Nanowires

Edited by

Paola Prete

Nanowires

Edited by Paola Prete

Republished by InTech

Janeza Trdine 9, 51000 Rijeka, Croatia

Technical Editor Teodora Smiljanic

Cover Designer Dino Smrekar

Additional hard copies can be obtained from orders@intechopen.com

Nanowires, Edited by Paola Prete
p. cm.
ISBN 978-953-7619-79-4

For EU product safety concerns:
IN TECH d.o.o., Prolaz Marije Krucifikse Kozulić 3, 51000 Rijeka, Croatia,
info@intechopen.com or visit our website at intechopen.com.

Preface

The study of quasi 1-dimensional (1d) semiconductor nano-crystals (so-called nanowires) represents the forefront of today's solid state physics and technology. Due to their many unique and fascinating physical properties (among others, superior mechanic toughness, higher carrier mobility and luminescence efficiency, and lowered lasing threshold), these systems are increasingly being considered as fundamental 'building blocks' for the realization of entirely new classes of nano-scale devices and circuits, with applications stretching from photonics to nano-electronics, and from sensors to photovoltaics. Free-standing semiconductor nanowires have been used to fabricate nanometer-scale field-effect transistors (FETs), bipolar junction transistors, and light-emitting diodes, nano-scale lasers, complementary inverters, complex logic gates, gas sensors, nano-resonators, nano-generators and nano-photovoltaic devices. Since the very first pioneering works carried out by K. Hiruma in Japan in the mid 90's, and by C.M. Lieber, P. Yang, and Z.L. Wang in USA by the end of 90's, successful demonstration of such a large variety of functional devices has lead to a rapidly growing interest in nanowire researches around the world and to a steep increase in the number of annual publications (the ISI-Thomson 'Web of Science' database reports over 800 papers published in the year 2009 under the combined keywords 'nanowires & semiconductors').

Such a breathtaking pace of research has allowed to explore the field in several directions within just a few years. The synthesis of free-standing nanowire structures now involves a large variety of methods and materials, including elemental and compound (both binary and multinary) semiconductors as well as complex modulated nanostructures. A seemingly vast amount of studies is being dedicated to understanding the physical-chemical and structural properties of these quasi 1d systems, and how they relate to the various synthesis mechanisms and parameters; this is being increasingly realized through the use of advanced nano-scale characterization tools and methods either directly on as-synthesized nanostructures or on nanowire-based devices. Still, a full understanding of nanowire physics cannot be achieved without a combination of these advanced characterizations with first principles (*ab initio*) calculations and methods to model their nano-scale characteristics and/or device performances.

This volume is intended to orient the reader in the fast developing field of semiconductor nanowires, by providing a series of self-contained monographs focusing on various nanowire-related topics. Each monograph serves as a short review of previous results in the literature and description of methods used in the field, as well as a summary of the authors recent achievements on the subject. Each report provides a brief sketch of the historical background behind, the physical and/or chemical principles underlying a specific nanowire fabrication/characterization technique, or the experimental/theoretical methods used to study a given nanowire property or device. Despite the diverse topics covered, the volume does appear as a unit. The writing is generally clear and precise, and the numerous illustrations provide an easier understanding of the phenomena described. The volume contains 20 Chapters covering altogether many (although not all) semiconductors of technological interest, starting with the IV-IV group compounds (SiC and SiGe), carrying on with the binary and ternary compounds of the III-V (GaAs, AlGaAs, GaSb, InAs, GaP, InP, and GaN) and II-VI (HgTe, HgCdTe) families, the metal oxides (CuO, ZnO, ZnCoO, tungsten oxide, and $PbTiO_3$), and finishing with Bi (a semimetal).

The selected reports can be grouped into four sections: the first one (Chapters 1-9) focuses on the synthesis of various semiconductor nanowires by bottom-up (self-assembly) methods; the second (Chapters 10-13) deals with the application of advanced characterisation tools/methods for studying semiconductor nanowires; a third section (Chapters 14-18) reports on *ab-initio* calculations and modelling studies of the nanowire electronic, optical and electrical properties, with emphasis on nano-scale effects. A last group of papers focuses on methods to integrate them for large-scale device fabrication (Chapters 19), and on 3-dimensional simulation of Si nanowire based FinFETs (Chapters 20).

A most promising way to the synthesis of free-standing semiconductor nanowire structures is by 'bottom-up' nano-technological approaches employing self-assembly methods. Among others, metal-catalyst assisted growth of nanowires, through the so-called vapour-liquid-solid (VLS) mechanism, is being considered a most straightforward and successful way for the synthesis of high-quality semiconductor quasi 1d nanostructures. VLS-based methods rely on alloying of a metal catalyst nanoparticle – most often Au – with atoms of the semiconductor materials, the latter supplied through the vapour. The as-formed alloy acts as nucleation site for the material and guides the nanowire growth, its diameter being controlled by that of the nanoparticle. The method has found applications to a variety of semiconductors and vapour growth techniques.

In this respect, Chapter 1 reports on the growth of SiC nanowires by chemical vapor deposition (CVD), using either VLS or catalyst-free methods, and their morphological, structural and optical characterizations, with a view to some of their possible applications to nano-devices.

Chapter 2 and Chapter 3 focus instead on the VLS growth and characterization of III-V compounds based nanowires. Chapter 2 discusses the growth of GaAs and GaAsSb nanowires by molecular beam epitaxy (MBE), a techniques that allows to change the crystallographic phase (zinc-blend or wurtzite) of the material during nanowire growth. This ensures to study the effects of controlled crystal phase changes on the materials band alignment within the nanowires. In Chapter 3 the attention is focused instead on nanostructures self-assembled by metalorganic vapour phase epitaxy (MOVPE). The latter is a well-proven technology for the III-V opto-electronics industry, and a most promising one for any industrial scale-up of future nanowire-based devices. The Chapter reports in particular, on nanowires based on the GaAs–AlGaAs system, a prototypical materials combination in the opto-electronic field; Au-catalyst fabrication issues, a crucial step to achieve strict control over the nanowire size, density and crystallographic alignment, are also discussed in depth.

Chapter 4 moves the attention to the Au-catalyzed self-assembly of HgTe-based nanowires by MBE. Here, different methods are experimented and discussed by the authors to growth HgTe nanowires, which self-assemble by a mechanism different from VLS, while the growth of HgCdTe nanowires remains elusive. Interestingly, Te and Au nanowires were also fabricated.

Chapters 5 to 9 focus on the synthesis and properties of nanowires made of transition metal oxides, a technologically important class of nanostructures for a variety of applications, including solid-state lighting, solar blind photo-detectors, gas sensors and nano-photovoltaic cells. Chapter 5 reviews on the synthesis of ZnO and CuO nanowires by thermal oxidation techniques. The authors present the thermodynamics of Zn and Cu oxidization reactions and discuss differences between the two metals; a detailed growth model of ZnO and CuO nanowires based on oxygen surface adsorption, subsequent metal oxidization and materials nucleation is then presented. Chapter 6 reports on ZnO nanowires grown by high-pressure pulsed laser deposition (PLD) and carbo-thermal evaporation methods, and on their optical and electrical characterizations. Both undoped and P-doped ZnO nanowires are then employed by the authors for FET fabrication, the letter in turn used to investigate the electrical character (whether p- or n-type) and doping stability of as-grown material. Chapter 7 addresses the growth and properties of ZnO-based surface nanowires and related 1-dimensional quantum-confined structures fabricated by PLD. Semiconductor surface nanowires fabricated on high-index substrate planes or patterned (V-grooved) substrates have been extensively investigated in the mid-90's of last century, but the properties of structures based on ZnO remain today much less known with respect to III-V compounds. In this respect, the paper discusses polar and non-polar growth on ZnO layers and outline differences between ZnO and GaAs surface nanowires. Surface nanowires on *M* nonpolar ZnO (10-10) planes are largely different from those on high-index GaAs surfaces, indicating that such surface nanowires results from a new bottom-up self-organized process. Remarkably, the optical and electrical properties of as-grown ZnMgO/ZnO surface nanowires are modulated by the anisotropic nature of surface morphology. Although these nanowire may show completely different properties from those of free-standing nanowires, Chapter 7 provides sufficient data to interested readers for a meaningful comparison between the two classes of nanowires.

Diluted magnetic semiconductors obtained by doping ZnO with magnetic elements (such as cobalt, Co) show room-temperature ferromagnetism (RTFM). Insertion of magnetic atoms into ZnO nanowires allow to study the effects of low-dimensionality on RTFM properties. Chapter 8 investigates the structure, optical, and magnetic properties of pure, Co-implanted and annealed ZnO nanowires, the latter obtained by VLS using vapor transport methods. Analyses indicate the role of oxygen vacancies (zinc interstitials) in enhancing ferromagnetic interaction between Co atoms in ZnO nanowires and the observation of super-paramagnetic behavior.

Among transition metal oxides, tungsten oxide ($W_{18}O_{49}$) nanowires show good sensing and field emission (FE) properties. Chapter 9 describes the synthesis of uniform and crystalline $W_{18}O_{49}$ nanowires on tungsten thin films by thermal annealing in ethane and nitrogen gas. Interestingly, the formation of tungsten carbide on the metal film surface enhances the nanowire formation, a mechanisms ascribed by the authors to structural defects and strain formation. Good stability of FE properties at atmospheric pressure of a diode device based on such nanowires is reported and discussed. Finally, the use of $W_{18}O_{49}$ nanowires in a micro-plasma reactor is described.

Chapter 10 discusses the use of advanced electron microscopy techniques for an in-depth characterization of complex semiconductor nanowire structures. The chapter is divided into two parts: in the first one, atomic-scale characterization of 1d nanostructures using aberration-corrected scanning transmission electron microscopy and electron energy loss spectroscopy techniques is discussed and major advantages pointed out. The second part of the chapter focuses instead, on electron tomography, an increasingly important electron microscopy technique for the 3-dimensional reconstruction of the morphology (shape) and inner crystalline properties of hetero-structured nanowires.

Chapter 11 explores the crystalline structure of MBE-grown GaN nanowires and the dynamics of nano-crystal phase changes by using *in-situ* low temperature X-ray diffraction and Rietveld analysis. As-grown nanowires have wurtzite structure at ambient temperatures, but start to develop the zinc-blend phase below 260K. A finite size model wherein the random phase distribution is utilized to describe the development of short range atomic ordering. The phase separation is found reversible upon temperature cycling, and occurs through interaction and exchange of size between characteristic ordered crystal domains within GaN nanowires.

Chapter 12 provides a review of Raman spectroscopy applied to nanowires; here, an overview of the selection rules, appearance of new modes and other related effects is given. The fundamentals for understanding Raman scattering in semiconductor nanowires and the basic physical principles behind specific phenomena related to nanowires are also presented; in particular, the Chapter reports on novel phenomena such as inhomogeneous heating, quantum confinement, Fano effect, the existence of surface and breathing phonon modes and the existence of novel crystalline phases. Chapter 13 reports then on the application of Raman scattering to the characterization of GaAs and InAs nanowires grown by MBE using either Au and Mn as catalysts. The phonon downshifts and asymmetrical broadenings found in the Raman spectra correlate to defects present within the nanowires. A phonon confinement model is used to calculate the average distance between defects and their density. Interestingly, the structural quality of Mn-catalyzed nanowires is comparable to that of Au-catalyzed ones, confirming that Mn is an interesting alternative metal-catalyst, especially for fabrication of dilute ferromagnetic III-V based nanowires.

The fully understand the huge wealth of experimental data currently provided by conventional and advanced materials characterization tools when applied to nanowires, requires to develop suitable theoretical models able to describe the peculiar nanowire physics; this is especially true if considering the 1-dimensional size effects on the nanowire electronic, electrical, and optical properties. Indeed, at size comparable to the mean free path of conduction electrons a mesoscopic transport regime is entered, leading to so-called finite-size effects; for even smaller size reaching the Fermi-wavelength, electronic wave function becomes confined and quantum-size effects are expected to occur, which affect both the transport and optical properties. This is exemplified in Chapter 14, focusing on both finite-size and quantum-size effects on charge transport properties of semi-metallic Bi nanowires fabricated by different techniques. Due to its large electron mean free path and Fermi wavelength, Bi is an ideal semi-metal to study nano-scale effects using

nanowires of relatively large diameters. The effect of strain on the electronic states and the piezo-resistance coefficients in single-crystal Si nanowires are instead studied in Chapter 15 by using first-principles calculations. Comparisons of first-principles predictions with piezo-resistance coefficients measured for p-doped single-crystal Si nanowires is then presented, the latter fabricated by electron beam lithography and reactive ion etching (top-down) methods.

Quantum confinement effects on the electronic states of free-standing hetero-structured nanowires are studied theoretically in Chapter 16, for GaP/GaAs/GaP and InP/InAs/InP axially hetero-structured, and $Si/Si_{1-x}Ge_x$ radially hetero-structured nanowires. The effect of nanowire diameter on changing the materials band-alignment along either the nanowire axis or in the radial direction is pointed out. Furthermore, it is shown that the existence of graded interfaces between materials within the nanowires can lead to significant fluctuations in the confinement energies and can even affect the excitonic properties of these systems. Chapter 17 presents instead, an optical model based on classical electrodynamics for analyzing the photoluminescence of single and multilayered (core-shell) nanowires; as concrete examples, the authors investigate the emissions from ZnO nanowires and nanotubes, and ZnO/silica nanocables. Optical properties are presented as functions of geometrical parameters and excitation polarization.

Chapter 18 highlights the ferro-electricity of $PbTiO_3$ nanowires, focusing on their surface and edge structures from the atomistic and electronic points of view by means of first principles density functional calculations. The coupling effects between mechanical deformation and electric properties here discussed for $PbTiO_3$ nanowires, are generally known as "multi-physics properties" and have been previously reported also for Si nanowires. Understanding these properties is clearly important in designing future electronic nano-materials and nano-devices.

Despite the many progress made to date in the synthesis, characterisation and physics comprehension of semiconductor nanowires, their exploitation to large-scale device fabrication requires to develop reliable and cost-affordable methods of nanowire integration into working nano-devices. To this purpose, Chapter 19 reviews the current state-of-the-art in the field, by focussing on various techniques proposed for the large-scale assembling of nanowire devices; these methods can be broadly divided into two main technologies: (i) transfer with alignment of pre-grown nanowires onto a surface; and (ii) direct growth of nanowires onto a pre-patterned substrate at desired positions. Benefits and drawbacks of each of the methods are presented and discussed in the Chapter, alongside with examples of nanowire devices fabricated by the given assembly techniques. As the authors point out, much remain however to be done in this field.

An example of the problem posed by the design and fabrication of future nanowire-based electronic devices is exemplified in Chapter 20, reporting on the use of a 3-dimensional process simulation tool to study the suitability of CAD technology for Si nanowire FinFETs process development. CAD technology predictability for FinFETs fabricated using a conventional CMOS-like process flow for novel strain-engineered Si nanowires is assessed. Effects of process-induced strain on the performance enhancement of Si nanowire FinFETs are also discussed.

This book is the result of the best of the scientific work and contributions of many researchers worldwide to whom a sincere thank is address for the care and skills lavished in the preparation of each Chapter. I wish the reader an enjoyable reading and a profitable use of this book in their future research work in this challenging field of science.

March 2010

Editor

Dr. Paola Prete
IMM-CNR
Lecce, Italy

Contents

1

Cubic SiC Nanowires: Growth, Characterization and Applications

Giovanni Attolini, Francesca Rossi, Filippo Fabbri, Matteo Bosi, Giancarlo Salviati and Bernard Enrico Watts
IMEM-CNR Institute
Italy

1. Introduction

Since the introduction of carbon nanotubes in 1992, the study of one-dimesional nanomaterials, which includes metallic, magnetic, semiconducting and oxide compounds, has attracted considerable interest, especially as regards nanowires (NW), nanobelts and nanorods [Kolasinski, 2006].

The main interests of this research are in the realization of nanoelectronic devices (e.g. nano field-effect transistors), nano-electromechanical systems, and nano-sensors exploiting high selectivity and compatibility with biological systems. Nanostructures may present very different characteristic and novel properties with respect to the corresponding bulk material, and they have important physical and chemical properties, in particular large specific surface/volume ratio and quantum size effects, which permit many applications such as nanoscale devices, sensors and scanning probes not possible with standard structures.

Silicon carbide (SiC) has gained importance as both a coating and a structural material for Micro Electro Mechanical Systems (MEMS) (Sarro, 2000; Mehregany et al, 2000). SiC is a wide bandgap semiconductor used for high temperature, high power applications and radiation hard environments. The high Si-C bond energy confers a high Young's modulus and hence mechanical toughness and high fracture strength (Li and Bhushan, 1999), moreover, it is chemically inert to the most corrosive and erosive chemicals and is biocompatible (Willander et al. 2006; Casady and Johnson 1996; Yakimova et al. 2007). More than 100 polytypes of SiC exist but the SiC cubic phase (3C-SiC) has drawn particular attention because it can be deposited on Si (Marshall et al. 1973).

The combination of these distinctive physical, chemical and mechanical properties of SiC and the possibility to synthesize SiC NW make this material an excellent candidate for the design and fabrication of nanodevices.

Surface functionalization introduces specific chemical functional groups onto a surface, in order to tailor its properties to specific needs. Functionalization of NW is nowadays a burgeoning field of activity, and motivates researchers involved in nanotechnology and related activities: defining a specific molecule/NW interface, suitable for selective bonding to a chosen chemical species, is a key step of the development of nano-objects tuned by the physical and chemical properties of molecules, and may have notable application in sensing and biosensing.

Functionalized 3C-silicon carbide NW have the potential to act as highly sensitive detector elements in bio-chemical field (Yakimova et al., 2007).
Many methods are currently being used to prepare SiC-NW (pure or with a SiO_2 shell) on silicon substrates using a catalyst, including chemical vapour deposition, vacuum evaporation of SiC, direct synthesis from Si and C powders.
In this paper we will present a brief review of growth methods used to obtain cubic silicon carbide NW, both with and without SiO_2 shell, and our results on the NW growth and characterization of morphological, structural and optical properties by SEM, TEM, CL and Raman.
Finally we will review some of the possible applications for nanodevices.

2. A brief review of NW growth methods

Different growth methods have been developed to prepare NW of different materials, and several theoretical models have been proposed to explain the growth mechanisms.
Semiconductor NW are generally synthesized via a Vapor-Liquid-Solid (VLS) process (Wagner and Ellis, 1964), a process that can be divided in three main steps: a) formation of a small liquid droplet on the surface of the substrate, b) supersaturation of the liquid by the incorporation of gaseous precursors and c) subsequent nucleation and growth of the NW from the liquid-solid interface.
Small metal clusters are deposited by different techniques on the substrate surface, forming nanosized dots.
In a second step, a gas of the proper precursors flows through the reaction tube and, when in contact with the metal droplets, the precursor deposits on the liquid surface and forms an alloy. A continuous incorporation of the precursor leads to a supersaturation of the desired compounds and as a consequence to the NW growth at the solid-liquid interface.
With other growth techniques it is possible to realize NW without a metal catalyst on the substrate surface, by thermal evaporation of a suitable source near its melting point and subsequent deposition at cooler temperature. This mechanism is called "vapor-solid" (VS) growth and has been mainly used to synthesize metal oxide and some semiconductor nanomaterials (Wang et al., 2008). It is often called self-catalytic growth, since in this case one component of the gaseous atoms might play the role of the catalyst.
3C-SiC/SiO_2 core-shell NW have been synthesized both by a direct heating method using WO_3 and graphite mixed powder as starting material and Ni as catalyst (Bark et al., 2006) and from a mixture of activated carbon and sol-gel derived silica embedded with Fe nanoparticles (Liang et al., 2000).
They are also obtained using iron catalyst by chemical vapor reaction in a mixture of milled Si and SiO_2 powders and C_3H_6 as raw materials in a graphite reaction cell (Meng et al., 2007).
Alternatively, a CVD method can be employed making use of Fe as catalyst and methane as precursor (Zang et al., 2002).
A high yield core/shell SiC/SiO_2 NW production method without the use of catalyst was developed starting from raw powders of Si via an oxide assisted thermal evaporation process (Khongwong et al., 2009).
Core-shell SiC NW have also been synthesized from carbon monoxide using Ni catalyst by carbothermal reduction method (Attolini et al., 2008).

Pure 3C-SiC NW, without shell and free from impurities with the exception of those related to the catalyst at the tip, have been prepared as follows:
3C-SiC NW were deposited on silicon substrate by metalorganic chemical vapor deposition (MOCVD) by using Ni, Au, Fe as catalysts and dichloromethylvinylsilane or methyltrichrolosilane as precursors (Kang et al., 2004; Takai et al., 2007; Yang et al., 2004; Choi et al.,2004; Seong et at. 2004).
The CVD approach consists in flowing the reactants with a carrier gas in a reaction tube inserted in a furnace where either (100) or (111) oriented silicon substrate are placed.
Polycarbosilane (PCS) was used as a precursor to grow porous silicon carbide ceramics with embedded β-SiC NW (Zhu et al., 2005).
Single β-SiC NW were grown through annealing of polycrystalline SiC layer in hydrogen atmosphere at 1150°C (Yang et al., 2006).
Table 1 gives a brief overview of the different methods with the main parameters used in SiC NW growth.

Methods	Chemical Vapor Deposition (including Metal Organic Vapor Phase Epitaxy, Chemical Vapor Reaction, Chemical Vapor Infiltration), Physical Vapor Deposition, Sputtering
Precursors	Methytrichlorosilane, Dichlorometilsilane, Methane, Propane, Silane, Diethylsilane
Starting materials	Si+SiO_2 with methane; Si+C; SiC powder;
Substrates	(100), (111) Silicon
Catalyst	Ni, Fe, Au, Pt, Pd, Fe/Co, Al
Growth temperatures range	From 1000 to 1400 °C

Table 1. Typical methods and conditions to prepare 3C-SiC NW

3. Experimental

In our laboratory we obtained 3C-SiC NW with three different growth methods:

1. core-shell NW were obtained using carbon oxide in an open tube;
2. SiC NW were synthesized in a heated quartz tube using carbon tetrachloride;
3. SiC NW were grown in a VPE reactor with silane and propane as precursors.

In this section, we will describe these growth procedures and the characterization of the NW in detail.
The morphological and optical characterizations of the as-grown NW were performed by acquiring secondary electron images in an S360 Cambridge Stereoscan Scanning Electron Microscope (SEM) equipped with a Gatan MonoCL2 system with photomultiplier detector to collect Cathodoluminescence (CL) spectra. The structural and compositional analyses were performed by Transmission Electron Microscopy (TEM) in a JEOL 2200FS working at 200 kV, equipped with an in-column Ω filter, a High Angle Annular Dark Field (HAADF) detector for Z-contrast imaging and an Energy-Dispersive X-Ray (EDX) detector for elemental mapping and EDX spectroscopy.
Micro-Raman scattering measurements were performed at room temperature (RT) with a 100x objective and a 532 nm excitation light. The spectrum resolution was about 0.2–0.3 cm^{-1}.

3.1 Growth and characterization of core-shell NW

The growth of 3C-SiC core-shell NW on Si substrates was performed in an open-tube configuration by flowing carbon oxide and nitrogen or argon as carrier gases.
The growth procedure is the following:

- (100) oriented silicon substrates are cleaned in organic solvents with an ultrasonic bath, dipped in a nickel-salt solution and dried in an oven at 60 °C before being placed into the reactor;
- the substrate is placed in an open tube inside a horizontal furnace, previously purged with inert gas to remove air. The central position of the furnace is selected as a zone of constant temperature;
- the temperature is raised to 1100°C and, after temperature stabilization, carbon oxide is introduced into the tube. The growth time was varied from 1 to 60 minutes, while the gas flow was kept constant in all the experiments.

The NW grow on the Ni-covered substrate areas and are arranged in dense forests (see representative SEM images in Fig. 1), with a quite narrow diameter distribution and with lengths up to several tens of microns. A round-shaped tip is observed on the NW, as discussed in the following.

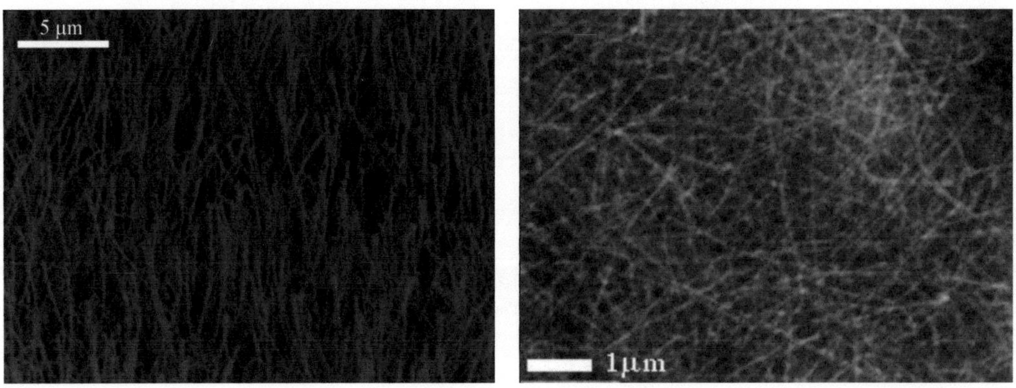

Fig. 1. SEM images of 3C-SiC/SiO_2 core-shell NW. a) 45°- tilted view, showing a good vertical alignment of the NW, b) planar view, showing the quite uniform NW diameter.

Compositional analyses performed by energy filtered TEM and HAADF imaging (see Fig. 2) highlighted the core/shell structure of the NWs. Fig. 2a reports a zero-loss filtered image of a typical wire, showing the crystalline core and the amorphous shell. Elemental mapping (see Fig. 2b-e) confirms the complementary distribution of carbon and oxygen, in the SiC core and in the oxide shell respectively. The oxygen to silicon ratio in the shell, as estimated by EDX point spectra, is very close to two, allowing to identify the shell as silicon dioxide. Further, EDX maps and HAADF images acquired in the tip region (see Fig. 3) confirm the presence of a high-Z nichel-containing particle on top of the NWs, consistently with a VLS growth process.
Structural studies were performed by HRTEM on the NW core. The symmetry of the crystal (inset in Fig.4a) and the lattice spacings identify the structure as 3C-SiC, with <111> growth axis. As reported in Fig. 4a, quite long segments grow almost free of planar defects.

However, the insertion of (111) stacking faults and the occurrence of local stacking sequences of 2H, 4H and 6H polytypes is observed in some areas (Fig. 4b). These results are consistent with the wide literature on planar defects and phase transitions in 3C-SiC whiskers(Seo et al., 2000; Yoshida et al., 2007).

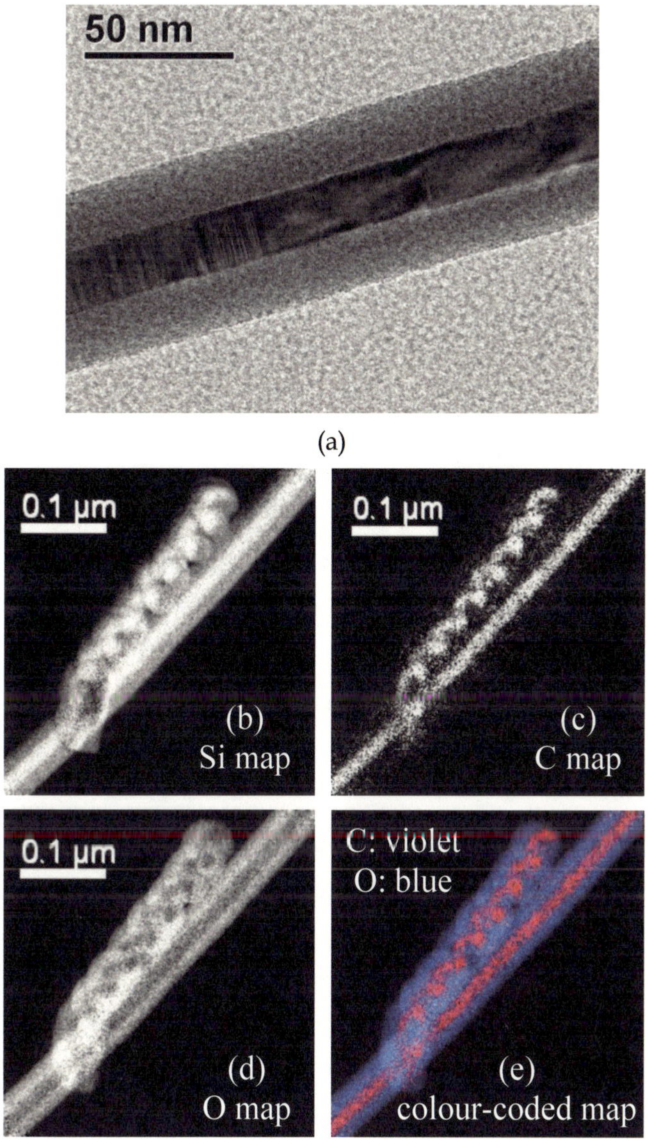

Fig. 2. (a) zero-loss filtered TEM image, evidencing the SiC/SiO_2 core/shell structure, (b)-(d) elemental maps computed from energy-filtered images with the 3 window method. The $L_{2,3}$ silicon edge and the K carbon and oxygen edges have been used for the energy filtering. The map in (e) is obtained by color-mix of the C and O maps.

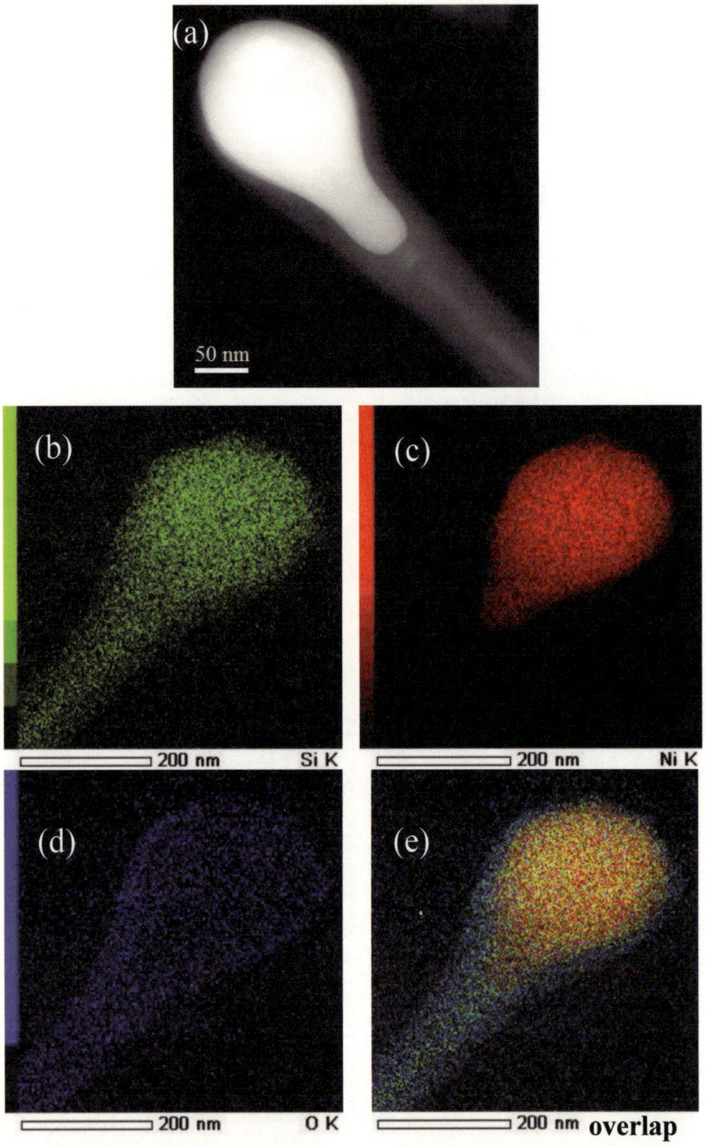

Fig. 3. HAADF image (a) and EDX maps (b-d) showing the distribution of silicon, nichel and oxygen in the tip region of a typical nanowire. (e) is obtained by colour-mix of the maps in (b)-(d).

X-Ray Diffraction (XRD) measurements (see Fig. 5) confirmed the presence of several 3C-SiC peaks, while no evidence of other polytypes phases was observed. A weak shoulder detected on (111) peak at 33.7° could be related to the presence of stacking faults. After a Reitveld refinement carried out using the MAUD program, a value of 4.361 Å has been calculated for the lattice parameter, in good agreement with the expected value for the β polytype.

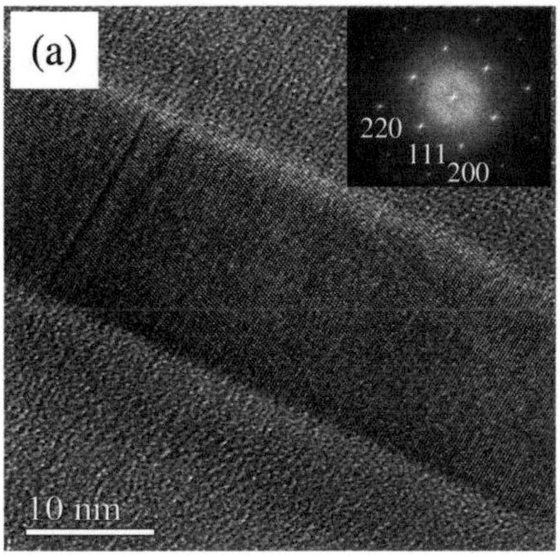

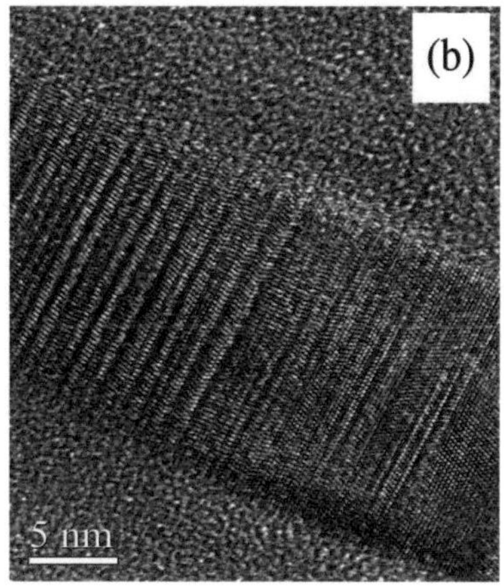

Fig. 4. HRTEM images in [110] zone axis of the 3C-SiC crystalline core: (a) segment almost free of planar defects, (b) defective area. Inset in (a) shows the fast Fourier transform.

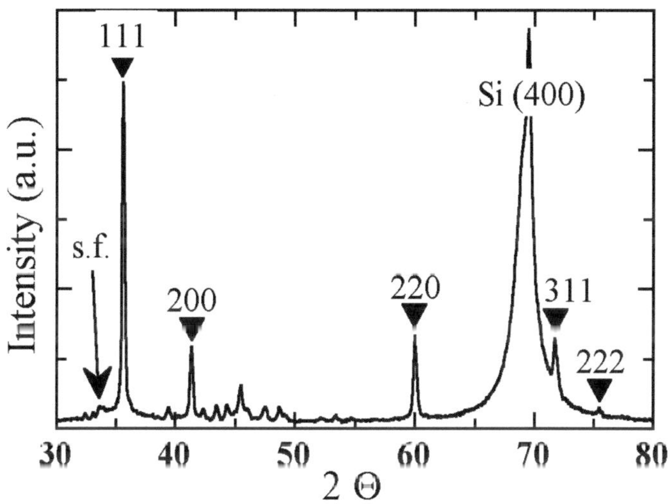

Fig. 5. XRD pattern of a NW sample on a 100 Si substrate. The 3C- SiC peaks (∇) and the extra stacking fault peak (s.f.) are indexed.

The lattice parameter was also analysed by micro-Raman experiments. As shown in Fig. 6, vibrational modes due to the shell and the TO and LO phonon modes of the β-SiC core were revealed. The latter are detected at 797.3 and 976.2 cm^{-1} respectively, slightly shifted from the expected 3C-SiC bulk values. The positions of Raman modes are known to shift because of lattice mismatch (Zhu et al., 2000) and it is possible to estimate the value of $\Delta a/ a_0$ from the following equations (Nakashima & Harima, 1997):

$$\omega(TO) = 796.5 + 3734 \cdot \left(\frac{-\Delta a}{a_0} \right) \tag{1}$$

$$\omega(LO) = 973.0 + 4532 \cdot \left(\frac{-\Delta a}{a_0} \right) \qquad (2)$$

We estimated $\Delta a/a_0$ of the order of -0.02% and -0.07% from the TO and LO mode respectively, corresponding to $-\Delta a$ of the order of 10^{-3} Å. The longitudinal variation is slightly higher than the transverse, consistently with the NW geometry. Some broadening of the SiC phonon modes was observed (e.g. TO-FWHM= 4 cm^{-1}) and ascribed to lattice imperfections.

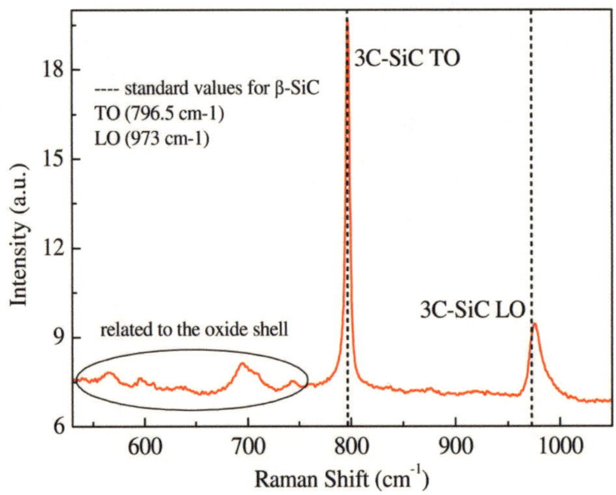

Fig. 6. Raman spectrum of core/shell NW.

Cathodoluminescence experiments were performed to investigate the optical properties of the core/shell NW. A broad room temperature emission, with an intense above-band-gap component, was detected (Fig. 7). By Gaussian deconvolution, the 3C-SiC near-band-edge (NBE) indirect transition was identified at 2.25±0.5 eV, while a second intense band was revealed at about 2.7 eV (blue band).

Above-band-gap luminescence from cubic SiC whiskers and nanocrystallites has been reported in the literature (Xi et al., 2006), but the contribution of a SiO_2 related emission in this spectral region (McKnight & Palik, 1980) has to be considered. In our samples, size-dependent CL studies on single NWs do not give clear evidence of quantum-confinement effects.

Electron beam irradiation effects (Fig. 7a-c) were studied to investigate the blue luminescence. The electron beam in the SEM was kept continuously impinging on the sample and CL spectra were acquired every minute until a total irradiation time of 1 hour. The results of this experiment are reported in Fig. 7c, showing the CL intensity evolution for the two main Gaussian components. The SiC-related emission stays almost constant, while the blue luminescence increases till saturation as a function of the irradiation time. This is consistent with an attribution of the 2.7 eV band to oxygen-deficiency-centres ODC(II) in the silicon dioxide (Skuja, 1998).

Further experiments on the core/shell NW geometry were aimed to study the selective removal of the silicon dioxide layer by etching. A long growth time (1 hour) was selected, to allow the synthesis of a larger amount of NW and the formation of a thicker shell. The sample was then submitted to etching treatments in HF:H_2O (1:3) solution, and room-temperature CL spectra were acquired at different etching steps (Fig. 8).

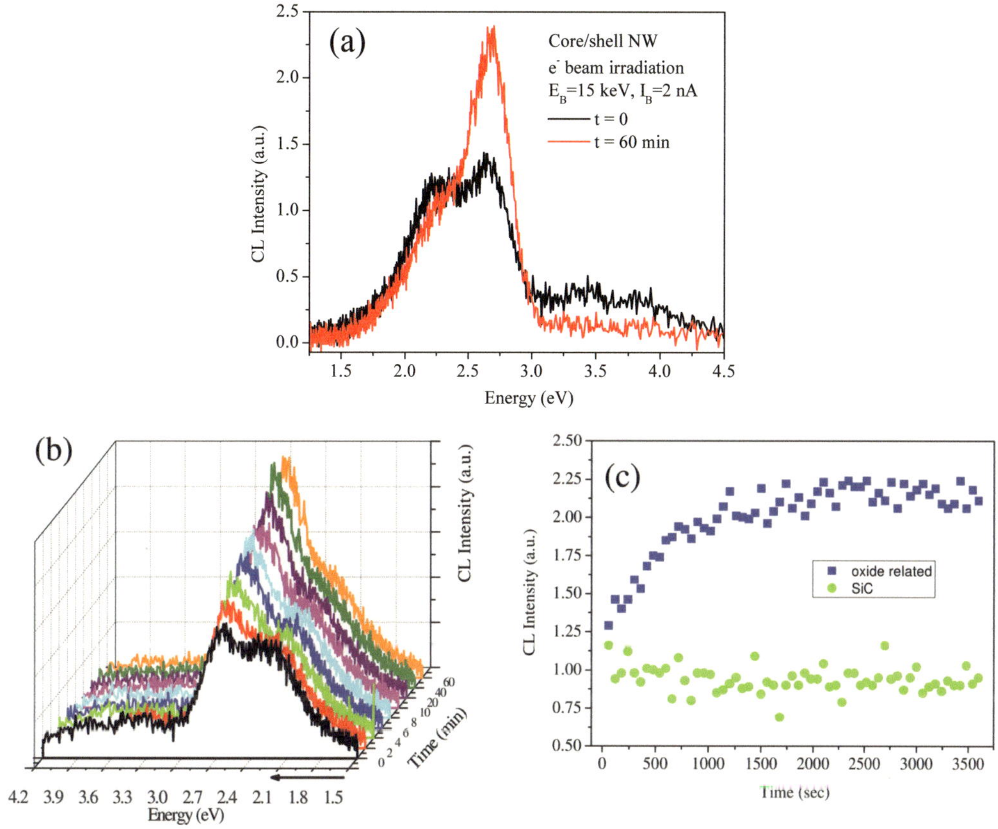

Fig. 7. Room temperature CL spectra of core/shell NW. The spectral evolution after 1 h exposure to the electron beam in the SEM at accelerating voltage of 15 KeV and beam current of 2 nA is reported in (a). Systematic spectra have been acquired as a function of the irradiation time (b), and the time-evolution of the CL intensity is reported in (c) for the two Gaussian components due to the SiC core (circles) and the shell layer (squares).

In these conditions the as-grown NW showed the SiC-NBE and the SiO_2-ODC related emissions discussed above, convoluted with additional bands peaked at about 1.98 eV, 2.55 eV, and 3.15 eV. In agreement with literature data, the 1.98 eV shoulder on the low energy side was assigned to the presence of substitutional oxygen on carbon site (O_C defects), unintentionally incorporated in the silicon carbide lattice [Gali et al, 2002; Kassiba et al., 2002). The high-energy emission at 3.15 eV could be related to the nanometric inclusions of hexagonal SiC polytypes with $E_{G\text{-}6H}$ = 3.15 eV, $E_{G\text{-}4H}$ = 3.24 eV (Bechstedt et al., 1997) as observed by high-resolution TEM (Fig. 4). As for the 2.55 eV band, it was tentatively ascribed to interface states related to carbon clusterization at the core/shell interface (Afanas'ev et al., 1996), suggested also by XPS data (not shown here).
A decrease of the whole CL emission was observed as an effect of the etching treatments (Fig. 8a). In particular, the SiO_2 related bands showed an exponential decay as a function of

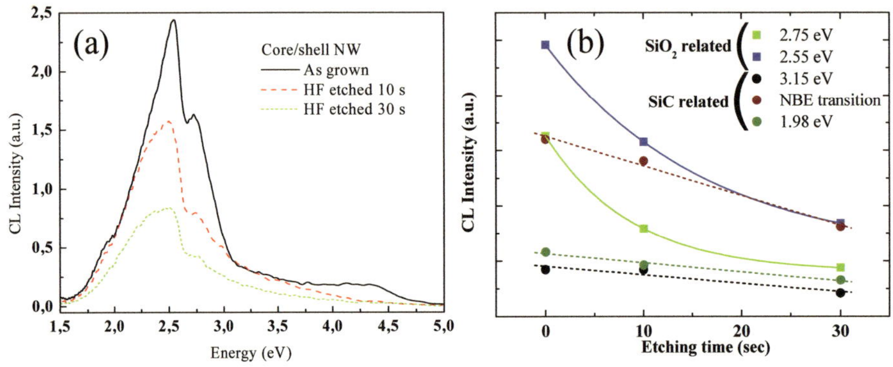

Fig. 8. CL spectra of single NW at different HF etching steps (a) and intensity evolution as a function of the etching time (b).The silicon dioxide and the silicon carbide related emissions show different decays.

the etching time, while the SiC related emissions showed a linear decrease (Fig. 8b). The SiO_2 related emissions decrease in intensity due to material removal. On the contrary, the chemical inertness of the silicon carbide to the hydrofluoric acid etching ensures that the decrease of the SiC related emissions has a different origin. A peculiar relationship between the removal of the silicon dioxide and the decrease of the core luminescence can be hypothesized, as discussed in the following.

It is possible to propose a model to explain why the presence of the silicon dioxide shell increases the radiative recombination in the silicon carbide core. A type I band alignment (Pistol & Pryor, 2008) of 3C-SiC and SiO_2 can be hypothesized, with conduction and valence band-offsets ΔE_C= 3.6 eV and ΔE_V = 2.9 eV (Afanas'ev et al, 1996). In this framework, the carriers generated by the electron beam in the shell diffuse into the core, and here recombine according to the allowed transitions in 3C-SiC. The diffusion of the carriers could be considered as an energy transfer from the shell to the core, an effect that has been observed for semiconductor nanoparticles (Louis et al., 2006) but not yet in NW. In our system, the amorphous shell results to be beneficial to enhance the luminescence intensity of the crystalline core, preferentially the SiC NBE radiative recombination. Besides the effectiveness as a carrier injector region, this could be partly related to the fact that the shell can act as a passivation layer to reduce the non-radiative recombination related to surface states, likewise in the case of entirely crystalline core/shell systems such as GaAs-based NWs (Skold et al. 2005; Jabeen et al. 2008; Tomioka et al., 2009).

3.2 Growth of NW using carbon tetrachloride

A fused silica reaction tube is placed inside an external liner and uniformly heated in a resistance furnace as shown in Fig. 9. The reaction tube is connected to a gas inlet through which a gaseous mixture of N_2-CCl_4 flows, obtained by flowing N_2 into a bubbler containing CCl_4 at 293 K. (100) oriented silicon substrates are cleaned in organic solvents with an ultrasonic bath, dipped in a nickel-salt solution, dried in an oven at 60 °C and put inside the reaction tube.

The air is completely purged by flowing nitrogen. The furnace is heated to 1000 °C while flowing the N_2-CCl_4 mixture and held at this temperature for a growth time of 30 minutes.

The furnace is then switched off, the tube is removed and cooled down to room temperature.

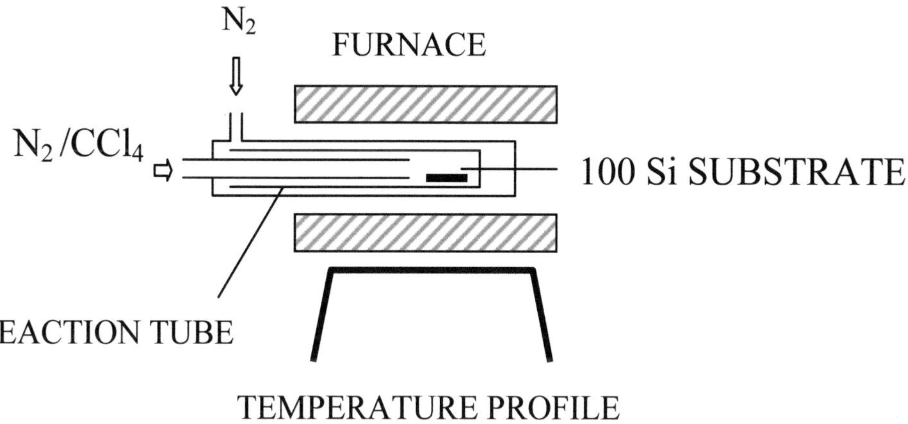

Fig. 9. Schematics of the experimental apparatus for the growth of NW with carbon tetrachloride.

In this case the NW synthesis is obtained by a three step process, consisting in:

1. thermal decomposition of CCl_4
2. reaction of chlorine decomposition products with silicon, both on substrate surface and in the vapor phase, to form silicon chlorides
3. reaction of silicon chlorides with carbon atoms to produce silicon carbide NW.

CCl_4 decomposition in the temperature range 767 - 1913 °C releases two chlorine atoms per decomposing molecule. In thermal plasmas in N_2 atmosphere the main species formed in the temperature range 477 - 1477 °C is chlorine (Huang & Guo, 2006). Either the main decomposition product is Cl_2 or Cl, a silicon extraction from the substrate is expected to occur due to the presence of halogens. This should be the result both of the surface etching (Dohnalek et al., 1994) and the reaction with the silicon vapor at the growth temperature.
All silicon chlorides, from SiCl to $SiCl_4$, are expected to be formed in the gas phase, with different concentrations (Chaussende et al, 1999). These species react with the C formed by CCl_4 decomposition, giving silicon carbide. The growth of NW is catalyzed by the nickel atoms present on the silicon surface, according to the preferential interface nucleation mechanism (Wacaser et al., 2009).
This process produces NW several microns long, with diameters below 80 nm (see the secondary electron image in Fig. 10a). They are arranged into dense networks randomly oriented on the silicon surface. Some craters (Fig. 10b) are opened in the substrate surface due to the etching action of the chlorine, as mentioned above for the reaction process.
A first investigation of the SiC crystalline structure performed by X-ray diffraction (not shown here) confirms the presence of the characteristic peaks of the cubic SiC polytype, in particular the most intense (111) diffraction at $2\Theta = 36.6°$. In order to investigate the structure of the single NW, a more accurate structural characterization is carried out by TEM analyses. A typical low-magnification image of a single wire, with uniform diameter of about 50 nm, is shown in Fig. 11a. As verified by SAED and high resolution studies, the lattice is cubic (see Fig. 11b) and the NW growth axis is parallel to the [111] direction.

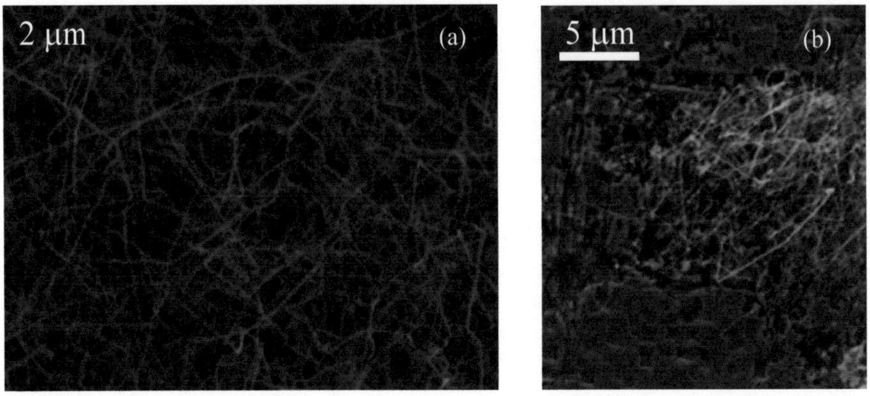

Fig. 10. SEM images showing (a) a typical NW ensemble, (b) the substrate surface etched by the chlorine.

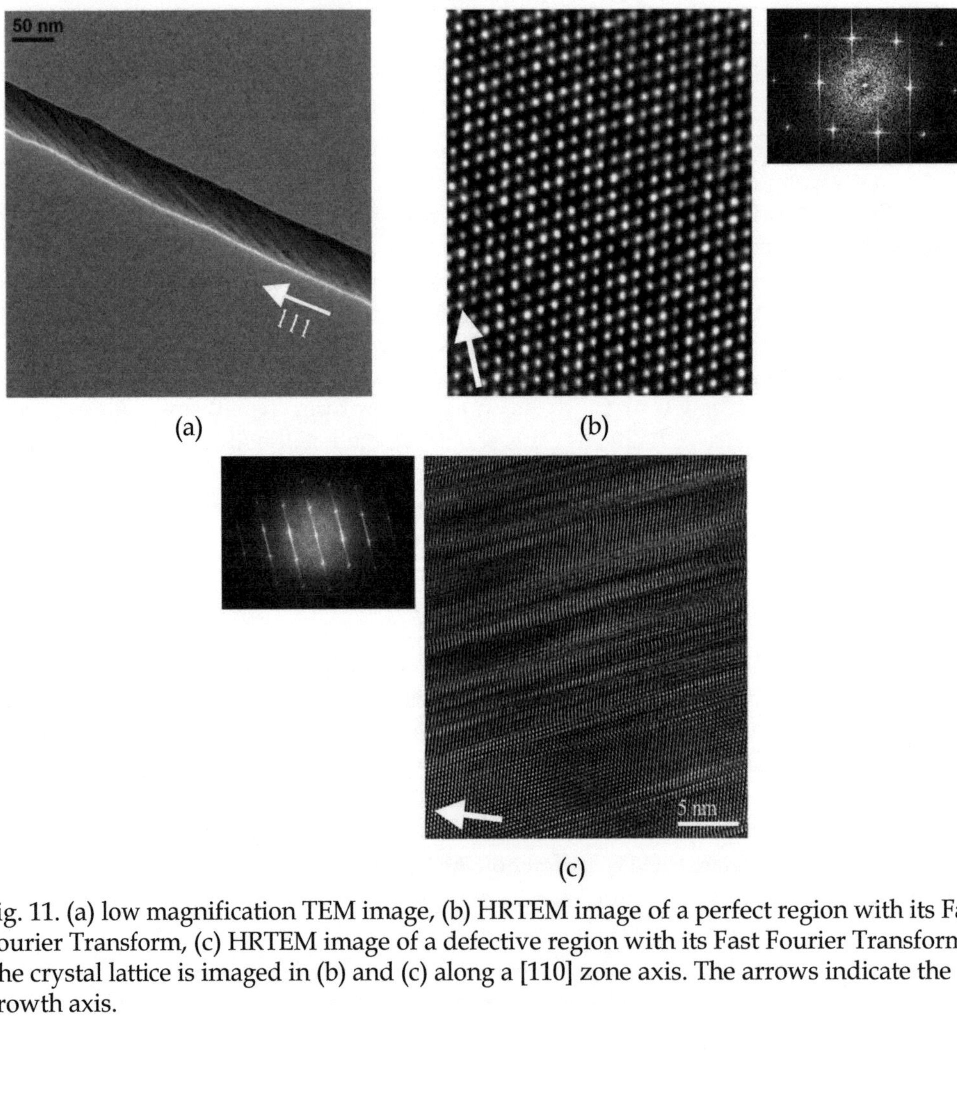

Fig. 11. (a) low magnification TEM image, (b) HRTEM image of a perfect region with its Fast Fourier Transform, (c) HRTEM image of a defective region with its Fast Fourier Transform. The crystal lattice is imaged in (b) and (c) along a [110] zone axis. The arrows indicate the growth axis.

Defective areas are observed, a detail of which is shown in Fig. 11c, together with the corresponding streaky pattern in the power spectrum. These results are consistent with common findings on 3C-SiC whiskers, as discussed above for core/shell NW.
The CL emission spectrum of the NW ensemble (see Fig. 12) is quite broad. A Gaussian fitting and deconvolution was performed in order to resolve the different components. The peak related to the indirect band gap emission of the 3C-SiC NW is detected at 2.20±0.02 eV. Two anomalous components, the band centered at 1.730±0.005 eV and the dominant band at 1.950±0.005 eV, are identified. A reference spectrum was acquired from a region of the substrate not covered by the NW, to eliminate possible contributions from porous silicon or silicon dioxide substrate. Since the luminescence from the substrate in this spectral region is about three orders of magnitude weaker, any spurious contribution can be excluded. In agreement with literature data, the emission at 1.73 eV can involve deep levels in 3C-SiC with activation energy of about 0.5 eV (Alfieri & Kimono, 2009; Itoh et al., 1997). The dominant red emission at about 1.95 eV can be tentatively assigned to the presence of substitutional oxygen on carbon site (O_C defects), unintentionally incorporated due to contaminants present in the open-tube growth (Gali et al., 2002). Finally, a high-energy emission is observed in the CL spectrum at 2.55±0.02 eV. Luminescence above the band-gap in cubic SiC has been previously debated in the literature: attributions to quantum confinement effects in nanowhiskers (Zhang et al., 2002), effects of morphology, orientation, defects and dangling bonds in NW facets (Chen et al., 2008), or to recombination involving surface/interfacial groups (Shim et al., 1997) have been proposed. Since we have a typical NW size too large to achieve appreciable quantum confinement, the weak blue emission detected in CL could be more likely related to surface modifications.

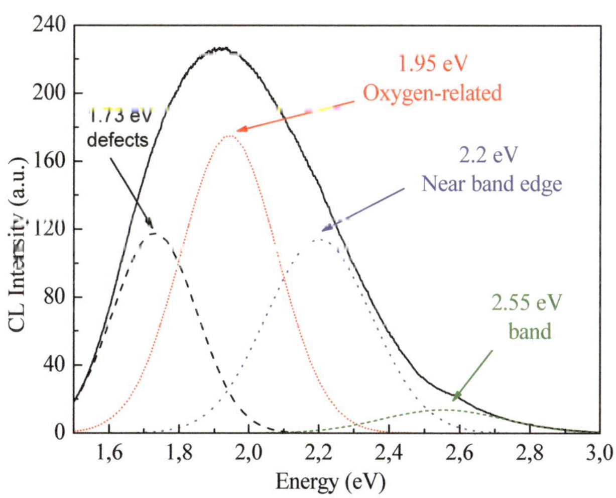

Fig. 12. CL spectrum at T=77 K of NW grown with CCl_4 precursor.

3.3 Growth of NW in VPE reactor

3C-SiC NW have been grown in a home-made induction heated Vapor Phase Epitaxy (VPE) reactor at atmospheric pressure (Fig. 13), by using diluted (3%) propane and silane as precursors, hydrogen as carrier gas and Ni as catalyst.

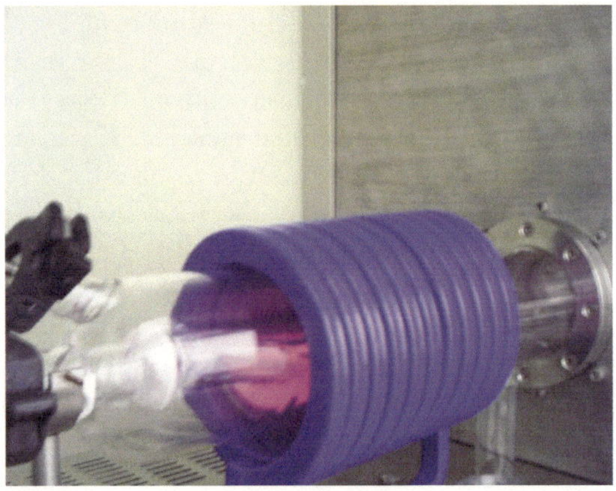

Fig. 13. Photo of the VPE reactor used to grow SiC NW.

A nominally 4 nm thick nickel film was deposited by e-beam evaporation system on a Si (100) substrate, previously etched in an HF solution for 60 s in order to remove the surface native oxide. For the VPE growth, a preheating of the nickel-deposited substrate at 1100°C for 2 minutes was performed in order to get catalyst dewetting, then the reagents were introduced into the growth chamber for a growth time of 5 minutes.
The process resulted in the formation of a NW network (Fig. 14a) in the nichel-covered substate areas. By spatially selective nichel deposition using conventional photolitographic techniques, peculiar geometries could be obtained (Fig. 14b).

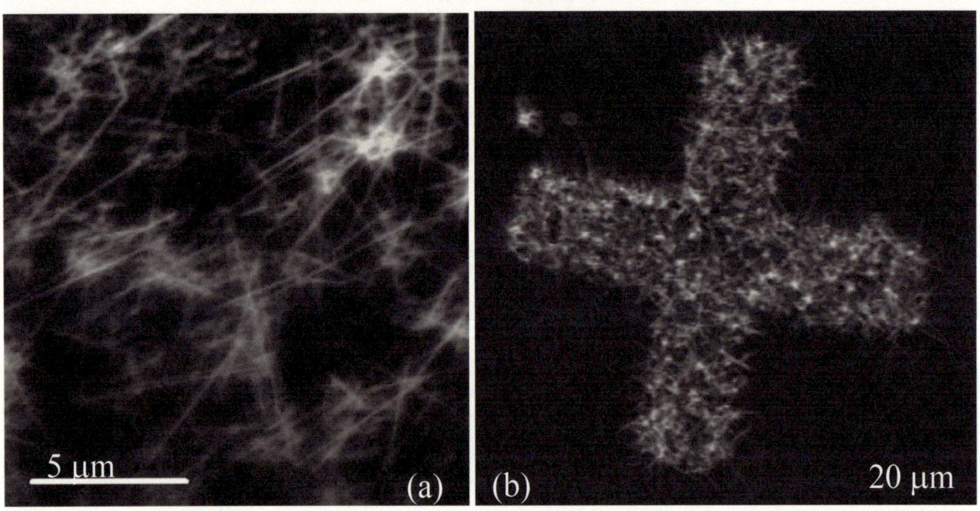

Fig. 14. SEM images of VPE grown NW. (a) NW network, b) selective NW growth in a cross-shaped area of the substrate.

Z-contrast analyses (Fig. 15) showed a high-Z (brightest) particle on top of the NW, indicating a metal-catalysed VLS growth mechanism. A tapering of the NW was observed: the average size decreases from about 80 nm in the center-base region to about 10 nm near the tip.
The lattice simmetry and structural quality of the as-grown NW were characterized by X-ray powder diffraction and TEM studies. The X-Ray profiles (not shown here) showed the characteristic peaks of β-SiC at $2\Theta = 35.6°$ (111), 41.4° (200), and 59.9° (220). The electron

diffraction patterns and the high-resolution TEM data confirmed the NW are predominantly 3C polytype with (111) growth axis, but stacking defects occur on (111) planes (Fig. 16a) and local stacking sequences characteristic of other polytypes (in particular 4H) are observed in the near-tip segments (Fig. 16b).

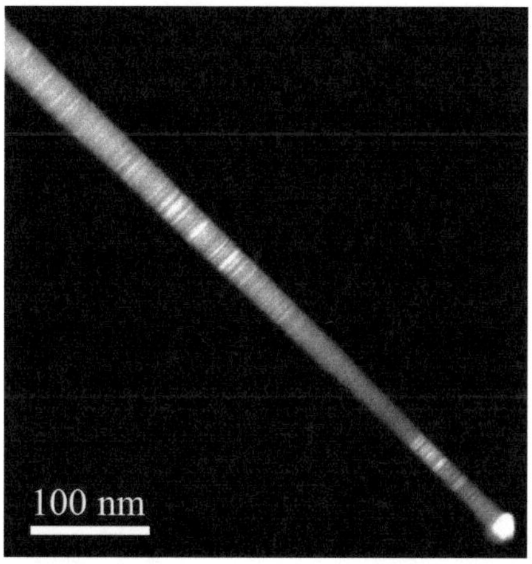

Fig. 15. TEM-HAADF (Z-contrast) image of a VPE grown NW.

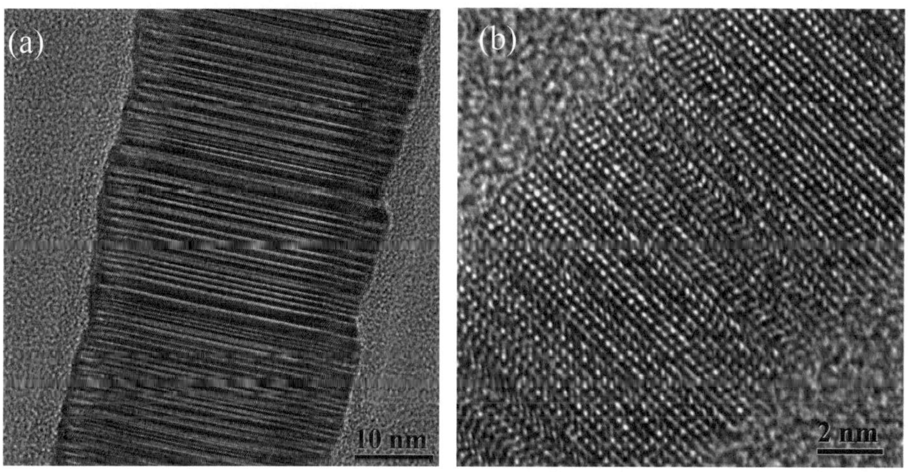

Fig. 16. HRTEM images in the center region (a) and in the tip region (b) of a VPE-grown NW.

The growth of NW was also tested on 3C-SiC substrates previously grown in the same reactor. The 3C-SiC film was deposited on 2″ p-type (001) Si substrates. The substrate was etched in 10% HF solution for 60 s before introduction in the growth chamber and heated at 1000 °C for 10 minutes to completely remove the Si native oxide layer. In order to relax the lattice mismatch between SiC and Si, a carburisation step was performed before the growth: the temperature was lowered to 400 °C, 3 sccm of propane were injected in the growth chamber and the temperature was raised to 1100 °C in about 2 minutes. This temperature was then held for 5 minutes to complete the carburisation process and form a thin monocrystalline SiC layer on the Si substrate. After the carburisation process the propane flow is interrupted, the temperature is raised to 1200 °C and SiH_4 and C_3H_8 are injected in the growth chamber for 10 minutes (Bosi et al., 2009).

A Ni film, nominally 4 nm thick, was deposited on the 3C-SiC/Si wafer and the wires were grown in the same manner as described above.
A VLS nichel assisted growth of NW was achieved. A typical side-view SEM image is shown in Fig. 17a. The lattice structure and spacings (Fig. 17b) correspond to those of 3C-SiC, with stacking defects on (111) planes. At the tip, a quite sharp interface was observed between the SiC wire and the catalyst particle (magnified detail in Fig. 17c). TEM observation shows that zones with a lower defect density are observed on SiC NW grown on 3C-SiC with respect to the ones grown directly on Si.

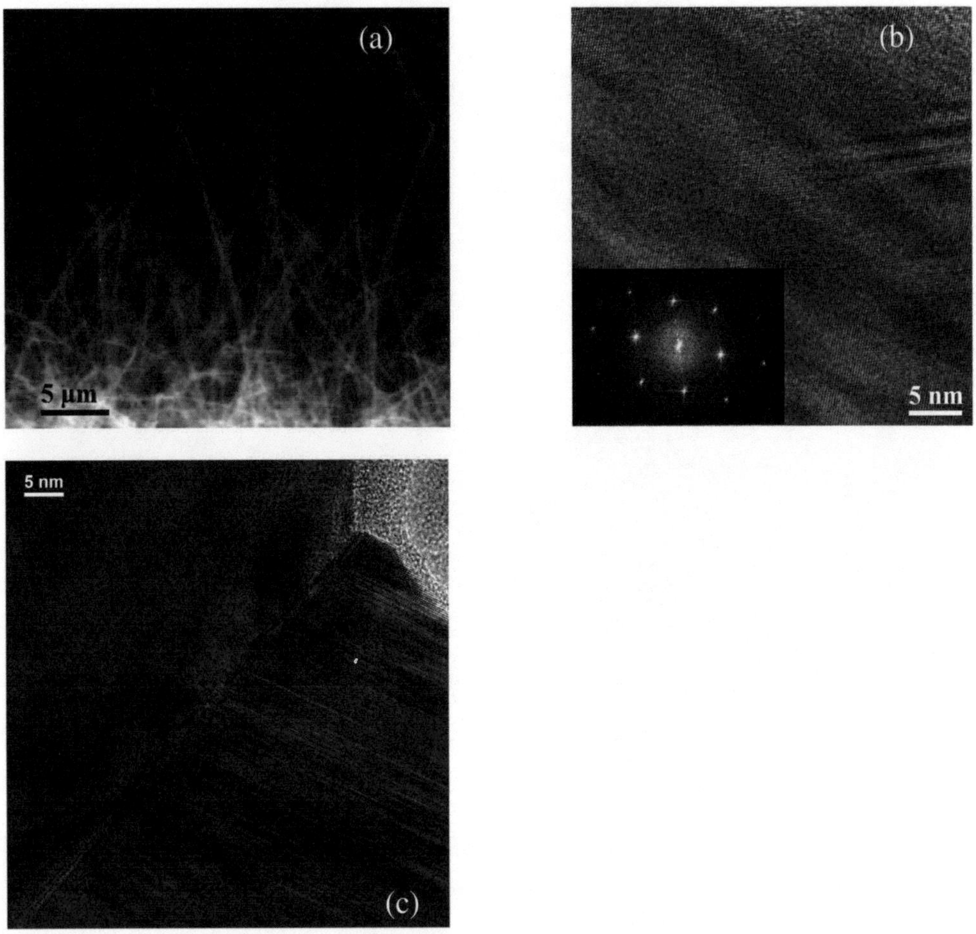

Fig. 17. (a) side-view SEM image of 3C-SiC NW grown on 3C-SiC substrate, (b) HRTEM image of the NW lattice and corresponding FFT analysis, (c) HRTEM image of the catalyst/wire interface at the tip.

4. Applications

The great interest that SiC has gathered in latest years is due the possibility to combine in one single material good semiconducting and chemical properties, biocompatibility and sensing potentiality. SiC chemical inertness, wide bandgap, tribological properties, hydroxyapatite-like osseointegration, and hemocompatibility make it a very promising candidate for biosensors, interface with biological tissues and lab-on-chip medical devices. Moreover, the possibility to realize nanostructures opens the possibility to assembly novel nanoscale devices and arrays by the bottom-up approach, and to enhance the sensing capabilities of the material thanks to the high surface/volume ratio of NW.

As it happens with other materials, low dimensional SiC may exhibit peculiar electronic, optical, thermal and chemical properties, as well as carrier quantum confinement. Many mono-dimensional structures have proven to exhibit better properties than the same material with bulk size, and the bottom-up approach would permit to combine appropriate building blocks to obtain unique functions, or combinations of functions, in integrated nanosystems, unavailable in the conventional top-down approach.

Several electronic devices, such as power diodes or field-effect transistors based on 3C-SiC (SiCFET) have already been developed (Zhou et al. 2006a, Chung et al., 2008) and SiC power diodes have already reached the market stage (Roussel, 2007). Several devices are fabricated by thermally growing SiO_2 layer on n-type Si wafers followed by deposition of parallel Au contacts, used as source and drain electrodes, while the n-type silicon is used as back gate. n-type 3C-SiC FETs were tested at different temperatures and were proven to have good performance at high temperatures, with larger current and higher carrier mobility. SiC nanostructures could be used to enhance these landmarks even more, as it happened with other compounds, but very few literature papers deal with SiC NW applications, even if there are already several "proof of concept" nanodevices realized with conventional semiconductors such as Si, GaAs, GaN NW (Xia et al. 2006, Lieber & Wang, 2007).

One of the challenges for SiC research for the next years would involve using the experience gathered on other compounds in order to exploit SiC possibilities in nanoscience.

Large-scale β-SiC NW coated with SiO_2 layer exhibit the characteristics of an excellent photocatalyst. The photocatalytic activity was evaluated by the photocatalytic decomposition of gaseous acetaldehyde accompanied by generation of carbon dioxide. SiC/SiO_2 core/shell NW present higher photocatlytic activity than the pure 3C-SiC NW, resulting from their stronger absorption of gaseous acetaldehyde by SiO_2 and holes remaining in the valence band of SiC (Zhou et al., 2006b).

Functionalization of the SiC surface with specific organic or inorganic molecules is a possible route to obtain specific surface properties such as high selectivity, biomolecular recognition and adhesion of species of biological interest. SiC, in all its polytypes, is suggested as a suitable material for biofunctionalization of H-terminated surfaces via hydrosilylation similar to silicon (Yakimova et al., 2007). Experimental study of pyrrole funtionalized Si- and C- terminated SiC have been made. This organic compound contain in its molecule NH group and two C-C double bond which may be used for further functionalizations (Preuss et al., 2006).

Other organic molecules with different functional groups, such as 3-aminepropyl triethoxysilane (APTES) having amine group or mercaptopropyl trimethoxysilane with thiol group, have been suggested for SiC NW surface functionalization.

In this framework we have used a Supersonic Molecular Beam Deposition approach (Milani & Iannotta, 1999) to functionalize 3C-SiC thick films and NW grown with our VPE reactor (Aversa et al., 2009). Porphyrins have been deposited onto the SiC surface and the core level/valence band emission have been studied with synchrotron radiation light. First experiments and analysis on SiC film show that both porphyrin macrocycle and the phenyl groups of the molecule have an interaction with the substrate: core level photoemission reveals significant changes on peak positions and lineshape. Functionalized 3C-SiC NW shows that oxidized and/or amorphous/oxycarbide components are present, while the Si_{2p} and C_{1s} bands show a strong lineshape change. The interaction of attached molecules is

more evident on SiC NWs than on films, and a role of F in "etching" the outer shell of NW could be hypothesized.
MEMS are a family of technologies and integrated devices that are becoming more and more important in modern life. Some areas in which these systems are already applied are shock sensors for airbag, inkjet printers, accelerometers and gyroscopes for boats and airplanes, entertainment, healthcare instruments, communication and information technologies, biology and biosensors. MEMS offer significant advantages over hybrid systems and devices because of their small dimensions, integration of different components and low power consumption (Godignon & Placidi, 2007). Although Si is presently the most used material for MEMS fabrication it has serious limitations for some applications, such as high temperature ($T > 300$ °C) and/or harsh environments with corrosive chemicals and biocompatibility, and SiC could be a viable alternative material. Micro cantilever of 3C-SiC with oscillating frequency in the order of 10^6 Hz and Q factors of 10^3 were realized and the resonators were used for high sensing applications such as mass detection for gas sensing devices in harsh environments (Zorman & Parro, 2008).
Combining MEMS and nanostructures peculiar characteristics such as high affinity to selected species via functionalization and high measurement sensitivity thanks to high resonant frequency and Q factors could open interesting possibility in detection devices for biological applications.
We have made some preliminary 3C-SiC NW depositions, starting from a patterned Si substrate on which structures such as micron-sized cantilever, beam and springs were fabricated. A Ni layer, nominally 2 nm thick, was deposited as catalyst on this microstructured Si substrate with an e-beam evaporator and the growth was performed in the VPE reactor, using the procedure described above. A SEM micrograph of the NW obtained on the spring is shown in fig. 18.

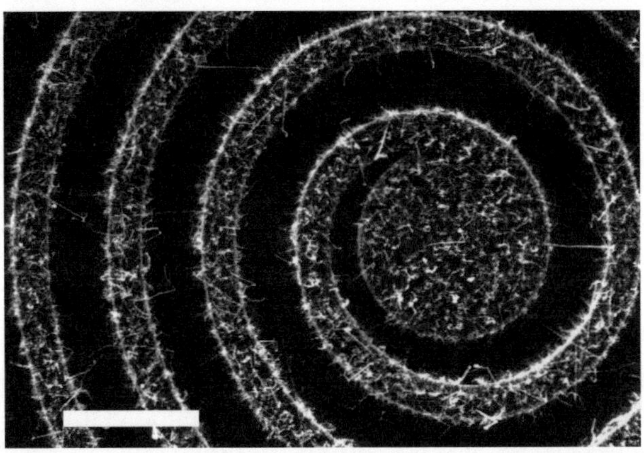

Fig. 18. SEM micrograph of 3C-SiC NW obtained on a Si spring (Si micromachined substrates were kindly provided by Dr. A. Roncaglia, IMM-Bologna, Italy). Marker is 20 μm.

5. Conclusions

Nanowires (NWs) open promising near-future perspectives for the design and fabrication of nano-scale devices.
In this paper we discussed the 3C-SiC NW preparation obtained by three different growth procedures:

- SiC/SiO_2 core/shell NWs grown by a Chemical Vapor Deposition (CVD) process on n-type Si (001) substrates
- 3C-SiC (without shell) NWs obtained by a chemical reaction using carbon tetrachloride (CCl4) as a precursor and Ni as catalyst.
- 3C-SiC NWs grown in an home-made Vapor Phase Epitaxy (VPE) reactor using propane and silane as precursors both diluted (3%) in hydrogen, and Ni as catalyst.

Selected morphological, structural and optical characterizations of the three NW sets, performed by Scanning Electron Microscopy, Transmission Electron Microscopy, X-Ray Diffraction and SEM-Cathodoluminescence, were presented.

We have presented some preliminary results about patterned NW growth on Si using standard photolitographic techniques to draw figures like crosses, NW functionalization using porphyrins and NW deposition on Si cantilever obtained by microstructuring a Si substrate.

6. Acknowledgements

We acknoledge the help of Mr. A. Motta for photolitographic work. Patterned SiC substrates were kindly provided by A. Roncaglia, IMM-CNR Bologna. We are grateful to Dr. L. Aversa, R. Verucchi and M. Nardi for synchrotron analysis.

7. References

Afanas'ev V. V.; Bassler M.; Pensl G. & Schulz M. J. (1996). Band offsets and electronic structure of SiC/SiO2 interfaces, *J. Appl. Phys.*, Vol. 79, issue 6 (15 March 1996), pag. 3108-3114, ISSN: 0021-8979.

Alfieri G. & Kimono T. (2009). Capacitance Spectroscopy Study of Midgap Levels in n-Type SiC Polytypes, *Mat. Sci. Forum*, Vol. 615 - 617 (March 2009) , pag. 389-392, ISSN: 0255-5476.

Attolini, G., Rossi, F., Bosi, M., Watts, B. E. & Salviati, G. (2008). Synthesis and characterisation of 3C-SiC NW. *J of non crystalline solid*, Vol. 354, issue 47-51 (1 December 2008) pag. 5227-5229, ISSN: 0022-3093

Aversa L. , Verucchi R., Nardi M., Bosi M., Attolini G., Rossi F., Watts B., Nasi L., Salviati G., Iannotta S.: "SiC functionalization by porphyrin supersonic molecular beams", 26th European Conference on Surface Science (ECOSS), Parma, 30/8 - 4/9/2009.

Bark, Y., Ryu, Y. & Yong K. (2006). Structural characterization of β-SiC nanowires synthesized by direct heating method. *Materials Science and Eng. C* Vol. 26, issue 5-7, Pag. 805-808, ISSN: 0928-4931

Bechstedt F.; Kackell P.; Zywietz A.; Karch K.; Adolph B.; Tenelsen K. & Furthmuller (1997). Polytypism and properties of silicon carbide, *Phys. Stat. Sol. B*, Vol. 202 Issue 1 (16 November 1997), pag. 35-62, ISSN: 1521-3951.

Bosi M.; Watts B. E.; Attolini G.; Ferrari C.; Frigeri C.; Salviati G.; Poggi A.; Mancarella F.; Roncaglia A.; Martnez O. & Hortelano V., 2009. Growth and Characterization of 3C-SiC Films for Micro Electro Mechanical Systems (MEMS) Applications, *Crystal Growth and Design*, DOI: 10.1021/cg900677c

Casady J. B. & Johnson R. W. (1996) Status of silicon carbide (SiC) as a wide-bandgap semiconductor for high-temperature applications: a review, *Solid State Electronics*, Vol. 39, issue 10, pag. 1409-1422, ISSN: 0038-1101

Chaussende, D. ; Manteil, Y.; Aboughe-nze, P.; Brylinski, C. & Bouix, J (1999). Thermodynamical calculations on the chemical vapour transport of silicon carbide, *Mat Sci Eng B*, Vol. 30 (July 1999), pag. 98-101, ISSN: 0921-5107.

Chen J.; Wu R.; Yang G.; Pan Y.; Lin J.; Wu L. & Zhai R. (2008). Synthesis and photoluminescence of needle-shaped 3C–SiC nanowires on the substrate of PAN carbon fiber. *J. of Alloy and Compounds*, Vol. 456, issue 1-2, (25 February 2004), pag. 320-23, ISSN: 0925-8388.

Choi, H-J.; Seong, H-K.; Lee, J-C. & Sung, Y-M (2004). Growth and modulation of silicon carbide NW. *Journal of crystal growth Vol.* 269, Issues 2-4 (1 September 2004), pag. 472-478, ISSN: 0022-0248.

Chung G. S.; Ahn J. H. & Han K. B. (2008). Fabrication of poly 3C-SiC thin film diodes for extreme environment applications, *IEEE International Symposium on Industrial Electronics, 2008. ISIE 2008* (30 June 2008), pag. 2576-2579, ISBN: 978-1-4244-1665-3.

Dohnalek, Z.; Gao, Q.; Choyke, W.J. & Yates Jr, J.T. (1994). Chemical reactions of chlorine on a vicinal Si (100) surface studied by ESDIAD, *Surf Sci*, Vol. 320, issue 3, 238-246, ISSN: 0039-6028.

Gali A.; Heringer D.; Deak P.; Hajnal Z.; Frauenheim Th.; Devaty R.P. & Choyke W.J. (2002). Isolated oxygen defects in 3C- and 4H-SiC: A theoretical study, *Phys. Rev. B*, Vol. 66, Issue 12 (27 September 2002), pag. 125208-125215, ISSN: 1550-235x.

Godignon P. & Placidi M. (2007). Recent improvements of SiC micro-resonators, *Phys. Stat. Sol. C*, Vol. 4, issue 4 (28 march 2007), Pag. 1548-1553, ISSN: 1610-1642.

Huang, J.; Guo, W. & Xu, P. (2007). Comparative Study of Decomposition of CCl4 in Different Atmosphere Thermal Plasmas, *Plasma Sci Technol.* Vol. 9, issue 1 (february 2007) pag. 76-79, ISSN: 1009- 0630.

Itoh H.; Kawasuso A.; Ohshima T.; Yoshikawa M.; Nashiyama I.; Tanigawa S.; Misawa S.; Okumura H. & Yoshida S. (1997). Intrinsic Defects in Cubic Silicon Carbide, *Phys. Stat. Sol. (a)*, Vol. 162, Issue 1 (16 november 2001), pag. 173-198. ISSN: 1862-6319.

Jabeen F.; Rubini S.; Grillo V.; Felisari L. & F. Martelli (2008). Room temperature luminescent InGaAs/GaAs core-shell nanowires, *Appl. Phys. Lett.*, Vol. 93, issue 8 (28 August 2008), pag. 083117 -083119, ISSN: 0003-6951.

Kang, B.-C., Lee, S.-B. & Boo, J.-H. (2004). Growth of β-SiC NW on Si(100) substrates by MOCVD using nichel as a catalyst. *Thin Solid Films* Vol. 464-465, (October 2004), pag. 215-219, ISSN: 0040-6090

Kassiba A.; Makowska-Janusik M.; Boucle J.; Bardeau J. F.; Bulou A. & Herlin-Boime N. (2002). Photoluminescence features on the Raman spectra of quasistoichiometric SiC nanoparticles: Experimental and numerical simulations, *Phys. Rev. B*, Vol. 66, issue 15 (14 October 2002), pag. 155317-155324, ISSN: 1550-235x.

Khongwong, W., Imai, M., Yoshida K. & Yano, T. (2009). Influence of raw powder size, reaction temperature, and soaking time on synthesis of SiC/SiO2 coaxial nanowires via thermal evaporation, *Journal of Ceramic Society of Japan*, Vol. 117, pag. 439-444, ISSN: 1348-6535

Kolasinski, K.W. (2006). Catalytic growth of nanowires: Vapor–liquid–solid, vapor–solid–solid, solution–liquid–solid and solid–liquid–solid growth. *Curr. Opin. Solid State Mater. Sci.* vol. 10, issue 3-4, pag. 182-191, ISSN: 1359-0286.

Li X. & Bhushan B. (1999). Micro/nanomechanical characterization of ceramic films for microdevices, *Thin Solid Films*, Vol. 340, issue 1, pag. 210-217, ISSN: 0040-6090.

Liang, C.H., Meng, G.W., Zhand, L.D., Wu, Y.C. & Cui, Z. (2000). Large-scale synthesis of β-SiC nanowires by using mesoporous silica embedded with Fe nanoparticles, *Chem. Phys. Lett.* Vol. 329, issue 3-4 (20 October 2000) pag. 323-328. ISSN: 0009-2614

Lieber C. M. & Wang Z. L. (2007). Functional Nanowires, *MRS Bull.*, Vol. 32 (february 2007), pag. 99-108, ISSN: 0883-7694.

Louis C. ; Roux S.; Ledoux G. ; Dujardin C. ; Tillement O. ; Cheng B. L. & Perriat P. (2006). Luminescence enhancement by energy transfer in core-shell structures, *Chem. Phys. Lett.* Vol. 429, issue 1-3 (29 september 2006) pag. 157–160, ISSN: 0009-2614.

M. Mehregany, C. A. Zorman, S. Roy, A. J. Fleischman, C. H. Wu, & N. Rajan (2000). Silicon Carbide for Microelectromechanical Systems, *Int. Mater. Rev.*, Vol. 45, issue 3, pag. 85–108, ISSN: 0950-6608

Marshall, R.C., Faust J.W., & Ryan, C.E.; (1973) Silicon carbide; *Proceedings of the International Conference on Silicon Carbide* Columbia : University of South Carolina Press, ISBN 0872493156.

McKnight S.W. & Palik E.D. (1980). Cathodoluminescence of Films, *J. Non-Crystal. Solids*, Vol. 40, issue 1-3, pag. 595-604, ISSN: 0022-3093.

Meng, A., Li, Z., Zhang, J., GaO, L. & Li, H. (2007). Synthesis and Raman scattering of β-SiC/SiO2 core–shell nanowires, *J. Cryst. Growth.* Vol. 308, issue 2, (15 October 2007) pag. 263-268, ISSN: 0022-0248

Milani P. & Iannotta S. (1999). *Cluster Beam Synthesis of Nanostructured Materials (Series in Cluster Physics, 1)* Springer, ISBN: 3540643702.

Nakashima S. & Harima H. (1997). Raman investigation of SiC polytypes, *Phys. Stat. Sol. A*, Vol. 162, Issue 1 (16 November 1997), pag. 39-64, ISSN: 1862-6319

Pistol M.E. & Pryor C.E. (2008). Band structure of core-shell semiconductor nanowires, *Phys. Rev. B*, Vol. 78, issue 11 (23 September 2008), pag. 115319, ISSN: 1550-235x.

Preuss, M.; Bechstedt, F.; Schmidt, W.G.; Sochos, J.; Schroter, B. & Richter, W. (2006). Clean and pyrrole-functionalized Si- and C-terminated SiC surfaces: First-principles calculations of geometry and energetics compared with LEED and XPS, *Phys. Rev.* B, Vol. 74, issue 23 (December 2006), pag. 235406 ISSN: 1550-235x.

Roussel, P, Semiconductor Today vol. 2, issue 5, 2007, pag.40 - http://www.semiconductor-today.com/features/Semiconductor%20Today%20-%20SiC%20power%20devices.pdf

Sarro P. M. (2000). Silicon carbon as a new MEMS technology, *Sens. Actuators A*, vol. 82, issue 1-3, pag. 210-218, ISSN: 0924-4247

Seo, W. -S., Koumoto, K. & Aria, S. (2000). Morphology and Stacking Faults of-SiC Whisker Synthezised by Carbothermal Reduction, *J. Am. Ceram. Soc.* Vol. 83, issue 10, pag. 2584-2592, ISSN 0002-7820.

Seong, H-K,; Choi, H-J.; Lee, S-K.; Lee, J-I. & Choi, D-J. (2004). Optical and electrical transport properties in silicon carbide NW. Applied Physics Letters, vol. 85, issue 7 (August 2004), pag. 1256-1258, ISSN: 0003-6951.

Shim H.W.; Kim K.C.; Seo Y.H.; Nahm K.S.; Suh E.K.; Lee H.J. & Hwang Y.G. (1997) Anomalous photoluminescence from 3C-SiC grown on Si(111) by rapid thermal chemical vapor deposition. *Appl. Phys. Lett.* Vol. 70, issue 13 (31 march 1997), pag. 1757-59. ISSN: 0003-6951.

Skold N.; Karlsson L.S.; Larsson M.W.; Pistol M.E.; Seifert W.; Tragardh J. & L. Samuelson (2005). Growth and Optical Properties of Strained GaAs–$Ga_xIn_{1-x}P$ Core–Shell Nanowires, *Nanolett.*, Vol. 5, issue 10 (13 September 2005), pag. 1943-1947, ISSN: 1530-6984.

Skuja L. (1998). Optically active oxygen-deficiency-related centers in amorphous silicon dioxide, *J. Non-Crystal. Solids,* Vol. 239, pag. 16-48, ISSN: 0022-3093.

Takai, S., Kohno, H. & Takeda, S. (2007). MOCVD growth of SiC NW aiming at the control of their shape. *Advanced Materials Research*, Vol. 26-28, Pag. 657-660, ISSN: 1022-6680

Tomioka K.; Kobayashi Y.; Motohisa J.; Hara S. & T. Fukui (2009). Selective-area growth of vertically aligned GaAs and GaAs/AlGaAs core–shell nanowires on Si(111) substrate, *Nanotechnol.* Vol. 20 Issue 14 (April 2009), pag. 145302-145305, ISSN:0957-4484.

Wacaser, B.A.; Dick, K.A.; Johansson, J.; Borgstrom, M.T.; Deppert, K. & Samuelson L. (2009). Preferential Interface Nucleation: An Expansion of the VLS Growth Mechanism for Nanowires, *Adv. Mater.* Vol. 21, Issue 2 (25 November 2008), pag. 153-165, ISSN: 1521-4095.

Wagner, R. S. & Ellis W.C. (1964). Vapor-liquid-solid mechanism of single crystal growth, *Appl. Phys. Lett.* Vol. 4 , pag. 89, ISSN: 0003-6951

Wang N., Cai Y., & Zhang R.Q. (2008). Growth of nanowires, *Mater. Sci. Eng. R.* Vol. 60, issue 1-6, pag. 1–51, ISSN: 0927-796X

Willander M., Friesel M., Whahab Q-U & Straumal B. (2006), Silicon carbide and diamond for high temperature device applications, *J. Mater. Sci: Mater. El.*, Vol. 17, issue 1, pag. 1-25, ISSN: 0957-4522

Xi G., Liu Y., Liu X., Wang X. & Qian Y. (2006). Mg-catalyzed autoclave synthesis of aligned silicon carbide nanostructures. *Phys. Chem. B*, Vol. 110, issue 29 (Jul 2006), pag. 14172-14178, ISSN: 1089-5647.

Xia B. Y.; Yang P.; Sun Y.; Wu Y.; Mayers B.; Gates B.; Yin Y.; Kim F. & Yan H. (2003). One-dimesnional nanostructures: synthesis, characterization and applications, *Adv. Mater.* Vol. 15. issue 5 (march 2004), pag. 353-389, ISSN: 1521-4095.

Yakimova R., Petoral R. M., Yazdi G. R., Vahlberg C., Petz A. L. & Uvdal K. (2007). Surface functionalization and biomedical applications based on SiC, Surface functionalization and biomedical applications based on SiC, *J. Phys. D: Appl. Phys.* Vol. 40, pag. 6435-6442. ISSN: 0022-3727

Yang, L.; Zhang, X.; Huang, R.; Zhang, G. & Xue, C. (2006). Formation of β-SiC NW by annealing SiC films in hydrogen atmosphere. *Physica E*, Vol. 35, Issue: 1 (October 2006), pag. 146-150, ISSN: 1386-9477.

Yang, W.; Araki, H.; Kohyama, A.; Thaveethavon S.; Suzuki, H. & Noda, T. (2004). Fabrication in-situ NW/SiC matrix composite by chemical vapour infiltration process. *Materials Letters*, Vol. 58, Issue 25 (October, 2004), Pag. 3145-3148, ISSN 0167-577X.

Yoshida H.; Kohno H; Ichikawa S.; Akita T. & S. Takeda (2007). Inner potential fluctuation in SiC nanowires with modulated interior structure, *Mat. Lett.* Vol. 61, Issue 14-15 (June 2007), pag. 3134-3137, ISSN: 0167-577X

Zhang Y.; Nishitani-Gamo M.; Xiao C. & Ando T. (2002) Synthesis of 3C–SiC nanowhiskers and emission of visible photoluminescence. J. Appl. Phys. Vol. 91, issue 9 (1 may 2002), pag. 6066-6070. ISSN: 0021-8979.

Zhang, H-F., Wang, C-M. & Wang, L-S. (2002). Helical Crystalline SiC/SiO2 Core− Shell Nanowires, *Nano Letters* Vol. 2, issue 9, (26 July 2002), pag. 941-944. ISSN: 1530-6984

Zhou, W.M.; Fang, F.; Hou, Z.Y.; Yan, L.J. & Zhang, Y.F. (2006a). *Fiel-Effect transistor based on β-SiC NW. IEEE Electron Device Letters*, Vol. 27, issue 6 (june 2006), pag. 463-465. ISSN: 0741-3106.

Zhou, W.; Yan, L.; Wang, Y. & Zhang, Y.; (2006b) Sic NW: a photocatilytic nanomaterial. *Appl. Phys. Lett.*, Vol. 89, issue 1 (5 july 2006), pag. 013105, ISSN: 0003-6951.

Zhu J.; Liu S. & Liang J. (2000). Raman study on residual strains in thin 3C-SiC epitaxial layers grown on Si(001), *Thin Solid Films*, Vol. 368, Issue 2 (15 June 2000, pag. 307-311, ISSN: 0040-6090.

Zhu, S.; Xi, H-A.; Li, Q. & Wang, R.; (2005) In situ growth of β-SiC NW in porous SiC ceramics. Journal Am. Ceram. Soc. Vol 88, Issue 9 (September 2005), pag. 2619-2621, ISSN 0002-7820

Zorman C.; & Parro R. (2008). Micro- and nanomechanical structures for silicon carbide MEMS and NEMS, *Phys. Stat. Sol (b)*, Vol. 245, issue 7 (July 2008), pag. 1404-1424, ISSN: 1521-3951.

2

Heterostructured III-V Nanowires with Mixed Crystal Phases Grown by Au-assisted Molecular Beam Epitaxy

D.L. Dheeraj, H.L. Zhou, A.F. Moses, T.B. Hoang, A.T.J. van Helvoort, B.O. Fimland and H. Weman
Norwegian University of Science and Technology
Norway

1. Introduction

Nanowires (NWs) with diameters in the range of few 10s of nanometers and lengths up to few micrometers long are one-dimensional materials that have attracted a great deal of interest recently in various fields, including photonics, electronics and medicine. These one-dimensional materials can offer improvement in the performance of optical and electronic devices due to their size-induced quantum confinement effects. In addition, their high surface area to volume ratio makes them potential candidates for various sensor applications. Several review articles reporting about the properties and applications of NWs made of different materials have already been published (Tian et al., 2009; Agarwal, 2008; Schmidt et al., 2009; Agarwal & Lieber, 2006; Lauhon et al., 2004). In this book chapter, we focus on heterostructured III-V semiconductor NWs with increased potential for use in photonic applications grown by molecular beam epitaxy.

Most optoelectronic devices are fabricated by using various III-V materials, because of their direct bandgap nature. The efficiency of these devices have improved substantially in the last 5 decades by reducing the dimensions of these materials, reducing defect density and changing the design using e.g. quantum wells (QW), multiple QWs, superlattice heterostructures and quantum dots (QDs) in the active layers. QWs consist of a very thin layer of a material with a lower bandgap sandwiched between materials with a higher bandgap. Such structures not only confine the carriers in the growth direction but also passivate surface states at the lower bandgap material surface. However, the choice of materials for the growth of defect free heterostructures is limited by the lattice mismatch of different materials. This is also one of the main reasons inhibiting the integration of III-V materials on Si. Such problems can be avoided in the growth of NWs, which provides the provision to relax in the radial direction without forming any dislocations. In fact, several successful NW based heterostructures, including structures with lattice mismatch as large as 7.8%, has been reported (Guo et al., 2006).

1.1 Vapor-Liquid-Solid growth mechanism

Fabrication of NWs by the bottom-up approach was first demonstrated by Wagner and Ellis in 1964, and the technique was named as vapor-liquid-solid (VLS) growth. This technique uses a metal particle (often Au) as a catalyst, which forms a liquid-alloy particle on reacting with source materials supplied in vapor form. On supplying the source material continuously to this liquid-alloy, the excess material dissolved in it precipitates in the form of a solid NW at the bottom of the particle (Wu & Yang et al., 2001). The diameter and position of the NW are defined by the Au particle. As shown in Fig. 1, the Ga adatoms hitting the substrate surface diffuse to the Au droplet along the NW sidewall. With this technique, it is possible to tailor the composition of the NW in both the axial and radial directions. The different steps involved in the VLS growth mechanism are shown in Fig. 1.

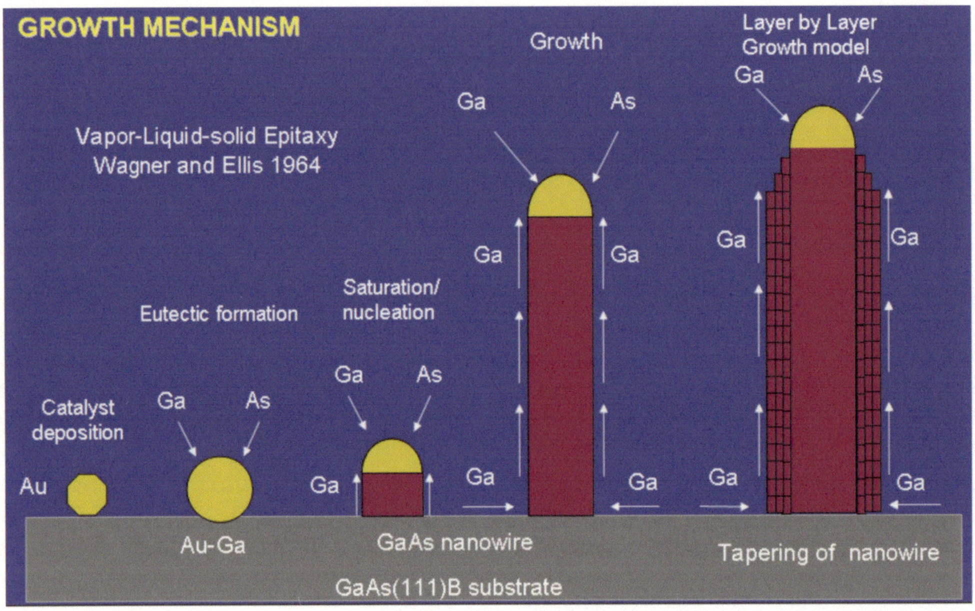

Fig. 1. Schematic diagram showing the various steps in the VLS growth of a GaAs NW.

As the NWs grow longer, tapering of the NWs is often observed. There are mainly two arguments explaining the reason for tapering: the first reason is based on the decrease in diameter of the Au droplet and hence the NW with its length due to the decrease of incoming Ga flux to the Au droplet (Harmand et al., 2005), while the second reason is explained by the (radial) layer-by-layer growth mechanism which would become effective when the length of the NW exceeds the diffusion length of Ga adatoms on the NW sidewall (Chen et al., 2006).

NWs have been grown by using the VLS growth technique in most III-V materials combinations, on native substrates as well as on silicon (Ihn et al., 2007; Tomioka et al., 2009), and with several epitaxial techniques, including metal organic vapor phase epitaxy (MOVPE), chemical beam epitaxy (CBE) and molecular beam epitaxy (MBE). Over the past decade, there has been considerable interest focused on the growth of heterostructured NWs (both axial and radial core-shell structures), followed by the first demonstrations of device prototypes, such as light-emitting diodes (Minot et al., 2007; Svensson et al., 2008;), high electron mobility devices (Jiang et al., 2007), laser-diodes (Duan et al., 2003), and photovoltaic cells (Czban et al., 2009).

1.2 Molecular beam epitaxial growth of NWs

MBE is one of the most controllable and widely used non-equilibrium growth techniques for growing thin, epitaxial films of a wide variety of materials. The deposition of material on the substrate is performed by evaporating the material from the effusion cells in a chamber maintained under ultrahigh vacuum. The deposition performed in an ultrahigh vacuum chamber not only makes it possible to grow highly pure materials but also to install in-situ characterization techniques such as reflection high-energy electron diffraction (RHEED). A deposition rate as low as 0.1 ML/s can be achieved by changing the temperature of the effusion cell. In addition, a flux of molecules or atoms towards the substrate can be abruptly released (or blocked) by controlling the shutters in front of the cell. This allows growing heterostructures with abrupt interfaces. Another major difference between the MBE growth method and other growth techniques is that it is far from thermodynamic equilibrium conditions, i.e., it is mainly governed by the kinetics of the surface processes.

The growth of NWs by the VLS mechanism in MBE was initially speculated to be impossible due to that the material is supplied in the form of atoms, where no catalytic activity of Au to decompose the metallic groups to atoms is required as in metal-organic chemical vapor deposition. However, these speculations were proven to be incomplete since several groups have demonstrated the growth of GaAs NWs by MBE (Wu et al., 2002; Cirlin et al., 2005; Harmand et al., 2005; Plante and LaPierre, 2006). Furthermore, remarkable progress has also been shown in self-catalyst growth of GaAs NWs by MBE (Fontcuberta i Morral et al., 2008).

The Au-assisted process of NW growth in MBE is explained to be a thermodynamic process due to which the group III adatoms diffuse from the substrate surface, which is at a higher chemical potential as compared to the Au droplet, which has lower chemical potential (Harmand et al., 2006). On the other hand, it has also been shown that the kinetic processes involved in the NW growth also play an important role during the NW growth, and are a bit different than those during the thin-film growth. The major kinetic processes involved during the growth of NWs by Au-assisted MBE are schematically shown in Fig. 2.

The dependence of the (axial) growth rate of NWs in the MBE process on the NW diameter is found to be different from that of the observations made on the NW growth by the CVD process (Givargizov, 1975). It is observed that the MBE growth rate of the NWs decreases with increase in the diameter of NWs (Plante & LaPierre, 2006). This was theoretically explained by Dubrovskii et al., 2005, attributing the diameter dependence of the NW growth to the diffusion of group III adatoms from the substrate surface to the Au particle.

The demonstrations of the growth of homogeneous binary compound NWs, including GaAs, InP, InAs and GaP by MBE (Cornet et al., 2007; Harmand et al., 2005; Ihn & Song, 2007) were followed by the growth of heterostructures involving ternary compounds such as AlGaAs, InGaAs, GaAsSb, InAsP and GaAsP (Chen et al., 2006; Chen et al., 2007; Cornet & LaPierre, 2007, Dheeraj et al., 2008a, Tchernycheva et al., 2007). However, the progress in the MBE growth of ternary compound NWs involving different group III elements, such as AlGaAs and InGaAs, has been inhibited due to difficulties in obtaining uniform composition along the length of the NW. This problem arises due to the difference in diffusion lengths of different group III elements along the NW sidewall. On the other hand, no variation in composition along the NW has been reported for the ternary compound NWs involving different group V elements such as InAsP and GaAsP. This could be due to the longer diffusion lengths of group V elements compared to that of the group III elements. The variation in composition could, however, be quite small in short NWs or inserts involving ternary structures.

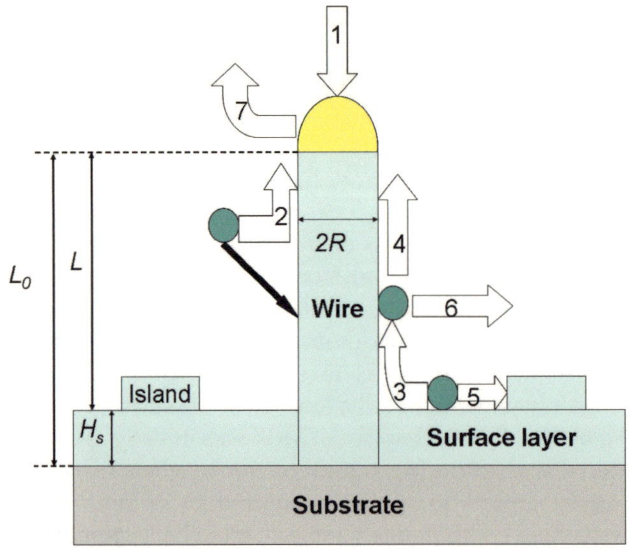

Fig. 2. The model of Au-assisted NW growth by MBE: The different pathways for adatoms are through; 1) direct impingement, 2) diffusion flux from the side walls to the droplet, 3) diffusion from the substrate to the sidewalls, 4) diffusion from the substrate along the sidewalls to the droplet, 5) nucleation on the surface, 6) desorption from the sidewalls, 7) desorption from the droplet (Dubrovskii & Sibirev, 2007).

Despite the fast progress in the growth of heterostructured NWs by MBE, the understanding and control of the crystal structure of these NWs is a challenging task. In fact, most of the III-V NWs grown by MBE adapt wurtzite (WZ) crystalline structure, even though the III-V materials exhibit zinc-blende (ZB) structure in bulk form. Since the WZ crystalline structure is not a stable form of these materials, the NWs often exhibit a high density of crystal defects (stacking faults, twins) irregularly within the NWs. A stacking fault in a WZ phase inherently forms a unit cell of ZB segment, while a twin in a ZB phase inherently forms a unit cell of a WZ segment (Caroff et al., 2009). It is very important to reduce the density of such crystal defects since they deteriorate the optical and electrical properties of the NWs.

Due to that the NWs have been found to form with different crystal phases depending on the growth conditions, this now allows for *heterocrystalline* band-structure engineering. This includes crystal phase-modulated as well as combined crystal phase- and crystal material-modulated heterostructures. This is because III-V materials in the ZB form have different band gap and different band offsets compared to that in the WZ form. Therefore, heterojunctions form in the same material wherever the material changes its crystalline phase. Recently, some research groups have shown the ability to control the crystal structure in MOCVD grown NWs (Caroff et al., 2009; Algra et al., 2008). Despite of a few successful attempts in controlling stacking faults in NWs grown by MBE, it is still a challenging task to understand the factors influencing the crystal structure (Shtrikman et al., 2009; Patriarche 2008).

Although the WZ crystal phase of GaAs in NWs has been achieved by many groups by now, little is known about its electronic and optical properties. Several works have attempted to determine the band gap in WZ GaAs NWs but the results are still inconsistent (Zanolli et al., 2007; Martelli et al., 2007; Moewe et al., 2008). As mentioned earlier, stacking faults in the

WZ phase inherently possess the ZB phase. Depending on the conduction and valence band offset energies in the ZB crystal structure with respect to WZ, even a low density of ZB segments could trap most of the photo-excited electrons and/or holes. As a result, one may observe that the dominating photoluminescence related emission in such NWs occur at energies close to (type I band offsets) or even below (type II band offsets) the band gap energy of ZB GaAs even if the band gap is higher for WZ GaAs.

1.3 Overview of the book chapter

In this book chapter, we will discuss the structural and optical properties of axial and radial heterostructured III-V NWs with mixed crystal phases grown by Au-assisted MBE.

Our GaAs NWs are observed to exhibit WZ crystalline structure and often stacking faults (SFs) are randomly distributed in the NW. Interestingly, by incorporating Sb in the GaAs NWs, we have found that GaAsSb NWs form in the ZB crystalline structure with very few twin defects (Dheeraj et al., 2008a). Recently, we have demonstrated the growth and characterized the optical properties of single and multiple SF-free ZB GaAsSb inserts in WZ GaAs NWs (Dheeraj et al., 2008b; Dheeraj et al., 2009). The growth of GaAs NWs with multiple GaAsSb inserts enabled us to conclude that the growth rate of NWs changes with growth time due to the inclined molecular beam in the growth chamber, and revealed the growth parameters influencing the formation of SFs as we will demonstrate within this chapter.

Micro-photoluminescence (μ-PL) characterization of SF-free WZ GaAs NWs reveals that the bandgap of WZ GaAs is ~ 29 meV larger than that of ZB GaAs (Hoang et al., 2009). Further, we demonstrate the growth procedure of GaAs/AlGaAs radial heterostructured core-shell NWs, with strongly enhanced photoluminescence efficiency from the GaAs core due to the shell passivation of surface states (Zhou et al., 2009). The structural characterization of these NWs by high resolution transmission electron microscopy (HRTEM) shows that the shell copies the crystalline structure of the core NW.

Finally, we discuss the growth of single GaAs/AlGaAs core-shell NWs with axial GaAsSb core-inserts. We demonstrate that it is possible to change the crystal phase of GaAs at the upper GaAsSb/GaAs interface by introducing a growth interruption. By changing the GaAs crystal phase above the insert from ZB to WZ, we observe dramatic changes in the photoluminescence properties, explained to be due to that the heterojunction band alignment is changed from type II to type I.

2. Experimental procedure

The NWs were grown in either a Varian Gen II Modular or a Riber 32 MBE system equipped with a Ga dual filament cell, an Sb cracker cell, and an As valved cracker cell, allowing to fix the proportion of dimers and tetramers. In the present study, the major species of arsenic and antimony were As_4 and Sb_2, respectively. The substrate surface was first deoxidized at 620°C, and then a 60 nm thick GaAs buffer, a 72 nm thick $Al_{0.33}Ga_{0.67}As$ film and a 36 nm thick GaAs film were grown under growth conditions producing an atomically flat surface (Yang et al., 1992). The AlGaAs film embedded in the GaAs buffer can be used as a marker to distinguish between the GaAs buffer and the 2D GaAs grown during the growth of the NWs. The GaAs/AlGaAs/GaAs heterostructure was capped with an amorphous As layer to avoid oxidation during its transfer in ambient air to an electron-beam evaporation system for gold deposition. The As cap was then desorbed at 280°C under high vacuum (10^{-7} Torr) in the electron-beam evaporation chamber (Bernstein et al., 1992). A ~ 1 nm thick Au film, as determined by a quartz crystal thickness monitor, was deposited on the sample surface. The sample was then transferred to and loaded into the MBE system. Under an As_4 flux of 6×10^{-6} Torr, the substrate temperature was increased to 540°C, a temperature suitable for GaAs NW growth. At this stage, nanoparticles containing Au alloyed with the substrate constituents were formed. GaAs NW growth was initiated by opening the shutter of the Ga effusion cell. The temperature of the Ga effusion cell was preset to yield a nominal planar growth rate of 0.7 ML/s. The Sb shutter was also opened to supply an Sb_2 flux of 6×10^{-7} Torr to grow GaAsSb NWs, whenever needed.

Morphological characterization of NWs was performed in a Zeiss Supra field-emission scanning electron microscopy (FE-SEM) operating at 5 kV. Crystalline structure was analyzed in either a Philips CM20, Philips CM30 or Jeol 2010F TEM operating at 200 kV and equipped with an Oxford instruments INCA energy dispersive X-ray (EDX) spectrometer for composition measurements. For TEM characterization, the NWs were scrapped off from the substrate and transferred to a Cu grid with a lacey carbon film. A more detailed report on the TEM techniques used for characterization of NWs can be found in Van Helvoort et al., 2009.

The crystallographic orientation of the NW ensembles on as-grown substrates was checked by X-ray diffraction (XRD) pole figure measurements (Largeau et al., 2008) carried out with a PANalytical X'Pert Pro MRD diffractometer with a point focus configuration defined by a poly capillary lens and crossed slits.

μ-PL measurements were carried out using an Attocube CFMI optical cryostat. Samples were placed in a He exchange gas and kept at a temperature of 4.4 K. For temperature dependent measurements, the sample can be heated up to 70 K. Single NWs were excited by a 633 nm laser line. The laser was defocused onto the NWs with an excitation density of approximately 1 $kWcm^{-2}$ using a 0.65 numerical aperture objective lens. The μ-PL from single NWs was dispersed by a 0.55 m focal length Jobin-Yvon spectrograph and detected by an Andor Newton thermo-electrically cooled Si CCD camera. The spectral resolution of the system is ~ 200 μeV. For single NW measurements, NWs were removed from their grown substrate and dispersed on a Si substrate with an average density of ~ 0.1 NW per μm^2.

3. Results

3.1 Growth of GaAsSb NWs

GaAs NWs were grown for 5 minutes and GaAsSb NWs for 20 minutes on GaAs(111)B substrates. Typical 45° tilted (Fig. 3(a)) and top-view (Fig. 3(b)) SEM images show the GaAsSb NWs with their Au catalyst and a growth axis normal to the GaAs(111)B surface. Diameters of the NWs range from 30 to 100 nm and are nearly uniform along the NW, except at the base which is broader than the rest of the NW. The top-view image (Fig. 3(b)) shows a hexagonal cross sectional shape of the NW as well as at the base region. The length and diameter of the NWs was measured from the SEM images. The NW length is observed to decrease with increase in NW diameter (Dheeraj et al., 2008a). This dependence of NW length on diameter confirms that growth is to a large extent fed by diffusion of adatoms from the surface of the substrate to the Au droplet at top of the NWs, as described by Dubrovskii & Sibirev, 2007.

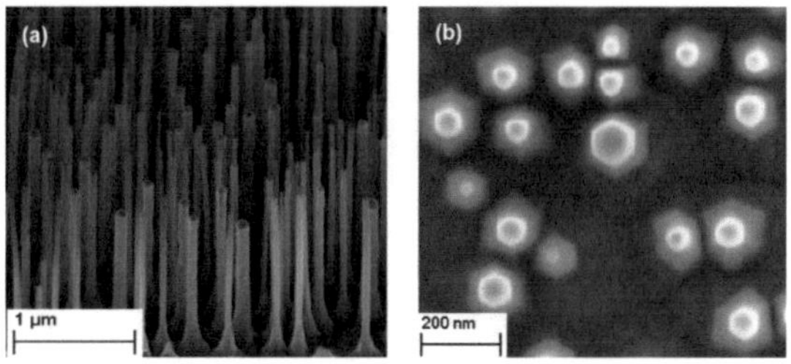

Fig. 3. (a) 45° tilted view and (b) top view SEM images of as-grown GaAsSb NWs grown on a GaAs(111)B substrate by Au-assisted MBE (Dheeraj et al., 2008a).

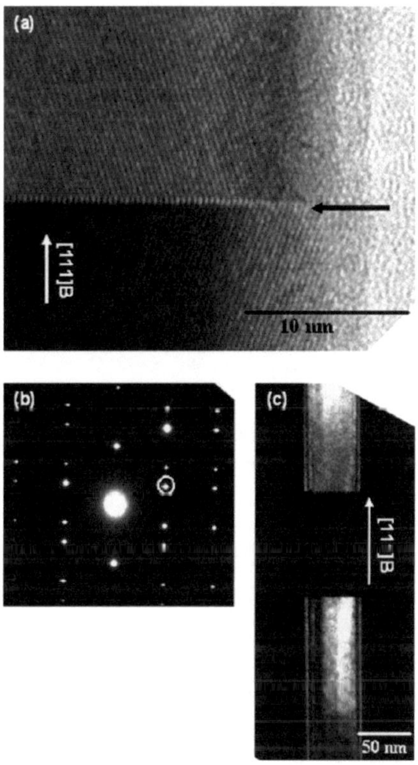

Fig. 4. (a) High resolution TEM image and (b) electron diffraction pattern showing the ZB structure with twins. (c) Dark-field image obtained by using the $1\bar{1}1$ diffraction spot, marked by a circle in Fig. 4(b) (Dheeraj et al., 2008a).

In Fig. 4, we show TEM images of a typical GaAsSb NW. The most striking feature of our experimental results is that our GaAsSb NWs adapt ZB structure, whereas the same growth conditions (except the Sb flux) produce GaAs NWs with WZ structure, as commonly observed for [111]B-oriented NWs grown by MBE. Fig. 4(a) shows a [110] zone axis HRTEM image revealing the ZB structure of the GaAsSb NW, with a twinning plane (indicated by black arrow) perpendicular to the [111]B growth axis. The crystal planes above the twinning plane (orientation A) is a rotation by 60° relative to the crystal planes below the twinning plane (orientation B) about the growth axis. The selected area electron diffraction (SAED) pattern in Fig. 4(b) shows additional spots related to the presence of both crystal orientations. Using the $1\bar{1}1$ diffraction spot marked by a circle in Fig. 4(b), we obtained the dark field image as shown in Fig. 4(c), where crystal orientation A appears bright and the crystal orientation B dark. By investigating the full length of several NWs, single-oriented ZB segments as long as 500 nm were observed. We have also demonstrated a technique to determine the average volumes of crystal orientation A and B parts of GaAsSb NW ensembles by XRD pole figure measurements, and found the volumes to be of equal ratio.

The XRD pole figure is a convenient characterization tool to obtain average data of NW ensembles. The pole figures of a sample with as-grown GaAsSb NWs were first recorded at the Bragg angles of the 111 and 220 GaAs reflections (Dheeraj et al., 2008a). We observed a central diffraction spot in the 111 GaAs XRD pole figure pattern, corresponding to the (111) planes parallel to the surface, and additional sharp spots located at an inclination angle of 70.5° in a sixfold symmetry pattern, corresponding to the other {111} planes. In the 220 XRD pole figure, spots were located at 35.3° in a similar sixfold symmetry pattern, corresponding to {220} planes. These features confirm that the NW growth axis was aligned with the $[\bar{1}\,\bar{1}\,\bar{1}]$ GaAs direction. Both pole figures show a sixfold symmetry, which must be due to the twinning of the ZB structure in the NWs, as a pure and single ZB phase has a threefold symmetry along <111> directions. θ -2θ scans (not shown here) were performed across each spot of the pole figures and we observed that the GaAsSb diffraction peaks of crystal orientation A and B parts had equivalent intensities. This means that the average volumes of GaAsSb segments with crystal orientation A and B, respectively, in the NW ensembles are equal.

In the 220 GaAs XRD pole figure pattern, we also observed a second set of six spots corresponding to hexagonal WZ $\{11\bar{2}0\}$ planes normal to the substrate surface. It is very likely that this hexagonal phase corresponds to the bottom part of the NWs, which consists of GaAs (no Sb) base of the NW as was mentioned in the growth procedure described above. This is consistent with our observation (as mentioned earlier) that pure GaAs NWs grown under our experimental conditions adopt the WZ structure, in accordance with previous reports on MBE-grown GaAs NWs. We should mention that the GaAs NW base of the present GaAsSb NWs was not observed by TEM. During sample scratching to collect NWs on the TEM grid, the NWs were probably broken above the GaAs part of the NW.

3.2 Growth of GaAs/GaAsSb/GaAs axial heterostructured NWs

GaAs/GaAsSb/GaAs axial heterostructured NWs were grown with 20 minutes, 30 seconds and 5 minutes growth times for each respective segment. The TEM image of such NW is shown in Fig. 5. The GaAs NW segment before the GaAsSb insert exhibited a pure WZ phase with almost no stacking faults. As can be seen in Fig. 5(a) and the HRTEM image of Fig. 5(c), the transition (marked as T1) from the GaAs WZ phase to the GaAsSb ZB phase insert is very abrupt. In about 1/3 of the studied NWs, a short (<5 nm) twinned ZB transition region is observed at the start of the GaAsSb ZB insert. The second transition, from the GaAsSb ZB insert to the upper GaAs NW segment (marked as T2), is shown by HRTEM in Fig. 5(b). In most of the NWs, the defect-free GaAsSb ZB insert was directly followed by a few nanometers of a twinned GaAs cubic phase (ZB/3C). Above this microtwin, a GaAs 4H polytype phase is formed, before the GaAs NW segment again returned to a WZ phase. The length of the GaAs 4H polytype phase is varying between 20 and 40 nm depending on the NW diameter. The WZ phase of the upper GaAs NW segment contains randomly spaced stacking faults, small regions of 4H and twinned ZB, as can be

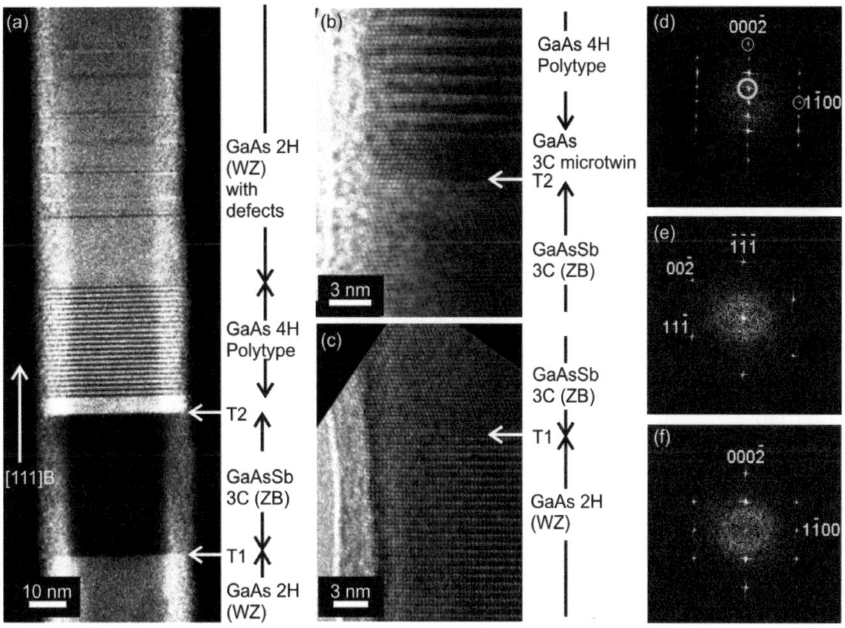

Fig. 5. Dark-field (a) and high-resolution (b, c) TEM images of a typical heterostructured GaAs/GaAsSb NW. (d), (e) and (f) are Fast Fourier Transforms (FFT) of HRTEM images of the GaAs 4H polytype above the insert, the GaAsSb ZB insert and the GaAs WZ phases below the insert, respectively (Dheeraj et al., 2008b). © 2008 American Chemical Society.

seen in Fig. 5(a). Fast Fourier transforms on HRTEM images of the 4H, ZB and WZ phases are shown in Figs. 5(d), 5(e), and 5(f), respectively. The appearance of a $1/4\,(0002)$ reflection in Fig. 5(d) (indicated by a thick circle) confirms a GaAs 4H polytype phase with the stacking sequence of ABCBABCB….

The 4H polytype structure is well known in bulk form in wide band gap materials such as SiC and AlN. The 4H polytype of SiC is studied in detail due to its superior electron mobility (Meyer et al., 2000). The understanding of the NW growth conditions favoring the formation of the 4H polytype in III-V materials that normally grow with a cubic ZB phase is thus an intriguing topic. The presence of a 4H polytype phase in GaAs and AlGaAs NWs was shown by Soshnikov et al., 2005. We believe that the 4H polytype is the intermediate phase between the ZB and WZ phases. As suggested by Dubrovskii et al., 2008 the gradual change in the supersaturation conditions could lead to the formation of intermediate structures such as cubic twins and a 4H polytype phase and delay the formation of the WZ phase. It is reasonable to believe that the supersaturation is gradually increasing after the shut off of the Sb flux and that this induces the 4H polytype crystal structure. However, no Sb was found above the insert, or in the gold particle by EDX, as mentioned above.

3.3 Effect of Sb on the crystal structure of NWs:

The understanding of the effect of Sb on the crystal structure of these NWs is very crucial. According to a recent model proposed by Glas et al., 2007, the crystalline structure of NWs depends on the orientational position (ZB or WZ) taken by the critical nucleus of each new monolayer. This is determined mainly by the following parameters: the stacking fault energy in the bulk ZB phase, the energy $\gamma_{\ell V}^{j}$ (j = ZB or WZ) of the lateral interface between the NW sidewalls and the vapor (V) phase, and the chemical potential difference (supersaturation) $\Delta\mu_{LS}$ of the liquid (L) phase with respect to the solid (S) phase. WZ formation in NWs of a given material requires two conditions: a facet energy lower for WZ than for ZB and a supersaturation $\Delta\mu_{LS}$ higher than a material-dependent critical value $\Delta\mu_c$. The latter scales with the stacking fault energy, which is about 20% higher in GaSb than in GaAs (Takeuchi & Suzuki, 1999). Moreover, according to the model of Glas et al., 2007, we expect $\Delta\mu_c$ to increase if the difference $\gamma_{\ell V}^{ZB} - \gamma_{\ell V}^{WZ}$ decreases. This difference should indeed be lower in GaAsSb, since the cohesive energy of GaSb is about 10% less than that of GaAs (Phillips, 1973). Hence, $\Delta\mu_c$ could be significantly larger in GaAsSb than in GaAs. This could be enough to offset the balance in favor of ZB nucleation in GaAsSb NWs, even if $\Delta\mu_{LS}$ does not change much. On the other hand, the addition of Sb in the vapor phase could induce a significant change of $\Delta\mu_{LS}$ in the liquid catalyst. If $\Delta\mu_{LS}$ decreases, this change would also be in favor of ZB nucleation. However, for the quaternary system (Au-Ga-As-Sb) considered here, there is no suitable thermodynamical data from which the change of $\Delta\mu_{LS}$ induced by Sb can be predicted. Therefore, at the present stage, we cannot define which phenomenon prevails to explain the ZB structure of [111]B-oriented GaAsSb NWs. It can be due to either an increase of $\Delta\mu_c$ or decrease of $\Delta\mu_{LS}$ (or both).

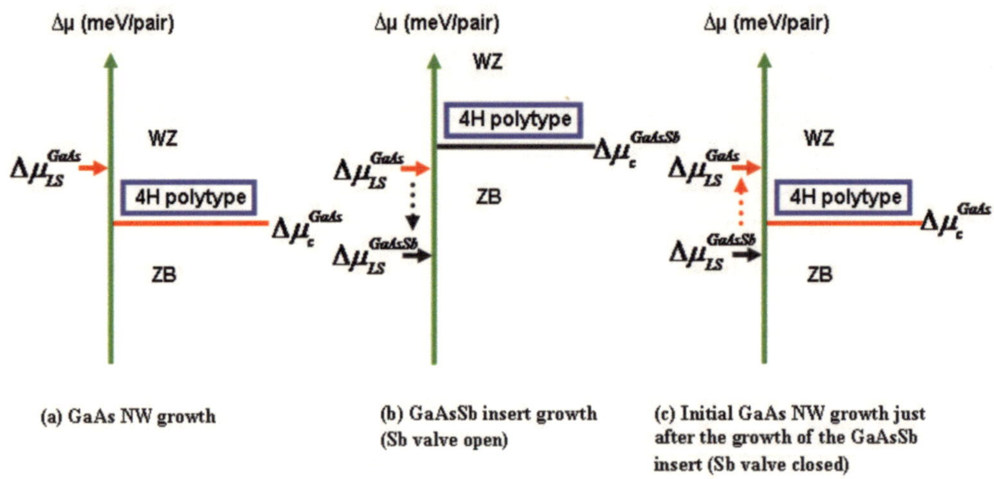

Fig. 6. Schematic illustrations of the possible configurations of the chemical potential difference $\Delta\mu_{LS}$ (red and black arrows) and material-dependent critical supersaturation $\Delta\mu_c$ (red and black horizontal lines) at different stages of NW growth (Dheeraj et al., 2008b). © 2008 American Chemical Society.

According to the HRTEM results, the most likely explanation for the phase transitions at the interfaces (T1 and T2 in Fig. 5) is that they are induced by the changing of both $\Delta\mu_c$ and $\Delta\mu_{LS}$ due to the introduction of Sb flux while growing. It should be noted that the chemical potential difference (supersaturation) $\Delta\mu_{LS}$ is gradually varying with time when the Sb flux is opened or closed, whereas the material-dependent critical supersaturation value $\Delta\mu_c$ is instantaneously changed between its corresponding values $\Delta\mu_c^{GaAsSb}$ and $\Delta\mu_c^{GaAs}$ for GaAsSb and GaAs, respectively. We believe that this is the main reason for forming two different interfaces at T1 and T2, and the detailed discussion is described in the following. According to the NW growth models by Glas et al., 2007, and Dubrovskii et al., 2008, the supersaturation value for the Au liquid droplet ($\Delta\mu_{LS}$) required to nucleate the 4H polytype phase is higher than for the ZB phase, and the WZ phase is higher than for the 4H polytype phase, as shown in Fig. 6. As the GaAs NW grows, the chemical potential difference $\Delta\mu_{LS}$ is $\Delta\mu_{LS}^{GaAs}$ in the WZ phase region (indicated with a red arrow in Fig. 6(a)). At the start of the GaAsSb insert growth, the critical supersaturation value $\Delta\mu_c$ instantaneously increases to the value required for the formation of the GaAsSb ZB phase ($\Delta\mu_c^{GaAsSb}$) (black horizontal line in Fig. 6(b)), whereas the chemical potential difference needs some time to gradually decrease to the equilibrium condition ($\Delta\mu_{LS}^{GaAsSb}$, indicated by a black arrow in Fig. 6(b)) for the formation of the GaAsSb ZB phase. However, the NWs should have ZB phase after introducing Sb into the MBE chamber due to that $\Delta\mu_{LS} < \Delta\mu_c^{GaAsSb}$, and hence a sharp interface occurs at T1. Reversely, after turning off the Sb flux, even though $\Delta\mu_c$ instantaneously goes back to $\Delta\mu_c^{GaAs}$, as shown by the red horizontal line in Fig. 6(c), $\Delta\mu_{LS}$ is gradually increasing to the supersaturation value $\Delta\mu_{LS}^{GaAs}$. This gradual change in the supersaturation condition could lead to the formation of intermediate structures such as cubic twins and 4H polytype, and thus delay the formation of the WZ phase. This is, however, just one possible explanation for why there are two different GaAs/GaAsSb interface transitions (at T1 and T2) based on our HRTEM observations, and does not yet give a complete description of the actual variations of $\Delta\mu_{LS}$ and $\Delta\mu_c$.

3.4 Growth of GaAs NWs with multiple GaAsSb inserts

We further demonstrate the growth of GaAs NWs with 4 GaAsSb inserts at an interval of growth time of 10 minutes on GaAs(111)B substrates. More than half a dozen of NWs were investigated by TEM and a typical NW is shown in Fig. 7. We could observe only three GaAsSb inserts in a NW by TEM. We believe that the first insert could have broken during the TEM sample preparation. The average growth rates of the three GaAsSb inserts and the lengths of the GaAs segments between each insert were measured from the TEM image in Fig. 7 and the results are shown in Fig. 8. It is observed that the average growth rate for the third insert is higher than for the second insert. TEM investigation on a number of these NWs show a similar increase in growth rate from the second to the third insert, while the growth rate of the fourth insert is observed to either increase, remain constant or decrease as compared to that of the third. The same trend is observed for the GaAs segments above these inserts.

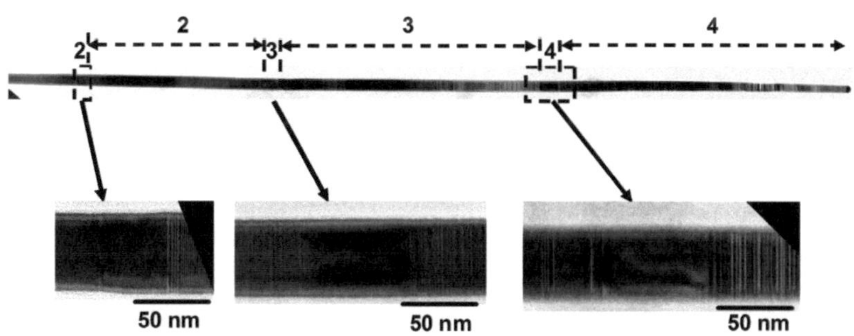

Fig. 7. (a) TEM image of a typical GaAs NW with three of four GaAsSb inserts visible (Dheeraj et al., 2009). © 2009 Elsevier.

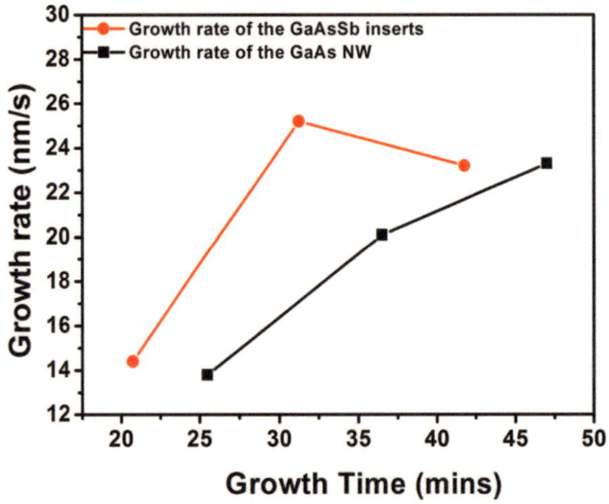

Fig. 8. Average growth rate of each GaAs segment and GaAsSb insert of the NW shown in Fig. 7 plotted versus the time at which half of the segment/insert is grown (Dheeraj et al., 2009). © 2009 Elsevier.

The crystalline structure of the GaAs segments is observed to be dominated by the WZ phase, whereas the GaAsSb inserts are observed to be of ZB phase. This result is consistent with the observations made before (Dheeraj et al., 2008a). The lower part of the GaAs NW has almost no stacking faults and the transition from the GaAs WZ phase to the GaAsSb ZB phase is always abrupt. The transition from the GaAsSb ZB phase of the first insert to the GaAs WZ phase above, however, exhibits a few nanometres of a 4H polytype phase followed by some stacking faults up to a few tens of nanometers, before maintaining pure WZ structure for a couple of hundred nanometers. This is a similar structure as depicted in Fig. 5. Further, with the increase in length of the NWs, the density of stacking faults was observed to increase in the GaAs segments. Interestingly, in spite of the higher density of stacking faults present in the upper GaAs segments, all the upper GaAsSb inserts were pure ZB without any twinning defects, and in addition the transition from the GaAs WZ phase to the GaAsSb ZB phase and vice-versa is abrupt (i.e, no 4H polytype). It is plausible that the growth rate of the NWs affects the kinetics of the change in supersaturation levels required for formation of ZB to WZ. Thus, we speculate that it is the slower growth rate near e.g. the position of the second insert that helps the formation of intermediate phases such as the GaAs 4H polytype, and as the growth rate of the NW increases, the change in supersaturation levels instead instantaneously leads to the formation of a WZ GaAs phase directly after the GaAsSb insert.

EDX investigations on the GaAsSb inserts revealed that the mole fraction of Sb was increasing with the insert's position along the NW. No significant relation between the Sb mole fraction and either the lengths of the GaAsSb inserts or the diameters of the NWs was observed. We believe the increase in Sb mole fraction with the NW length could be either due to increase in growth rate of the NW or due to decrease in the NW temperature with its length. However, further investigation is needed to clarify which mechanisms that are responsible for the change in Sb mole fraction along the NW.

3.5 Growth kinetics of GaAs NWs and the formation of stacking faults

According to the VLS growth model proposed by Dubrovskii & Sibirev 2007, the most predominant Ga adatom diffusion paths favouring the growth of NWs by MBE were considered to be (a) diffusion of Ga adatoms on the substrate surface to the NW base and (b) diffusion of Ga adatoms from the NW base to the Au droplet along the NW sidewall. Part of the Ga adatoms diffusing on the substrate could contribute to the growth of a 2D layer and part of the Ga adatoms diffusing along the NW sidewall contribute to the radial growth of the NW. We should point out here that, due to the design of the MBE growth chamber with the incident angles of molecular beams being typically 57° to the substrate surface in our case, the number of Ga adatoms hitting the NW sidewall increases proportionally to the length of the NW until shadowing from adjacent NWs sets in. This could lead to the increase in the probability of Ga adatoms either diffusing to the Au droplet or adding to the radial growth of the NW. As we hardly observed any radial growth of the NWs, an increase in axial growth rate of the NWs is quite plausible, which explains the increase in growth rate from the second to the third GaAsSb insert (and likewise for the GaAs segments above these inserts) , as was observed in the NW of Fig. 7. As mentioned above, after a certain critical average NW length, the growth rates of NWs were observed to either increase further, become constant or decrease, which we believe is dependent on the amount of shadowing of the molecular beam from adjacent NWs. The amount of shadowing, which is time-averaged due to the rotating substrate, will be different from NW to NW due to the random distribution of NW positions and the variation in length of adjacent NWs. In general, shadowing will affect the growth rate relatively more for shorter NWs than for longer NWs.

It has been shown that the WZ phase is favoured above a certain critical liquid supersaturation and that the ZB phase is prevailing below (Glas et al., 2007). The liquid supersaturation is quite sensitive to the rate of incoming adatoms to the Au droplet and to the rate of NW formation under the Au droplet, which could vary during the formation of every monolayer of the NW, leading to fluctuations in supersaturation levels. If the supersaturation is close to the critical supersaturation, these fluctuations could lead to the bistable conditions favouring both WZ and ZB phase, leading to the formation of stacking faults. The formation of the high density of stacking faults in the GaAs segment after the third and fourth GaAsSb insert could be attributed to fluctuations in the levels of supersaturation caused by turning on and off the Sb flux. However, this argument is not sufficient to explain the reason for the increase in the density of stacking faults with the length of the NW.

Cornet et al., 2007, suggested that a decrease in the rate of incoming In adatoms to the Au particle after a certain length of the InP NWs, lead to the formation of stacking faults due to the decrease in supersaturation to bistable conditions where both WZ and ZB exists. Conversely, we noticed an increase in the growth rate and density of stacking faults in GaAs NWs with its length, indicating that the formation of stacking faults is possible due to the increase in growth rate of NWs. This is further corroborated by the presence of a high density of stacking faults in the GaAs NWs grown under higher Ga flux equivalent to a nominal planar growth rate of 1.8 MLs^{-1} (not shown here). Although little can be said about the effects of increase in growth rate of NWs on change in supersaturation levels, these results suggest that the formation of stacking faults introduced in the GaAs segment after the third and fourth GaAsSb insert could be due to the coincidence of a change in supersaturation conditions by turning on and off the Sb flux and the increase in the growth rate of the NW. These results suggest that long and defect-free WZ GaAs NWs can be obtained at a lower growth rate of NWs.

3.6 Growth of GaAs/AlGaAs radial heterostructured NWs

Here, we demonstrate the growth procedure of GaAs/AlGaAs radial heterostructured core-shell NWs, with strongly enhanced photoluminescence efficiency from the GaAs core due to the shell passivation of surface states (Zhou et al., 2009). GaAs NWs were first grown for 25 minutes in the MBE chamber at the growth conditions mentioned earlier. The growth of AlGaAs was initiated by introducing the Al flux towards the substrate by opening the shutter in front of the effusion cell. The temperature of the Al effusion cell was preset to what would yield an $Al_xGa_{1-x}As$ layer with x equal to 0.3 or 0.6 on a GaAs(001) substrate. The V/III flux ratio was kept constant for both Al fluxes used by adjusting the As_4 flux. For the (nominal) $Al_{0.3}Ga_{0.7}As$ shell samples, the shell growth times were 1 min, 5 min and 15 min, whereas for the (nominal) $Al_{0.6}Ga_{0.4}As$ shell samples, the shell growth times were 0.5 min, 2 min and 7 min. The Al, Ga and As_4 fluxes were shut down simultaneously at the end of the AlGaAs shell growth and the substrate temperature immediately ramped down to room temperature. The as-grown NWs were characterized by SEM to systematically investigate the morphology of the GaAs core NWs and GaAs/AlGaAs core-shell NWs with different growth times and Al contents. 45°-tilted view SEM images of the GaAs core NWs and the GaAs/AlGaAs core-shell NWs are shown in Figs. 9 (a-c). For determination of the geometric sizes of the as-grown NWs, 40 NWs for each growth condition were randomly chosen from the SEM images.

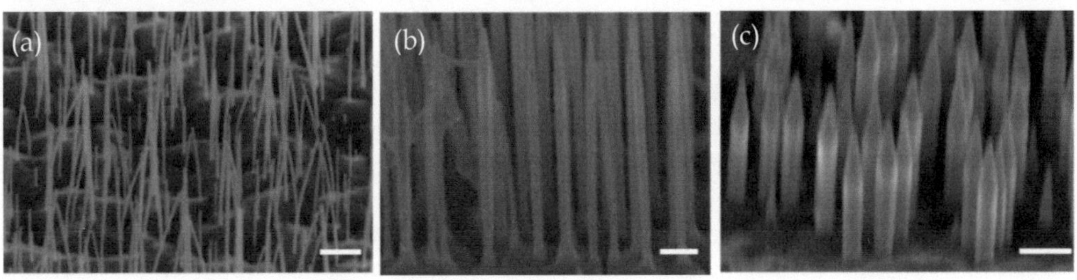

Fig. 9. 45°-tilted SEM images of the GaAs core NWs grown for 20 min (a), with $Al_{0.3}Ga_{0.7}As$ shell grown for 15 min (b), and with $Al_{0.6}Ga_{0.4}As$ shells grown for 7 min (c). Scale bar is 200 nm (Zhou et al., 2009).

The GaAs core NWs grew predominantly perpendicular to the substrate surface ([111]B direction) with a constant diameter over its length. The length of the NW is defined from the GaAs surface of the buffer to the tip of the NW. The diameters of the GaAs core NWs are varying from 10 nm to 25 nm and the lengths of the NWs are ranging from 240 nm to about 1.2 μm. The lengths of the GaAs core NWs were observed to decrease with increasing diameter. It confirms that growth of the NWs is fed by diffusion of adatoms from the surface of the substrate and/or from the sidewalls to the Au droplet at top of the NWs, as described by Dubrovskii & Sibirev, 2007. However, this behavior did not retain after the initiation of AlGaAs shell growth, which could be due to the deposition of Ga and Al adatoms directly on the NW sidewalls by a vapor solid (VS) growth mechanism.

No tapering was observed in any of the investigated GaAs NWs, whereas tapering was observed in the GaAs/AlGaAs core-shell NWs. Tapering is often observed when the length of the NWs exceeds the diffusion length of group III adatoms impinging on the NW sidewall as mentioned in 1.1. It is well known that Al is very reactive. Consequently, the effective surface diffusion barrier is higher for Al adatoms (~1.74 eV for AlAs grown on AlAs (001)) than for Ga adatoms (~1.58 eV for GaAs grown on GaAs (001)) (Shitara et al., 1993). This results in a shorter adatom diffusion length for Al compared to Ga, whereby radial sidewall deposition becomes more pronounced leading to a tapered morphology for GaAs/AlGaAs core-shell NWs. The tapering of the core-shell NWs becomes much more pronounced with increasing shell growth time and higher Al content. Also, we observed that both the mean diameter and the mean length of the NWs are increasing with increasing shell growth time, which shows that the AlGaAs shell grows in both axial and radial directions.

Comparing the $Al_{0.3}Ga_{0.7}As$ and $Al_{0.6}Ga_{0.4}As$ shell growth, we find an increasing radial and a decreasing axial growth rate of the AlGaAs shell with an increased Al content. At the same time, the mean volume of the GaAs/AlGaAs core-shell NWs quickly increases when the AlGaAs shell is formed, e.g. about 90 and 60 times for samples with 30% Al and 60% Al, respectively, compared to the mean volume of the NWs without AlGaAs shell. This is consistent with an observed decrease in the growth rate of the 2D layer with increase in shell growth time, where the 2D layer height is measured by the distance between the $Al_{0.3}Ga_{0.7}As$ marker layer in the buffer and the top surface of the sample in the cross-sectional SEM image.

The observed rapid increase of the NW volume with the AlGaAs growth time could be due to the shorter diffusion length of the Al adatoms as well as to the increasing collection area of the sidewalls of NWs with the increasing shell growth time. NWs collect Ga and Al adatoms in three different ways; (1) diffusion from the substrate surface, (2) direct impingement on the NW sidewalls, and (3) direct impingement on the gold droplet. A larger area of the NW sidewalls (effective collection area = *length* × *diameter*) is obtained when NWs get longer, whereby the NWs can collect more Al and Ga adatoms. At the same time, the shorter diffusion length of the Al also contributes to trap more and more Al and Ga adatoms migrating on the sidewalls of the NWs. Both of these two factors enhance each other, which results in a very fast increase of the NW volume. After the AlGaAs shells start to grow, the total flux impinging on the sample surface should remain constant. The AlGaAs either deposits on the NWs or on the substrate surface. Once NWs can capture more AlGaAs with a larger area of NW sidewalls, a slower growth rate of the 2D layer should be expected. A decrease in the growth rate of the 2D layer with the increase in growth time of the AlGaAs shell is also indeed what is observed (Zhou et al., 2009).

Finally, lower density of NWs was also found with increasing AlGaAs shell growth time, e.g. it is about 90 NWs/μm^2 for the GaAs core NW sample and roughly 50 NWs/μm^2 for the sample with $Al_{0.3}Ga_{0.7}As$ shell grown for 15 minutes. When the AlGaAs growth takes place, there is a possibility that some of the shorter NWs (e.g. the GaAs core NWs ~250 nm) might be buried by the 2D layer at the AlGaAs shell growth stage. We observed that both the radial growth rate of the AlGaAs NW shell and the growth rate of 2D AlGaAs increase with increased Al flux, whereas the NW axial AlGaAs growth rate is not significantly changed for average size NWs. The ratio between the NW axial AlGaAs growth rate and the 2D AlGaAs growth rate, of importance for NW burial, thus reduces with increase in Al flux. The ratio will be even further reduced for shorter NWs by the fact that shorter GaAs/AlGaAs core-shell NWs will get less Ga and/or Al adatoms compared to longer NWs due to both the smaller sidewall collection area and the shadowing effect. The shadowing effect is induced by the inclined molecular beam with respect to the substrate normal (here ~37°). Due to this effect, shorter NWs will be shadowed by longer neighboring NWs, which will also lead to a reduced growth rate. Shorter NWs may therefore be completely buried by the 2D layer, leading to a decrease in NW density.

Fig. 10(a) shows the [110] zone axis HRTEM image of a typical GaAs core NW (14 nm diameter). A HRTEM image of a typical GaAs/AlGaAs core-shell NW ($Al_{0.3}Ga_{0.7}As$ shell grown for 15 minutes) is shown in Fig. 10(b). SAED patterns in Fig. 10(a) and 10(b) reveal a wurtzite (WZ) crystal structure both in the GaAs core and in the AlGaAs shell. By comparing the HRTEM (Fig. 10(c)) and scanning-TEM images (Zhou et al., 2009) of the same NW, it indicates that apart from radial AlGaAs growth, an axial AlGaAs is also grown on top of the GaAs core. It can also be seen from HRTEM image (Fig. 10(c)), that stacking faults appear at the region where the axial AlGaAs starts to grow (by VLS) on top of the GaAs core. At the same time, it also shows that the radial AlGaAs shell exactly follows the crystal structure (including stacking faults) of the GaAs core. The interface between the GaAs core and the radial AlGaAs shell is relatively sharp and the thickness of the radial shell along the NW axis is highly uniform until the tapered AlGaAs NW region at the top. We have no indication that Al is diffusing into the GaAs NW core for neither of the two studied shell compositions. For the core-shell NW shown in Fig. 10(b) ($Al_{0.3}Ga_{0.7}As$ shell grown for 15 minutes), the diameter of the GaAs NW core is ~15 nm and for the whole core-shell NW ~ 60 nm, which is consistent with the observations from SEM.

In addition, a high angle annular dark field STEM characterization performed on these NWs (see Zhou et al., 2009) revealed that the AlGaAs NW grown on the top of the GaAs NW is a long (~ 800 nm) AlGaAs "core" with a radial AlGaAs "shell", with different Al compositions in the "core" and "shell", respectively. The observation of forming self organized core-shell structures is consistent with previous published results by Chen et al., 2007.

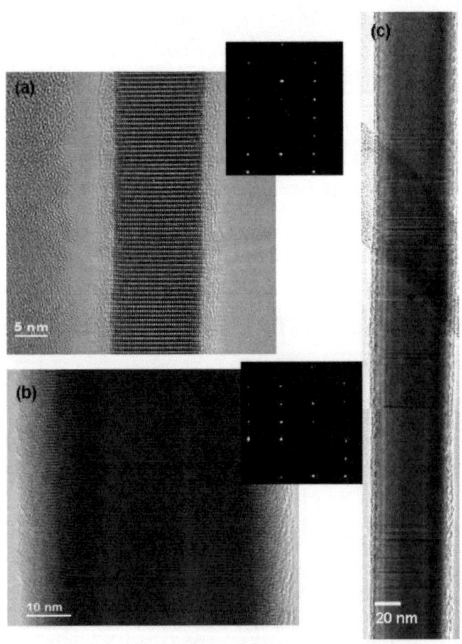

Fig. 10. HRTEM images of (a) GaAs core and (b), and (c) a typical GaAs/$Al_{0.3}Ga_{0.7}As$ core-shell NW, together with the selected area electron diffraction patterns (inset of image (a) and (b) (Zhou et al., 2009).

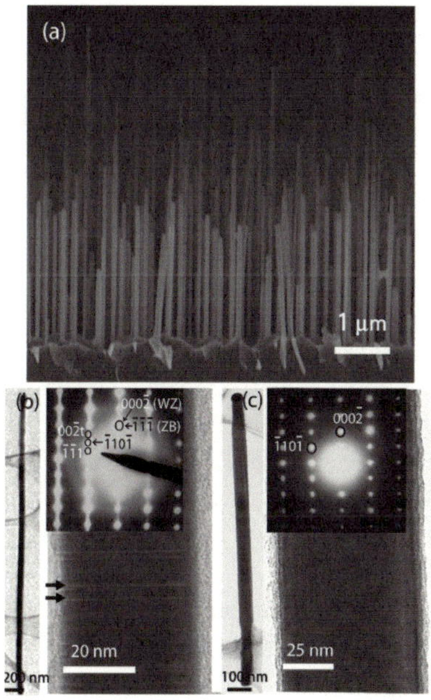

Fig. 11. a) Side view SEM image of the as-grown GaAs NWs. TEM (left) and HRTEM (right) image of a GaAs NW with (b) and without (c) stacking faults (Hoang et al., 2009). © 2009 American Institute of Physics.

3.7 Band gap of WZ GaAs

The newly observed WZ phase of the GaAs NWs raises questions about fundamental physical parameters such as the band gap energy, exciton binding energy, carrier effective masses and phonon energies. As mentioned earlier, it is necessary to grow the NWs without any stacking faults to determine the bandgap energy of WZ material. The SEM image of the GaAs NWs shown in Fig. 11 (a) depicts that there are two typical sets of wires: short wires (~ 2 µm, and large diameter ~ 60 nm) and long wires (~ 4 µm, and smaller diameter ~ 30-40 nm). This is in agreement with the model proposed by Dubrovskii & Sibirev, 2007, which shows that the NWs with smaller diameter grow faster than NWs with large diameter at the same flux.

From HRTEM characterization, it appears that the narrow and long NWs are associated with stacking faults while the wide and short NWs are free of stacking faults. These observations are in agreement with our previous report where we attributed the formation of stacking faults to the faster growth rate of NWs (Dheeraj et al., 2009). Figs. 11(b) and 11(c) show TEM images of a GaAs NW with and without stacking faults, respectively. The right panel in Fig. 11(b) shows a magnified image of a small portion of the same NW shown in the left panel with clear appearance of stacking faults and is complemented with the electron diffraction pattern shown in the inset of Fig. 11(b). The right panel in Fig. 11(c) shows a magnified image of the stacking fault free NW, indicating a pure WZ crystal structure for this NW. From the TEM images of these two representative NWs and several others studied, we conclude that narrow (and long) NWs appear to have a low density of stacking faults associated with short ZB segments (~ 2 nm) sandwiched in between a dominating WZ structures while wide (and short) NWs show a pure WZ crystal structure free of stacking faults or ZB segments.

Low temperature μ-PL measurements from single WZ GaAs NWs show a PL emission band at an energy of ~ 29 meV higher in comparison with the ZB GaAs free exciton emission energy (1.515 eV). Some NWs exhibited PL emissions from both ZB and WZ phases, at ~1.515 and 1.544 eV, respectively. The results are consistent with the HRTEM observation that short ZB segments appear in the GaAs NWs that contain stacking faults.

Fig. 12 shows PL spectra from four single WZ GaAs NWs at 4.4 K. The 633 nm cw excitation laser was defocused (spot size ~ 5 μm) so that individual NWs were entirely excited. The PL emissions from all these NWs exhibited strong emission bands centered at ~ 1.535 eV. The PL spectra display a broad emission range with several features which we now discuss in detail. All NWs (wires 1 - 4) show an emission peak at 1.544 eV (in wire 2 the peak appears as a shoulder), 29 meV higher in comparison with the ZB GaAs free exciton emission energy (1.515 eV). We suggest that this is the free exciton emission for WZ GaAs. There exist a few recent results on the PL emission for WZ GaAs nanostructures but the results are inconsistent. Martelli et al., 2007, reported the free exciton emission peak in WZ GaAs NWs to be at an energy 7 meV higher compared with the ZB GaAs free exciton energy, while Moewe et al., 2008, reported a 10 meV lower energy for WZ GaAs nano-needles. Our result (~ 29 meV higher), however, is close to the results reported in some earlier theoretical works (Mujica et al., 1995; Murayama & Nakayama, 1994; Yeh et al., 1992) where the WZ GaAs band gap energy was predicted to be ~ 30 - 33 meV higher than for ZB GaAs.

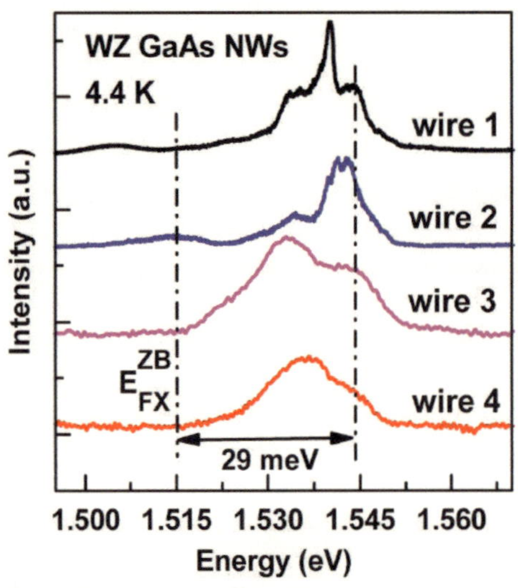

Fig. 12. μ-PL spectra from four different NWs labeled 1-4 (Hoang et al., 2009). © 2009 American Institute of Physics.

Wires 1 and 2 exhibit narrow emission lines at ~ 1.54 eV, while a broader peak at lower energy (~ 1.532 eV) is observed in all wires 1 – 4. The origin of these emission bands are not clear at the moment, but the emissions most probably result from excitons bound to defects such as morphological irregularities or are impurity related. We note that in the PL spectra of wires 1 and 2 there are also emission peaks at 1.505 and 1.515 eV, respectively. These peaks are close to the exciton emission energies in ZB GaAs NWs (Titova et al., 2006). However, in wires 3 and 4, there are no indications of emission in this energy range. We suggest that these weak emissions are associated with the ZB GaAs segments that are observed in the WZ GaAs NWs with stacking faults. Again, this result is consistent with HRTEM results shown in Figs. 11 (b) and (c), where some GaAs NWs exhibit pure WZ free of stacking faults (and ZB segments), while other NWs exhibit a low density of stacking faults, which inherently possesses ZB phase.

In order to gain further insight of the observed emission peaks from these WZ GaAs NWs, we have measured the temperature dependence of the PL spectra of wire 1 from 4.4 K to 70 K. Several observations were obtained from the temperature dependence of the various peaks. The sharp emission peak at 1.54 eV (labeled 2) dominates at low temperature while the GaAs WZ free exciton emission peak (labeled 1) only appears as a shoulder. As the temperature increases, the sharp peak (2) decreases in intensity and disappears by 40 K, above which only the free exciton emission peak remains.

The temperature dependence of the emission energies of peaks 1 and 2 is plotted in Fig. 13, and is compared with the emission energy of a ZB GaAs/AlGaAs core-shell NW as reported by Titova et al., 2006. The solid lines are fits to the data of peaks 1 and 2 as well as the ZB NWs from Titova et al., 2006, using a modified form of the Varshni equation (Cardona et al., 2004). We note that the emission energies of the WZ free exciton (peak 1) and defect-related (peak 2) peaks have a similar temperature dependence as the exciton emission energy of the ZB GaAs NW. The defect-related emission energy closely follows the free exciton energy and quenches at ~ 40 K, suggesting that the emissions are not related to deep levels but are rather excitonic in nature.

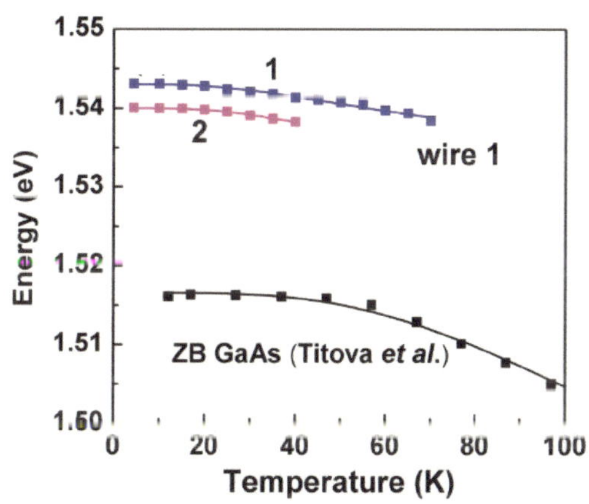

Fig. 13. PL energies of peaks 1 and 2 (see text) as a function of temperature (Hoang et al., 2009). © 2009 American Institute of Physics.

Hence, we conclude that PL measurements on single NWs at low temperatures reveal a peak related to the WZ GaAs free exciton emission at an energy ~ 29 meV higher in comparison with the free exciton emission energy in ZB GaAs. The WZ/ZB GaAs band alignment is believed to be type II with the ZB conduction band edge energy lower than for WZ, in which case PL emission can be expected anywhere in the energy range from the WZ GaAs band gap to even below the ZB GaAs band gap depending on the density, size and distribution of the ZB segments. Our observation of weak emission peaks at energies close to the free exciton energy in ZB GaAs for some NWs, in combination with our HRTEM images of some of the WZ NWs where stacking faults (inherently possessing ZB phase) are observed, corroborates this view.

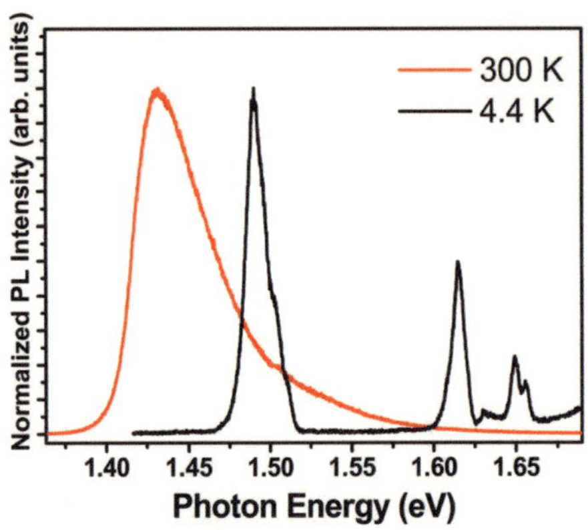

Fig. 14. Room and low (4.4 K) temperature μ-PL emission spectra from a single GaAs/AlGaAs core-shell NW (Zhou et al., 2009).

3.8 Optical properties of GaAs/AlGaAs core-shell NWs

Fig. 14 shows RT and low temperature (4.4 K) μ-PL emission spectra from the same GaAs/$Al_{0.3}Ga_{0.7}As$ core-shell NW. For the investigated bare GaAs core NWs, there is no detectable PL signal at RT. The RT spectrum of the core-shell NW shows only GaAs related emission at 1.431 eV (full width at half maximum (FWHM) is ~49 meV). In contrast, at 4.4 K, the spectrum of the core-shell NW shows both GaAs emission at 1.489 eV (FWHM is ~12 meV) and several PL peaks above 1.6 eV.

We note here that the emission energy (1.489 eV) for the GaAs core is lower than the free exciton energy (1.515 eV) of ZB GaAs. This is probably due to the recombination from the type II band alignment that can occur due to stacking fault created GaAs ZB segments sandwiched in between a dominating GaAs WZ structure (Pemasiri et al., 2009; Hoang et al., 2009). It is important to note that no PL emission was observed from bare GaAs core NWs at RT. As has already been reported by several authors (Titova et al., 2006; Chen et al., 2007), the shell in the GaAs/AlGaAs core-shell NW helps to increase the radiative recombination efficiency by approximately two orders of magnitude through the suppression of non-radiative surface recombination at the GaAs surface. As can be seen in Fig. 14, PL emissions from both GaAs and AlGaAs are observed at low temperature.

The PL peaks above 1.6 eV are most likely related to recombination from carriers localized in the AlGaAs region (~1 μm) at the top of the NW. The occurrence of several PL peaks is believed to be due to some local composition variations (10-15%) in this AlGaAs region. It has already been reported that for the axial VLS growth of AlGaAs on top of GaAs, a self-formed AlGaAs core-shell structure is formed with a radial variation in the Al composition (Chen et al., 2007). The existence of AlGaAs related PL peaks from the GaAs/AlGaAs core-shell NWs suggests that there is room for improvement of the growth, especially by reducing the axial growth of AlGaAs

3.9 Optical properties of GaAs/AlGaAs core-shell NWs with GaAsSb insert

The GaAs/GaAsSb/GaAs axial heterostructured NWs are very interesting due to that band-structure in these materials is engineered not just by changing the crystal material but also by the crystal phase. Here, we show that the crystal structure of the GaAs barrier on the upper interface of GaAsSb insert can be altered by introducing a growth interruption (GI)

immediately after the growth of the GaAsSb insert. The optical properties of such structures are discussed. In addition, these NW have a AlGaAs shell as discussed in 3.7.
Fig. 15 (a) shows the dark field TEM image of a GaAs/AlGaAs core-shell NW with a GaAsSb insert where no GI was employed after insert growth. The GaAsSb insert has a ZB crystal structure, whereas the GaAs core has WZ crystal structure. The AlGaAs shell copies the crystal structure of the WZ or ZB core as discussed before. HRTEM images of the top and bottom GaAs/GaAsSb interfaces are shown in Fig. 15(b) and (c), respectively. Due to the absence of the GI in this NW, a ZB GaAs segment (~5 nm) appears after the growth of the GaAsSb insert, denoted by a double headed arrow in Fig. 15(b). In Fig. 16(a), the PL emission from a single GaAs/AlGaAs core-shell NW with a GaAsSb insert is shown. The μ-PL emission peak related to the GaAsSb insert is observed at ~1.27 eV. The ZB segment in the GaAs barrier at the top interface of the GaAsSb insert causes a type II band alignment. This results in the spatial separation of electrons and holes in the adjacent ZB GaAs and ZB GaAsSb layers, respectively. Fig. 16(b) shows the power dependent spectra from the GaAsSb related peak at 1.27 eV. With an increase of the excitation power with a factor of ten the PL emission is broadened and the PL peak is blue shifted ~100 meV.

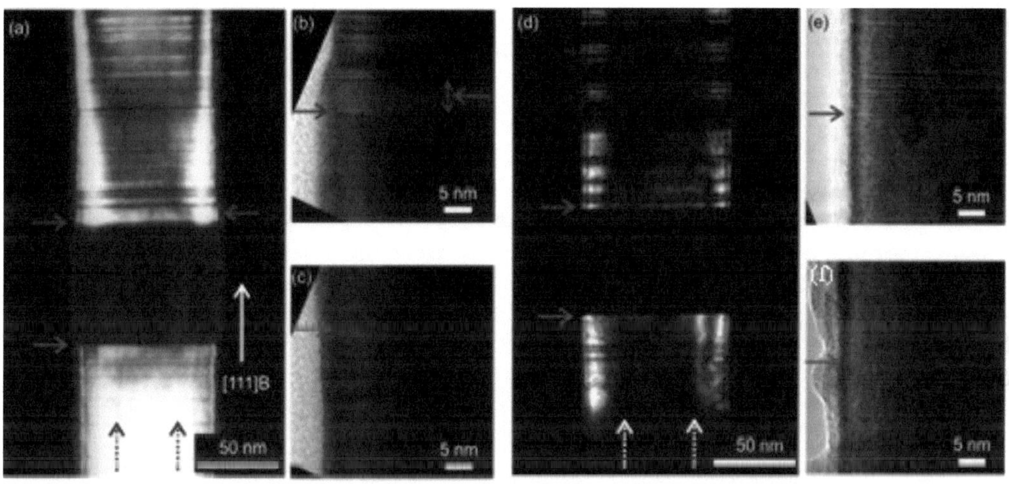

Fig. 15. Dark field TEM of a WZ GaAs/AlGaAs core-shell NW with ZB GaAsSb insert without (a) and with (d) GI, with dotted black/white arrows indicating core size. HRTEM image of (b,e) the top and (c,f) the bottom interface of the GaAsSb insert in Fig. (a,d) respectively. Arrow pointing to the left indicates ZB GaAs segment. Arrows pointing to the right indicate GaAsSb/GaAs interfaces (Moses et al., 2009). Arrows pointing to the left indicate a twin.

The large blue shift can be attributed to the band filling and band bending effect as the excitation intensity increases (Chiu et al., 2002; Dinu et al., 2003). The spatially separated electrons and holes create an electric field at the GaAsSb/GaAs interface that results in band bending as the excitation density increases. The broadening of the PL peak at higher excitation intensities is probably due to the electron state filling in the ZB GaAs segment.
A dark field TEM image of a GaAs/AlGaAs core-shell NW with GaAsSb insert where GI after the growth of GaAsSb insert is employed is shown in Fig. 15 (d). The HRTEM images

of the top and bottom GaAs/GaAsSb interfaces are shown is Fig. 15 (e) and (f), respectively. By GI, the formation of ZB GaAs segment at the upper interface of the GaAsSb insert can be avoided, which can be observed in Fig. 15(e). Hence the ZB GaAsSb insert has a well defined WZ GaAs barrier on both the interfaces. However, the WZ GaAs barrier above the GaAsSb insert has some stacking faults, the density of which is less when compared to the sample grown without GI.
In Fig. 17(a), PL emission from a single GaAs/AlGaAs core-shell NW with GI after the GaAsSb insert growth is shown. The GaAsSb related PL emission peak is observed at ~1.39

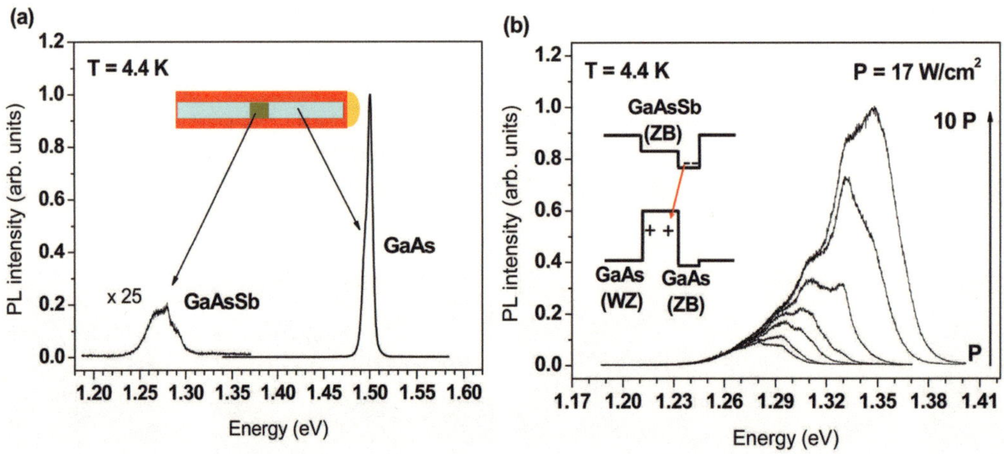

Fig. 16. (a) PL spectrum from a single GaAs/AlGaAs core-shell NW without GI after the GaAsSb insert. (b) Power dependent spectra from the GaAsSb related PL peak. Inset shows schematically the type II band alignment when the upper GaAs barrier contains a few nm thin ZB GaAs segment (Moses et al., 2009).

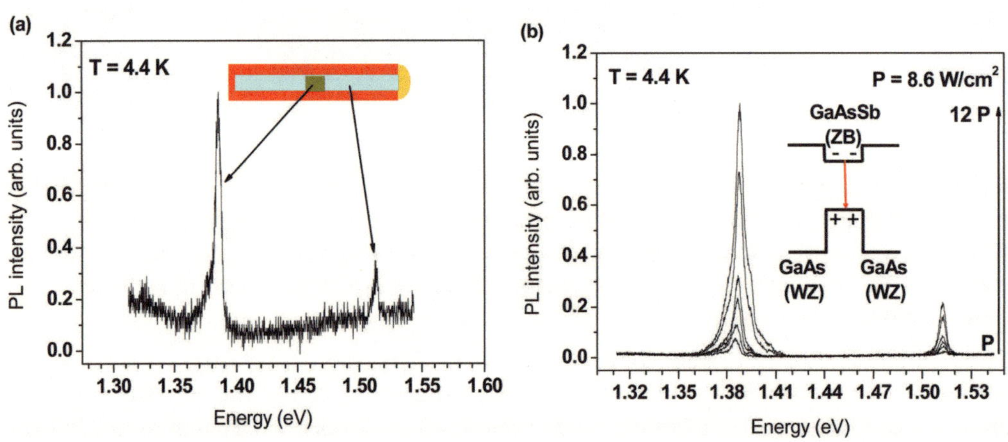

Fig. 17. (a) PL spectrum from a single GaAs/AlGaAs core-shell NW with GI after the GaAsSb insert. (b) Power dependent spectra of a GaAs/AlGaAs core-shell NW with GI after the GaAsSb insert. Inset shows schematically the type I band alignment and recombination therein (Moses et al., 2009).

eV. The WZ GaAs barriers in the GaAs/AlGaAs core-shell NW with GaAsSb insert will lead to a weak type I band alignment with both electrons and holes confined in the GaAsSb core-insert. In Fig. 17(b), the power dependent spectra of a single GaAs/AlGaAs core-shell NW with GI after the GaAsSb insert growth are shown. There is a small blue shift (~2 meV) of the GaAsSb related PL peak as the excitation intensity increases.

The power dependent PL spectra of this NW are dramatically different from the GaAsSb core-insert NW that contains a thin ZB GaAs segment at the upper GaAs/GaAsSb interface (Fig. 16(b)). We believe this is a strong indication that a type I band alignment has been achieved around the ZB GaAsSb insert with a WZ GaAs barrier at both GaAs/GaAsSb interfaces.

In all NW samples, the GaAs related PL emission is observed at ~1.50-1.52 eV, well below the free exciton PL emission observed in stacking fault free WZ GaAs NWs. This lower value is typical for WZ GaAs NWs that contain stacking faults and short ZB segments. This gives rise to type II PL emission where the stacking faults and short ZB GaAs segments act as electron traps as dicussed in 3.7.

4. Conclusions

We have demonstrated the MBE growth of ZB structured GaAsSb NWs, WZ GaAs NWs with single and multiple ZB GaAsSb inserts, and AlGaAs shells around such NWs. HRTEM studies show the GaAsSb inserts to be defect free displaying a ZB phase. Also, the crystal phase transition was observed to be abrupt at the WZ GaAs to ZB GaAsSb lower interface, whereas an intermediate 4H polytype phase was observed in the GaAs segment above the GaAsSb upper interface. We suggest that the formation of ZB GaAsSb could be due to either an increase of the material-dependent critical supersaturation value $\Delta\mu_c$ or a decrease of the supersaturation $\Delta\mu_{LS}$ in the Au particle (or both).

We showed that the growth rate of NWs in MBE increases with NW length, due to the increase in NW sidewall area (i.e. the increase in sidewall collection area for the incoming molecular beam). After a certain critical average NW length, the growth rates of NWs were observed to either increase further, become constant or decrease, which we believe is dependent on the amount of shadowing of the molecular beam on a NW from its adjacent NWs. Further, the density of stacking faults in the WZ GaAs NWs is found to be dependent on the growth rate of the NWs. These results suggest that the formation of stacking faults can be minimized by decreasing the growth rate of the GaAs NWs.

We have demonstrated the growth of stacking fault-free WZ GaAs NWs and determined the free exciton emission energy. PL measurements on single NWs at low temperatures reveal a peak related to the WZ GaAs free exciton emission at an energy ~ 30 meV higher in comparison with the free exciton emission energy in ZB GaAs. Further, we have shown the importance of AlGaAs shell to increase the radiative recombination efficiency through the suppression of non-radiative surface recombination at the GaAs surface.

Finally, we have characterized GaAs/AlGaAs core-shell NWs with a single GaAsSb insert grown without/with growth interruption (GI) after the GaAsSb insert, with ZB and WZ GaAs barrier, respectively, at the top interface of the GaAsSb insert. GaAsSb heterostructure with either ZB GaAs or WZ GaAs is believed to exhibit type II or type I band-alignment, respectively. We confirmed this by performing power dependent PL measurements, which show a strong (~100 meV) or a weak (~2 meV) blue shift in the samples grown without or with GI, respectively. We attributed the strong blue shift and line width broadening to band bending and band filling due to a type II band alignment; while the weak blue shift is attributed to a type I band alignment. The band alignment is depending on controllable crystal phase variations in the NW and opens up new possibilities for band structure engineering of NWs.

5. Acknowledgements

The authors would like to thank J.C. Harmand, G. Patriarche, F. Glas, and L. Largeau for many fruitful discussions. The authors would also like to thank G. Patriarche and L. Largeau for their help with some of the TEM, and XRD pole figure measurements. Part of this work was supported by the 'NANOMAT' program (Grant No. 182091) and the Norwegian-French 'AURORA' program (Grant No. 187692) of the Research Council of Norway.

6. References

Agarwal, R. & Lieber, C.M. (2006). Semiconductor nanowires: optics and optoelectronics *Appl. Phys. A: Mat. Sci. Processing*, 85, 209-215.

Agarwal, R. (2008). Heterointerfaces in semiconductor nanowires. *Small*, 4, 1872-1893.

Akiyama, T.; Sano, K.; Nakamura, K & Ito, T. (2006). An empirical potential approach to wurtzite–zinc-blende polytypism in group III–V semiconductor nanowires, *Jap. J. Appl. Phys.*, 45, L275 - L278.

Algra R.E.; Verheijen, M.A.; Borgström, M.T.; Feiner, L.F.; Immink, G.; van Enckevort, W.J.P.; Vlieg, E. & Bakkers, E.P.A.M. (2008). Twinning superlattices in indium phosphide nanowires *Nature*, 456, 369.

Bernstein, R.W.; Borg, A.; Husby, H.; Fimland, B.O. & Grepstad, J.K. (1992). Capping and decapping of MBE grown GaAs(001), $Al_{0.5}Ga_{0.5}As$(001), and AlAs(001) investigated with ASP, PES, LEED, and RHEED, *Appl. Surf. Sci.*, 56-58, 74-80.

Cardona, M.; Meyer, T.; & Thewalt, M. (2004). Temperature dependence of the energy gap of semiconductors in the low-temperature limit, *Phys. Rev. Lett.*, 92, 196403.

Caroff, P.; Dick, K.A.; Johansson, J.; Messing, M.E.; Deppert, K. & Samuelson, L. (2009). Controlled polytypic and twin-plane superlattices in III-V nanowires, *Nat. Nanotechnology*, 4, 50-55.

Chen, C.; Plante, M. C.; Fradin, C. & LaPierre, R. R. (2006). Layer-by-layer and step-flow growth mechanisms in GaAsP/GaP nanowire heterostructures, *J. Mat. Res.*, 21, 2801-2809.

Chen, C.; Shehata, S.; Fradin, C.; LaPierre, R.; Couteau, C. & Gregor, W. (2007). Self-directed growth of AlGaAs core-shell nanowires for visible light applications, *Nano Lett.*, 7, 2584-2589.

Chiu, Y.S.; Ya, M.H.; Su, W.S. & Chen, Y.F. (2002). Properties of photoluminescence in type-II GaAsSb/GaAs multiple quantum wells, *J. Appl. Phys.*, 92, 5810-5813.

Cirlin, G. E.; Dubrovski, V. G.; Sibirev, N. V.; Soshnikov, I. P.; Samsonenko, Y.B.; Tonkikh, A. A. & Ustinov, V. M. (2005).The diffusion mechanism in the formation of GaAs and AlGaAs nanowhiskers during the process of molecular-beam epitaxy, *Semiconductors*, 39, 557–564.

Cornet, D.M. & LaPierre, R.R. (2007a). InGaAs/InP core-shell and axial heterostructure nanowires, *Nanotechnology*, 18, 385305.

Cornet, D. M.; Mazzetti, V.G.M. & LaPierre, R. R. (2007b). Onset of stacking faults in InP nanowires grown by gas source molecular beam epitaxy *Appl. Phys. Lett.*, 90, 013116.

Czban, J.A.; Thompson, D.A. & LaPierre, R. (2009). GaAs core−shell nanowires for photovoltaic applications. *Nano Lett.*, 9, 148-154.

Dheeraj, D.L.; Patriarche, G.; Largeau, L.; Zhou, H.L.; van Helvoort, A.T.J.; Glas, F.; Harmand, J.C.; Fimland, B.O. & Weman, H. (2008a). Zinc blende GaAsSb nanowires grown by molecular beam epitaxy, *Nanotechnology*, 19, 275 605.

Dheeraj, D.L.; Patriarche, G.; Zhou, H.; Hoang, T.B.; Moses, A.F.; Grønsberg, S.; van Helvoort, A.T.J.; Fimland, B.O. & Weman, H. (2008b). Growth and characterization of wurtzite GaAs nanowires with defect-free zinc blende GaAsSb inserts, *Nano Lett.* 8, 4459-4463.

Dheeraj, D.L.; Patriarche, G.; Zhou, H.; Harmand, J.C.; Weman, H. & B.O. Fimland, (2009). Growth and structural characterization of GaAs/GaAsSb axial heterostructured NWs, *J. Crystal Growth*, 311, 1847-1850.

Dinu, M.; Cunningham, J.E.; Quochi, F. & Shah, J. (2003). Optical properties of strained antimonide-based heterostructures, *J. Appl. Phys.*, 94, 1506-1512.

Duan, X.; Huang, Y.; Agarwal, R.; & Lieber, C.M. (2003). Single-nanowire electrically driven lasers, *Nature*, 421, 241-245.

Dubrovskii, V.G.; Cirlin, G.E.; Soshnikov, I.P.; Tonkikh, A.A.; Sibirev, N.V.; Samsonenko, Y.B. & Ustinov, V.M. (2005). Diffusion-induced growth of GaAs nanowhiskers during molecular beam epitaxy: Theory and experiment, *Phys. Rev. B*, 71, 205325.

Dubrovskii, V.G. & Sibirev, N.V. (2007). General form of the dependence of nanowire growth rate on the nanowire radius, *J. Cryst. Growth*, 304, 504-513.

Dubrovskii, V.G. & Sibirev, N.V. (2008). Growth thermodynamics of nanowires and its application to polytypism of zinc blende III-V nanowires, *Phys. Rev. B*, 77, 035414.

Fontcuberta i Morral, A.; Spirkoska, D.; Arbiol, J.; Heigoldt, M., Ramon Morante, J. & Abstreiter, G. (2008). Prismatic quantum heterostructures synthesized on molecular-beam epitaxy GaAs nanowires, *Small*, 4, 899-903.

Givargizov, E.I. (1975). Fundamental aspects of VLS growth, *J. Cryst. Growth*, 31, 20-30.

Glas, F.; Harmand, J.C. & Patriarche, G. (2007). Why does wurtzite form in nanowires of III-V zinc blende semiconductors? *Phys. Rev. Lett.*, 99, 146101.

Guo, Y.N.; Zou, J.; Paladugu, M.; Wang, H.; Gao, Q.; Tan, H.H.; & Jagadish, C. (2006). Structural characteristics of GaSb/GaAs nanowire heterostructures grown by metal-organic chemical vapor deposition. *Appl. Phys. Lett.*, 89, 231917.

Harmand, J.C.; Tchernycheva,M.; Patriarche, G.; Travers, L.; Glas, F. & Cirlin, G. (2007). GaAs nanowires formed by Au-assisted molecular beam epitaxy: Effect of growth temperature, *J. Cryst. Growth*, 301–302, 853–856.

Harmand, J. C.; Patriarche, G.; Péré-Laperne, N.; Mérat-Combes, M.N. ; Travers, L.; & Glas, F. (2005) Analysis of vapor-liquid-solid mechanism in Au-assisted GaAs nanowire growth, *Appl. Phys. Lett.* 87, 203101.

Hoang, T. B.; Moses, A.F.; Zhou, H.L.; Dheeraj, D.L.; Fimland, B.O. & Weman, H. (2009). Observation of free excition photoluminescence emission from single wurtzite GaAs nanowires, *Appl. Phys. Lett.*, 94, 133105.

Ihn, S.G.; Song, J.I.; Kim, T.W.; Leem, D.S.; Lee, T.; Lee, S.G.; Kooh, E.K. & Song, K. (2007). Morphology- and orientation-controlled gallium arsenide nanowires on silicon substrates *Nano Lett.*, 7, 39-44.

Ihn, S.G. & Song, J.I. (2007). InAs nanowires on Si substrates grown by solid source molecular beam epitaxy, *Nanotechnology*, 18, 355603.

Jiang, X.; Xiong, Q. Nam, S.; Qian, F.; Li, Y. & Lieber C.M. (2007). InAs/InP radial nanowire heterostructures as high electron mobility devices. *Nano Lett.*, 7, 3214-3218.

Largeau, L.; Dheeraj, D.L.; Tchernycheva, M.; Cirlin, G.E. & Harmand, J.C. (2008). Facet and in-plane crystallographic orientations of GaN nanowires grown on Si(111), *Nanotechnology*, 19, 155704.

Lauhon, L.J.; Gudiksen, M.S.; & Lieber, C.M. (2004). Semiconductor nanowire heterostructures, *Phil. Trans. R. Soc. Lond.* A, 362, 1247–1260.

Martelli, F.; Piccin, M.; Bais, G.; Jabeen, F.; Ambrosini, S.; Rubini, S.; & Franciosi, A. (2007). Photoluminescence of Mn-catalyzed GaAs nanowires grown by molecular beam epitaxy, *Nanotechnology*, 18, 125603.

Meyer, B.K.; Hofmann, D.M.; Volm, D.; Chen, W.M.; Son, N.T.; Janzén, E.; (2001). Optically detected cyclotron resonance investigations on 4*H* and 6*H* SiC: Band-structure and transport properties, *Phys. Rev. B*, *61*, 4844-4849.

Minot, E.D.; Kelkensberg, F.; Van Kouwen, M.; Van Dam, J.A.; Kouwenhoven, L.P.; Zwiller, V.; Borgström, M.T.; Wunnicke, O.; Verheijen, M.A. & Bakkers, E.P.A.M. (2007). Single quantum dot nanowire LEDs. *Nano Lett.*, 7, 367-371.

Moewe, M.; Chuang, L.; Crankshaw, S.; Chase, C. & Chang-Hasnain, C. (2008). Atomically sharp catalyst-free wurtzite GaAs/AlGaAs nanoneedles grown on silicon, *Appl. Phys. Lett.*, 93, 23116.

Moses, A.F.; Hoang, T.B.; Dheeraj, D.L.; Zhou, H.L.; van Helvoort, A.T.J.; Fimland, B.O. & Weman, H. (2009). Micro-photoluminescence study of single GaAsSb/GaAs radial and axial heterostructured core-shell nanowires, IOP Conf. Series: Materials Science and Engineering 6, 012001.

Mujica, A.; Needs, R. & Munoz, A. (1995). First-principles pseudopotential study of the phase stability of the III-V semiconductors GaAs and AlAs, *Phys. Rev. B*, 52, 8881.

Murayama, M. & Nakayama, T. (1994). Chemical trend of band offsets at wurtzite/zinc-blende heterocrystalline semiconductor interfaces, *Phys. Rev. B*, 49, 4710-4724.

Pemasiri, K.; Montazeri, M.; Gass, R.; Smith, L.M.; Jackson, H.E.; Yarrison-Rice, J.; Paiman, S.; Gao, Q.; Tan, H.H.; Jagadish, C.; Zhang, X. & Zou, J. (2009). Carrier dynamics and quantum confinement in type II ZB-WZ InP nanowire homostructures, *Nano Lett.*, 9, 648-654.

Patriarche, G.; Glas, F.; Tchernycheva, M.; Sartel, C.; Largeau, L. & Harmand, J.C. (2008). Wurtzite to zinc blende phase transition in GaAs nanowires induced by epitaxial burying, *Nano Lett*, 8,1638-1643.

Phillips, J.C. (1973). Bonds and bands in semiconductors (Academic Press, New York and London, 1973).

Plante, M.C. & LaPierre, R.R. (2006). Growth mechanisms of GaAs nanowires by gas source molecular beam epitaxy, *J. Cryst. Growth,* 286, 394–399.

Schmidt, V.; Wittemann, J.V.; Senz, S.; & Gösele, U. (2009). Silicon nanowires: A review on aspects of their growth and their electrical properties. *Adv. Mater.* 21, 2681-2702.

Shitara, T.; Neave, J.H.; & Joyce, B.A. (1993). Reflection high-energy electron diffraction intensity oscillations and anisotropy on vicinal AlAs(001) during molecular-beam epitaxy, *Appl. Phys. Lett.*, 62, 1658.

Shtrikman, H.; Popovitz-Biro, R.; Kretinin, A. & Heiblum, M. (2009). Stacking-faults-free zinc blende GaAs nanowires, *Nano Lett.*, 9, 215-219.

Soshnikov, I. P.; Cirlin, G. E.; Dubrovskii, V. G.; Veretekha A. V.;, Gladyshev, A.G. & Ustinov, V.M. (2005). Formation of GaAs nanowhisker arrays by magnetron sputtering deposition, *Phys. Solid State*, 48, 786–791.

Svensson, C.P.T.; Mårtensson, T.; Trägårdh, J.; Larsson, C.; Rask, M.; Hessman, D.; Samuelson, L. & Ohlsson, J. (2008) Monolithic GaAs/InGaP nanowire light emitting diodes on silicon. *Nanotechnology,* 19, 305201.

Takeuchi, S. & Suzuki, K. (1999). Stacking fault energies of tetrahedrally coordinated crystals, *Phys. Status Solidi. A*, 171, 99-103.

Tchernycheva, M.; Cirlin, G.E.; Patriarche, G.; Travers, L.; Zwiller, V.; Perinetti, U. & Harmand, J.C. (2007). Growth and characterization of InP nanowires with InAsP insertions, *Nano Lett.*, 7, 1500-1504.

Tian, B.; Kempa, T.J.; & Lieber, C.M. (2009). Sinlge nanowire photovoltaics. *Chem. Soc. Rev.*, 38, 16–24.

Titova, L.; Hoang, T.; Jackson, H.; Smith, L.; Yarrison-Rice, J.; Kim, Y.; Joyce, H.; Tan, H. & Jagadish, C. (2006). Temperature dependence of photoluminescence from single core-shell GaAs–AlGaAs nanowires, *Appl. Phys. Lett.*, 89, 173126.

Tomioka, K.; Kobayashi, Y.; Motohisa, J.; Hara, S. & Fukui, T. (2009). Selective-area growth of vertically aligned GaAs and GaAs/AlGaAs core–shell nanowires on Si(111) substrate, *Nanotechnology,* 20, 145302.

Van Helvoort, A.T.J.; Dheeraj, D.L.; Grønsberg, S.; Zhou, H.L.; Fimland, B.O. & Weman, H. (2009). Dark field transmission electron microscopy techniques for structural characterization of nanowire heterostructures, accepted for publication in *J. Phys: Conf. Series (EMAG2009).*

Wagner, R.S. & Ellis, W.C. (1964). Vapor-liquid-solid mechanism of single crystal growth. *Appl. Phys. Lett.*, 4, 89.

Wu, Y. & Yang, P. (2001). Direct observation of vapor-liquid-solid nanowire growth *J. Am. Chem. Soc.*, 123, 3165-3166.

Yang, K. & Schowalter, L.J. (1992). Surface reconstruction phase diagram and growth on GaAs(111)B substrates by molecular beam epitaxy, *Appl. Phys. Lett.* 60, 1851-1853.

Yeh, C.Y.; Lu, S.; Froyen, S. & Zunger, A. (1992). Zinc-blende-wurtzite polytypism in semiconductors, *Phys. Rev. B* 46, 10086-10097.

Zanolli, Z.; Fuchs, F.; Furthmüller, J.; von Barth, U. & Bechstedt, F. (2007). Model GW band structure of InAs and GaAs in the wurtzite phase, *Phys. Rev. B*, 75, 245121.

Zhou, H. L.; Hoang, T.B.; Dheeraj, D.L.; van Helvoort, A.T.J.; Liu, L.; Harmand, J.C.; Fimland, B.O. & Weman, H. (2009). Wurtzite GaAs/AlGaAs core-shell nanowires grown by molecular beam epitaxy, *Nanotechnology*, 20, 415701.

3

MOVPE Self-Assembly and Physical Properties of Free-Standing III-V Nanowires

Paola Prete[1] and Nicola Lovergine[2]
[1]*IMM-CNR, Lecce Research Unit, Lecce,*
[2]*Dept. of Innovation Engineering, University of Salento, Lecce,*
Italy

1. Introduction

Quasi 1-dimensional (quasi-1D) semiconductor nano-crystals (so-called nanowires) represent the forefront of today's solid state physics and technology. These systems, having two of their dimensions comparable to the wavelengths of the electronic or phononic wave-functions, are expected to show a variety of quantum confinement effects, such as density of state singularities, molecular-like states extending over large distances, high luminescence efficiencies and low lasing threshold: these properties have led to consider quasi-1D nanostructures as the fundamental building blocks for the realization of novel types of photonic and electronic nano-devices. The high surface-to-volume ratio of these nano-structures allows also to exploit the role of surface states (and their ambient-driven changes) in determining the nanostructure carrier transport and optical excitation/recombination properties.

Despite the abovementioned attractive physical and technological assets, until very recently, not too many studies were performed on quasi-1D semiconductor systems, whilst two-dimensional (2D) structures – quantum wells – have been under study already for more than three decades and quasi zero-dimensional (0D) objects – quantum dots, QDs – have been in the focus of researchers for nearly two decades. The main reason for such discrepancy resides in the difficulty of fabricating these nanostructures. In the recent past, quasi-1D nanostructures of III-V compound semiconductors have been generally grown on three-dimensional patterned (V-grooved) substrates (Kapon, 1994), or on high-index (such as {n11}-oriented) substrate planes (Nötzel & Ploog, 2000). In such cases the major nanowire dimension was running in the substrate plane, with the consequence, however, that interactions with the substrate may dominate over 1D effects. These difficulties have lead to consider free-standing quasi-1D nano-crystals having negligible interaction with the substrate.

Free-standing nanowires based on III-V compound semiconductors are nowadays in the focus of intense research activities throughout the world; this in reason of the prominent role of these compounds in the optoelectronic arena. Demonstration of resonant tunnelling diodes (Bjork et al. 2002; Wensorra et al., 2005), single electron transistors (Thelander et al., 2003; Thelander et al., 2005), and photodetectors (Pettersson et al., 2006) based on quasi-1D nanostructures of III-V compounds have been reported. Also intense single photon sources have been fabricated by embedding InGaP quantum dots in free-standing GaP nanowires

(Borgström et al., 2005). Nanowires of InP (Goto et al., 2009) and GaAs (Colombo et al., 2009) were also recently shown to have photovoltaic (PV) properties, with photon energy conversion efficiencies exciding a few percent; theoretical considerations suggest that these structures may be ideal building blocks for the realization of high efficiency PV cells.

Relatively short free-standing semiconductor nanowires (nanorods) can be synthesized in colloidal form by reactions of suitable chemical precursors in coordinating solvents (Shieh, 2005), and afterwards collected on a suitable solid support (glass, silica) for study and/or device fabrication. However, crystallographically-oriented (i.e., epitaxial) quasi-1D nanostructures and their arrays can be grown (monolithically) on single-crystal substrates, provided their major dimension is running out of the substrate plane. The fabrication of free-standing quasi-1D semiconductor nanostructures of III-V compounds is currently investigated by exploiting different combinations of nano-technological methods and advanced epitaxial growth techniques, among the latter, molecular beam epitaxy (MBE), metalorganic vapour phase epitaxy (MOVPE) and other vapour phase growth methods; a major advantage of epitaxial technologies is that they promise precise and reproducible control over the nanowire size, shape, material composition and/or intentional doping along either the nanowire length or the lateral directions. Moreover, vapour epitaxy methods (in special MOVPE) represent the industry standard process for the mass production of most III-V based semiconductor devices; their exploitation in achieving the controlled self-assembly of III-V nanowires would then allow an easy integration of future nanowire-based devices with the current planar-device III-V technology.

In the following, we will briefly review the different approaches being currently explored for the fabrication of free-standing quasi-1D nanostructures of III-V compound semiconductors utilizing the MOVPE technology: major limitations and advantages will be discussed. In particular, we will focus on the self-assembly of semiconductor nanowires by the so-called metal-catalyst assisted — or VLS — mechanism. The latter is currently considered a most promising technology for the realization of high quality quasi-1D nanostructures. Examples of this approach will be given based on results obtained in the author's laboratory using low pressure MOVPE to growth nanowire structures of III-As compounds.

2. "Top-down" versus "bottom-up" technologies for III-V nanowire fabrication

The fabrication of nanostructured materials is generally achieved by two well-known alternative nano-technological approaches: (i) subtractive (so called "top-down") methods, where excess material is physically or chemically removed from an otherwise bulk-like or epitaxially micro-structured material; or (ii) additive (so called "bottom-up") methods, relying on the ability to self-assemble the nanostructure atom-by-atom (or molecule-by-molecule) up to the final dimensions.

A typical "top-down" approach to the fabrication of epitaxial quasi-1D nanostructures would utilize nano-lithographic techniques, such as electron beam lithography (EBL) and proximal probe patterning, in combination with dry-etching methods to remove excess material from planar epitaxial structures (Wensorra et al., 2005). However, top-down methods do not warrant a sufficiently low dimensionality, the spatial resolution limit of nano-lithographic techniques being typically above the size (20 nm) below which quantum-confinement effects appear in these nanostructures. Moreover, the resulting nano-crystals may suffer from process damages and surface roughening that could seriously deteriorate their electronic and/or optical properties.

"Bottom-up" methods achieving quasi-1D nanocrystals through self-assembly growth methods allow a superior crystalline perfection and size down to a few tens of nm; moreover, these methods allow the fabrication of quasi-1D nanostructures with compositional modulations along the nanowire axis (quasi-1D heterostructures) or in the radial direction (so-called core/shell nanostructures), which cannot be easily achieved by "top-down" approaches or chemical solution methods. In the field of MOVPE, two alternative bottom-up technologies for fabrication of III-V nanowires are being considered: (i) selective area epitaxy (SAE) (Motohisa et al., 2004), and (ii) metal-catalyst assisted self-assembly (Hiruma et al., 1995), through the so-called Vapour-Liquid-Solid (VLS) mechanism.

In the case of SAE, a (111)-oriented (III-V or Si) substrate is first covered with a thin (usually 20 nm) SiO_2 mask and circular openings with diameters of 50–200 nm are then fabricated in the mask by standard EBL and plasma etching processes. At relatively high temperatures (650÷750°C) III-V nanowires can then be grown by MOVPE within the mask openings with size comparable to the opening diameters and lengths up to several microns; the nanowires grow preferentially along the substrate (111) direction, i.e. with their major direction normal to the substrate plane (Motohisa et al., 2004). A clear advantage of this method is its ability to control the exact nanowire position on the substrate. Controlling the nanostructure positions and density across the substrate surface could pave the way to the realization of large and dense periodic arrays of almost identical quasi-1D nanostructures, the latter having potential applications in the field of 2-dimensional photonic crystal cavities and waveguides. Highly ordered arrays of GaAs (Noborisaka et al., 2005a) and InP/InAs/InP core-multishell (Mohan et al., 2006) nanowires, have been recently demonstrated. However, the method appears restricted to the use of (111)-oriented substrates, while the diameters of as-grown III-V nanowires and quasi-1D heterostructures are still limited to above 50 nm by the spatial resolution limits of both EBL and plasma etching techniques (Motohisa et al., 2004) used to pattern the SiO_2 mask.

Many of the above limitations are overcome in the case of nanowires self-assembled by the VLS mechanism (Wagner & Ellis, 1964; Givargizov & Chernov, 1973). Such mechanism rely on the ability of a suitable metal catalyst (usually Au) nanoparticle (NP) to react and alloy at relatively low temperatures with one or more of the III-group metal elements. The catalyst NP acts thus as a solvent for the metal(s) in a manner similar to the case of liquid phase epitaxy. It is assumed that, when in contact with a nutrient vapour phase (referred to as the Vapour in the VLS acronym) and for temperatures above the Au/III-group eutectic composition melting point, the NP forms a supersaturated liquid alloy droplet (the Liquid), finally leading to the solid state precipitation, nucleation and subsequent growth of the semiconductor nanowire (the Solid). A detailed study on the VLS growth of InAs nanowires indicates however, that the Au/In alloy droplet remains solid during the process (Dick et al., 2005), suggesting that the Au-catalysed growth of III-V nanowires may proceed through a Vapour-Solid-Solid (VSS) mechanism rather than through the VLS one. Despite these uncertainties, the acronym VLS is often retained in the nanowire literature to generally indicate a metal-catalyst assisted growth process.

A general characteristic of the VLS self-assembly of nanowires using MOVPE is that it relies on the existence of a kinetic hindrance to (planar) epitaxial growth (Seifert et al., 2004), a condition satisfied by growth at sufficiently low temperatures (typically below 500°C).

Noteworthy is also that the diameter of a VLS-grown semiconductor nanowire mimics the size of the original metal catalyst NP: nanowire diameters below 10 nm can be obtained by this method, provided that NPs of comparably small size are used in the process. No nano-lithographic processing of the substrate is thus required by this technology, unless precise patterning of the nanowires onto the substrate is desired. In this respect, a major technological issue is the ability to precisely and reproducibly control the actual size and dispersion of the metal catalyst NPs, along with their density and homogeneity across the substrate surface.

3. Au-catalyst NP technology

A critical step in the fabrication of semiconductor nanowires by VLS self-assembly is the controlled preparation of the Au catalyst NPs on various substrates. Three methods have been reported to date in the literature: (a) Volmer-Weber self-organization from a thin metal film; (b) chemical synthesis in the form of colloidal solutions and subsequent deposition on the substrate; and (c) synthesis of aerosol Au NPs, their precise size-selection using electrostatic methods, and subsequent deposition on the substrate (Magnussson et al., 1999). The former method was first used by Hiruma and his colleagues at Hitachi Central Research Laboratory (Japan) for the VLS growth of GaAs and InAs nanowires (Hiruma et al., 1995). The second technique relies on the well-know and widely used methods of synthesizing metal-NP colloids based on solution chemical reactions (Jana et al., 2001; Walker et al., 2001). Finally, the third method is being employed by Samuelson and colleagues at the University of Lund (Sweden) for the VLS growth of III-V nanowires by chemical beam epitaxy and MOVPE (Ohlsson et al., 2001; Seifert et al., 2004). Examples of the first two methods are given in the following for preparation of Au NPs on crystalline semiconductor substrates.

3.1 Au NPs from thin film self-organisation

This is a straightforward method often used in the literature to fabricate Au NPs by the high temperature annealing of very thin (a few nm or less) UHV-evaporated films. Minimization of the material (surface plus strain) energy leads to the dewetting of the metal film and to the Volmer-Weber self-organization of the deposit into a dense distribution of NPs. A schematic of this method is shown in Fig. 1(a): a thin (typically 0.1÷10 nm) Au film is evaporated under UHV conditions on suitably prepared substrates and afterwards annealed under inert gas flow. For relatively high annealing temperatures Au atoms gain sufficient surface mobility to rearrange the metal film into a dense array of NPs, thus minimizing its total energy. As the Au surface mobility depends on the interaction of the atoms with the underlying substrate material, the optimal annealing temperatures and the NP size and density will vary with the type of substrates. The case of Au on Si is specially interesting to study, as Si is a technologically important substrate for III-V nanowire fabrication (allowing the integration of future nanowire-based devices with Si microelectronics).

We studied the self-assembly process of Au thin fims on Si substrates. To this purpose, p-type (100)Si wafers were first degreased in iso-propanol vapours for 1h for removal of any surface organic contaminants, subsequently etched at room temperature (RT) in a NH_4F:HF solution for 6 min to eliminate the native Si-oxide layer from the substrate surface, rinsed in deionised water, and blown-dry under pure N_2. The substrates were then immediately loaded in the UHV chamber of a Joule evaporator for deposition of the Au thin film. The evaporation rate was 0.02 nm/s and the final Au film thickness was 2 nm. After evaporation

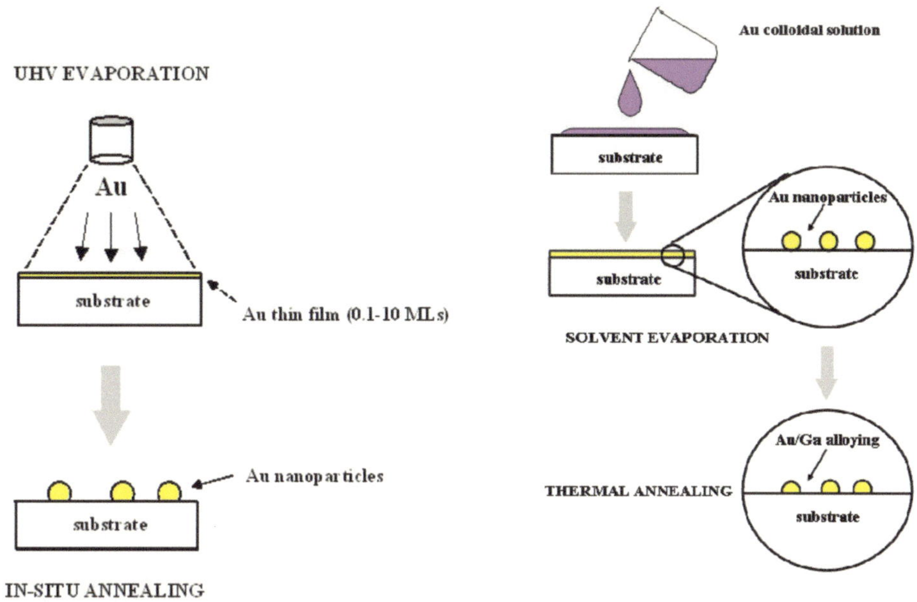

(a) (b)

Fig. 1. Schematic of the Au NP fabrication/deposition steps onto a semiconductor substrate: (a) self-assembly from a UHV evaporated thin Au film; and (b) from colloidal Au solution.

the samples were loaded into the quartz chamber of a horizontal tubular resistance furnace for annealing under 6.0N pure N_2 atmosphere. Annealing experiments were performed for 20 min between 350°C and 814 °C, i.e. either just below or well above the reported melting point (363°C) of the Au-Si eutectic composition; the sample temperature was increased from RT to the final value at a rate of about 10°C/sec. This treatment allows the Au film to self-assemble into a dense and uniform array of NPs across the entire Si surface. Fig. 2 shows the sample surface morphology before and after annealing at 814°C. The as-deposited thin film evidences a granular morphology, but after annealing it self-organises into distinct NPs with average size in the 10÷30 nm range and surface densities around 10^{10}÷10^{11} cm^{-2}; the NP shape appears droplet-like, as shown by the 40°-tilted view field emission scanning electron microscopy (FE-SEM) image reported in the insert of Fig. 2(b). Glancing incidence X-ray diffraction (GIXRD) analyses of the as-annealed samples have shown that the NPs are composed of pure Au in its fcc phase, while diffraction experiments performed in the Bragg geometry showed that the Au NPs self-assembled at 814°C are mostly (111)-oriented with respect to the (100)-planes of Si (Spadavecchia et al., 2005). High resolution transmission electron microscopy (HR-TEM) analyses further clarified that the NPs are single crystalline lens-shaped nanoislands and, for samples annealed at 814°C, the epitaxial relationships between the nanoislands and Si are uniquely defined as [001]Au $\|$ $[0\bar{1}1]$Si and [220]Au $\|$ [311]Si (Piscopiello et al., 2008). In contrast, at lower annealing temperatures the nanoislands turn out to be randomly oriented. In all cases however, they appear partially embedded into the Si substrate, the Au/Si interface being defined by smooth crystallographic facets, corresponding to the (111), (311), (711) and $(\bar{1}11)$ Si-planes (see Fig. 3). For the epitaxially-grown nanoislands these planes are almost parallel to the (410), (110) and (160) planes of Au, respectively. As a consequence, the Au lattice cell is rotated by 45°

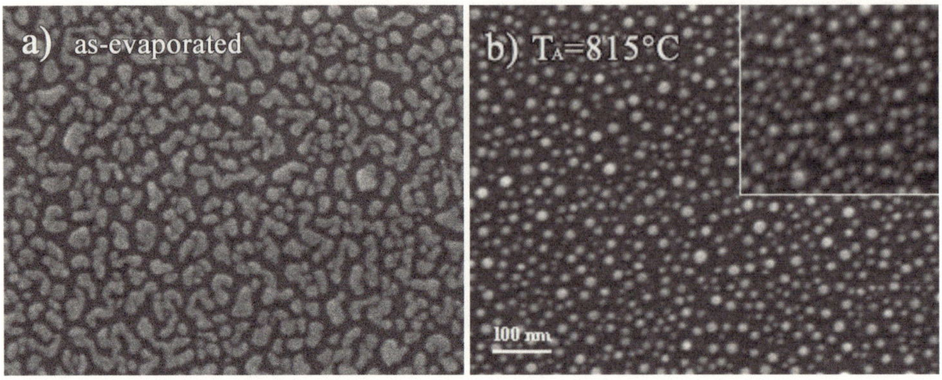

Fig. 2. FEG-SEM micrographs (150,000× magnification, plan-view) of 2-nm thin Au films on (100)Si (a) as-evaporated; and (b) after annealing at 814°C (inset represents a 40°-tilted view of the same sample). The marker in (b) represents 100 nm for both micrographs.

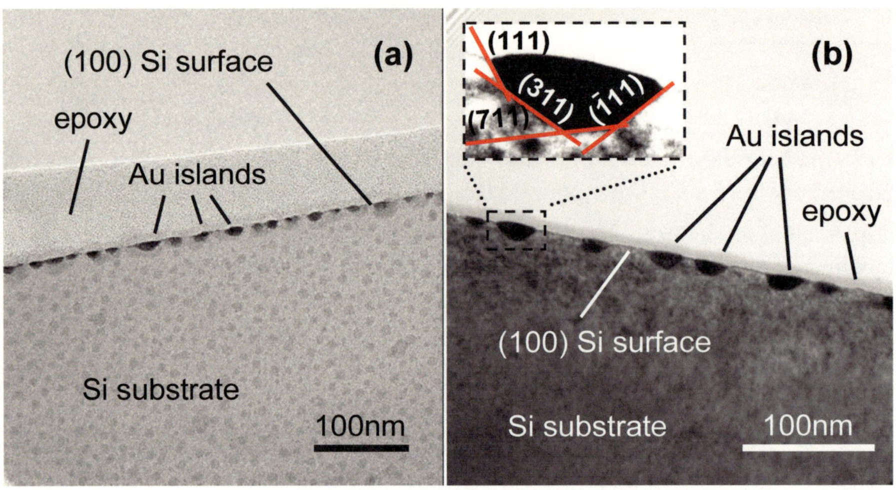

Fig. 3. Bright field cross-sectional TEM images of Au/Si samples annealed at (a) 350°C, and (b) 814°C, respectively. Au nanoislands are well visible (darker contrast) on the Si substrate surface. The inset in (b) shows an enlarged image of a Au nanoisland bound by well-defined crystallographic facets, corresponding to the (111), (311), (711) and ($\bar{1}$11) Si-planes.

and tilted by 21° with respect to the Si lattice. The formation of such epitaxial interface minimizes the huge lattice mismatch (25%) between Au and Si down to about 4%; this, along with the small nanoisland size, allows the easier elastic relaxation of the Au lattice.
The growth of partially embedded Au nanoislands can be explained by the formation of a liquid Si-Au alloy droplet during annealing at temperatures above the Au-Si eutectic melting point.[1] The presence of residual oxygen and water (<1 ppm) in the N_2 gas used in

1 In this respect, we stress the fact that the melting temperature of a nano-sized Au-Si eutectic alloy may well decrease below 350°C, in the same way the melting temperature of pure Au is reduced from its bulk value in the case of metal nano-crystals below a few tens of nanometres.

our experiments leads to oxidation of the Si atoms at the droplet surface and its evaporation in the form of SiO_x molecules; in order to keep the Si concentration in the droplets constant (in accordance with the Au-Si phase diagram equilibrium conditions) more Si atoms must diffuse from the substrate into the droplets. This process make the Au-Si droplets sinking into the substrate. Upon cooling of the sample the droplets get supersaturated and re-deposition of crystalline Si occurs, leading to the crystallisation of the pure Au nanoislands. During this step the facetted Au-Si interface of Fig. 3(b) thus forms. It is noteworthy that this process is expected for any substrate orientation, although the actual shape (faceting) of the resulting Au-Si nano-interface would be different from that observed for (100)Si. Furthermore, even the use of ultra-pure (to the level typically utilised for MOVPE) H_2 may not completely prevent this phenomenon, if a residual level of moisture is somehow still present in the reactor chamber during the high temperature annealing stages.

Despite the complex process leading to the formation of Au NPs on Si, their size and distribution show a relatively weak dependence on annealing temperature. The average Au NP size and statistical dispersion across the (100)Si substrate surface can be determined quantitatively by the analysis of the sample FE-SEM micrographs. Fig. 4(a) reports a typical count histogram of the apparent Au NP diameters, as measured from a series of plan-view micrographs, for one of the Au/(100)Si samples after annealing at 815°C. Taking D as the Au NP diameter, its statistical distribution is best-described by the Gamma distribution function:

$$P_{D_0,M}(D) = \frac{1}{\Gamma(M)}\left(\frac{M}{D_0}\right)^M \exp\left(-\frac{M}{D_0}D\right)D^{M-1} \tag{1}$$

where D_0 is the most probable NP diameter, M is the distribution shape parameter and Γ is Euler's function. The distribution standard deviation is then given by $\sigma=D_0/M^{1/2}$. Fig. 4(b) shows that for all samples D_0 varies between 15 nm and 35 nm, but the distribution relative standard deviations remain at ~29%.

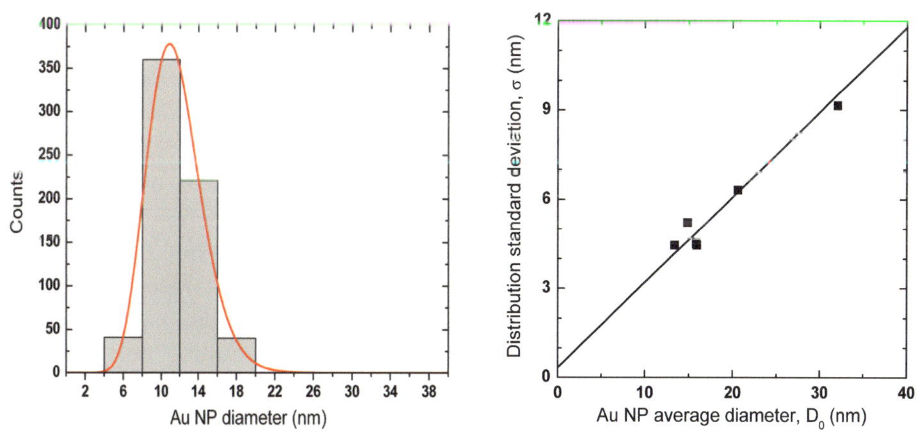

Fig. 4. (a) Count histogram of Au NP diameters for a 2-nm thin Au/(100)Si sample annealed at 815°C. The red curve is the Gamma distribution function [Eq. (1)] best-fitting the data, with fitting parameters D_0=(11.6±0.02) nm and M=16.1±0.2; (b) the distribution standard deviations (σ) as function of D_0 for various samples annealed at 815°C. The slope of the best-fitting solid line corresponds to a 29% average relative standard deviation.

3.2 Au NPs from colloidal solution

The preparation of noble metal NPs in colloidal form can be obtained by a variety of chemical reaction methods (Turkevich et al., 1951; Jana et al., 2001; Walker et al., 2001). In particular, colloidal Au NPs can be prepared in aqueous solution by the reaction of tetrachloro-aurate ($HAuCl_4$) with sodium citrate (Turkevich et al., 1951) or other reagents containing more Au-philic molecules (such as thiols, etc.). The latter act as capping agents for the process, their concentration in the solution controlling the actual size of as-synthesized Au NPs. As a matter of fact, changing the molar ratio between $HAuCl_4$ and sodium citrate allows one to control the size of colloidal Au NPs anywhere within the 10-100 nm range. Colloidal NPs usually show a relatively narrow size dispersion and have been successfully used in the literature for the VLS growth of III-V nanowires by MOVPE (Khorenko et al., 2005; Paiano et al., 2006). In the following we report on the use of colloidal Au NPs in aqueous solution (synthesized as described above) for the VLS self-assembly of GaAs nanowires. The use of aqueous solutions is particularly suitable for applications with advanced epitaxial technologies, as the solvent (water) can be readily evaporated during preparation stages, moreover it is compatible with most standard wet-etching processes used for surface preparation of semiconductor wafers before growth.

Fig. 5(a) reports the linear optical absorption coefficient spectrum measured at RT for one such Au colloidal solution (with NP size around 23.4 nm and volume density around 10^{11} cm^{-3}) with the strong plasmonic resonance of Au NPs peaked at around 518 nm. The resonance shifts its peak wavelength position and width with the actual size of the NPs (Njoki et al., 2007); however, it is difficult to determine both the average size and dispersion of the NPs from the optical absorption spectra of colloidal solutions, especially for NP diameters in the 10÷30 nm range. The size dispersion of colloidal Au NPs can be more precisely studied by FE-SEM observations upon deposition of the NPs on a solid support. A typical result of the FE-SEM analysis is shown in Fig. 5(b): as in the case of Au NPs self-assembled onto Si (Sec. 3.1 above), the diameters of colloidal Au NPs appear dispersed according to a Gamma distribution function [Eq. (1)] (Nagao et al., 2004). By choosing the right $HAuCl_4$ to sodium citrate molar ratio, Au NPs with average size down to 16 nm can be easily synthesised; the NP distribution relative dispersions range instead around 13-18%.

A schematic of the deposition of colloidal Au NPs onto a substrate is illustrated in Fig. 1(b): the colloidal solution may be spin-dropped at RT on the surface of a freshly etched substrate. The solvent (water) is then evaporated by holding the sample on a hot (70—80°C) plate (in air) for a few minutes. In order to attach the catalyst NPs on the surface, this is often treated by thoroughly rinsing in a poly-L-lysine aqueous solution (Hochbaum et al., 2005). The role of poly-L-lysine is to favour the adherence of Au NPs onto the substrate surface: the polymer possesses a net positive charge in aqueous solution at neutral pH and hence adsorbs onto the substrate surface. Consequently, the resulting poly-L-lysine film presents a positively charged group to the negatively charged (citrate anion capped) Au NPs, which attracts them to the surface. Rinsing the as-treated substrate into the colloidal solution finally leads to an even attachment of Au NPs onto the substrate surface. A possible drawback of poly-L-lysine is however, an increased carbon contamination of the surface.

As the deposition of colloidal NPs onto the substrate is generally realised in ambient atmosphere, this may easily lead to re-oxidation of the surface; the presence of surface oxides may prevent the crystallographic alignment of nanowires to the underlying substrate lattice. For this reason an *in-situ* (i.e., inside the MOVPE reactor chamber) H_2 heat cleaning is

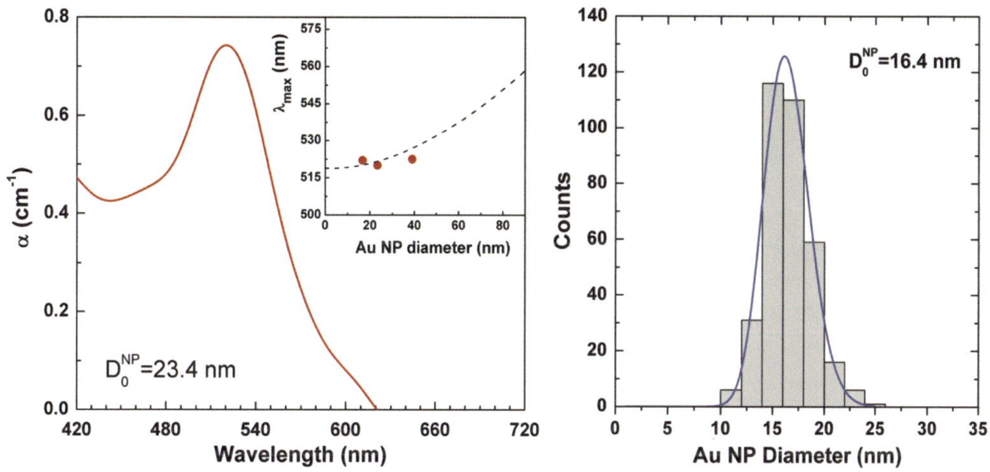

Fig. 5. (a) RT linear optical absorption coefficient of a colloidal Au NPs in aqueous solution around the plasmon resonance spectral region; the dashed curve in the inset shows the plasmon resonance peak wavelength as function of NP size (Njoki et al., 2007) along with present experimental data; (b) count histogram of Au NP diameters from one such colloidal solution, as-measured from FE-SEM observations. The blue solid curve is the best-fitting Gamma distribution function [Eq. (1)], with D_0=(16.40±0.07) nm and M=57.4±3.6. The distribution relative standard deviation is 13.2%.

performed on the substrate before nanowire growth. This treatment additionally ensures the efficient pyrolysis and elimination of any organic contaminant (NP capping agents, poly-L-lysine, etc.) brought onto the growth surface during its preparation with colloidal NPs. The actual temperature and vapour conditions of the H_2 annealing step depends on the chemical nature of the substrate and thermal stability of its surface oxides. In the case of Si, H_2 heat cleaning at temperatures well above 900°C are necessary to fully desorb any residual silicon dioxide from the substrate surface (Moore et al., 2007), although annealing temperatures in the 600-700°C interval have been most often reported (Mårtensson et al., 2004; Khorenko et al., 2004) before III-V nanowire growth. In any case, these temperatures are well above the Au-Si eutectic melting point; therefore, phenomena similar to those observed for Au NPs self-assembled from a thin film (Sec. 3.1) should occur also for colloidal NPs annealed together with the Si substrate. On the contrary, the native oxides of Ga and As are much less stable than silicon dioxide and H_2 treatment of GaAs at temperatures around 600-650°C are usually sufficient to reduce any trace of residual oxides from the substrate surface. GaAs annealing is usually performed under excess As in the vapour to prevent the material non-congruent evaporation. It has been claimed that under these As-excess conditions GaAs is stable against Au interaction (Dick et al., 2005); therefore, the behaviour of Au NPs on the surface of GaAs under high temperature annealing is expected to be entirely different from that observed for Si.

In order to gain insights on this behaviour we studied the evolution of the colloidal Au NP size on the surface of GaAs upon annealing at 625°C. To this purpose ($\bar{1}\bar{1}\bar{1}$)B-oriented GaAs wafers were first degreased in iso-propanol vapours for 1h, subsequently etched at ~40°C in a H_2SO_4:H_2O_2:H_2O (4:1:2) solution for 8 min, thoroughly washed in deionised water, and finally dried under pure N_2. Colloidal NPs were then deposited on the freshly etched substrates as described above. To keep the GaAs surface as clean as possible, no poly-L-lysine was used to adhere the NPs onto the substrate. After loading the sample to the reactor chamber, its temperature was raised to 625°C under a H_2+tertiarybutylarsine (TBAs) atmosphere and annealing was then performed for 10 min. The sample cool-down was also performed under the same H_2+TBAs flow. FE-SEM micrographs [Fig. 6(a)] of colloidal Au NPs after annealing were again studied for a quantitative analysis of their size dispersion. Fig. 6(b) shows a typical count histogram of the diameters for as-annealed NPs: noteworthy is that much larger diameters and larger distribution widths than those of initial colloidal NPs (Fig. 5) are observed. This effect can be ascribed to the well-known 'Ostwald ripening' mechanism, where smaller and less stable NPs shrink their size (and may even disappear) by losing Au atoms in favour of larger and more stable ones. As the mechanism is driven by the surface diffusion of Au atoms, sufficiently high annealing temperatures are required. Systematic experiments have shown that the effect sets in for temperatures above about 500°C.

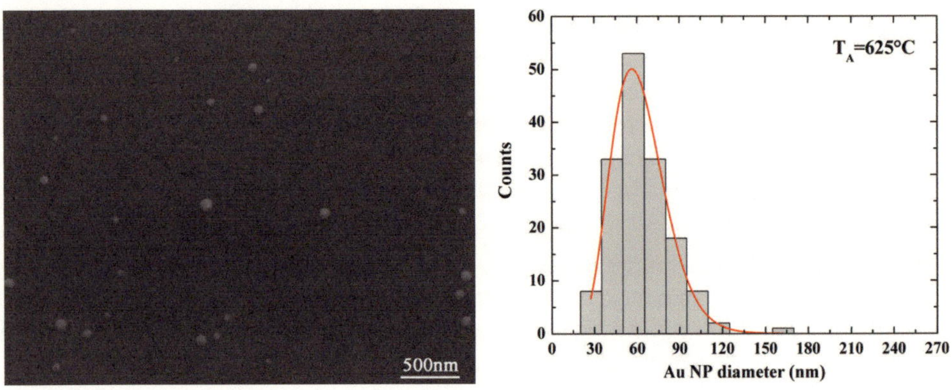

(a) (b)

Fig. 6. (a) FE-SEM plan-view micrograph of colloidal Au NPs on the surface of a $(\bar{1}\bar{1}\bar{1})$B-GaAs substrate after annealing at 625°C; (b) count histogram of NP diameters, as measured from the micrograph in (a); the red curve in (b) represents the Gamma function [Eq. (1)] best-fitting the experimental data, with D_0=(62.4±0.8) nm and M=10.8±0.8.

3.3 Comparison between Au NPs from colloidal solutions and thin films

It is useful to compare the average size and density of Au NPs observed after annealing on (100)Si and $(\bar{1}\bar{1}\bar{1})$B-GaAs. In order to make a meaningful comparison we begin with analysing the characteristics of NPs self-assembled on the two substrates upon deposition of similar surface doses of Au atoms (equivalent to a film thickness around 1-2 nm). Fig. 7 reports both the average diameters and surface densities for Au NPs self-assembled as described in Secs. 3.1 and 3.2 – i.e. upon annealing at temperatures in the 500-625°C interval (for GaAs) and 815°C (for Si). It appears that for the same deposited thickness much larger and less dense NPs self-assemble on GaAs than on Si, despite the annealing temperatures used for Si are hundreds of degrees higher. This result can be ascribed to the higher surface diffusion of Au atoms on the surface of GaAs with respect to Si, leading to a much more efficient Ostwald ripening of Au NPs for the former substrate. We showed elsewhere that Au NPs on (100)Si do show however, Ostwald ripening (Prete et al., 2007), but the strong

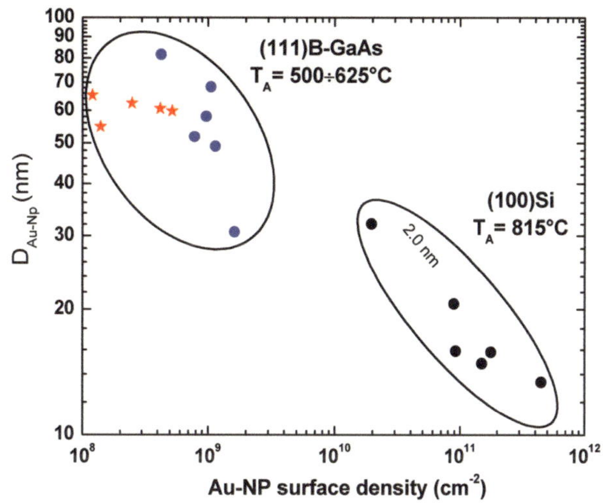

Fig. 7. Average Au NP diameters as function of surface density on $(\bar{1}\bar{1}\bar{1})$B-GaAs and (100)Si substrates after annealing at temperatures between 500-625°C for GaAs, and at 815°C for Si. In the diagram, black and blue solid points refer to 1-2 nm Au deposition on GaAs (•) and Si (•), respectively, while the stars (★) refer to colloidal NPs deposited onto GaAs.

interaction of Au NPs with Si discussed in Sec. 3.1 clearly slows down the process with respect to GaAs. The Ostwald ripening appears also to affect the behaviour of colloidal Au NPs onto GaAs (Sec. 3.2), such that substrate annealing under standard MOVPE conditions results into huge changes of the NP size, size distribution and densities with respect to initial (as-deposited) values. As this phenomenon is intrinsic of the Au-GaAs interaction, the effect should hold as well for aerosol-produced Au NPs annealed under similar conditions.

4. III-V nanowires grown by Au-catalyst assisted MOVPE

In the mid 90's Hiruma and colleagues at Hitachi Central Research Laboratory (Japan) (Hiruma et al., 1995) were the first to demonstrate the VLS growth of III-V nanowires using Au NPs self assembled from UHV evaporated thin metal films. Samuelson and co-workers of Lund University (Sweden) later on demonstrated the growth of free-standing GaAs nanowires using chemical beam epitaxy and size-selected aerosol Au NPs as metal catalysts (Ohlsson et al., 2001). The Au-catalysed MOVPE of GaAs, GaP, InAs, and InP nanowires was similarly demonstrated by the same group a few years later (Seifert et al., 2004). Today the MOVPE self-assembly of III-V nanowires is studied by several research groups, which however, in most cases use colloidal Au NPs as catalysts for the VLS growth.

To date the Au-catalyst assisted MOVPE of III-V nanowires has been extended to ternary alloys, such as GaAsP (Svensson et al., 2005), GaInP (Sköld et al., 2005), and InGaAs (Regolin et al., 2006), while only few data can yet be found on the catalyst-assisted MOVPE of AlGaAs nanowires (Lim et al., 2008). For these ternary alloys a major challenge is to growth nanowires with uniform composition along both their major axis and in the radial direction (Regolin et al., 2006; Kim et al., 2006; & Lim et al., 2008). The MOVPE growth of nanowires with modulation of the material composition in either the axial or radial direction has been also demonstrated for a number of III-V systems. In particular, the growth of axially

modulated nanostructures has been reported for the GaP/GaAsP and InP/InAsP systems, for whom examples of double-hetero-junction structures were reported (Svensson et al., 2005), while few data still exist on the growth and properties of AlGaAs/GaAs (Ouattara et al., 2007) and InGaAs/GaAs quasi-1D nanostructures. In these latter cases, the occurrence of group-III metal memory effects and instabilities within the catalyst NPs during growth lead to difficulties in achieving sharp compositional changes at the various interfaces within these structures (Kim et al., 2006; & Ouattara et al., 2007).
The application of GaAs nanowires to novel nanophotonic devices is hindered by the proclivity of GaAs to readily oxidise in air, leading to the formation of surface states that reduce carrier lifetimes and quench the material optical emission. The growth of a shell using a large band-gap material around GaAs nanowires reduces the effects of surface states and enhances the radiative efficiency of GaAs nanowires; besides, changing the band-gap of the shell material allows to control the carrier confinement within the GaAs core and may also improve the optical confinement of the photons by acting as an optical cavity. The Au-catalysed growth and optical properties of GaAs/GaInP core-shell nanowires were first reported (Sköld et al., 2005), showing that alloy ordering of GaInP induces compositional and band-gap variations in the shell, leading to potential fluctuations within the shell and different strain states in the core. For these reasons, emphasis has been placed in the recent literature on the growth of GaAs/AlGaAs core-shell nanowire structures (Samuelson et al., 2004; Titova et al., 2006; Prete et al., 2008). The latter are of particular interest, as the GaAs-AlGaAs materials combination represents a prototypical system in semiconductor opto-electronics. Also, the use of AlGaAs as shell material leads to almost strain-free nanowires, simplifying the analysis of their electronic/optical properties.
Despite the amount of work reported in the literature, systematic studies on the VLS self-assembly of III-V nanowires by MOVPE under different epitaxy conditions (substrate material and orientations, metal-catalyst preparation, type of precursors and purity, growth parameters, etc.) are still largely lacking. In particular, almost exclusive use of arsine (AsH_3) and phosphine as precursors of group-V elements was made in most of the studies above. Substitution of the toxic hydrides with alkyl-substituted arsine and phosphine molecules, besides leading to a safer process, may also improve the materials electronic properties (i.e. by reducing carbon and other unintentional impurity incorporation into the material).
In the following, we report on the Au-catalysed growth of GaAs nanowires and GaAs-AlGaAs core-shell nanostructures using the liquid precursor tertiarybutylarsine (TBAs).

4.1 Growth of GaAs nanowires using TBAs

The MOVPE growth of GaAs nanowires using TBAs has been previously reported in conjunction with either triethylgallium (TEGa) or trimethylgallium (TMGa), and pure N_2 as carrier gas (Khorenko et al., 2005). Growth temperatures were limited to over ~430°C in the case of TEGa and over 470°C for TMGa, i.e. above those (≥400°C) reported using AsH_3 (Seifert et al., 2004). As the VLS self-assembly of nanowires rely on the existence of a kinetic hindrance to (planar) growth, whose onset for GaAs is ~480-500°C, lower temperatures are essential to guarantee a good control over nanowire growth and shape. Furthermore, GaAs nanowires grown using TBAs+TEGa in N_2 were heavily affected by bending and kinking (Khorenko et al., 2004; Khorenko et al., 2005), an effect ascribed to growth instability at the catalyst/nanowire interface. It is therefore of great importance to clarify the role of MOVPE conditions on the morphology and defects of as-grown GaAs nanowires using TBAs.

GaAs nanowires were grown under H_2 ambient by low (50 mbar) pressure MOVPE using an Aixtron reactor model AIX200 RD. TMGa and TBAs were used as Ga and As precursors, respectively. Semi-insulating $(\bar{1}\bar{1}\bar{1})$B-oriented (within ±0.01°) GaAs wafers were used as substrates. GaAs pieces were cut from wafers by cleavage along ⟨110⟩ in-plane directions. Colloidal Au NPs, synthesized in aqueous solution, were deposited onto the freshly-etched substrates as described in Sec. 3.2. A rough estimate of as-deposited Au dose gives an equivalent Au film thickness around 0.1-0.5 nm; we notice that these values compare well with the thickness of UHV-evaporated Au films previously reported in the literature for the VLS growth of III-V nanowires (Hiruma et al., 1995). Immediately before growth the as-deposited substrates were annealed as described in Sec. 3.2. After ramping down the sample temperature to the final growth value, varied between 375°C and 525°C, TMGa was admitted to the reactor chamber and the growth initiated. All growth runs were carried out under a H_2 flow of 7.0 sl/min, a TMGa molar fraction in the vapour of 3×10^{-5}, and a As:Ga precursor molar ratio around 7:1.
The above growth conditions allowed us to growth GaAs nanowires down to less than 400°C (Paiano et al., 2006). Fig. 8 shows the morphology of as-grown nanowires for different temperatures. The nanowires consist of straight GaAs segments epitaxially well-aligned in the substrate $[\bar{1}\bar{1}\bar{1}]$B-direction, as confirmed by X-ray diffraction (XRD) measurements (see Sec. 4.2 below) and surface densities in the $(1–6)\times10^{8}$ cm^{-2} interval. Furthermore, each nanowire shows a droplet at its tip, with size comparable to that observed for the Au NPs after their annealing at 625°C. This confirms that the growth occurred through the VLS mechanism. Noteworthy, nanowires grown above 425°C show a substantial degree of tapering along the $[\bar{1}\bar{1}\bar{1}]$B growth direction, while constant diameter nanowires are obtained at lower temperatures. Plan-view FE-SEM observations of nanostructures grown above 425°C indicate that the tapering gives rise to a nanowire hexagonal section (Fig. 9); by comparing the directions of the hexagonal base edges with that of the substrate ⟨110⟩ and $\langle 1\bar{1}0 \rangle$ cleaved edges, it turned out that each couple of parallel base edges is aligned normal to one of the three equivalent $\langle \bar{2}11 \rangle$ in-plane GaAs directions.
Systematic FE-SEM plan-view observations of the samples have allowed a quantitative evaluation of the nanowire diameter D (taken as the major diagonal length of the nanowire hexagonal base). Fig. 10 reports a count histogram of base diameter values obtained from the analysis of several FE-SEM micrographs of a sample grown at 425°C. It appears that the nanowires have a relatively narrow base diameter distribution, which can be best-fitted by a Gamma-distribution function [Eq. (1)]; this is no surprise as the nanowire dimensions mimic that of the Au-catalyst NPs that have generated them. A best-fit of the experimental data in Fig. 10 gives D_0=(60.6±0.4) nm and M=28.2±0.9, leading to a standard deviation of the base diameter distribution σ_{NW}=11.4 nm (corresponding to a 18.8% relative standard deviation). Significantly, the average diameter of the distribution closely matches that [D_{Au}=(62.4±0.8) nm] of Au NPs after annealing: this confirms that negligible tapering of nanowires occurs at 425°C. The same quantitative study was performed on nanowires grown between 375°C and 500°C, allowing the precise estimate of the sample average base diameter D_0 as function of temperature. The results of such analysis are reported in Fig. 11 for 10 min grown samples. It appears that growing below about 425°C ensures nanowires without tapering (i.e., with diameters almost coinciding with that of the Au NPs), while D_0 increases exponentially with temperature above 425°C. This latter finding indicates a kinetics-limited growth regime of the nanowire sidewalls. Fig. 10(b) further shows that tapered nanowires tends to have a larger diameter distribution, which linearly scales with the degree of tapering.

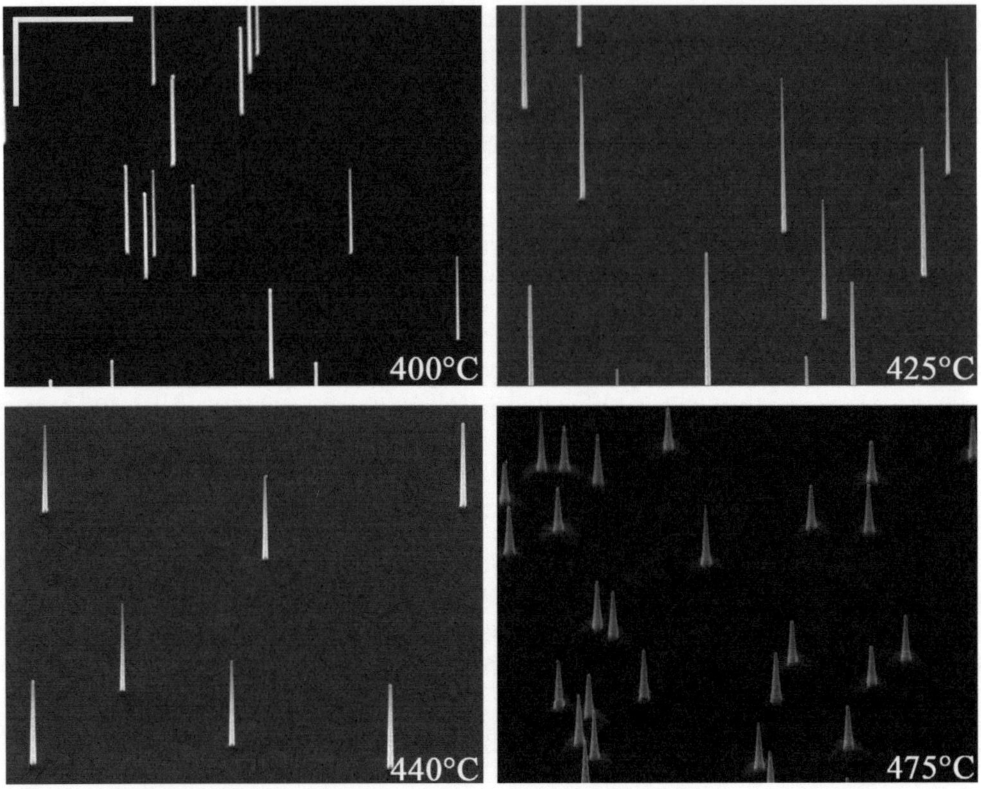

Fig. 8. FE-SEM micrographs (15,000× magnification, 45° tilt view) of GaAs nanowires grown on ($\bar{1}\,\bar{1}\,\bar{1}$)B GaAs at different temperatures. The growth time is 20 min for samples grown at 400°C and 425°C, and 10 min for those grown at 440°C and 475°C. Markers in the upper left micrograph represent 1 μm.

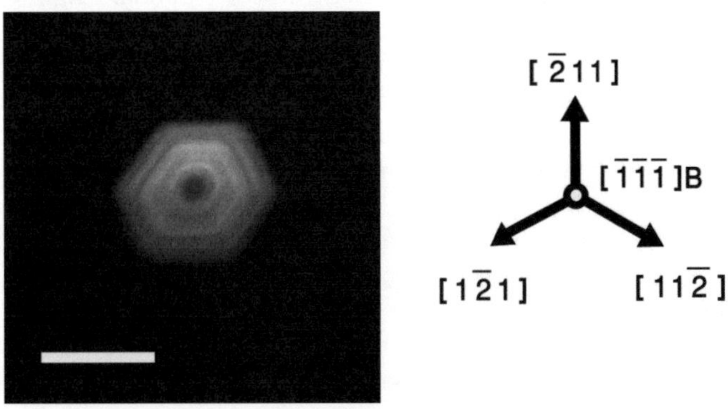

Fig. 9. FE-SEM micrograph (80,000× magnification, plan-view) of a [$\bar{1}\,\bar{1}\,\bar{1}$]-aligned GaAs nanowire grown at 500°C. The nanowire shows a hexagonal base, each couple of parallel base edges being aligned normal to one of the three equivalent ⟨$\bar{2}$11⟩ substrate directions (as shown in the diagram on the left). A 60 nm diameter droplet is also observed at the nanowire tip. Marker in the micrograph represents 200 nm.

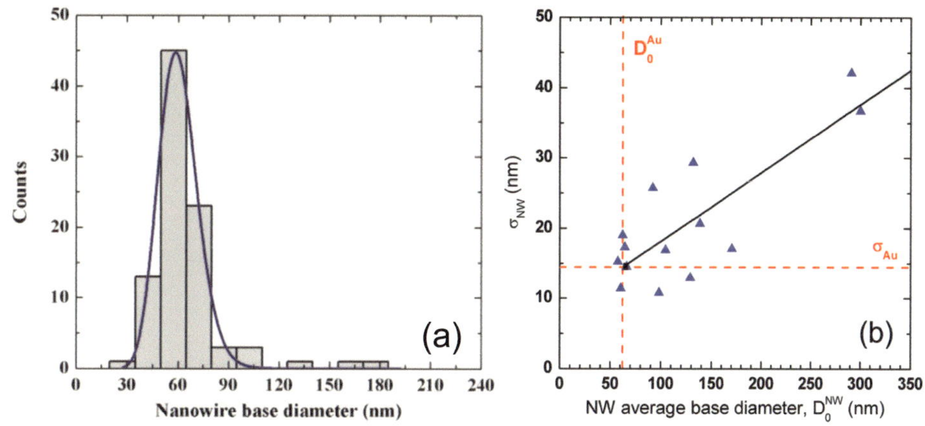

Fig. 10. (a) Count histogram of GaAs nanowire base diameters for a sample grown at 425 °C. The solid curve represents the Gamma-distribution function [Eq. (1)] best-fitting the data, with D_0=(60.6±0.4) nm and shape parameter M=28.2±0.9. (b) Variation of the base diameter distribution standard deviation with the nanowires average base diameter for samples grown at different temperatures. The red dashed lines indicate the (run-to-run) average diameter and distribution standard deviation of Au NPs annealed on $(\bar{1}\bar{1}\bar{1})$B-GaAs.

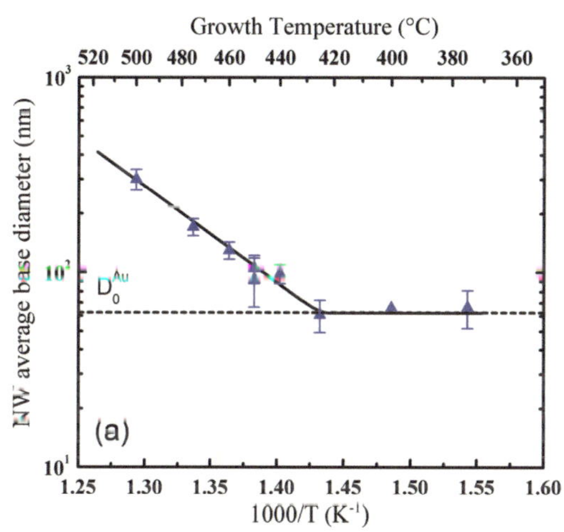

Fig. 11. Arrhenius plot of the nanowire average base diameters as function of the growth temperature. The horizontal dashed line represents the (run-to-run) average diameter D_{Au}=62.4 nm of Au NPs after annealing at 625°C. Solid lines are only guides for the eye.

To compare the growth kinetics of the material along the nanowire sidewalls with that in the $[\bar{1}\bar{1}\bar{1}]$B (axial) direction we measured the nanowire lengths of the same samples above. However, as the VLS growth rate of a nanowire along its axis is a non-linear function of the diameter (Johansson et al., 2005), we selected only those nanowires having tip diameter values around D_{Au}=62 nm (within ±8 nm). The results are plotted in Fig. 12, where each data point is averaged over several tens of nanowires. It is first of all interesting to compare the as-measured axial nanowire growth rates with that of planar GaAs epilayers on the $(\bar{1}\bar{1}\bar{1})$B

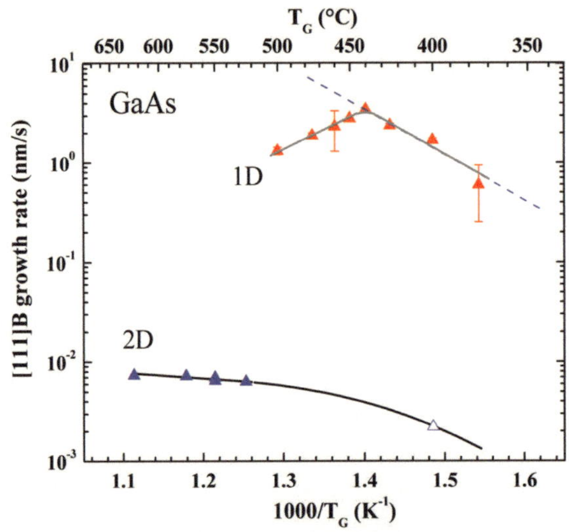

Fig. 12. Axial (▲) (for nanowires) and planar (▲) [on the $(\bar{1}\,\bar{1}\,\bar{1})$B-oriented substrate surface] growth rate of GaAs as function of inverse growth temperature (Arrhenius plot). The open triangle (Δ) indicates the planar growth rate estimated at 400°C. Solid lines are only an aid for the eye.

surface under identical vapour conditions: planar growth in the $[\bar{1}\,\bar{1}\,\bar{1}]$B direction turns to be over two orders of magnitude lower than that of nanowires, confirming the existence of a kinetic hindrance to the GaAs planar growth also for TBAs. The nanowire growth in the axial direction shows a maximum of 3.44 nm/s at around 440°C; above this value the axial growth rate of the nanowires decreases due to the competing growth on the nanostructure sidewalls (tapering). Instead, below about 425°C almost no growth occurs at those surfaces. Under these conditions the growth appears thermally activated with an activation energy of (20.7±3.2) kcal/mol. This value is slightly above what reported for Au-catalysed GaAs nanowires grown using TMGa+AsH_3 (Borgström et al., 2005). However, in both cases the maximum growth rates occur at temperatures for which almost complete decomposition of TMGa is expected (DenBaars et al., 1986). Clearly, the use of TBAs in place of AsH_3 does not greatly affect the pyrolysis efficiency of TMGa, if H_2 is used as carrier gas; we notice however, that TBAs allows nanowire growth with good morphological quality for As:Ga molar ratios in the vapor (7:1) much lower than those (>27:1) typically reported for AsH_3 (Seifert et al., 2004), a result of the lower decomposition temperatures of TBAs. Furthermore, our results indicate a substantial decrease of nanowire growth temperatures with respect to using N_2 as carrier gas (Khorenko et al., 2005). In a purely surface-kinetics controlled growth regime, H_2 has no role on the pyrolysis of TMGa, as the amount of adsorbed hydrogen on GaAs surface is known to be negligible (Qi et al., 1996); instead, H_2 may enhance TMGa dissociation in the vapour through well-known hydrogen radical attack reactions.

4.2 Crystallography of GaAs nanowires on (111)GaAs and (111)Si substrates

The perfect epitaxial relationships of GaAs nanowires grown on $(\bar{1}\,\bar{1}\,\bar{1})$B-GaAs substrate is evidenced by the high-angle XRD spectrum reported in Fig. 13 for a sample grown at 450°C; indeed, the spectrum shows only the (111), the quasi-forbidden (222) and the (333)

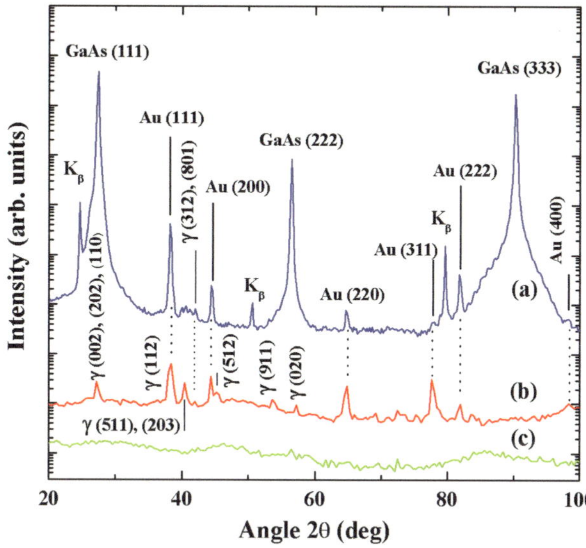

Fig. 13. (a) θ-2θ XRD, and (b) GIXRD spectra of a GaAs nanowires sample grown at 450°C; (c) GIXRD spectrum of a bare $(\bar{1}\bar{1}\bar{1})$B-oriented GaAs substrate. Diffraction peaks belonging to GaAs, Au and the Au_2Ga alloy (γ phase) are indicated in the diagram. The spectra have been shifted along the intensity axis for clarity sake.

diffraction peaks of cubic GaAs, confirming that the nanowires are epitaxially well-aligned along the substrate $[\bar{1}\bar{1}\bar{1}]$B-direction. No peak signatures relating to the wurtzite phase of GaAs can be observed, indicating that the nanowires crystallise in the zinc-blend phase solely. Besides, several peaks which can be assigned to the fcc phase of Au, are also observed in the spectrum, most of the Au being located at the nanowire tips. The Au peak positions in the θ-2θ spectrum agree with the angular values reported in the ICPDS powder diffraction database (ICPDS, 2000). However, the relative peak intensities are very different from that of powder-like material: the intensities of the Au (111) and (222) peaks are much enhanced, indicating that the (111)Au and (111)GaAs crystallographic planes are parallel to each other in most cases.

In addition, other very weak peaks occur in the angular region 39°<2θ<42° that cannot be attributed to either GaAs or Au. The exact crystallographic phase originating these peaks can be identified by the analysis of the GIXRD spectra. Noteworthy, the GIXRD pattern of the nanowire sample [curve (b) in Fig. 13] exhibits several weak diffraction peaks, not observed in the GIXRD pattern of the bare $(\bar{1}\bar{1}\bar{1})$B GaAs substrate taken as reference [curve (c)]. Besides the expected relatively intense diffraction peaks of Au, other peaks corresponding to the most intense of the diffraction peaks of the orthorhombic Au_2Ga (ICPDS, 2000) appear in the spectrum at 27.2°, 40.4°, 42.0°, 45.3°, 53.6° and 57.2°, a few others being likely shadowed by the superposition of Au peaks. The Au_2Ga alloy (the so called γ phase) is one of the Ga-rich alloys that may form in the Au-Ga binary system. Clearly, the alloy must be located at the Au-NP/nanowire interface.

The γ phase is stable at temperatures below 349°C and its formation has been reported upon cooling below 370°C of a Au-Ga melted alloy (Zeng et al., 1987). This suggests that the catalyst NP at the tip of the nanowire is liquid at 450°C, a temperature lower than that (470-

480°C) generally expected for the formation of a Au-Ga melted alloy (Zeng et al., 1987; Dick et al., 2005). The formation of a saturated solid Au-Ga alloy (about 10 at-% Ga, α phase) at the Au-GaAs interface has been instead claimed (Dick et al., 2005) during nanowire growth up to 475°C. On this basis, these authors proposed a VSS process, as the driving mechanism of the Au-catalyst assisted growth of III-V nanowires. One should however mention that the concentration of Ga reached in the Au-Ga alloy at the tip of the nanowire depends on the overall dynamic balance between Ga supply rate from the vapour, diffusion rate through the alloy and nanowire growth rate, which may change depending on actual MOVPE growth parameters. As a matter of fact, GIXRD measurements performed on nanowire samples grown at 400°C did not show the γ phase peaks, suggesting that at lower temperatures the Au-Ga droplet may indeed remain solid.

GaAs nanowires were also grown on (111)Si wafers to compare their crystallography with that observed on ($\bar{1}\bar{1}\bar{1}$)B-GaAs; to this purpose p-type (111)Si substrates were cleaned and etched as described in Sec. 3.1 and afterwards deposited with colloidal Au NPs having average size around 16.4 nm; as-prepared samples were then annealed at 625°C under pure H_2 flow for 20 min. GaAs nanowires were grown at 450°C under MOVPE conditions similar to those used for the samples in Fig. 8. The typical morphology of GaAs nanowires grown on (111)Si is shown in Fig. 14: as expected, each nanowire appears tapered along its axis, indicating a substantial sidewall growth of the material; however, the nanowires show no preferential direction with respect to the substrate, few of them being also heavily kinked. Detailed study of the samples by FE-SEM observations do show that only a small fraction (1-2%) of the nanowires is aligned with their axis normal to the substrate plane. Similar results have been also reported in the literature for GaAs nanowires grown on Si using TBAs and N_2 as carrier gas (Khorenko et al., 2005). This is in striking contrast with other results reporting on the growth of well-aligned GaP (Mårtensson et al., 2004) and GaAs (Bao et al., 2008) nanowires on (111)Si. Clearly, actual conditions of Si surface preparation, the size of Au NPs and different procedures for their subsequent deposition and annealing may strongly influence the complex Au-Si interaction discussed in Sec. 3.1, so to adversely affect the VLS growth of III-V nanowires. More research is needed however, on this topic.

Fig. 14. FE-SEM micrograph (15,000× magnification, 45° tilt view) of GaAs nanowires grown at 450°C on a (111)Si substrate.

4.3 Growth of GaAs-AlGaAs core-shell nanowires

The growth of core-shell nanowires structures by MOVPE technology is easily realised by overgrowing the desired shell material around a nanowire core by conventional vapour epitaxy. This requires growth temperatures high enough to suppress the VLS mechanism in favour of a conformal (around the nanowire) deposition of the material. We have grown GaAs-AlGaAs core-shell nanowires on ($\bar{1}\bar{1}\bar{1}$)B-GaAs substrates. In addition to TMGa and TBAs, trimethylaluminum (TMAl) was used as Al precursor. In these experiments the growth temperature of the GaAs core was kept at 400°C to ensure constant diameter nanowires throughout their entire lengths (see Sec. 4.1). Once the growth of the GaAs core was terminated, the substrate temperature was increased to 650 °C under a H_2+TBAs for the AlGaAs shell overgrowth; to this purpose the Al fraction of the III-group metal precursors in the vapour x^v=[TMAl]/([TMAl]+[TMGa]) was fixed at x^v=0.50. The shell growth time ranged between 4 and 10 min. The nanostructures were grown by keeping the precursors V:III ratio in the vapour fixed to 5:1 for both core and shell and under a total H_2 flow through the chamber of 7.0 sl/min. Samples were cooled under a continuous H_2+TBAs flow. Fig. 15(a) shows the morphology of as-grown GaAs/AlGaAs core–shell nanowires. They appear straight (kink-free) segments, with their major dimension running normal to the substrate ($\bar{1}\bar{1}\bar{1}$)B surface. As expected, the nanostructures have a constant diameter throughout their entire lengths, but for a small tapered section close-by their upper ends, where the original Au NPs (used to grow the GaAs core) are clearly visible. Plan-view FE-SEM observations showed that all core–shell nanowires have a hexagonal section (Fig. 16), with their side facets normal to one of the three equivalent in-plane [110] substrate directions. A similar result has been reported also for GaAs/AlGaAs core–shell nanowires grown by SAE-MOVPE on SiO_2-patterned ($\bar{1}\bar{1}\bar{1}$)B-GaAs (Noborisaka et al., 2005b).
Fig. 15(b) shows a high-angle annular dark field (Z-contrast) image of a single GaAs-AlGaAs core-shell nanowire recorded in a scanning transmission electron microscope. The contrast variation across the nanowire diameter is due to the different Ga composition of the

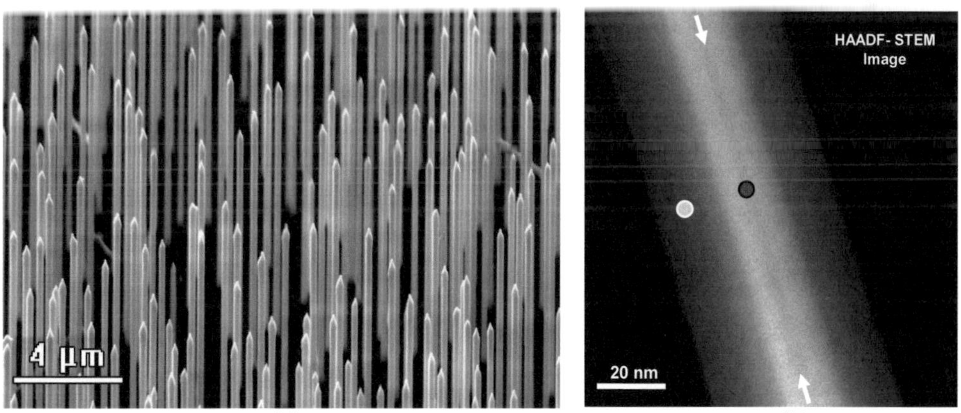

(a) (b)

Fig. 15. (a) FE-SEM micrograph (21,000× magnification, 45°-tilt view) of a dense ensemble of GaAs–AlGaAs core–shell nanowires as-grown on the ($\bar{1}\bar{1}\bar{1}$)B GaAs substrate. (b) Z-contrast image of a single GaAs-AlGaAs core-shell nanowire. The white and grey circles indicate regions from where EDS measurements were performed. The white arrows in the image indicate a central darker area inside the GaAs core region.

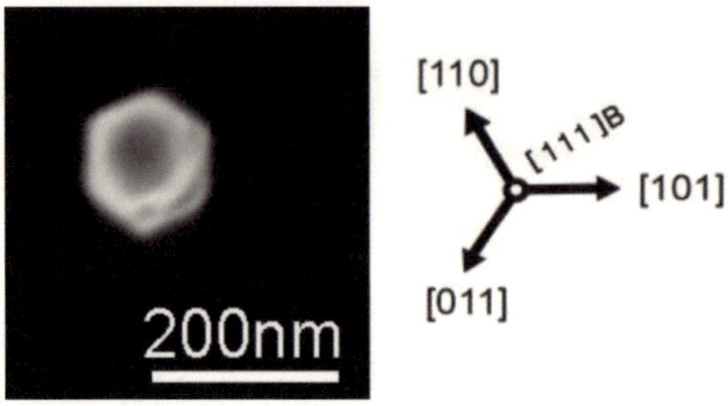

Fig. 16. FE-SEM micrograph of a single core–shell nanowire (150,000× magnification, plan-view) showing its hexagonal section.

core (brighter) and the shell (darker) regions, as also indicated by X-ray energy dispersion spectroscopy (EDS) measurements performed at different positions along the same nanowire [indicated in Fig. 15(b)]. Analysis of EDS spectra collected from the shell region indicated an Al composition of the AlGaAs alloy x_{Al}=0.32±0.02. The Z-contrast image shows the presence of a darker region at the center of the GaAs core [indicated by the arrows in Fig. 15(b)]; in addition, a higher carbon (C) concentration was observed by EDS in the nanowire core than in the shell, most likely responsible for the lower Z-contrast signal observed in Fig 15(b). C is thus more efficiently incorporated into the GaAs nanowire core during VLS growth than in the AlGaAs shell under conventional MOVPE growth mode.

A quantitative analysis of FE-SEM plan-view images of as-grown nanowires recorded for the various samples allowed to estimate the GaAs core diameter and the corresponding AlGaAs shell thickness for each sample. Fig. 17(a) shows the size distribution of Au NPs sitting at the tips of the core-shell nanowires for a typical sample. Best-fitting the experimental data with the Gamma distribution function [Eq. (1)] gives a NP average diameter not far from the run-to-run average values reported in Sec. 4.1 for annealed Au NPs. As the Au-catalysed growth of GaAs at 400°C leads to cylindrical (untapered) GaAs nanowires, the above Au NP values are thus assumed as representative of the nanostructure core size and

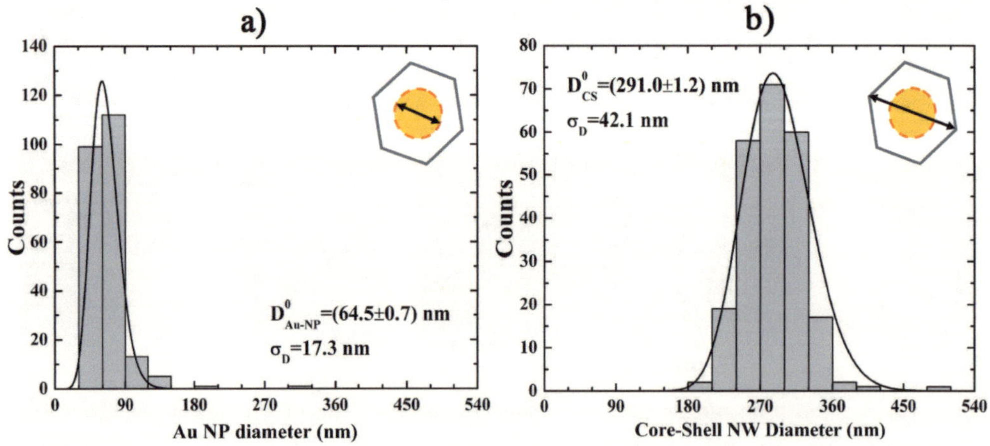

Fig. 17. Diameter count histograms of (a) Au NPs at the tips of core-shell nanowires, and (b) core-shell nanowires. The solid curves in both diagrams represent the Gamma distribution functions (Eq. 1) best-fitting the experimental data (fitting parameters shown in the graphs).

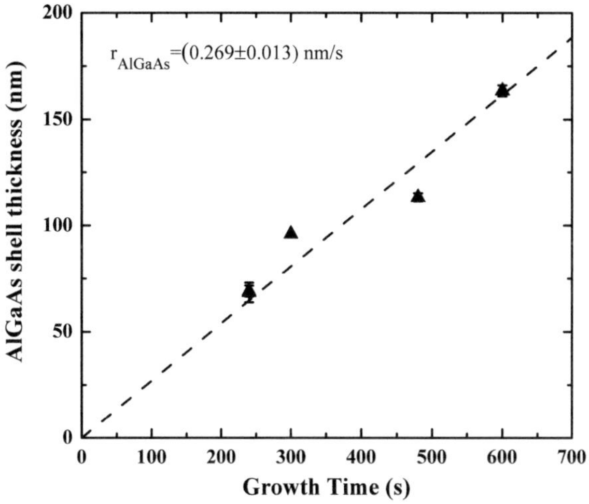

Fig. 18. Average thickness (▲) of the AlGaAs shell as function of growth time. The dashed line is the linear best-fit of experimental points.

distribution width, respectively. Fig. 17(b) shows the diameter distribution of the same core-shell nanowires as in Fig. 17(a) along with its best-fitting Gamma distribution function: the nanowires, whose AlGaAs shell was grown for 8 min, have an average diameter D_0^{NW} =(291.0±1.2) nm and a distribution standard deviation σ_D=42.1 nm. The shell thickness for each core-shell nanowire can be then easily estimated. For the sample in Fig. 17 the average shell thickness is then $h_{AlGaAs}=\left(D_0^{NW}-D_0^{NP}\right)/2$ =(113 ±1) nm. Fig. 18 reports the average thickness of the AlGaAs shell as a function of the alloy growth time for different core-shell samples (shell thickness values were estimated in the same way as above): the AlGaAs thickness increases linearly with time, resulting in an alloy growth rate of 0.269 nm/s. This value must be compared with the 0.667 nm/s growth rate of an AlGaAs epilayer deposited under identical conditions.

4.4 Luminescence properties of GaAs-AlGaAs core-shell nanowires

The radiative properties of GaAs-AlGaAs core-shell nanowires were determined by studying the photoluminescence (PL) of dense nanowire ensembles without removing them from their GaAs substrate. Fig. 19 shows the PL emissions measured at 7K under 532 nm laser excitation from a GaAs/AlGaAs core–shell sample, whose core and shell materials were grown for 20 and 10 min, respectively. In order to discriminate between the nanowire emission and that of the substrate, PL spectra were recorded from different regions of the sample; in particular, the spectrum labelled (1) in the Figure refers to a region where no Au NPs were deposited, consisting of a (unintentionally-grown) planar AlGaAs/GaAs hetero-structure, the AlGaAs epilayer being a few tenths of micron thick; spectrum (2) originates, instead, from a sample area where a dense ensemble of core–shell nanowires was grown. Given the laser beam spot size, this allowed to probe the collective PL emission of 10^6–10^7 nanowires. The planar AlGaAs/GaAs region shows an intense PL emission due to radiative recombination from the GaAs substrate. Noteworthy, no signal was detectable above the GaAs band-edge, indicating negligible emission from the thin AlGaAs layer, which acts solely as passivation for the underlying GaAs crystal.

An entirely different PL signal was recorded from the dense core-shell nanowire region: the PL emission, whose intensity is more than one order of magnitude weaker than that detected for the planar AlGaAs/GaAs region, is dominated by a broad band peaked at 1.49 eV, showing a very broad tail on its low-energy side. Moreover, some additional broad and weak luminescence contributions can be observed between 1.52 and 2.05 eV; a Gaussian line-shape fitting of the PL spectrum [curve (3) in Fig. 19] demonstrates that this high-energy emission is made up of two very broad bands at around 1.67 and 1.90 eV, followed on their high-energy side by a relatively narrower (FWHM~49 meV) peak at 1.997 eV. These emissions weaken with decreasing the AlGaAs shell thickness in the nanowires; the dominant broad band below 1.52 eV is observed instead for all samples, although its exact peak position and width may slightly change from sample to sample. High resolution cathodoluminescence (CL) measurements performed at 10K on these core-shell nanowires after removal from their substrate confirmed the spectral features above, further allowing to compare the emission from dense ensembles of core-shell nanowires with that of single/few nanostructure(s) (Prete et al., 2008); also, comparison between secondary electron and monochromatic CL images of single nanowires further allowed to spatially resolve the main luminescence features, demonstrating that the 1.49 eV emission originates from the GaAs core region of the nanowire, while those at 1.67 eV, 1.90 eV and 1.997 eV are associated to the AlGaAs shell.

Temperature-dependent PL measurements (not reported here) have shown that the spectral feature at 1.997 eV weakens and red-shifts with temperature until it vanishes above around 60 K (i.e. for k_BT~5–6 meV). We assign this peak to the band-edge (excitonic) emission of the AlGaAs shell. This allows to determine the Al molar fraction in the alloy as $x_{Al}=0.34\pm0.01$ (Pavesi & Guzzi, 1994; Wasilewski et al., 1997) in very good agreement with EDS

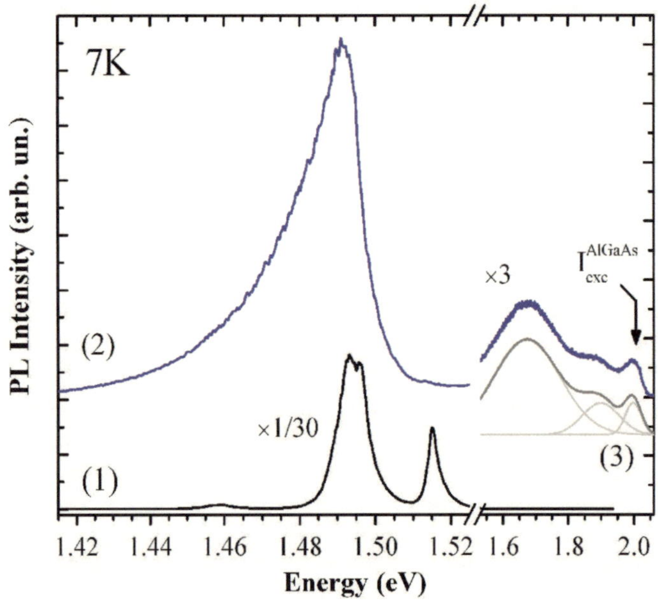

Fig. 19. PL spectra recorded at 7K from the GaAs/AlGaAs core-shell nanostructures [curve (2)] and that of an ‚equivalent' AlGaAs/GaAs planar structure [(curve (1)]. All spectra were recorded under a 532 nm laser line excitation. Excitation power for curve (2) was ~5 mW.

measurements (Sec. 4.3). Preliminary Raman scattering experiments performed on single nanowires from the same samples confirmed that the AlGaAs composition is around 33%. This strongly points towards the very good sample-to-sample and within-sample compositional homogeneity of the AlGaAs shell in present nanowires.

Noteworthy is that neither the broad band at around 1.67 eV nor the one at 1.90 eV can be observed under He–Ne laser excitation, i.e. for a laser photon energy just below the band-gap of the $Al_{0.33}Ga_{0.67}As$ alloy indicating that both these bands originate from photo-excitation of the same $Al_{0.33}Ga_{0.67}As$ alloy. The energy position of the weak 1.90 eV band coincides fairly well with that of the so-called D_1A donor-acceptor pair recombination band observed in lightly Si-doped AlGaAs layers (Pavesi & Guzzi, 1994), suggesting that the shell may contain a residual concentration of Si donors. Secondary ion mass spectrometry analysis of AlGaAs epilayers grown on (110)-oriented GaAs substrates under MOVPE conditions identical to those used for the nanowire shells confirmed that Si is indeed unintentionally incorporated in the material. The origin of Si is likely ascribable to the TMAl source used for present growth experiments, whose residual Si contamination is ~0.8 ppm.

Finally, low temperature CL measurements performed on single nanostructures suggest that each nanowire has a slightly different core luminescence, its spectral peak position shifting from nanowire to nanowire within the same sample over a relatively narrow (~50 meV) energy interval (Prete et al., 2008). These energy shifts suggest the existence of a built-in electric field at the nanowire core-shell interface: indeed, such electric field may bend the materials conduction and valence bands, leading to the spatial separation of photo-generated carriers within the nanowire and to large energy shifts of the luminescence (Su et al., 2008). These effects require intense electric fields, corresponding to large space charge densities at the core-shell interface. The observation of a relatively high C signal within the nanowire core (Sec. 4.3) suggests that substantial carbon doping may occur into the GaAs core; this, together with the unintentional Si doping of the AlGaAs shell, may lead to the required space charge build-up at the GaAs-AlGaAs interface.

5. Conclusions

We briefly reviewed the current interests in free-standing III-V nanowires and major technological approaches being investigated for the synthesis, along with their limitations and advantages. In particular, self-assembly of these nanostructures by employing Au-catalyst assisted (VLS) MOVPE is currently considered a most promising technology in reason of its potentials in terms of materials control and scalability to industrial mass production of future nanowire-based devices.

Several methodologies are employed for the synthesis of Au catalyst NPs onto the surface of semiconductor substrates, among others self-assembly from thin metal films and deposition from colloidal Au solutions. These methods depend on metal catalyst surface diffusion on and interaction with the underlying substrate material; therefore, NP properties (i.e., size distribution, density, etc.), important to the VLS growth of nanowires, critically vary with preparation parameters. Examples of these effects have been reported for Au NPs on Si and GaAs substrates, annealed at temperatures similar to those used during MOVPE.

The growth of nanowires and related quasi-1D nanostructures has been demonstrated in the literature using Au-catalyst assisted MOVPE for most of III-V compounds. Despite the amount of work reported to date, a full understanding of the fundamental mechanism of

VLS self-assembly of III-V nanowires and the role of growth parameters on the physical properties of as-grown nanostructures is still largely lacking.

In this Chapter, the growth and some of the physical properties (size, morphology, structure, luminescence, impurities) of GaAs and GaAs-AlGaAs core-shell nanowires using colloidal Au NPs as catalysts and TBAs in substitution of AsH_3 is reported as case study. Good control over the size, shape and compositional uniformity of these nanostructures is demonstrated for epitaxially oriented nanowires; core-shell nanowires show strong luminescence from both their core and shell materials, but their origins remain to be fully understood, along with the role of MOVPE conditions and unintentional impurities (especially carbon) on the observed properties.

6. Acknowledgements

The authors would like to acknowledge F. Marzo, I. Miccoli, and P. Paiano (of University of Salento, Lecce, Italy), E. Piscopiello, and L. Tapfer (of Centro Ricerche ENEA, Brindisi, Italy), as well as L. Lazzarini, and G. Salviati (of IMEM-CNR, Parma, Italy) for their collaboration to this work and the many discussions. The partial support of the Italian Ministry of Foreign Affairs (MAE) through the Italy-Japan Significant Bilateral Project: "Nanocharacterization of nanowires, nanomagnets and laser diodes for sensors, opto-electronics and data storage" is also acknowledged.

7. References

Bao, X.-Y.; Soci, C.; Susac, D.; Bratvold, J.; Aplin D.P.R.; Wei, W.: Chen, C.-Y.; Dayeh S.A.; Kavanagh K.L. & Wang, D. (2008). Heteroepitaxial growth of vertical GaAs nanowires on Si (111) substrates by metal−organic chemical vapor deposition. *Nano Lett.*, 8, 11, (November 2008) 3755–3760, ISSN: 0028-0836.

Bjork, M.T.; Ohlsson, B.J.; Thelander, C.; Persson, A.I.; Deppert, K.; Wallenberg, L.R. & Samuelson, L. (2002). Nanowire resonant tunneling diodes, *Appl. Phys. Lett.*, 81, (December 2002) 4458-4460, ISSN: 0003-6951.

Borgström, M.; Deppert, K.; Samuelson, L. & Seifert, W. (2004). Size- and shape-controlled GaAs nano-whiskers grown by MOVPE: a growth study. *J. Cryst. Growth* 260 (January 2004) 18-22, ISSN: 0022-0248.

Borgström, M.T.; Zwiller, V.; Müller, E. & Imamoglu, A. (2005). Optically Bright Quantum Dots in Single Nanowires. *Nano Lett.* 5, (June 2005) 1439-1443, ISSN: 0028-0836.

Colombo, C.; Hei, M.; Grätzel, M. & Fontcuberta i Morral, A. (2009). Gallium Arsenide p-i-n Radial Structures for Photovoltaic Applications. *Appl. Phys. Lett.*, 94, (April 2009) 173108,1-3, ISSN: 0003-6951.

DenBaars, S. P.; Maa, B. Y.; Dapkus, P. D.; Danner, A. D. & Lee, H. C. (1986). Homogeneous and heterogeneous thermal decomposition rates of trimethylgallium and arsine and their relevance to the growth of GaAs by MOCVD. *J. Cryst. Growth* 77 (September 1986) 188-193, ISSN: 0022-0248.

Dick, K.A.; Deppert, K.; Karlsson, L.S.; Wallenberg, L.R.; Samuelson, L. & Seifert, W. (2005). A new understanding of Au-assisted growth of III-V semiconductor nanowires. *Adv. Funct. Mater.*, 15, (October 2005) 1603-1610, ISSN: 1616-301X.

Givargizov, E. I. & Chernov, A.A. (1973). Rate of whisker growth by the vapor-liquid-crystal mechanism and the role of surface energy. *Sov. Phys. Crystall.*, 18, (July-August 1973) 89-92, ISSN: 0038-5638.

Goto, H.; Nosaki, K.; Tomioka, K.; Hara, S.; Hiruma, K.; Motohisa, J. & Fukui, T. (2009). Growth of core–shell InP nanowires for photovoltaic application by selective-area metal organic vapor phase epitaxy. *Appl. Phys. Expr.*, 2, (February 2009) 035004, 1-3, ISSN: 1882-0778 (print), 1882-0786 (on-line)

Hiruma, K.; Yazawa, M.; Katsuyama, T.; Ogawa, K.; Haraguchi, K.; Koguchi, M.; & Kakibayashi, H. (1995). Growth and optical properties of nanometer-scale GaAs and InAs whiskers. *J. Appl. Phys.*, 77, 2, (January 1995) 447-462, ISSN: 0021-8979 (print), 1089-7550 (online).

Hochbaum, A. I. ; Fan, R.; Hui, R. & Yang, P.(2005). Controlled growth of Si nanowire arrays for device integration. *Nano Lett.* 5 (January 2005) 457- 460, ISSN: 0028-0836.

ICPDS-International Centre for Powder Diffraction Data (2000). PDF Cards no. 04-0784 & no. 29-0619.

Jana, N. R.; Gearheart, L. & Murphy, C. J. (2001). Wet chemical synthesis of high aspect ratio cylindrical gold nanorods. *J. Phys. Chem. B* 105 (April 2001) 4065-4067, ISSN (printed): 1089-5647. ISSN (electronic): 1520-5207.

Johansson, J.; Svensson, C. P. T.; Mårtensson, T.; Samuelson, L. & Seifert, W. (2005). Mass transport model for semiconductor nanowire growth. *J. Phys. Chem. B* 109 (June 2005) 13567-13571, ISSN (printed): 1089-5647. ISSN (electronic): 1520-5207.

Kapon, E. (1994). Lateral patterning of quantum well heterostructures by growth on non-planar substrates, in: *Semiconductors and Semimetals, Epitaxial Microstructures*, 40, Gossard A.C. (Ed.), 259-336, Academic Press, Inc. Published by Elsevier Ltd, ISBN: 9780127521404, Boston (USA).

Khorenko, V.; Regolin, I.; Neumann, S.; Prost, W. & Tegude, F.-J. (2004). Photoluminescence of GaAs nanowhiskers grown on Si substrate. *Appl. Phys. Lett.*, 85, 26, (December 2004) 6407-6408, ISSN: 0003-6951.

Khorenko, V.; Regolin, I.; Neumann, S.; Do, Q.; Prost, T.W. & Tegude, F.-J. (2005). Characterisation of GaAs nanowhiskers grown on GaAs and Si substrates, *Proceedings of 2005 Intern. Conf. on Indium Phosphide and Related Materials*, IEEE Conf. Procs. 363-366, ISBN: 0-7803-8891 -7 , Glasgow, Scotland, (UK), May 2005, Institute of Electrical and Electronics Engineers Inc.

Kim, Y.; Joyce, H. J.; Gao, Q.; Tan, H. H.; Jagadish, C.; Paladugu, M.; Zou, J. & Suvorova, A. A. (2006). Influence of nanowire density on the shape and optical properties of ternary InGaAs nanowires. *Nano Lett.* 6, (February 2006) 599-604, ISSN: 0028-0836.

Lim, S.K.; Tambe, M.J.; Brewster, M.M. & Gradečak, S. (2008). Controlled growth of ternary alloy nanowires using metalorganic chemical vapor deposition. *Nano Lett.*, 8, 5, (April 2008) 1386-1392, ISSN: 0028-0836.

Magnusson, M.H.; Deppert, K.; Malm, J.-O.; Bovin, J.-O. & Samuelson L. (1999) Size-selected gold nanoparticles by aerosol technology. *Nanostruct. Mater.*, 12, 1-4, (October 1999) 45-48, ISSN: 1359-6454.

Mårtensson, T.; Svensson, C.P.T.; Wacaser, B.A.; Larsson, M.W.; Seifert, W.; Deppert, K.; Gustafsson, A.; Wallenberg, L.R. & Samuelson, L. (2004). Epitaxial III-V nanowires on silicon. *Nano Lett.*, 4, 10, (September 2004) 1987-1990, ISSN: 0028-0836.

Mohan, P.; Motohisa, J. & Fukui, T. (2006). Fabrication of InP/InAs/InP core-multishell heterostructure nanowires by selective area metalorganic vapor phase epitaxy. *Appl. Phys. Lett.*, 88, (March 2006) 133105,1-3, ISSN: 0003-6951.

Moore, J.C.; Skrobiszewski, J.L. & Baski A.A. (2007). Sublimation behavior of SiO_2 from low- and high-index silicon surfaces. *J. Vac. Sci. Technol. A*, 25, 4, (July 2007) 812-815, ISSN: 0734-2101 (print) 1520-8559 (online).

Motohisa, J.; Noborisaka, J.; Takeda, J.; Inari, M. & Fukui T. (2004). Catalyst-free selective-area MOVPE of semiconductor nanowires on (111)B oriented substrates. J. Cryst. Growth, 272 (December 2004), 180-185, ISSN: 0022-0248.

Nagao, O.; Harada, G.; Sugawara, T.; Sasaki, A. & Ito, Y. (2004). Small-angle X-ray Scattering method to determine the size distribution of gold nanoparticles chemisorbed by thiol ligands. *Jpn. J. Appl. Phys.*, Part 1, 43, (November 2004) 7742-7746, ISSN (print): 0021-4922, ISSN (electronic): 1347-4065.

Njoki, P.N.; Lim, I-Im S.; Mott, D.; Park, H.-Y.; Khan, B.; Mishra, S.; Sujakumar, R.; Luo, J. & Zhong C.-J. (2007). Size correlation of optical and spectroscopic properties for gold nanoparticles. *J. Phys. Chem. C*, 111, 40, (September 2007) 14664–14669, ISSN (print): 1932-7447, ISSN(electronic): 1932-7455.

Noborisaka, J.; Motohisa, J.; & Fukui, T. (2005a). Catalyst-free growth of GaAs nanowires by selective-area metalorganic vapor-phase epitaxy. *Appl. Phys. Lett.*, 86, (May 2005) 213102, 1-3, ISSN: 0003-6951.

Noborisaka, J.; Motohisa J.; Hara, S. & Fukui, T (2005b). Fabrication and characterization of freestanding GaAs/AlGaAs core-shell nanowires and AlGaAs nanotubes by using selective-area metalorganic vapor phase epitaxy. *Appl. Phys. Lett.* 87, (August 2005) 093109, 1-3, ISSN: 0003-6951.

Nötzel, R. & Ploog, K.H. (2000). Quantum wires and quantum dots on high-index substrates. *Phys. E*, 8, (August 2000) 117-124, ISSN: 1386-9477.

Ohlsson, B.J.; Björk, M.T.; Magnusson, M.H.; Deppert, K.; Samuelson, L. & Wallenberg, L.R. (2001). Size-, shape-, and position-controlled GaAs nano-whiskers. *Appl. Phys. Lett.*, 79, 20, (November 2001) 3335-3337, ISSN: 0003-6951.

Ouattara, L.; Mikkelsen, A.; Skölld, N.; Eriksson, J.; Knaapen, T.; Ćavar, E.; Seifert, W.; Samuelson, L. & Lundgren, E. (2007). GaAs/AlGaAs nanowire heterostructures studied by scanning tunneling microscopy. *Nano Lett.*, 7, 9, (August 2007) 2859-2864, ISSN: 0028-0836.

Paiano, P.; Prete, P.; Lovergine, N. & Mancini, A. M. (2006). Size and shape control of GaAs nanowires grown by metalorganic vapor phase epitaxy using tertiarybutylarsine. *J. Appl. Phys.* 100 (November 2006) 094305, 1-4, ISSN: 0021-8979 (print), 1089-7550 (online).

Paiano, P.; Prete, P.; Speiser, E.; Lovergine, N.; Richter, W.; Tapfer, L. & Mancini A.M. (2007). GaAs nanowires grown by Au-catalyst assisted MOVPE using tertiarybutyl-arsine as Group-V Precursor. *J. Cryst. Growth*, 298 (January 2007) 620-624, ISSN: 0022-0248.

Pavesi, L. & Guzzi, M. (1994). Photoluminescence of $Al_xGa_{1-x}As$ alloys. *J. Appl. Phys.* 75, 10, (May 1994) 4779-4842, ISSN: 0021-8979 (print), 1089-7550 (online).

Pettersson, H.; Trägårdh, J. Persson A. I., Landin L., Hessman D., and Samuelson, L. (2006). Infrared photodetectors in heterostructure nanowires. *Nano Lett.* 6 (January 2006) 229-236, ISSN: 0028-0836.

Piscopiello, E.; Tapfer, L.; Paiano, P.; Prete, P. & Lovergine, N. (2008). Formation of epitaxial gold nanoislands on (100) silicon. *Phys. Rev. B*, 78, 3, (July 2008) 035305, 1-7, ISSN 1098-0121 (print), 1550-235X (online).

Prete, P.; Lovergine, N. & Tapfer, L. (2007). Nanostructure size evolution during Au-catalysed growth by carbo-thermal evaporation of well-aligned ZnO nanowires on (100)Si. *Appl. Phys. A*, 88, 1, (April 2007) 21-26, ISSN: ISSN: 0947-8396 (print) ISSN: 1432-0630 (electronic).

Prete, P.; Marzo, F.; Paiano, P.; Lovergine, N.; Salviati, G.; Lazzarini, L. and Sekiguchi, T. (2008). Luminescence of GaAs/AlGaAs core-shell nanowires grown by MOVPE using tertiarybutylarsine. *J. Cryst. Growth*, 310, (January 2008) 5114-5118, ISSN: 0022-0248.

Qi, H.; Gee, P. E. & Hicks, R. F. (1996). Sites for arsine adsorption on GaAs(001). *Surf. Sci.* 347, 3, (February 1996) 289-302, ISSN: 0039-6028.

Regolin, I.; Khorenko, V.; Prost, W.; Tegude F.-J.; Sudfeld, D.; Kästner, J. & Dumpich, G. (2006). Composition control in metal-organic vapor-phase epitaxy grown InGaAs nanowhiskers. *J. Appl. Phys.*, 100, 7, (October 2006) 074321, 1-5, ISSN: 0021-8979 (print), 1089-7550 (online).

Samuelson, L.; Thelander, C.; Björk, M.T.; Borgström, M.; Deppert, K.; Dick, K.A.; Hansen, A.E.; Mårtensson, T.; Panev, N.; Persson, A.I.; Seifert, W.; Sköld, N. Larsson, M.W.; Wallenberg, L.R. (2004). Semiconductor nanowires for 0D and 1D physics and applications. *Phys. E*, 25 (July 2004) 313- 318, ISSN: 1386-9477.

Seifert, W.; Borgström, M.; Deppert, K.; Dick, K.A.; Johansson, J.; Larsson, M.W.; Mårtensson, T.; Sköld, N.; Svensson, C.P.T.; Wacaser, B.A.; Wallenberg, L.R. & Samuelson, L. (2004). Growth of one-dimensional nanostructures in MOVPE. *J. Cryst. Growth*, 272, (October 2004) 211-220, ISSN: 0022-0248.

Shieh, F.; Saunders, A.E. & Korgel, B.A. (2005). General shape control of colloidal CdS, CdSe, CdTe quantum rods and quantum rod heterostructures. *J. Phys. Chem. B*, 109, (December 2005) 8538-8542, ISSN (printed): 1089-5647. ISSN (electronic): 1520-5207, and references therein.

Sköld, N.; Karlsson, L.S.; Larsson, M.W.; Pistol, M.-E.;Seifert, W.; Trägårdh, J. & Samuelson, L. (2005). Growth and optical properties of strained GaAs-$Ga_xIn_{1-x}P$ core-shell nanowires. *Nano Lett.* 5 (September 2005) 1943-1947, ISSN: 0028-0836.

Spadavecchia, J.; Prete, P.; Lovergine, N.; Tapfer, L. & Rella, R. (2005). Au nanoparticles prepared by physical method on Si and sapphire substrates for biosensor applications. *J. Phys. Chem. B*, 109, (August 2005) 17347-17349, ISSN (printed): 1089-5647. ISSN (electronic): 1520-5207.

Su, W.S.; Chen, T.T.; Cheng, C.L.; Fu, S.P.; Chen, Y.F.; Hsiao, C.L. & Tu, L.W.(2008). Built-in surface electric field, piezoelectricity and photoelastic effect in GaN nanorods for nanophotonic devices. *Nanotechnol.* 19, 11, (June 2008) 235401, 1-5, ISSN: 0957-4484 (print), ISSN 1361-6528 (online).

Svensson, C. P. T.; Seifert, W. ; Larsson, M. W.; Wallenberg, L. R.; Stangl, J.; Bauer, G. & Samuelson, L. (2005). Epitaxially grown GaP/$GaAs_{1-x}P_x$/GaP double hetero-structure nanowires for optical applications. *Nanotechnol.* 16, 6, (April 2005) 936-939, ISSN: 0957-4484 (print), ISSN 1361-6528 (online).

Thelander, C.; Mårtensson, T.; Bjork, M.T.; Ohlsson, B.J.; Larsson, M.W.; Wallenberg, L.R. & Samuelson, L. (2003). Single-electron transistors in heterostructure nanowires. *Appl. Phys. Lett.* 83, (September 2003) 2052-2054, ISSN: 0003-6951.

Thelander, C.; Nilsson, H.A.; Jensen, L.E. & Samuelson, L. (2005). Nanowire single-electron memory. *Nano Lett.* 5, (February 2005) 635-638, ISSN: 0028-0836.

Titova, L.V.; Hoang, T.B.; Jackson, H.E.; Smith, L.M.; Yarrison-Rice, J.M.; Kim, Y.; Joyce, H.J.; Tan, H.H.; and Jagadish, C. (2006). Temperature dependence of photoluminescence from single core-shell GaAs–AlGaAs nanowires. *Appl. Phys. Lett.* 89, (October 2006) 173126, 1-3, ISSN: 0003-6951.

Turkevich, J.; Cooper Stevenson, P. & Hillier, J. (1951). A study of the nucleation and growth processes in the synthesis of colloidal gold. *Discussions Farad. Soc.*, 11, (December 1951) 55-75, ISSN: 03669033.

Wagner R. S. & Ellis W.C. (1964). Vapor-Liquid-Solid mechanism of single crystal growth. *Appl. Phys. Lett.*, 4, (March 1964) 89-91, ISSN: 0003-6951.

Walker, C. H.; St. John, J. V. & Wisian-Neilson P. (2001). Synthesis and size control of gold nanoparticles stabilized by poly(methylphenylphosphazene). *J. Am. Chem. Soc.*, 123, 16, (April 2001) 3846- 3847, ISSN: 0002-7863 (print), 1520-5126 (on line).

Wasilewski, Z.R.; Dion, M.M.; Lockwood, D.J.; Poole, P.; Streater, R.W.; Spring Thorpe, A.J. (1997). Composition of AlGaAs. *J. Appl. Phys.*, 81, 4, (February 1997) 1683-1694, ISSN: 0021-8979 (print), 1089-7550 (online).

Wensorra J.; Indlekofer, K.M.; Lepsa, M.I.; Förster, A. & Lüth, H. (2005). Resonant tunneling in nanocolumns improved by quantum collimation. *Nano Lett.*, 5, (October 2005) 2470-2475, ISSN: 0028-0836.

Wu, Z.H.; Mei, X.Y.; Kim, D.; Blumin, M. & Ruda, H. (2002). Growth of Au-catalyzed ordered GaAs nanowire arrays by molecular-beam epitaxy. *Appl. Phys. Lett.*, 81, 27, (December 2002) 5177-5179, ISSN: 0003-6951.

Zeng, X.-F.; Chung, D.D.L. & Lakhani, A. (1987). Effect of heating on the structure of Au/GaAs encapsulated with SiO_2. *Solid St. Electr.* 30, 12, (December 1987) 1259-1266, ISSN: 0038-1101.

4

Molecular Beam Epitaxy Growth of Nanowires in the $Hg_{1-x}Cd_xTe$ Material System

Randi Haakenaasen and Espen Selvig
Norwegian Defence Research Establishment
Norway

1. Introduction

The size of electronic components keeps decreasing as more computing power is packed into the volume of a personal computer. There are also clear advantages to shrinking sensors and other electronic devices; they will be lighter, smaller and require less power. More functionality can then be added to portable instruments, whether they are cellular phones or uniforms for soldiers. In the quest for miniaturization, nanotechnology is an obvious field of study. Nanostructures can have properties that differ from those of the bulk material, for example size-tunable effective band gap or high sensitivity to surface preparation due to the large surface-to-volume ratio. This can lead to miniaturized components with completely new properties.

Nanowires are today grown in numerous material systems such as GaAs, Si, GaP, InP, ZnS, CdSe, ZnTe or GaAsSb (Olsson *et al.*, 2001; Duan & Lieber, 2000; Shan *et al.*, 2005; Janik *et al.*, 2006; Dheeraj *et al.*, 2008). Most of these are grown with vapor phase epitaxy techniques using the vapor-liquid-solid (VLS) or vapor-solid-solid (VSS) mechanism, in which the component fluxes go through or around a gold catalyst particle and the nanowire grows underneath this particle (Wagner & Ellis, 1964; Persson *et al.*, 2004). Many groups have also successfully grown segmented or heterostructured nanowires (Björk *et al.*, 2002; Wu *et al.*, 2002; Gudiksen *et al.*, 2002). Various nanowire devices have been demonstrated, for example a *pn* junction, field-effect transistor, photodetector, polarized light emitting diode (LED), laser, single electron transistor, optical switch, and detectors for biological and chemical molecules.

$Hg_{1-x}Cd_xTe$ is an alloy between the semimetal HgTe and the semiconductor CdTe, and it has a direct band gap that is tunable from -0.26 eV (HgTe) to 1.61 eV (CdTe) at 77 K, covering the entire infrared (IR) region. The small effective electron mass of $Hg_{1-x}Cd_xTe$ (minimum of about 0.02 m_0 for $Hg_{0.66}Cd_{0.34}Te$) leads to a quantum upshift for larger structures than in other materials and enables size-tunable electrical and optical properties. In HgTe particles there should, for example, be quantum effects for diameters smaller than 80 nm and a positive band gap below 18 nm (Green *et al.*, 2003). HgCdTe is mostly used for high performance IR detectors, but the small lattice mismatch in this material system (maximum 0.3% between CdTe and HgTe), facilitates growing heterostructures, including quantum wells, with good crystallinity (Tonheim *et al.*, 2008). Hg(Cd)Te also has a number of other special properties which make it appealing to explore nanostructures in this material system: it is piezoelectric and it has high electron mobility, high dielectric constant, large *g*-factor, large spin-orbit coupling, and Rashba effect. Many of these properties are strongly correlated with the fraction of HgTe in the material. HgTe is, for example, a candidate for spintronics applications, and a quantum spin Hall effect has been discovered in HgTe quantum wells (König *et al.*, 2007). HgTe is also a thermoelectric material, and recently nanomaterials have shown potential in improving thermoelectric materials. Enhanced thermoelectric performance in one-dimensional segmented nanowire systems has been demonstrated in $Bi/Bi_{1-x}Sb_x$ nanowires (Dresselhaus *et al.*, 2003). Possible interesting HgTe and HgCdTe nanowire electro-optical applications include polarization-sensitive detectors, detectors with reduced cooling requirements, lasers and LEDs in the infrared region of the spectrum.

Recently, HgTe nanorods have been synthesized in solution (Song *et al.*, 2004; Qin *et al.*, 2007) or formed within single wall carbon nanotubes (Carter *et al.*, 2006), and HgTe nanotubes have been prepared by spray deposition of solvothermally synthesized iodine-doped HgTe nanoparticles (Ranga Roa & Dutta, 2006). HgCdTe nanorods (up to 300 nm long) have been made by adding Hg^{2+} ions into a solution containing CdTe nanorods (Tang *et al.*, 2007). Networks or nanowires of HgCdTe nanoparticles have also been produced by dissolving CdTe nanoparticles in a solution containing Hg^{2+} ions (Yang *et al.*, 2009). As the Hg is added by in-diffusion in solution, a gradient in the composition within the nanowire is expected. Room temperature absorption and PL spectra showed peaks in the range 700 - 1200 nm. Photo detection was demonstrated as an incident light (λ = 785 nm) changed the conductivity of the network with a rise and fall time of 2 s.

The motivation for using molecular beam epitaxy (MBE) to grow HgTe and HgCdTe nanowires is that, although MBE growth of Hg(Cd)Te is both complicated and challenging, its high precision and flexibility can give good control over the growth of thin layers and abrupt junctions, which may be an advantage in future nanostructure devices. Furthermore, MBE growth of HgTe and HgCdTe thin films occurs near thermal equilibrium (~200 °C), and the Hg atoms re-evaporate easily while the Te atoms are very mobile on the surface (Selvig *et al.*, 2008a). This offers the potential for self-organized growth of crystals for deposition rates lower than during thin film growth.

Three different techniques have been tested to make HgTe and HgCdTe nanowires, namely self-organization (without a catalyst), VLS and using SiO_2 as a mask material for selective growth. Three types of HgTe nanowires have been grown: **I**) thick, polycrystalline HgTe wires, **II**) thin, twisted, <111>-oriented single crystal HgTe nanowires, and **III**) thin, straight, segmented HgTe<111>/Te<001> nanowires (Hadzialic, 2004; Selvig *et al.*, 2006; Haakenaasen *et al.*, 2008a; Haakenaasen *et al.*, 2008b). The HgTe wires do not grow by the VLS mechanism. Instead, small Au particles function as nucleation sites for the wires, which subsequently self-organize and grow laterally on the substrate, without epitaxial coupling to the latter. Te and Au nanowires were also fabricated. Attempts at growing HgCdTe nanowires by MBE have not been successful (Haakenaasen *et al.*, 2008a). A comparison with the phase diagrams for the GaAs material system offers an explanation for why HgCdTe/HgTe nanowires were not obtained by the VLS technique. SiO_2 functions as a mask material for HgTe, but not for HgCdTe. Conductive atomic force microscopy has been used to measure the resistivity of the polycrystalline type I HgTe nanowires.

2. Experimental

Si(100), GaAs(100) or $Cd_{0.96}Zn_{0.04}Te$ (100), (111)B and (211)B substrates were used. The substrates were generally wet-etched to remove oxides from the surface. Si and GaAs substrates were etched in hydrofluoric acid (1:10, 48% HF:H_2O) for 3 s and CdZnTe substrates were etched in a (1:100, Br:methanol) solution for 2 min. In the attempts at self-organized growth without a catalyst, some of the substrates were scratched with either a clean q-tip or a q-tip dipped in an 1-µm-diamond-slurry before wet-etching, while some were just etched. Some samples were partly covered by a sputtered layer of SiO_2. For the attempts at VLS growth, the substrates were etched and loaded into a sputtering machine where a ~ 5-60 Å thick layer of Au was deposited. The substrates were then loaded into a Riber 32P MBE machine and heated to 340°C for 10 minutes, resulting in break-up of the gold film into particles of diameter 2-50 nm. Some Si substrates were covered with a photoresist mask (for example an array of 3.5-µm-diameter holes) and the HF etch repeated before Au sputtering, resulting in Au particles in defined areas only.
MBE growth of HgTe and HgCdTe thin films of high crystalline quality requires controlling the substrate temperature to within ± 1°C of the optimum growth temperature, which is just below the Te-phase limit (onset of Te precipitation) (Colin & Skauli, 1997; Selvig *et al.*, 2007; Selvig *et al.* 2008a; Selvig *et al.*, 2008b). In our MBE machine this is 193°C for HgTe and 198°C for $Hg_{0.75}$ $Cd_{0.25}Te$. The growth rate is determined by the Te flux while the Hg flux is kept constant. The growth conditions for the nanowire experiments were similar to those used for HgTe or HgCdTe films grown on CdZnTe(211)B substrates except for lower growth temperatures (178, 186, or 188 °C for HgTe, 192 - 193°C for $Hg_{0.75}$ $Cd_{0.25}Te$, 335°C for CdTe) and much lower deposition rates (would give equivalent thin film growth rates in the range 1-100 Å/min). The equivalent film thickness deposited was varied from 20 Å to 0.5 µm in order to investigate how the nanowires evolve during growth. Below we will take 'deposition rate' to mean 'equivalent thin film growth rate'.
The grown samples were investigated with scanning electron microscopy (SEM), energy dispersive x-ray spectroscopy (EDX), x-ray photoelectron spectroscopy (XPS), and transmission electron microscopy (TEM). In the case of the thick type I nanowires, TEM samples were prepared by transferring a water droplet containing nanowires from the substrate onto a carbon-coated copper grid. This procedure did not work very well for the thin nanowires. Furthermore, for the nanowires that grew in patterns on the substrate, it was necessary to identify where in the pattern a given section of nanowire had grown. It was therefore necessary to develop a new specimen preparation technique for plane-view studies of undisturbed surfaces using TEM (Foss *et al.*, 2010). This technique could be useful in general for studying grown nanowires or quantum dots, defects or patterns of objects/structures on a sample surface. Electrical measurements were performed using an atomic force microscope (AFM) with a conductive tip.

3. Results

3.1 Growth without a catalyst

Several attempts were made to induce self-organized growth without a catalyst (Hadzialic, 2004). In these we used plain or scratched Si and GaAs substrates as described above. Deposition rates varied from 2.5 - 45 Å/min. During HgTe deposition a few crystals grew along scratches, but mostly there was no growth on the substrates. When a Cd flux was added, we obtained varying degrees of overgrown, polycrystalline HgCdTe layers, with more crystals along the scratches. No nanowires were found. Self-organized growth without a catalyst therefore does not seem to work.

3.2 Polycrystalline HgTe nanowires

VLS growth was then attempted using both Si, GaAs and CdZnTe substrates with Au particles. Fig. 1(a) shows a SEM image of HgTe nanowires (and platelets) grown on a Au-sputtered Si substrate. The uneven surfaces of the nanowires already hint at their polycrystalline nature, even though they look straight on a larger scale. The deposition rate was ~ 40 Å/min, the equivalent thin film thickness deposited 0.50 μm, and the growth temperature 178 °C. The wires are 100-500 nm wide and 1- 4 μm long, and they grow laterally on the surface with random orientation in the surface plane, which indicates that there is no epitaxial coupling to the substrate. This is supported by the fact that the wires are not firmly attached to the surface, as they were swept away by an AFM tip in contact mode. Furthermore, there is no visible Au droplet at the ends of the wires, as seen in TEM

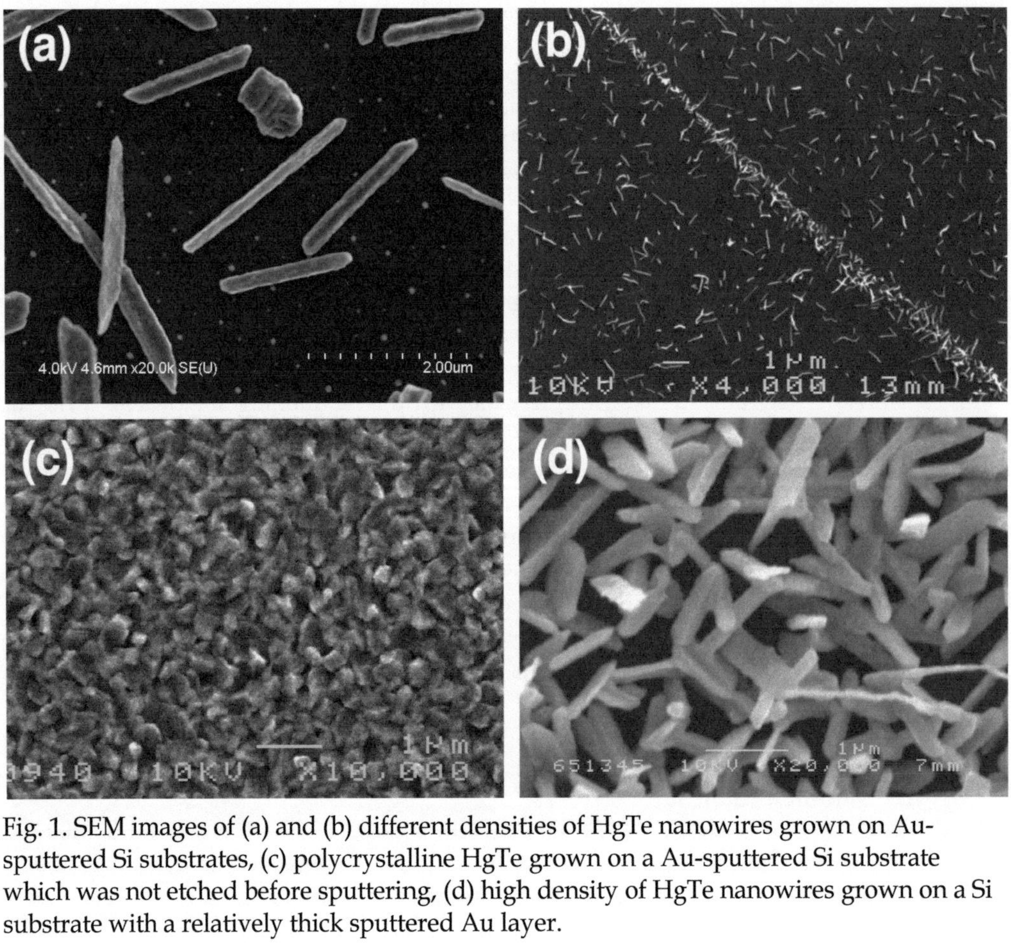

Fig. 1. SEM images of (a) and (b) different densities of HgTe nanowires grown on Au-sputtered Si substrates, (c) polycrystalline HgTe grown on a Au-sputtered Si substrate which was not etched before sputtering, (d) high density of HgTe nanowires grown on a Si substrate with a relatively thick sputtered Au layer.

or SEM images of nanowires grown by the VLS or VSS techniques in other materials. However, as described above, no wires formed on substrates without Au. We conclude that the wires have not grown by the VLS mechanism; instead we believe the Au particles function as nucleation centers for the nanowires. Possibly the Au particles can 'hold' Hg atoms until Te atoms arrive and nucleate a HgTe crystal. Once the crystal has nucleated, it grows by self-organization into a wire. These results are similar to the observations of Zhang *et al.* who used a Ti layer on a Si substrate for growth of IrO_2 nanowires (Zhang *et al.*, 2005). They did not observe Ti particles on the nanowires after growth and concluded the Ti just helped with nucleation.

The density of HgTe nanowires can vary a lot. A higher density of nanowires is observed along scratches in the substrate surface, as seen in Fig. 1(b). This is reasonable, as the surface energy necessary to nucleate a crystal in a scratch is lower. It is also possible that there is more Au in the scratch than outside. The scratches clearly help make nucleation sites. Most of the Au particles on the flat parts of the surface do not nucleate wires. This could be due to the size of the particle, or that the nucleation is very sensitive to the geometry of the particle.

On some substrates we did not remove the native oxide before Au sputtering; otherwise the normal growth process was performed on them. This resulted in a polycrystalline HgTe film, as seen in Fig. 1(c). Etched and Au sputtered substrates placed next to these substrates during growth had HgTe nanowires and no overgrown layer, similar to the wires on Figs. 1(a) and 1(b). Tests described below show that HgTe does not grow on SiO_2 at these growth conditions. We therefore believe that the oxide prevents diffusion of the Au, leaving Au nucleation sites all over the surface.

For etched and Au sputtered substrates we found a strong correlation between the density of wires and how many days the substrate had been kept in air after Au sputtering and before loading into the MBE machine. The sputtered Au layers are thin and porous, most likely allowing oxygen to react with the silicon surface. Also here the oxide probably impedes diffusion of Au during heating, resulting in more nucleation sites (or sites with a more favorable geometry) and a higher wire density. Other factors that affected the wire density were the Au layer thickness, where higher thickness gave higher density, as seen in Fig. 1(d), the trace of the RHEED beam on the surface (gave a different density of wires, but the density could be either lower or higher), and a rough ion milled surface before depositing the Au (gave a polycrystalline layer, probably due to impeding the Au diffusion). When the density is high, the wires also grow out from the substrate surface, as seen in Fig. 1(d).

A TEM image of some wires and their diffraction pattern are shown in Figs. 2(a) and 2(b), respectively. These wires were grown with a 10 °C higher substrate temperature than the wires in Fig. 1(a). The electron diffraction pattern shows that the nanowires are polycrystalline with lattice constants corresponding to HgTe with a sphalerite type crystal structure. The EDX spectrum from a small area of a different wire is shown in Fig. 2(c) (The Al peak comes from the sample holder). EDX analysis of nanowires on the Si substrates showed no Hg or Te between the wires. XPS analysis of a large substrate area containing wires gave a Hg:Te ratio of ~ 1:1.

Several variations in growth conditions were attempted in order to reduce polycrystallinity. Reduction of the deposition rate to ~ 1/3 resulted in no wires on the sample, which meant the incoming Te flux was too low compared to the Te re-evaporation rate. During three growth runs, attempts were made to nucleate the wires for 1 min or 5 min

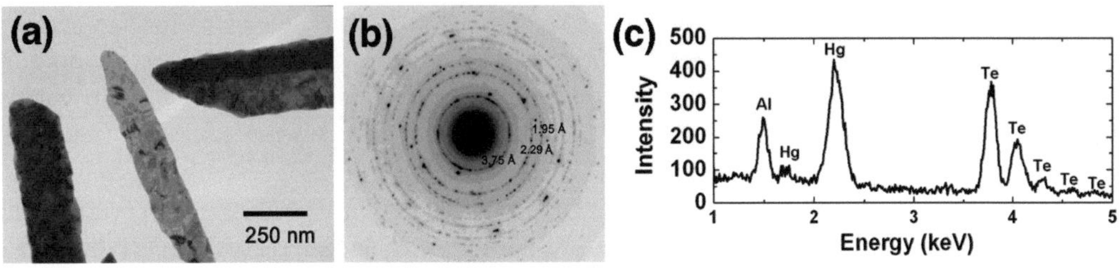

Fig. 2. (a) TEM image of HgTe nanowires, (b) electron diffraction pattern from the nanowires in (a), (c) EDX spectrum from a point on another HgTe nanowire that was transferred to a TEM carbon-coated Cu grid. (a) and (b) are taken from (Selvig *et al.*, 2006).

with normal Te flux, then turn down the flux to 50% or 70% for the rest of the growth. The idea was to inhibit nucleation of new grains by reducing the flux. This resulted in no (for 50 % flux) or fewer (70 % flux) wires that were still polycrystalline, and with random orientations on the substrate surface.

Raman scattering measurements were made at room temperature, as described in (Haakenaasen *et al.*, 2008a). In terms of lattice vibrations the alloy HgCdTe exhibits classic 'two-mode' behavior, which means that it behaves as though there were two sublattices present with their own phonon structures which evolve in frequency with x-value. The Raman spectrum from several wires clearly showed only the HgTe phonon modes. The scattering was strong, which indicates the crystal has well-defined electronic states. Furthermore, a relative sharpness of the longitudinal optical phonon indicates that the material has good crystallinity in the short range. As the wires are polycrystalline, this means that the crystallinity is good within the individual grains.

HgTe was also grown on Au-sputtered GaAs and CdZnTe substrates. Growth on GaAs substrates resulted in lateral HgTe nanowires and platelets similar to the ones on Si, but with a smaller average aspect ratio and slightly more rounded ends, as shown in Fig. 3(a). Growth on CdZnTe substrates resulted in a high density of nanocrystals with nanowires sticking out of the surface plane, as shown in Fig. 3(b). These nanowires grew out from the nanocrystals on the surface, and no epitaxial coupling to the substrate could be found.

The high density of crystals probably had something to do with substrate preparation, as occasional areas had a low density of lateral wires, similar to on Si or GaAs. Again, the explanation could be that Au did not diffuse as easily on the CdZnTe substrates.

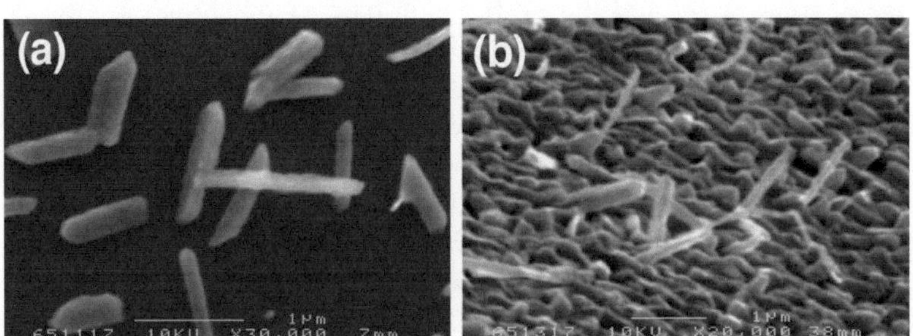

Fig. 3. (a) HgTe nanowires and platelets grown on a Au-sputtered GaAs substrate, (b) HgTe grains and nanowires grown on a Au-sputtered CdZnTe substrate.

3.3 Twisted, single crystal HgTe nanowires

Figs. 4(a) and 4(b) are SEM images of nanowires grown at 1/6 of the time of the wires in Fig. 1(a) (deposition rate 40 Å/min, equivalent film thickness 0.083 µm, growth temperature 186 °C). The wires or ribbons are 20-50 nm wide and 0.5-1 µm long, but not quite straight. These wires are thin enough that a positive band gap may have opened up, in which case they would be very interesting for IR devices. A bright field TEM image is shown in Fig. 4(c).

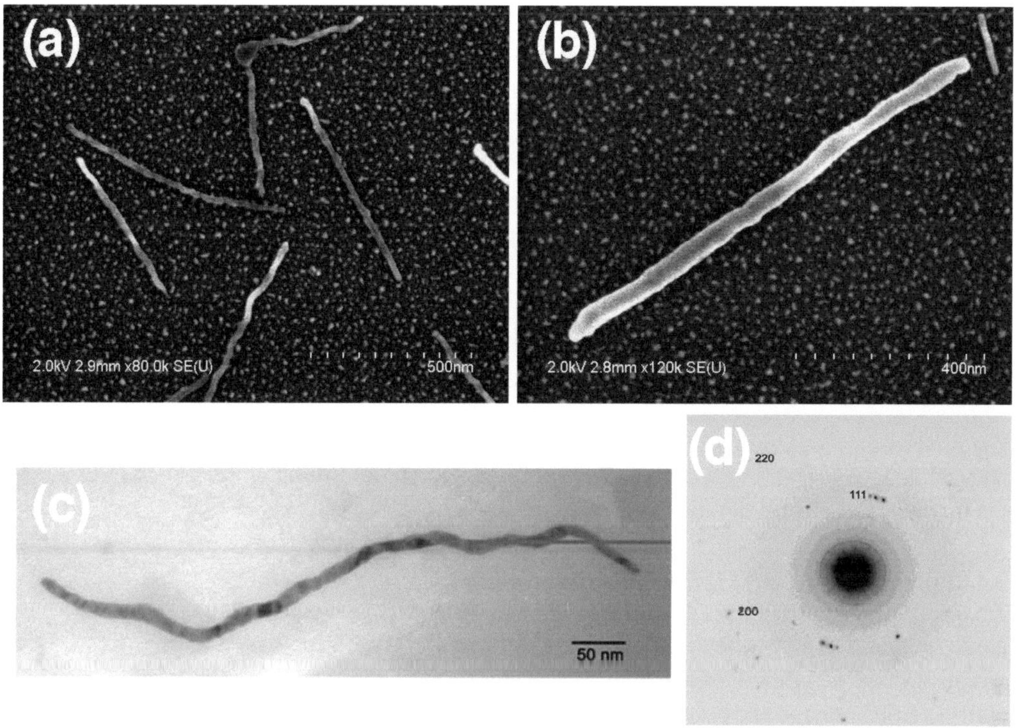

Fig. 4. (a) and (b): SEM images of thin, twisted nanowires grown on Au-sputtered Si substrates, c) Bright field TEM image of such a nanowire, d) electron diffraction pattern from the wire in c) plus another wire close by. The lattice constant (and extinction rules) is consistent with HgTe. a) is taken from (Haakenaasen *et al.*, 2008a), c) and d) are taken from (Selvig *et al.*, 2006).

A TEM diffraction pattern from the wire in Fig. 4(c) and a nearby wire is shown in Fig. 4(d). What at first looks like several grains in the wire, turns out on closer inspection to be bending contrast which moves when the wire is tilted with respect to the incoming electron beam. The wires seem to consist mainly of single crystal HgTe with the <111> direction oriented along the wire, but with bends in the wire (Haakenaasen *et al.*, 2008a).

Fig. 5 shows a TEM dark field image of a wire from a sample grown with 1/3 of the growth time used in Fig. 1(a); equivalent thin film thickness deposited was 0.17 µm. Lamellar twins by rotation around a <111> axis are clearly observed at a kink in the wire. This is commonly observed in MBE-grown HgCdTe (Selvig *et al.*, 1995).

The lateral nanowires can grow over each other, as seen in Figs. 1(a) and 3(a). This means that material can diffuse along the wire so that it grows in length. The height of both the thin, twisted wires and the thicker polycrystalline wires is smaller than the total equivalent thin film thickness for the given growth, so clearly not all material deposited is incorporated into the wires. As a growth mechanism for the wires, we suggest that after nucleation at Au particles, only the atoms that land on the wires are incorporated. These atoms can diffuse both to the ends, making the wire longer, and fill in the 'corners' so that the wires get both wider and seemingly straighter with time. They add new grains instead of making old grains larger. In this way the wire side walls can become reasonably straight even though the wires are not single crystal.

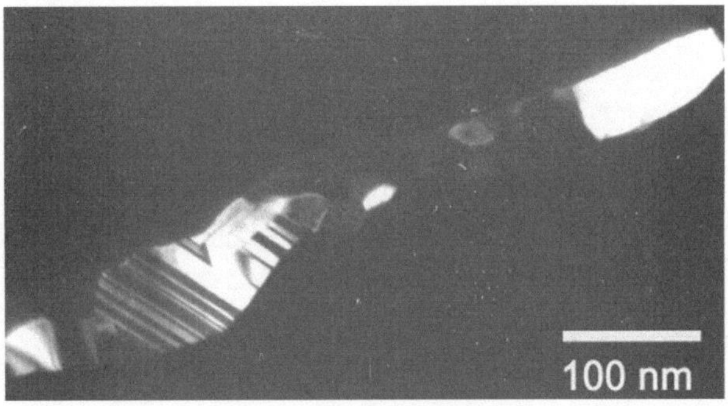

Fig. 5. TEM dark field image of a nanowire grown for approximately 1/3 of the time of the nanowires in Fig. 1(a). Twin defects can be observed. The 111 reflection was used to form the image. Taken from (Selvig *et al.*, 2006).

3.4 Segmented HgTe<111>/Te<001> nanowires

To test if the bending of the thin wires was due to interference between the growing nanowires and the Au particles, the Si substrates were patterned so that the Au particles were confined to certain areas, such as the array of 3.5-μm-diameter circles shown in Fig. 6(a). The SEM image shows nanowires that have nucleated and grown on and out from the Au particle areas. The growth conditions were the same as for the thin type II nanowires in Section 3.3. Fig. 6(b) is a higher magnification SEM image of a wire that grew from the inside to the outside of such an area. The wires are typically 15-70 nm wide and 0.5-1.5 μm long, and some of them may therefore have positive band gap. Wires that grew within the Au particle regions were uneven and not quite straight, while wires that grew out and away from the Au areas were straight and smooth, showing that the Au particles must indeed somehow block the straight growth of the wires. The straight wires were also often a little thicker at the end and had random orientation in the surface plane. These characteristics were common for the wires grown outward from various Au particle patterns. No wires grew entirely outside the Au particle areas, as the Au particles are necessary to nucleate the growth.

In order to find out if the sections of wire growing inside and outside a Au particle region were different, it was necessary to retain the surface pattern of Au particles and wires. Samples were prepared according to the inclined pseudo-plane-view specimen preparation technique developed for this purpose, as described in (Foss *et al.*, 2010). The TEM image of a wire growing on the border of a Au particle area is shown in Fig. 7(a) and a selected area

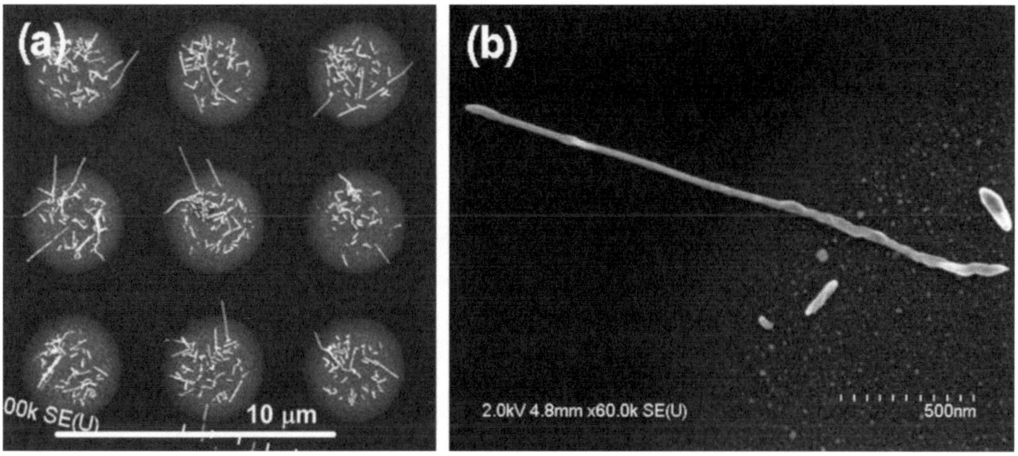

Fig. 6. SEM images of (a) HgTe nanowires that have grown on circular areas with Au particles on a Si substrate, and (b) a nanowire that has grown from the inside to the outside of such a Au particle area. (b) is taken from (Haakenaasen *et al.*, 2008b).

diffraction (SAD) pattern from the wire in Fig. 7(b). Although there are some Au particles where the wire has grown, they are very small and the wire is quite straight and relatively smooth. We see from the SAD pattern, which is from the right half of the middle of the wire (including a little of the dark region at the tip), that the hexagonal Te <001> direction is parallel to the cubic HgTe <111> direction and parallel to the wire axis. This was confirmed in all our TEM observations. EDX spectra were taken from 100 nm large areas along the wire. The spectrum from the dark tip is shown in Fig. 7(c). It contains peaks from Hg and Te (in addition to Si from the substrate and Cu from the grid) and is consistent with HgTe. A spectrum from the middle of the wire is shown in Fig. 7(d). It shows only Te peaks and thus is consistent with elemental Te. (The small peaks at 2.1 and 9.7 keV are due to Au from the Au particles). We notice that the tip of the wire is darkest, consistent with HgTe since Hg is a stronger scatterer than Te. The opposite end of the wire gives a similar spectrum to the one in Fig. 7(c), providing evidence for HgTe also there. This end seen in the image is not necessarily the true tip of the wire as the tip may have been ion milled away during TEM preparation (it is right next to the ion milled hole in the Si substrate). Fig. 7(e) shows another wire that has grown out of a Au particle area. The left one third of the wire is inside the Au particle area, the rest outside. EDX spectra show that also this wire has HgTe at the two ends. Between the ends the spectra indicate alternating sections of Te and HgTe.

The wires thus consist of HgTe at the ends and mostly Te segments (but some HgTe segments) in the middle. The segmentation is consistent with the Hg-Te phase diagram if the average composition is too low in Hg. According to this phase diagram, HgTe is a line compound with no other phases between elemental Te and HgTe. It is interesting that the bulk phase diagram seems to apply also for nanowires. Since the wires grow mainly at the ends, material that lands on the wires must diffuse to the ends before being incorporated into the crystal lattice. During MBE growth of Hg(Cd)Te there is a very large flux of Hg, as the Hg atoms have a very low sticking coefficient. It is likely that Hg atoms re-evaporate

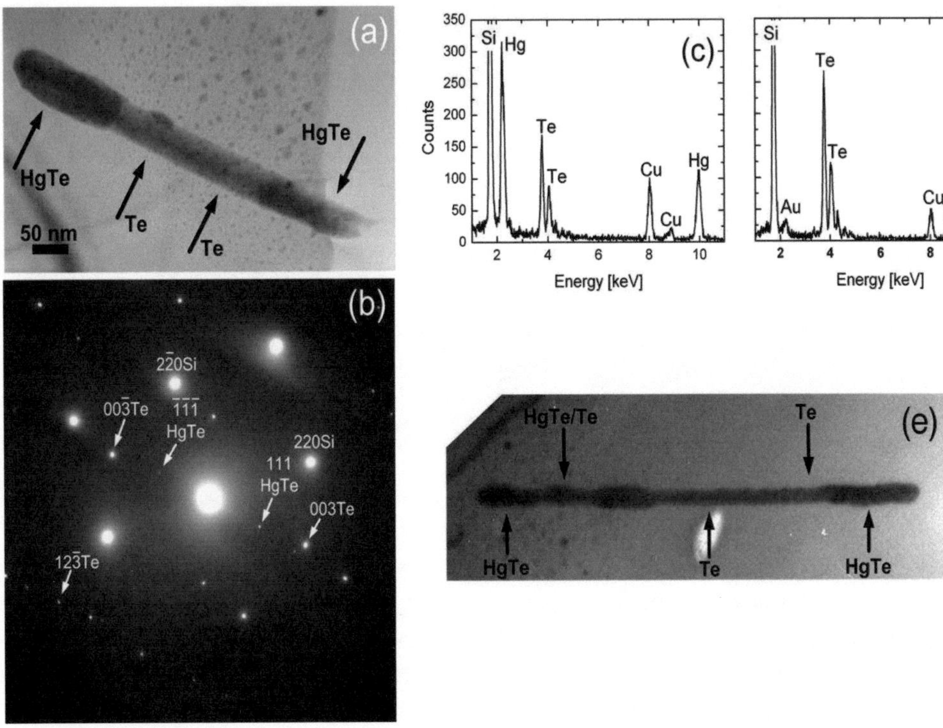

Fig. 7. (a) Bright field TEM image of a nanowire that has grown on the border of a Au particle area, (b) SAD pattern from the right half of the middle of the wire (including some of the darker region at the tip), (c) EDX spectrum from the dark tip on the left side of the wire, (d) EDX spectrum from the middle region of the wire, and (e) Bright field TEM image of a different, 1 µm long nanowire that has grown from the inside to the outside of a Au particle area. All figures taken from (Haakenaasen *et al.*, 2008b).

quickly rather than diffuse, whereas Te atoms are known to diffuse far on a Hg(Cd)Te film surface during growth. We therefore believe that the Te atoms diffuse along the wires to the ends, where there is Hg available to grow HgTe. The lack of Hg in parts of the wire could be due to less Hg being incorporated during growth of nanowires than thin films or that Hg evaporates from already grown sections of the wire during further growth of the wire.

The lattice constant of the cubic HgTe is 6.46 Å, while for hexagonal Te the lattice constants are a = 4.46 Å and c = 5.93 Å. Although HgTe and Te have completely different crystal structures (this is not like the stacking fault alternation of the cubic zincblende and hexagonal wurtzite crystal structures that is seen in many compound semiconductor nanowires grown by the VLS mechanism), both the (111) plane of HgTe and the (001) plane of Te have a six-fold symmetry. The d_{220} spacing of HgTe is 2.28 Å while the d_{110} spacing of Te is 2.23 Å, resulting in a lattice mismatch of ~ 2%. Although this would be a large. mismatch for thin films, heterostructured nanowires can accommodate larger mismatch and still be free from dislocations or other defects because the small lateral dimension allows efficient strain relaxation (Thelander *et al.*, 2004). The segments of HgTe and Te therefore can accommodate good epitaxial match.

3.5 Au and Te nanowires

On a few of the samples there were small areas with a different type of nanowire. While the HgTe wires looked black in the optical microscope, these wires looked white. SEM micrographs and EDX analysis revealed that these wires were very straight and smooth and consisted of Te, Figs. 8(a) and 8(b). They could appear both in a 'bird's nest' or as individual lateral nanowires on the substrate. It is not clear what caused these Te nanowires to grow instead of the HgTe nanowires.

For a thicker Au sputter layer (60 Å), the growth resulted in long (10-40 µm) Au wires along the <110> directions, in addition to the HgTe wires, Fig. 8(c). The Au wires are the result of Au diffusing to steps on the Si(100) surface, as shown by fewer Au particles in the immediate vicinity of these wires. There was generally no growth of HgTe nanowires from the long Au wires. It is possible that the geometry was not conducive to nucleation or that the volume of these wires is so large that the concentration of dissolved Hg was too low for nucleation.

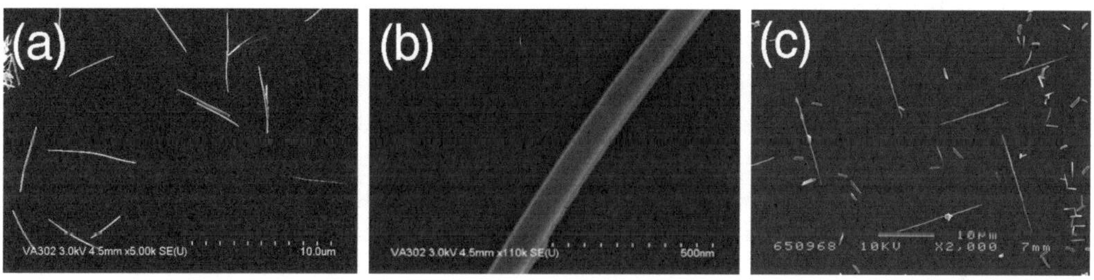

Fig. 8. SEM images of (a) and (b)Te nanowires grown on a Au-sputtered Si substrate, and (c) long Au wires in addition to HgTe nanowires grown on a Au-sputtered Si substrate. (b) is taken from (Haakenaasen *et al.*, 2008a).

3.6 Attempts to grow HgCdTe nanowires

Several attempts to grow $Hg_{0.75}Cd_{0.25}Te$ nanowires on Au-coated Si substrates were made with varying Au thickness, deposition rate (22 - 44 Å/min) and total deposited thickness (1760 - 5280Å). All growths resulted in a polycrystalline HgCdTe layer and no nanowires. A low deposition rate (10 Å/min) of CdTe at 340°C only resulted in Au particles and no growth of CdTe.

3.7 Explanation why VLS does not work

To try to understand why HgTe or HgCdTe nanowires did not grow by the VLS or VSS techniques, a comparison to the growth of Au-catalyzed GaAs nanowires was made (Tibballs, 2005). Persson *et al.* grow their GaAs wires at 540°C, but they have found that the Au catalyst particle is solid during growth (Persson *et al.* 2004). The VSS technique still works as enough Ga and As can diffuse through and around the Au particle. There is no phase diagram available for HgTe-Au or HgCdTe-Au, but we can get some information from the Cd-Au and Hg-Au phase diagrams (Morfatt, 1977). In Au-Hg, the melting point depression gives a solidus temperature of 419°C at 20 at.% Hg. At ~200°C, which is the Hg(Cd)Te MBE growth temperature, a 97 at.% Hg concentration in Au is needed for the Hg-Au solution to be fully liquid. However, at 200°C Au can dissolve at most 15 at.% Cd and Hg, and supersaturation will cover the Au particle with intermetallic phases, effectively capping the particle. This will probably prevent more Hg or Cd dissolving in the Au, so that the particle does not reach the high Hg concentration needed for this solution to become liquid. The VLS mechanism can therefore be excluded. Furthermore, intermetallic phases usually reduce diffusion, so the capping layer may prevent or at least impede further transport of Hg or Cd through the Au.

There is no published data on diffusion coefficients of Cd or Hg in Au, but employing the activation energies for Zn and Sn in (Mortlock & Rowe, 1965) as a guide, because these elements exhibit similar atomic size and solubility in Au, we note that one expects a six-order-of-magnitude reduction in diffusivity when going from 540°C to 200°C. We therefore believe the diffusion of Hg and Cd through Au is too slow even with the highest concentration gradient consistent with not precipitating intermetallics on the surface. We would need a growth temperature of approximately 420°C to attain a diffusivity comparable with the GaAs experiment, but this temperature is not compatible with the Hg fluxes available in the MBE chamber. Several other groups have reported growth of other II-VI VLS nanowires (without Hg) at around 400°C (Shan *et al.*, 2005; Janik *et al.*, 2006). An additional problem is that Au and Te form a solid phase at 200°C, which will compete with growth of HgTe. An additional problem for HgCdTe is that there is a large difference in stability between HgTe and CdTe, which means it will be very difficult to control the ratio between Hg and Cd in the grown crystal. There would have to be a very low Cd concentration in Au in equilibrium with HgCdTe; otherwise all Te would be bound to Cd and the crystal would become almost entirely CdTe. But such a low concentration would exclude enough transport of Cd through the Au particle. Looking for other possible catalyst materials, we find that Ag is not much better than Au, while In or Pb probably will precipitate a phase with Te. The only possibility at 200°C seems to be thallium, which we do not want in our growth chamber. Excluding thallium, we conclude that it is not possible to make HgTe or HgCdTe nanowires via VLS or VSS at the MBE growth temperature.

3.8 Selective growth using a SiO_2 mask

To investigate the possibility of selective growth, HgCdTe and HgTe were grown on a series of Si and CdZnTe substrates with and without a SiO_2 mask (Hadzialic, 2004). The HgCdTe and HgTe deposition rates were varied by a factor 100, from 1.3 to 135 Å/min, and EDX and XPS were used to determine if any growth had occurred. HgCdTe grew on SiO_2 for all except the lowest deposition rates (incompatible with realistic growth rates) while HgTe did not grow on SiO_2 for any deposition rate. SiO_2 can therefore be used as a mask for selective growth of HgTe, but not for HgCdTe.

3.9 Electrical characterization using conductive atomic force microscopy

To perform electrical measurements on a single nanowire, it is necessary to have two or more contacts to the wire. This is no trivial task since the wires are so small. Electron beam lithography (EBL) has been used on wires made in other materials, but HgTe will degrade or evaporate at the normal temperatures used for baking of EBL resists. An alternative method has been developed to obtain the circuit illustrated in Fig. 9(a). Large, sputtered Au electrodes were deposited, using optical lithography, on a sample with randomly distributed, polycrystalline type I HgTe nanowires, shown in Fig. 9(b). Some wires were partly covered by the Au electrode. A conductive AFM tip could then be used as a second, movable electrode to measure current-voltage (I-V) curves at several positions L along such

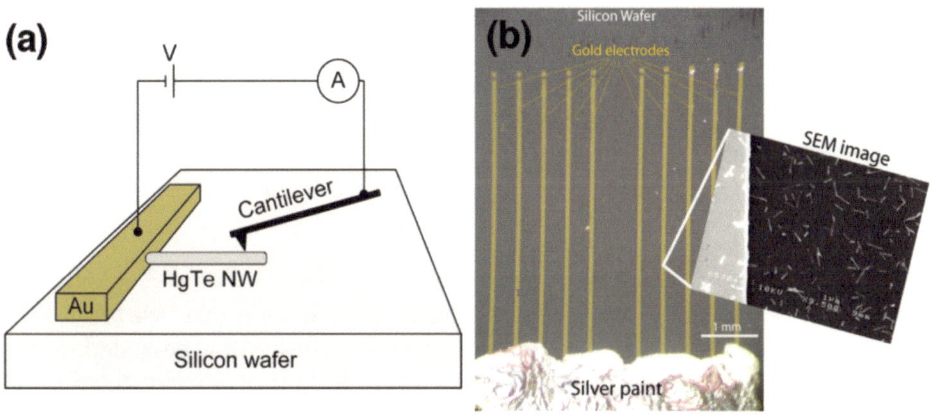

Fig. 9. (a) Schematic drawing of the nanowire-conductive AFM electrical circuit. (b) Optical microscope image of sputtered gold electrodes on a Si substrate with HgTe nanowires. Inset is an enlarged SEM image which shows randomly distributed HgTe nanowires, some of which are partly covered by the gold electrode.

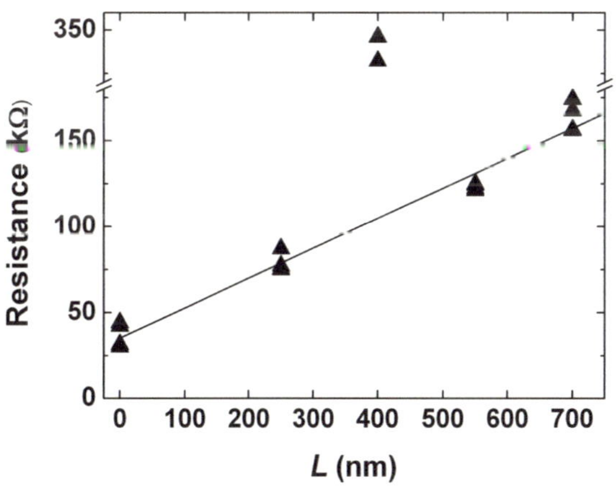

Fig. 10. The resistance values from the measured I-V curves for one nanowire, and a linear fit to the data (excluding the data points at L = 400 nm). Taken from (Gundersen et al., 2010).

wires. The I-V curves were linear and the resistance was found from the slope of linear fits to the data. Then all these measured resistance values were plotted vs. L, as shown in Fig. 10. Also these data were very well fitted with straight lines (excluding the points at L = 400 nm, where the tip had poor contact to the wire), and the slope dRmeasured/dL was extracted for each wire.

In the measurement circuit it was only the drift resistance of the wire that varied with L; therefore the wires were found to be diffusive (in the ballistic regime the resistance is independent of wire length). From the slopes dRmeasured/dL and the cross-section of the wires, resistivities were calculated for two wires. These were ~ 250/130 times larger than the resistivity 4.4×10^{-4} Ωcm of an MBE-grown 3 μm thick HgTe film, as measured by the van der Pauw technique. The difference is attributed to scattering at the grain boundaries and the surface. These measurements are described in more detail in (Gundersen, 2008; Gundersen et al., 2010).

4. Further work

To obtain straight, single crystal HgTe nanowires, the growth conditions should be varied to obtain a higher Hg content in the wires growing away from the Au particle areas. Electrical and optical characterization of the different types of wires is the next step. Resistivity measurements should be made on the type II and type III HgTe/Te wires described above. A Schottky curve could be an indication of a band gap or barrier in the wire. Measurements should be performed at small *L* to try to observe ballistic transport. The substrate could be used as a common back-gate electrode to modulate the carrier density of the wires, thereby providing information about the doping type of the wire. If two contacts could be made by EBL, the sample could be mounted in a cryostat. If the effective width of a wire in the ballistic regime could be controlled by a gate voltage, then conductance quantization could perhaps be observed at lower temperatures. Photoluminescence or photoconductive measurements could determine the effective band gap of Te nanowire segments and reveal whether a positive band gap has opened up in thin HgTe wires. This would open up for electro-optic devices in the infrared region of the spectrum. To stabilize the wires, a thin CdTe passivation layer could be grown *in-situ*. Spintronics and thermoelectric applications could also be considered.

5. Conclusions

HgTe nanowires nucleated by Au particles have been grown on Si and GaAs substrates by molecular beam epitaxy. The wires grow laterally on the surface, but they are not epitaxially coupled to the substrate. During the initial growth, Au particles on the surface block the straight, forward growth of the wires, resulting in twisted, but mostly single crystal, <111>-oriented wires. During further growth, atoms can diffuse along the wires, both making them longer and filling in corners to make them wider and straighter. New grains are added so that the larger wires end up polycrystalline. Raman scattering experiments show that the crystallinity is good within the individual grains. When the growing wires are unimpeded by Au particles, they become straight and smooth, and they consist of segments of cubic <111>HgTe and hexagonal <001>Te parallel to the wire axis.

Attempts at including Cd to grow HgCdTe on Au-coated substrates resulted in a polycrystalline layer and no nanowires. By comparing to Au-catalyzed growth of GaAs nanowires, we conclude that the diffusion coefficients of Cd and Hg through Au are too low for the VLS mechanism to function at the low HgCdTe MBE growth temperature. SiO_2 can be used as a mask for selective growth of HgTe, but not for HgCdTe. A new sample preparation technique for plane-view studies of undisturbed surfaces using TEM has been developed. The conductive tip of an AFM was used as a movable electrode to measure the resistivity of the thick, polycrystalline HgTe nanowires. Two nanowires were found to have diffusive electron transport, with resistivities two orders of magnitude larger than that of a HgTe film. The difference can be explained by scattering at the rough surface walls and at the grain boundaries in the wires.

6. Acknowledgements

The work was funded by Norwegian Ministry of Defence. Support for nanowire characterization was provided by the AFOSR under grant number FA9550-06-1-0484, P00002.

7. References

Björk, M. T.; Ohlsson, B. J.; Sass, T.; Persson, A. I.; Thelander, C.; Magnusson, M. H.;Deppert, K.; Wallenberg, L. R. & Samuelson, L. (2002). One-dimensional heterostructures in semiconductor nanowhiskers. *Appl. Phys. Lett.* 80, 6, (Feb. 2002), 1058-1060, 0003-6951.

Carter, R.; Sloan, J.; Kirkland, A. I.; Meyer, R. R.; Lindan, P. J. D.; Lin, G.; Green, M. L. H.;Vlandas, A.; Hutchinson, J. L. & Harding, J. (2006). Correlation of structural and electronic properties in a new low-dimensional form of mercury telluride. *Phys. Rev. Lett.* 96, 21, (Jun. 2006), 215501, 0031-9007.

Colin T. & Skauli, T. (1997). Applications of thermodynamical modelling in molecular beam epitaxy of $Cd_xHg_{1-x}Te$. *J. Electron. Mater.* 26, 6, (Jun. 1997), 688-696, 0361-5235

Dheeraj, D. L.; Patriarche, G.; Largeau, L.; Zhou, H. L.; van Helvoort A. T. J.; Glas, F.; Harmand, J. C.; Fimland B. O. & Weman H. (2008). Zinc blende GaAsSb nanowires grown by molecular beam epitaxy. *Nanotechnology* 19, 27, (Jul. 2008), 275605, 0957-4484.

Dresselhaus, M. S.; Lin, Y.-M.; Rabin, O. & Dresselhaus, G. (2003). Bismuth nanowires for thermoelectric applications. *REPORTS: Nanoscale and Microscale Thermophysical Engineering*, 7, 3, (Jul. 2003), 207-219, 1089-3954.

Duan, X. & Lieber, C. M. (2000). The general synthesis of compound semiconductor nanowires. *Adv. Mater.* 12, 4, (Feb. 2000), 298-302, 0935-9648.

Foss, S.; Taftø, J. & Haakenaasen, R. (2009). A specimen preparation technique for plane-view studies of surfaces using transmission electron microscopy. *J. Electron Microscopy*, 59, 1, (Feb. 2010), 27-31, 0022-0744.

Green, M; Wakefield, G. & Dobson, P. J. (2003). A simple metalorganic route to organicallypassivated mercury telluride nanocrystals. *J. Mater. Chem.* 13, 5, (May 2003), 1076-1078, 0959-9428.

Gudiksen, M. S.; Lauhon, L. J.; Wang, J.; Smith, D. C. & Lieber, C. M. (2002). Growth of nanowire superlattice structures for nanoscale photonics and electronics. *Nature* 415, 6872, (Feb. 2002), 617-620, 0028-0836.

Gundersen, P. (2008). *Electrical characterization of HgTe nanowires using conductive atomic force microscopy*, Master's thesis, Dept. of Physics, Norwegian University of Science and Technology, Trondheim.

Gundersen, P.; Kongshaug, K. O.; Selvig E. & Haakenaasen, R. (2010). Electrical characterization of HgTe nanowires using conductive atomic force microscopy. In preparation.

Hadzialic, S. (2004). *Growth and Characterization of* $Cd_xHg_{1-x}Te$ *Nanowires*, Cand. Scient. thesis, Dept of Physics, University of Oslo, Oslo.

Haakenaasen, R.; Selvig, E.; Hadzialic, S.; Skauli, T.; Hansen, V.; Tibballs, J. E.; Trosdahl-Iversen, L.; Steen, H.; Foss, S.; Taftø, J.; Halsall, M. & Orr, J. (2008a). Nanowires in the CdHgTe material system. *J. Elec. Mater.* 37, 9, (Sep. 2008), 1311-1317, 0361-5235.

Haakenaasen, R.; Selvig, E.; Foss, S.; Trosdahl-Iversen, L. & Taftø, J. (2008b). Segmented nanowires of HgTe and Te grown by molecular beam epitaxy. *Appl. Phys. Lett.* 92, 13, (Mar. 2008), 133108-1 - 133108-3, 0003-6951.

Janik, E.; Sadowski, J.; Dluzewski, P.; Kret, S.; Baczewski, L. T.; Petroutchik, A.; Lusakowska, E.; Wrobel, J.; Zaleszczyk, W.; Karczewski, G.; Wojtowicz, T. & Presz, A. (2006). ZnTe nanowires grown on GaAs(100) substrates by molecular beam epitaxy. *Appl. Phys. Lett.* 89, 13, (Sep. 2006), 133114, 0003-6951.

Konig, M.; Wiedmann, S.; Brune, C.; Roth, A.; Buhmann, H.; Molenkamp, L. W.; Qi, X. L.; and Zhang, S. C. (2007). Quantum spin hall insulator state in HgTe quantum wells. *Science* 318, 5851, (Nov. 2007), 766-770, 0036-8075.

Morfatt, W. G. (1977). *The handbook of binary phase diagrams,* General Electric Company, 0-931690-00-5, Schenectady.

Mortlock, I. & Rowe, A. H. (1965). Atomic diffusion of mercury in gold. *Philos. Mag.* 11, 114 (1965) 1157-1164, 1478-6435.

Ohlsson, B. J.; Björk, M. T.; Magnusson, M. H.; Deppert, K.; Samuelson, L. & Wallenberg, L. R. (2001). Size-, shape-, and position-controlled GaAs nano-whiskers. *Appl. Phys. Lett.* 79, 20, (Nov. 2001), 3335-3337, 0003-6951.

Persson, A. I.; Larsson, M. W.; Stenstrom, S.; Ohlsson, B. J.; Samuelson, L. & Wallenberg, L. R. (2004). Solid-phase diffusion mechanism for GaAs nanowire growth. *Nature Mater.* 3, 10, (Oct. 2004), 677-681, 1476-1122.

Qin, A.-M.; Fang, Y.-P. & Su, C.-Y. (2007). Hydrothermal synthesis of HgTe rod-shaped nanocrystals. *Mater. Lett.* 61, 1, (Jan. 2007), 126-129, 0167-577X.

Ranga Roa A. & Dutta, V. (2006). Nanotubes in spray deposited nanocrystalline HgTe:I thin films. *Mater. Res. Symp. Proc.* 901E, 0901-Ra11-19 - Rb11-19.1- 0901-Ra11-19 - Rb11-19.6, 0272-9172.

Selvig, E.; Gjønnes, K.; Olsen, A.; Colin, T. & Løvold, S. (1995). TEM study of (Cd,Hg)Te grown by molecular beam epitaxy. *Inst. Phys. Conf. Ser.* 146, pp. 317-320, 0951-3248, Oxford 1995, IOP Publishing Ltd., Bristol.

Selvig, E.; Hadzialic, S.; Haakenaasen, R.; Skauli, T.; Steen, H.; Hansen, V.; Trosdahl-Iversen, L.; van Rheenen, A. D. & Lorentzen, T. (2006). Growth of HgTe nanowires. *Phys. Scripta* T126, (2006), 115-120, 0031-8949.

Selvig, E.; Tonheim, C. R.; Kongshaug, K. O.; Skauli, T.; Lorentzen, T. & Haakenaasen, R. (2007). Defects in HgTe grown by molecular beam epitaxy on (211)B-oriented CdZnTe substrates. *J. Vac. Sci. Technol. B* 25, 6, (Nov./Dec. 2007), 1776 -1784, 1071-1023.

Selvig, E.; Tonheim, C. R.; Kongshaug, K. O.; Skauli, T.; Hemmen, H.; Lorentzen, T. & Haakenaasen, R . (2008a). Defects in CdHgTe grown by molecular beam epitaxy on (211)B-oriented CdZnTe substrates. *J. Vac. Sci. Technol. B* 26, 2, (Mar./Apr. 2008), 525-533, 1071-1023.

Selvig, E.; Tonheim, C.R.; Lorentzen, T.; Kongshaug, K. O.; Skauli, T. & Haakenaasen, R. (2008b). Defects in HgTe and CdHgTe grown by molecular beam epitaxy. *J. Elec. Mater.* 37, 9, (Sep. 2008), 1444 – 1452, 0361-5235.

Shan, C. X.; Liu, Z. & Hark, S. K. (2005). Highly oriented zinc blende CdSe nanoneedles. *Appl. Phys. Lett.* 87, 16, (Oct. 2005), 163108, 0003-6951.

Song, H.; Cho, K.; Kim, H.; Lee, J. S.; Min, B.; Kim, H. S.; Kim, S.-W.; Noh, T. & Kim, S. (2004). Synthesis and characterization of nanocrystalline mercury telluride by sonochemical method. *J. Cryst. Growth* 269, 2-4, (Sep. 2004), 317-323, 0022-0248.

Tang, B.; Yang, F.; Ciu, Y.; Zhuo, L.; Ge, J. & Cao, L. (2007). Synthesis and characterization of wavelength-tuneable, water-soluble, and near-infrared-emitting CdHgTe nanorods. *Chem. Mater.* 19, 6, (Feb. 2007), 1212-1214, 0897-4756.

Thelander, C.; Björk, M. T.; Larsson M. W.; Hansen A. E.; Wallenberg, L. R. & Samuelson L. (2004). Electron transport in InAs nanowires and heterostructure nanowire devices. *Solid State Commun.* 13, 9-10, (Sep. 2004), 573-579, 0038-1098.

Tibballs, J. E. (2005). Scandinavian Institute of Dental Materials, Haslum. Private communication.

Tonheim, C. R.; Selvig, E; Nicolas, S.; Gunnæs, A. E.; Breivik, M. & Haakenaasen, R. (2008). Excitation density dependence of the photoluminescence from $Cd_xHg_{1-x}Te$ multiple quantum wells. *J. Phys: Conf. Series.* 100, 042024, 1742-6588, Stockholm June 2007, IOP Publishing Ltd., Bristol.

Wagner, R. S. & Ellis, W. C. (1964). Vapor-liquid-solid mechanism of single crystal growth. *Appl. Phys. Lett.* 4, 5, (Mar. 1964), 89-90, 0003-6951.

Wu, Y.; Fan, R. & Yang, P. (2002). Block-by-block growth of single-crystalline Si/SiGe superlattice nanowires. *Nano. Lett.* 2, 2, (Feb. 2002), 83-86, 1530-6984.

Yang, J; Zhou, Y.; Zheng, S.; Liu, X.; Qiu, X.; Tang, Z.; Song, R.; He, Y.; Ahn, C. W. & Kim J. W. (2009). Self-reorganization of CdTe nanoparticles into near-infrared $Hg_{1-x}Cd_xTe$ nanowire networks. *Chem. Mater.* 21, 14, (Jul. 2009), 3177-3182, 0897-4756.

Zhang, F.; Barrowcliff, R.; Stecker, G.; Wang, D. & Hsu, S.-T. (2005). IrO_2 nano structures by metal organic chemical vapor deposition, *NSTI-Nanotech 2005 Conf. Proc. 2,* pp. 623-626, 0-9767985-4-9, Anaheim, May 2005, Nano Science and Technology Institute, Cambridge.

5

Metal-oxide Nanowires by Thermal Oxidation Reaction Technique

Supab Choopun, Niyom Hongsith and Ekasiddh Wongrat
Department of Physics and Materials Science, Faculty of Science, Chiang Mai University Chiang Mai 50200 and ThEP Center, CHE, Bangkok 10400, Thailand

1. Introduction

The metal-oxides are very interesting materials because they possess wide and universal properties including physical and chemical properties. For example, metal-oxides exhibit wide range of electrical property from superconducting, metallic, semiconducting, to insulating properties (Henrich & Cox, 1994). The wide ranges of properties makes metal-oxide suitable for many applications including corrosion protection, catalysis, fuel cells, gas sensor, solar cells, field effect transistor, magnetic storage (Henrich, 2001), UV light emitters, detectors, piezoelectric transducers, and transparent electronics (Hsueh & Hsu, 2008) etc.

Recently, nanostructures of metal-oxide such as nanowire, nanorod, nanobelt, nanosheet, nanoribbon, and nanotube have gained a great attention due to their distinctive and novel properties from conventional bulk and thin film materials for new potential applications. These unique properties cause by quantum confinement effect (Manmeet et al., 2006), lower dimensionality (Wang et al., 2008), change of density of state (Lyu et al., 2002), and high surface-to-volume ratio (Wangrat et al., 2009).

Nanowires can be regarded as one-dimensional (1D) nanostructures which have gained interest for nanodevice design and fabrication (Wang et al., 2008). As an example of metal-oxide nanowires, the materials are focused on zinc oxide (ZnO) and copper oxide (CuO). ZnO which is n-type semiconductor has been widely studied since 1935 with a direct band gap of 3.4 eV and large exciton binding energy of 60 meV at the room temperature (Coleman & Jagadish, 2006). ZnO has a wurtzite structure, while CuO, which is p-type semiconductor with narrow band gap of 1.2 eV , has a monoclinic crystal structure (Raksa et al., 2009).

ZnO and CuO can be synthesized by various techniques such as pulse laser deposition (PLD) (Choopun et al., 2005), chemical vapor deposition (VD) (Hirate et al., 2005), thermal evaporation (Jie et al., 2004; Ronning et al., 2004), metal-catalyzed molecular beam epitaxy (MBE) (Wu et al., 2002; Chan et al., 2003; Schubert et al., 2004), chemical beam epitaxy (CBE) (Björk et al., 2002) and thermal oxidation technique (Wongrat et al., 2009). Thermal oxidation technique is interesting because it is a simple, and cheap technique. Many researchers have reported about the growth of ZnO and CuO by thermal oxidation technique with difference conditions such as temperature, time, catalyst, and gas flow. The list of metal-oxide nanowires synthesized by thermal oxidation is shown in Table 1.

Materials	Temperature (°C)	Oxidation Time	Morphology	Diameter (nm)	Growth direction	Ref.
CuO	400-700	2-4 h	nanowire	30-100	$[\bar{1}11]$ and $[111]$	(Jiang et al., 2002)
CuO	300-500	0.5-24 h	nanowire	500	-	(Chen et al., 2008)
CuO	600	6 h	nanowire	100-400	[110]	(Raksa et al., 2008)
CuO	400	2 h	nanowire	60	$[\bar{1}11]$	(Nguyen et al., 2009)
CuO	400	4 h	nanowire	40-100	-	(Zeng et al., 2009)
CuO	400-800	4 h	nanowire	50-100	[010]	(Manmeet et al., 2006)
CuO	500	1.5 h	nanowire	100	$[\bar{1}11]$	(Hansen et al., 2008)
ZnO	300	5 min	Nanowire and nanoflake	100-150	-	(Hsueh & Hsu, 2008)
ZnO	600	24 h	nanowire	100-500	-	(Wongrat et al., 2009)
ZnO	300-600	1 h	nanoneedle	20-80	[0001]	(Yu & Pan, 2009)
ZnO	500	1 h	nanoplate	200-600	$[11\bar{2}0]$	(Kim et al., 2004)
ZnO	200-500	30 min	nanowire	30-350	[0001]	(Schroeder et al., 2009)
ZnO	300-600	1 h	nanowire	12-52	$[11\bar{2}0]$	(Fan et al., 2004)
ZnO	< 400	30 min	nanowire	20-150	$[11\bar{2}0]$	(Ren et al., 2007)
ZnO	600	1.5 h	nanowire	30-60	[0001]	(Sekar et al., 2005)
ZnO	400-600	1 h	nanowire and nanorod	20	$[2\bar{1}\bar{1}0]$	(Liang et al., 2008)

Table 1. List of metal-oxide nanowires synthesized by thermal oxidation.

2. Oxidation reaction

The thermal oxidation technique can be used to fabricate nanowires that are low-cost and high quality. A growth condition can be controlled with difference temperature and time. The oxidation reaction of metal on the surface yields metal-oxide semiconductor. However, the metal-oxide may be bulk or nanosize depending on the growth conditions such as the temperature, time, metal-catalyst and gas atmosphere.

In our previous report, we have successful synthesized ZnO and CuO nanowires by thermal oxidation technique without metal-catalyst under normal atmosphere. Zn metal was screened on the alumina substrate and sintered at the temperature of 500-700°c for 24 hours. It was found that the oxidation reaction occurred with two mechanism; above melting point of Zn metal and lower melting point of Zn metal. In phase diagram as shown in Fig. 1, Zn in liquid phase forms to ZnO nanowires at temperature more than 419.6°c while less than 419.6°c Zn solid phase can be oxidized to form ZnO nanowires on the substrate. The oxidation reaction of Zn solid (below melting point) can be expressed as:

$$Zn(s) + \frac{1}{2}O_2(g) \rightarrow ZnO(s) \tag{1}$$

While the chemical reaction of Zn liquid is shown as:

$$Zn(l) + \frac{1}{2}O_2(g) \rightarrow ZnO(s) \tag{2}$$

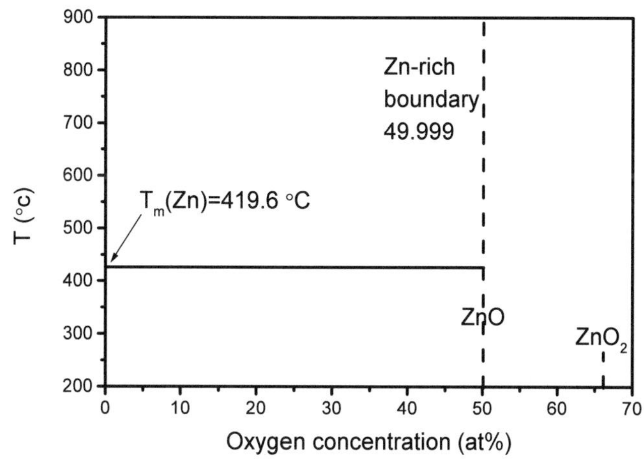

Fig. 1. Phase diagram of Zinc-Oxygen at the difference temperature (Ellmer & Klein, 2008).

Many researchers including us have reported about the synthesis of nanowires by thermal oxidation techniques. The mechanism of metal oxidation has also been reported by Wagner (Wagner & Grunewald, 1938). The oxide growth rate depends on transport properties such as diffusion coefficient that can be proved from Fick's first law at the steady state as following:

$$J = -D\frac{\partial c}{\partial x} \tag{1}$$

where J is the concentration gradient, c is the concentration of oxygen, D is the diffusion coefficient and x is the displacement. Integrating Eq. (1) when J/D is constant and the boundary conditions are given as

$$c(x=0,t) = c_1;\, c(x=T,t) = c_2 \tag{2}$$

, we get:

$$-\frac{JT}{D} = (c_2 - c_1) \tag{3}$$

where c_2 is concentration in the interface regions of oxide-oxygen molecule, while c_1 is in regions of metal-oxide interface as shown in Fig 2.

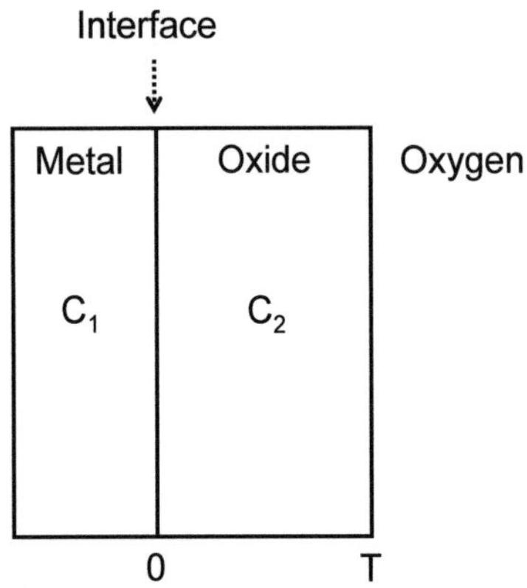

Fig. 2. Schematic diagram of metal-oxide interface and oxide-gas.

we assume c_1 is negligible since the number of oxygen is tiny. Therefore, Eq. (3) can be reduced to the following

$$-\frac{JT}{D}=c_2 \tag{4}$$

For conservation of mass (Belousov, 2007), Fick' s first law is indicated as:

$$J=c^*\frac{dT}{dt} \tag{5}$$

where c^* is the oxygen concentration in the oxygen product. From Eq. (4) and Eq. (5), we get

$$J=c^*\frac{dT}{dt}=-\frac{c_2D}{T} \tag{6}$$

Integrating equation (6), the solution is formed as

$$T^2=\frac{2c_2Dt}{c^*} \tag{7}$$

where

$$k=\frac{2c_2D}{c^*} \tag{8}$$

is a parabolic rate constant. From Eq. (7), it was found that the oxide thickness is proportional to the square root of time. However, parabolic rate constant is different for various metal-oxides since the diffusion in oxidation reaction is caused by many mechanisms such as metal vacancy, oxygen vacancy, metal interstitial and oxygen interstitial.

First, the oxidation reaction of metal occurs at the surface where metal lose electron to form M^{++} ions. Then, the electron from metal moves to the surface. The oxygen molecule and the electron reacts to form adsorbed oxygen (oxygen ion) on the surface. The adsorbed oxygen ions include O^{2-}, O^- and O_2^- which has reaction as (Martin & Fromm, 1997)

$$\left.\begin{array}{l}\frac{1}{2}O_2 + 2e \rightarrow O^{2-} \\ O_2 + e \rightarrow O_2^- \\ \frac{1}{2}O_2 + e \rightarrow O^-\end{array}\right\} \quad (9)$$

The diffusion across oxide layer owing to metal ion or oxygen ion is depended on the domination of transport. The transportation of ions can be considered for four possible mechanisms as shown in Fig. 3. First, the transportation of ions by oxygen is due to interstitial mechanism that the oxygen ions are more mobile than metal ions and pass from one interstitial site to one of its nearest-neighbor interstitial site without permanently displacing any of the matrix atom (Shewmon, 1989). Therefore, the new metal-oxide is formed at the metal-oxide interface as seen in Fig. 3(a). Second, if there is the vacancy of oxygen in the matrix atom or called unoccupied site (Shewmon, 1989), the nearest-neighbor oxygen ions can move from one to vacancy. The new metal-oxide is also formed at the metal-oxide interface as in Fig. 3(b).

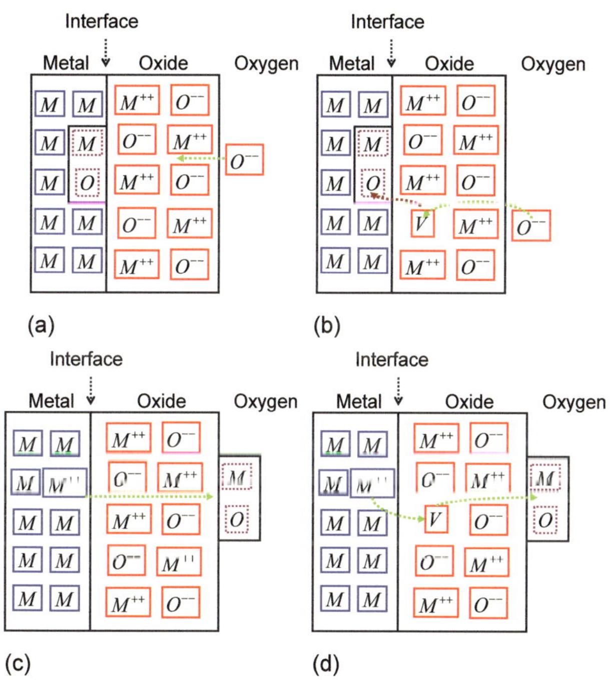

Fig. 3. Schematic diagrams of four possible mechanisms of ion transport in oxidation reaction, (a) the transportation of oxygen ions by oxygen interstitial mechanism (b) the transportation of oxygen ions by oxygen vacancy mechanism (c) the transportation of metal ions by metal interstitial mechanism (d) the transportation of metal ions by metal vacancy mechanism.

Third, on the contrary, it is for the case that the metal ions are more mobile than oxygen ions and pass from one site to the nearest-neighbor site without permanently displacing any of the matrix atom (Shewmon, 1989). The new metal-oxide is formed at oxide-oxygen interface as in Fig. 3 (c). Fourth, for the case that it has the unoccupied site of metal ions on the lattice, the metal ions can jump from one to the nearest-neighbor unoccupied site. The new metal-oxide is also formed at the oxide-oxygen interface as in Fig. 3(d). The new metal-oxide in Fig. 3 (a) and (b) is in MO form, while in Fig. 3 (c) and (d) is in M_2O form (Martin & Fromm, 1997). Moreover, the oxidation reaction of metal forms two layers of oxide and metal. The ratio of molar volume of oxide to molar volume of metal called "Pilling-Bedworth ratio" which is the indicator of whether an oxide layer is protective. The Pilling-Bedworth ratio can be written by (Barsoum, 2003):

$$P - Bratio = \chi = \frac{V_{MO_{Z/2}}}{V_M} = \frac{MW_{MO_{Z/2}} \rho_M}{MW_M \rho_{MO_{Z/2}}} \tag{10}$$

where V_M, $V_{MOz/2}$ is the molar volume of the metal and of the metal-oxide, respectively. MW and ρ is the molecular weights and densities of the metal and oxide, respectively. For metals having a P-B ratio less than unity, the metal-oxide tends to be porous and unprotective because metal-oxide volume is not enough to cover the underlying metal surface. For ratios larger than unity, compressive stresses build up in the film, and if the mismatch is too large, (P-B ratio>2), the oxide coating tends to buckle and flake off, continually exposing fresh metal, and is thus nonprotective. The ideal P-B ratio is 1, but protective coatings normally form for metals having P-B ratios between 1 and 2 (Barsoum, 2003).

3. Gibb free energy

From the chemical oxidation reaction of metal, the two possible chemical reactions are

$$M(s) + \frac{1}{2}O_2(g) \rightarrow MO(s) \tag{11}$$

and

$$M(l) + \frac{1}{2}O_2(g) \rightarrow MO(s) \tag{12}$$

where M is defined by metal. The difference between the two chemical reactions is that metal solid can be formed by oxidation reaction process under a melting point while metal liquid can be formed above melting point. However, both reactions are spontaneous reactions. An important parameter to obtain physical insight into the materials aspect of thermodynamics potential is Gibb free energy. This section will be described about the Gibb free energy in term of thermodynamic parameter. The Gibb free energy of a phase is given by:

$$G = H - TS \tag{13}$$

In thermodynamics, the parameters such as enthalpy, entropy and Gibb free energy are normally absolute value. So, it can be identified by a relative value with convention at standard state. The standard state is assigned to compare with the reference state at pressure of 1 atm and 298 K. Firstly, the enthalpy of chemical reaction is typically described by the change of two state; initial state and final state. So, the enthalpy of compound in chemical reaction at standard state is simply given by:

$$\Delta H^0 = \sum \Delta H^0_{products} - \sum \Delta H^0_{reac\tan ts} \tag{14}$$

$$\Delta H^0 = \sum \Delta H^0_{MO} - \frac{1}{2}\sum \Delta H^0_{O_2} - \sum \Delta H^0_{M} \tag{15}$$

Similarly, entropy changes in chemical reactions can be obtained in a same way and the sum of standard entropy for chemical reaction can be written as:

$$\Delta S^0 = \sum \Delta S^0_{products} - \sum \Delta S^0_{reac\tan ts} \tag{16}$$

$$\Delta S^0 = \sum \Delta S^0_{MO} - \frac{1}{2}\sum \Delta S^0_{O_2} - \sum \Delta S^0_{M} \tag{17}$$

Since, Gibb free energy is defined by Eq. (13), the Gibb free energy of chemical reaction process at various temperatures is given as:

$$\Delta G^0 = \Delta H^0 - T\Delta S^0 \tag{18}$$

On the other hand, if standard Gibb free energy is known, it can be simply calculated by:

$$\Delta G^0 = \sum \Delta G^0_{products} - \sum \Delta G^0_{reac\tan ts} \tag{19}$$

4. Thermodynamics equilibrium

Typically, equilibrium system is a stable system with time and the certain properties of the system are uniform throughout with the same temperature and pressure. The various phases can co-exist without driving force. The variation of Gibb free energy through temperature and pressure can be indicated as

$$dG = -SdT + VdP \tag{20}$$

For a system undergo a process at constant temperature, Eq. (20) becomes:

$$dG = VdP \tag{21}$$

The surrounding oxygen, which is assumed to be ideal gas can react with metal to form metal-oxide. Eq. (21) can be rewritten as

$$dG = \frac{RT}{P}dP \tag{22}$$

Integrating between state 1 and 2,

$$\Delta G = RT\ln\frac{P_2}{P_1} \tag{23}$$

Since the reference state is 1 atm and 298 K, Eq. (23) is reduced as following:

$$G = G^0 + RT\ln P \tag{24}$$

It was found that Eq. (24) of Gibb free energy is a function of pressure and temperature. However, the system is not ideal system that the relation in Eq. (24) is non-linear. Thus, for real gas the new function of fugacity (f) is introduced and defined as:

$$dG = RTd\ln f \tag{25}$$

Integrating Eq. (25) with initial state and final state yields:

$$\Delta G = RT\ln\frac{f_2}{f_1} \tag{26}$$

The activity is defined by:

$$a = \frac{f}{f^0} \tag{27}$$

where f^0 is fugacity in standard state. Replacing Eq. (26) with Eq. (27)

$$\left.\begin{aligned} G - G^0 &= RT\ln a \\ G &= G^0 + RT\ln a \end{aligned}\right\} \tag{28}$$

From the chemical oxidation reaction of metal, the Gibb free energy in the system can be written by:

$$\Delta G = G_{MO} - G_M - \frac{1}{2}G_{O_2} \tag{29}$$

Substituting Eq. (28) to Eq. (29):

$$\Delta G = \left(G^0_{MO} + RT\ln a_{MO}\right) - \left(G^0_M + RT\ln a_O\right) - \frac{1}{2}\left(G^0_{O_2} + RT\ln a_{O2}\right) \tag{30}$$

when

$$\Delta G^0 = G^0_{MO} - G^0_M - \frac{1}{2}G^0_{O_2} \tag{31}$$

Eq. (30) can be given as:

$$\Delta G = \Delta G^0 + RT\ln\frac{a_{MO}}{a_M a_{O_2}^{1/2}} \tag{32}$$

The activity is approximately one in solid. While activity in oxygen is an oxygen pressure in the system so Eq. (32) can be reduced to:

$$\Delta G = \Delta G^0 - \frac{1}{2}RT\ln P_{O_2} \tag{33}$$

The important parameter from Eq. (33) is Gibb free energy at standard state which is the function of temperature and pressure. The kinetic of oxide nucleation to form the nanostructure depend on the diffusion of oxygen ions and metal ions which is determined by the change of Gibb free energy. Nevertheless, the simple method to consider oxidation reaction at various temperature and pressure can be performed by Ellingham diagram that will be discussed in the next section.

5. Ellingham diagram

The oxidation reaction of metal is related to temperature and pressure in Eq. (33). Ellingham have presented a simple method to consider oxidation reaction. The temperature and pressure can oxidize metal to form oxide layer at the equilibrium of oxygen, metal, and metal-oxide (ΔG=0). At the equilibrium, Eq. (33) can be reduced to:

$$\Delta G^0 = \frac{1}{2} RT \ln P_{O_2} \tag{34}$$

For 1 mole of oxygen molecule in chemical oxidation reaction, Eq. (33) turns to

$$\Delta G^0 = RT \ln P_{O_2} \tag{35}$$

The relation between Gibb free energy, temperature and the pressure of oxygen can be plotted in Fig. 4. The linear relation between Gibb free energy and temperature for chemical reaction of Cu to Cu_2O and CuO is plotted together in Fig. 4. This linear relation passes zero point of Gibb free energy at absolute zero temperature with slope of $RlnP_{O2}$. An intersection between two lines shows the equilibrium between Cu and Cu_2O at the pressure and temperature that can form metal-oxide. For example, Ellingham diagram for some metal oxidation reaction is given in Fig. 5. for given control pressure and temperature.

6. Metal-oxide nanowires by thermal oxidation reaction technique

Metal-oxide nanowires have been successfully synthesized by thermal oxidation technique. This technique has been successfully used for synthesizing ZnO or CuO by simply heating

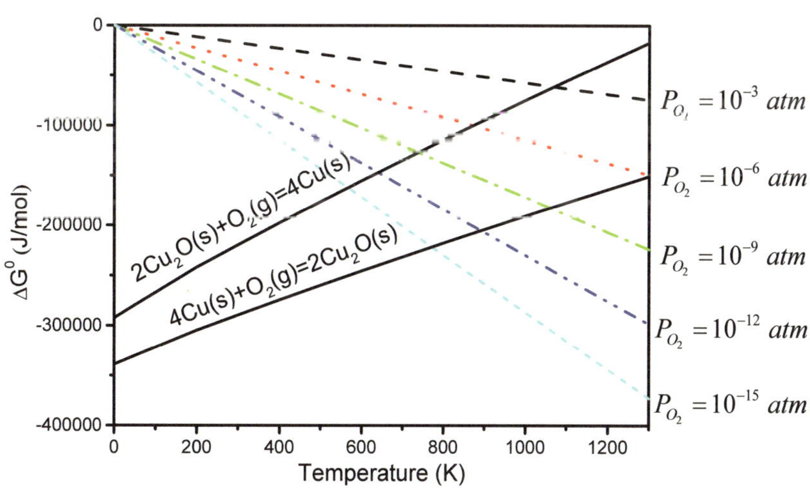

Fig. 4. Superposition between ΔG^0 versus T for oxidation reaction and for oxygen pressure at $\Delta G^0 = RT \ln P_{O_2}$.

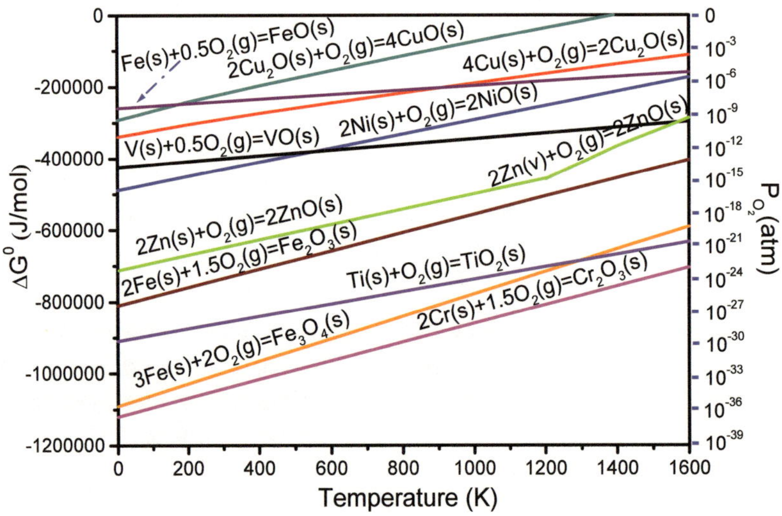

Fig. 5. Ellingham diagram for some metal-oxide.

pure Zn and Cu material sources, respectively. The process is usually conducted in a cylindrical furnace. The morphology of ZnO nanowires and CuO nanowires by thermal oxidation reaction technique revealed by field emission scanning electron microscopy (FE-SEM) are shown in Fig. 6 (a) and (b), respectively. ZnO nanowires were performed by heating zinc powder (purity 99.9%) that was screened on the alumina substrate. The oxidation process was sintered in a horizontal furnace in alumina crucible in air at atmospheric pressure at temperature 600°C for 24 hr. The ZnO nanowires have the diameters ranging from 60-180 nm and the lengths ranging from 5-10 μm. The wire-like structure of ZnO is clearly observed from TEM image, and the associative selected area electron diffraction pattern (SADP) as shown in Fig. 6 (c). The SADP shows a spot pattern, indicating a singlecrystalline property of the nanowire corresponding to the hexagonal structure of ZnO with the lattice constants, $a = b = 3.2$ Å, $c = 5.2$ Å. From the trace analysis,, it was found that the ZnO nanowire grew along the $<\bar{2}110>$ direction on (0001) plane..

Likewise, CuO nanowires were performed by heating copper plate (~99% of purity) at temperature 600°C for 24 hr. Clearly, the oxidized products exhibited nanowire structure with the diameter ranging from 100–300 nm and the length from 10-50 μm. In addition, at higher heating temperature leads to larger diameter of nanowires. Fig. 6 (d) shows TEM bright field image of CuO nanowires with its associated SADP. The wire-like structure can be observed from TEM image. The SADP of the nanowire shows a spot pattern, indicating a single-crystalline property of the nanowire which corresponds to the monoclinic structure of CuO with the lattice constants, $a = 4.7$ Å, $b = 3.4$ Å and $c = 5.1$ Å, and from the trace analysis, the spots can be also confirmed that the growth direction of CuO nanowires is <110>.

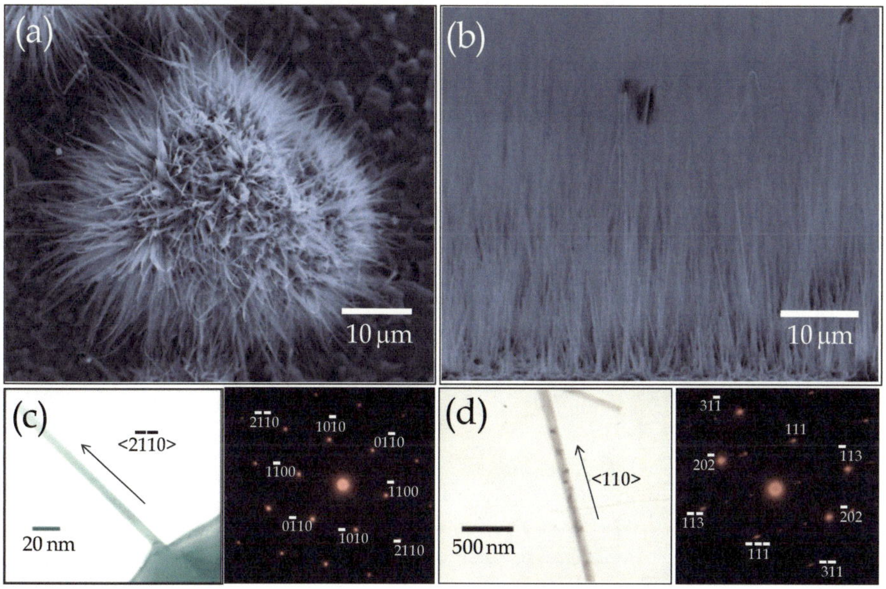

Fig. 6. FE-SEM images of (a) ZnO nanowires and (b) CuO nanowires, which prepared by thermal oxidation technique and TEM bright field image with the associated SADP for (c) ZnO nanowires and (d) CuO nanowires, respectively.

7. Growth mechanism of metal-oxide nanowire

Understanding the growth mechanism is critical in controlling and designing nanostructures. In 1950-1960, the growth mechanism of whisker that can be considered as a wire in micrometer size had been widely studied. V-S growth mechanism, V-L-S growth mechanism, and Frank dislocation growth mechanism were proposed for whisker growth (Dai, et al., 2003). However, the growth mechanism of metal-oxide nanowire should be different from whisker growth mechanism. Thus, in this section, we have proposed a possible growth mechanism that may be occurred in the formation of nanowire. The growth mechanism can be explained in term of Gibb free energy of oxidization process.

The growth mechanism of metal-oxide nanowire is proposed as in the following four steps: 1. oxygen adsorption, 2. surface oxidization to form nuclei, 3. nuclei arrangement and 4. nanowire formation as shown in Fig. 8.

Step 1. oxygen adsorption: Typically, oxygen molecules in air are adsorbed on the metal surface with diffusion process as described in Section 2. There are many reports to describe the mechanism of O_2 interaction with the transition metal surface (Martin & Fromm, 1997).

Step 2. surface oxidization to form nuclei: the metal-oxide nucleation was formed by diffusion of metal ions and/or oxygen ions in the oxide layer, and the reactions of metal ions with oxygen ions to form metal-oxide as seen in section 2.

To form nuclei of metal-oxide in the oxide layer (as in Fig. 7), metal-oxide nucleus is formed by agglomeration between metal ion and oxygen ion due to the minimization of surface energy. This phenomenon is similar to the coalescence behavior of two droplet of water when they are connected and then forming bigger one droplet instead of staying separately. In order to explain the growth mechanism, Gibb free energy change per nucleus (not per mole) of metal-oxide with radius r and volume V is introduced and defined as

$$\Delta G_N = V\Delta G^0 + A_f\gamma_f \tag{36}$$

where A_f and γ_f is surface area and surface energy, respectively. For a given ΔG^0 which is usually negative, the magnitude of ΔG_N depends on only two terms: a volume energy and surface energy. Volume energy is propotional to $-r^3$, but surface energy is propotional to r^2. Thus, ΔG_N as a function of r exhibits maximum value at some critical radius, r^*, as seen in Fig. 7 (c).

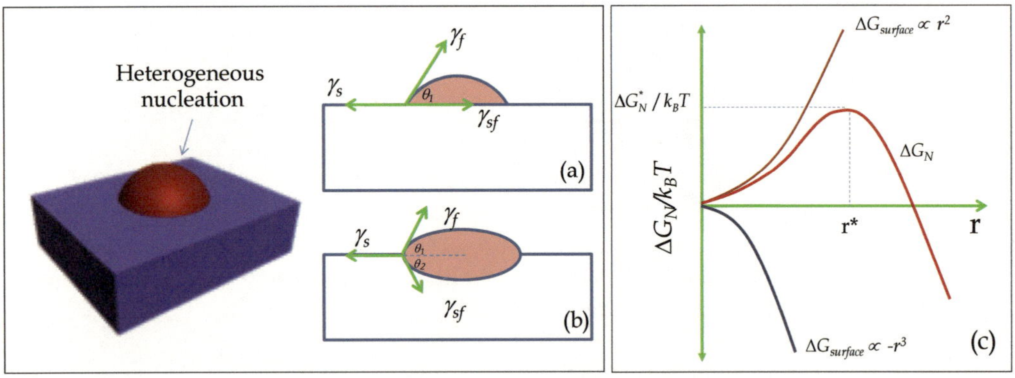

Fig. 7. Model of the metal-oxide nucleaion for (a) non-reactive and (b) reactive nucleation and (c) plot of ΔG_N as a function of radius.

For example in spherical homogeneous nucleation, the critical radius, r^*, can be simply obtained at $d(\Delta G_N/dr) = 0$ and can be written as

$$r^* = \frac{-2\gamma_f}{\Delta G^0} \text{ and } \Delta G_N^* = \frac{(16/3)\pi\gamma_f^3}{(\Delta G^0)^2} \tag{37}$$

Moreover, critical radius also implies the stability of nucleation. The nucleation will be stable at radius more than critical radius and proceed to grow spontaneously. In the other word, the spherical nucleation of stable phase forms with radius r when it can overcome the maximum energy barrier, ΔG^*_N or $\Delta G_N(r^*)$. However, the nucleation shape by oxidation process is different from spherical shape. Normally, nucleation shape by oxidation process is likely to be two shapes of nucleation: non-reactive and reactive nucleation as shown in Fig. 7 (a) and (b). The non-reactive reactive nucleation are metal-oxide nuclei that non-react and react with metal (or substrate), respectively. This is called heterogeneous nucleation. The non-reactive nucleation as seen in Fig. 7 (a) can be considered as the cluster of radius R_1 and contact angle θ_1 forming on a non-reactive substrate. In contrast, reactive nucleation as seen in Fig. 7 (b) can be considered as a double cap-shaped cluster with the upper cap having radius R_1 and contact angle θ_1, and the bottom cap having radius R_2 and contact angle θ_2. Similar to spherical homogenous nucleation, ΔG^*_N for heterogenous nucleation can be derived and given by

$$\Delta G_N^* = \frac{(16/3)\pi\gamma_f^3}{(\Delta G^0)^2} f(\theta_1) \tag{38}$$

and

$$\Delta G_N^* = \frac{(16/3)\pi\gamma_f^3}{\Delta G_0^2} h(\theta_1,\theta_2) \tag{39}$$

, for the non-reactive and reactive nucleation (Zhou, 2009), respectively. Where $f(\theta)$ and $h(\theta_1, \theta_2)$ normally called "shape factor" and given by

$$f(\theta) = \frac{(2+\cos\theta)(1-\cos\theta)^2}{4} \tag{40}$$

and

$$h(\theta_1,\theta_2) = \frac{\chi}{\chi-1}\left(\frac{\chi-1}{\chi} + \frac{\sin\theta_2}{\chi\sin\theta_1} \bullet \frac{\gamma_{sf}}{\gamma_f}\right)^3 f(\theta_1) \tag{41}$$

where χ is the volume ratio between the metal and oxide. The volume ratio χ can be related to the Pilling–Bedworth ratio, $\chi = Vox/Vm$, where Vm and Vox are the molar volume of the metal and oxide, respectively (Zhou, 2009). It should be noted that althought ΔG_N^* for heterogenous nucleation have a function of shape factor, the characterisic curve is not different compared with the spherical homogeneous nucleation. The shape factor just describes the magnitude decreasing of energy barrier ΔG_N^*.

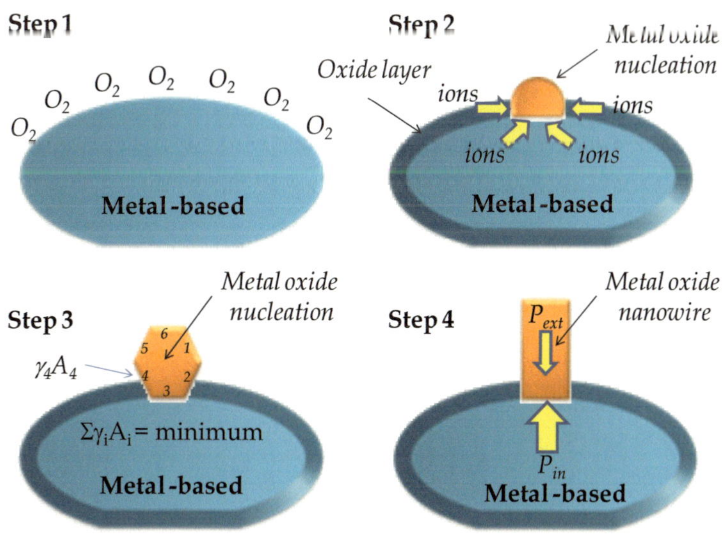

Fig. 8. Four steps of metal-oxide nanowire growth mechanism, Step 1 - oxygen adsorption, Step 2 - surface oxidization to form nuclei, Step 3 - nuclei arrangement and finally, Step 4 - nanowire formation.

Step 3. nuclei arrangement: the metal-oxide nuclei arrangement was designed based on the nuclei probability in term of the minimization of surface energy. As discuss above, a nucleation will form when it can overcome the maximum energy barrier, ΔG_N^*. Thus, we can write the probability of nucleation, P_N, and given

$$P_N = \frac{N}{N_0} = \exp(-\Delta G_N^* / k_B T) \tag{42}$$

where N is the surface concentration of nuclei and N_0 is the possible surface concentration of nuclei on metal surface, k_B is Boltzmann constant and T is temperature. It can be seen that the magnitude of P_N depends on three parameters; temperature, surface energy, and ΔG^0.

For temperature effect, it can be seen that the probability increases when the temperature increases at constant surface energy and ΔG^0. The increment of nucleation probability suggests that the nucleation frequency of formation is increased as seen in Fig 9 (a). Compare to experimental results by varying the temperatures of oxidation process, the ZnO nanowires were started to observe for heating temperature of 400°C (Fig. 9 (b)), and a lot of nanowires were observed for heating temperature of 500-800°C (Fig. 9 (c)-(f)). However, the nanowires could not be observed for heating temperature of 900°C (Fig. 9(g)). It was also found that the diameter of nanowire and the length was highest at 600°C. Therefore, the number of nanowires can be explained in term of nucleation frequency.

Fig. 9. (a) The increase of nucleation frequency of formation at higher temperature and (b)-(g) FE-SEM images of ZnO nanowires at various oxidation temperatures.

Structure	Example	Low-γ facets
Body-centered cubic (bcc)	Cr, Fe	{110}
Face-centered cubic (fcc)	Au, Al	{111}
Hexagonal close-packed (hcp)	Zn, ZnO, Mg	{0001}
Diamond	Si, Ge	{111}
Zinc blende	GaAs, ZnSe	{110}
Fluorite	MgF_2, CaF_2	{111}
Rock salt	NaCl, PbTe	{100}

Table 2. Facets of the lowest surface energy for various crytal structures (Smith, 1995).

After the metal-oxide nuclei was formed, the arrangement of nuclei will try to adjust itself to maximize nucleation probability or to minimize total surface energy that is

$$\Sigma \gamma_i A_i = \text{minimum} \tag{43}$$

Subscript *i* denotes terms corresponding to the nucleation free surface, the interface to the substrate, and the substrate free surface, respectively. In the case of liquid or amorphous nuclei, which have no γ anisotropy, there is only one term of $\gamma_f A_f$. In the common case of crystalline nuclei, these surface terms include all of the various exposed atomic planes or facets. So, the metal-oxide nuclei arrangement must be designed based on the high nuclei probability or low total surface energy. For example, the facets of lowest surface energy for various crystal structures are shown in Table 2.

For our work, we have specified that γ of ZnO on the $\{0001\}$, $\{11\bar{2}0\}$ and $\{10\bar{1}0\}$ is about 1.2 J/m^2, 1.4 J/m^2 and 1.6 J/m^2, respectively (Jiang et al., 2002). Therefore, the nucleation arrangement will just control the growth direction of metal nanowire. It can be seen that the growth direction of nanowires, both ZnO and CuO, has a few directions which have low miller index due to the lowest surface energy.

Step 4. nanowire formation: nanowire was grown by the driving force from the difference of the surface pressure between metal-oxide nucleation and substrate (or metal based). First, let consider the pressure inside the particles, which have radius *r* and surface energy γ as given by Laplace equation (Stolen et al., 2004) as

$$P_i = P_g + \frac{2\gamma_i}{r_i} \tag{44}$$

where P_g is the external pressure in the surrounding, subscript *i* referred to the solid or liquid phase. We will now consider a case where metal-oxide nucleation as a solid phase with radius r_f adsorbs on metal-based substrate with radius r_l. At interface between metal-oxide nucleation and substrate, which has surface energy γ_{sf}, the different pressure at interface is

$$P_s - P_l = \frac{2\gamma_s}{r_s} - \frac{2\gamma_l}{r_l} \tag{45}$$

In our case, we assumed that the substrate or metal-based has a large radius r_l, so the second term in Eq. 45 should be neglected. Therefore, we can write the differential form of Eq. 45 as

$$d\left(P_s - P\right)_l = d\left(\frac{2\gamma_{sf}}{r_s}\right) \tag{46}$$

Integrating from a flat interface ($r = \infty$), we obtained

$$\Delta P = P_s - P_l = \left(\frac{2\gamma_{sf}}{r_s}\right)\Bigg|_{r_s=\infty}^{r_s=r_s} = \left(\frac{2\gamma_{sf}}{r_s}\right) \tag{47}$$

It can be seen that the different pressure is inversely propotional to nucleation radius at certain surface. As an example, the different pressure at interface between the metal surface and nucleation surface is plotted as a function of the nucleation radius for a case of ZnO in Fig. 10 by using surface energy of about 1.2 J/m^2.

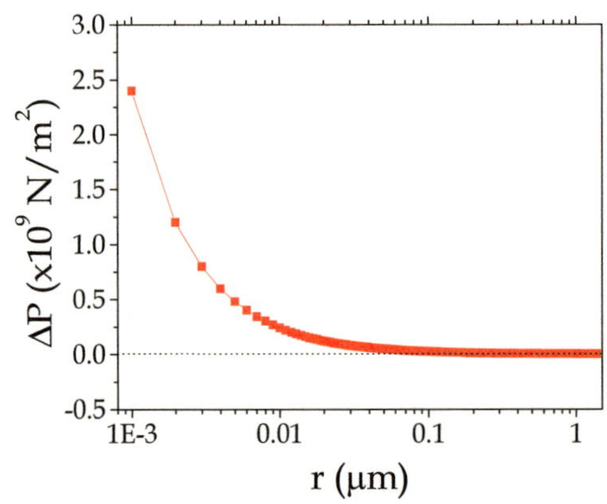

Fig. 10. Plot of the different pressure at interface between the metal surface and nucleation surface as a function of the nucleation radius for a case of ZnO.

Typically, pressure can be related to force. Thus, the increasing of different pressure at the interface suggests that there is an increasing of driving force with decreasing of the size of nucleation in order to push metal-oxide out to form nanowire. Moreover, the driving force begins to increase rapidly when the size of nucleation below 100 nm.

In the case of ZnO, at oxidation temperature of 600°C that over the melting point of Zn, oxygen molecules in air is dissociated and then adsorbed on the surface of Zn melt for ready to oxidized as discuss in Step 1. After that, the nucleation of ZnO was formed on the metal surface by the coalescence behavior of oxygen ions and metal ions. The nucleation probability depends on the temperature, surface energy and ΔG^0 as discussed in Step 2. And next step, to minimize total surface energy, the arrangement of ZnO nuclei was designed based on the nuclei probability in term of surface energy. The high possible of arrangement of ZnO nuclei is <0001>, $<11\bar{2}0>$ and $<10\bar{1}0>$ direction for their surface energy about of 1.2 J/m^2, 1.4 J/m^2 and 1.6 J/m^2, respectively (Jiang et al., 2002). Finally, the size of ZnO nucleation will be control and/or produce the driving force for nanowire growth based on the Laplace different pressure at interface between the ZnO nucleation and substrate. Since at below 100 nm the driving force increases rapidly, it can be related to FE-SEM image of ZnO nanowires that shows the size or diameter ranging in 100-200 nm. However, when the oxidation temperature is high (more than 800°C), the probability of nucleation is high too resulting in the high surface density of ZnO nuclei. Therefore, the size of nucleation is larger and there is no driving force at interface for nanowire growth.

However, CuO case is different from ZnO case because the oxidation temperature of 600°C is significantly lower the melting point of Cu metal. First, the Cu was easily oxidized to form Cu_2O layer due to the Cu_2O has lower oxidization Gibb free energy than CuO as showed in Fig 11. So, we will consider the Cu_2O as the substrate for creation CuO nucleation. Similar to ZnO case, the CuO nuclei were formed and arranged based on the minimization of total surface energy. After that, the driving force for CuO nanowires was produced by the small size of CuO nucleation. The growth mechanism is confirmed by the cross-section FE-SEM image of CuO nanowire on Cu substrate as shown in Fig. 11. It can be seen that the thick Cu_2O layer is firstly formed on copper plate, followed by the formation of CuO layer and then CuO nanowires was finally formed on CuO layer.

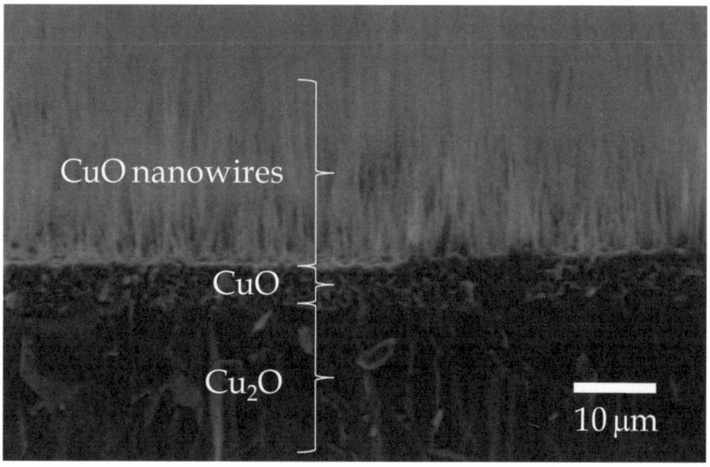

Fig. 11. Cross section FE-SEM image of CuO nanowires by heating copper plate at 600°C. Three layers can be observed for Cu_2O, CuO and CuO nanowires, respectively.

8. Applications device based on Metal-oxide nanowires

Metal-oxide nanowires are already widely used and indeed a key element in many industrial manufacturing processes. Recently, ZnO nanowires and CuO nanowires can be successfully applied as gas sensor and dye-sensitized solar cell (DSSC) (Choopun et al., 2009, Raksa et al., 2009). There are several reports that gas sensors and DSSC based on ZnO and CuO nanowire exhibited better performance than that of bulk material. For example, For gas sensor applications, normally, sensor response or the sensitivity of sensor strongly depends on the surface morphology of sensor material. From our report, nanowires were prepared by thermal oxidation of Zn powder on alumina substrate under normal atmosphere at various temperature. From sensor results, it was found that the sensitivity of nanowires sensor is improved compared with bulk sample (Wongrat et al., 2009).

9. Conclusions

This article reviews the metal-oxide nanowires among the group of ZnO and CuO which have been successfully synthesized by thermal oxidation technique. In addition, we have proposed a possible growth mechanism that may occur in the formation of nanowire. The growth mechanism can be explained in term of the Gibb free energy of oxidization process. Four steps of growth mechanism was proposed and discussed including oxygen adsorption, surface oxidization to form nuclei, nuclei arrangement and nanowire formation is considered. This growth model can be explored to explain other metal-oxide.

10. Acknowledgments

This work was supported by Thailand Research Fund (TRF). Also, we would like to thank P. Raksa, Department of Physics and Materials Science, Chiang Mai University for useful discussion on the parts of CuO nanowires. N. Hongsith would like to acknowledge the Development and Promotion of Science and Technology Talents Project (DPST) scholarship and E. Wongrat would like to acknowledge the Graduate School, Chiang Mai University for financial support.

11. References

Barsoum, M.W. (2003). *Fundamentals of ceramics*, IOP Publishing Ltd, ISBN 0-7503-0902-4, Institute of Physics Publishing Bristol and Philadelphia.

Belousov, V.V. (2007). Mechanisms of Accelerated Oxidation of Copper in the Presence of Molten Oxides. *Oxid. Met.*, 67, 235-250, ISSN 0030-770X.

Björk, M.T., Ohlsson, B.J., Sass, T., Persson, A.I., Thelander, C., Magnusson, M.H., Deppert, K., Wallenberg, L.R. & Samuelson, L. (2002). One-dimensional heterostructures in semiconductor nanowhiskers. *Appl. Phys. Lett.*, 80, 1058-1060, ISSN 0003-6951.

Chan, Y.F., Duan, X.F., Chan, S.K., Sou, I.K., Zhang, X.X. & Wang, N. (2003). ZnSe nanowires epitaxially growth on GaP(111) substrates by molecular-beam epitaxy. *Appl. Phys. Lett.*, 83, 2665-2667, ISSN 0003-6951.

Chen, J.T., Zhang, F., Wang, J., Zhang, G.A., Miao, B.B., Fan , X.Y., Yan, D. & Yan, P.X. (2008). CuO nanowires synthesized by thermal oxidation route. *J. Alloy. Compd.*, 454, 268-273, ISSN 0925-8388.

Choopun, S., Hongsith, N., Wongrat, E., Kamwanna, T., Singkarat, S., Mangkorntong, P. & Mangkorntong, N. (2008). Growth Kinetic and Characterization of RF-Sputtered ZnO:Al Nanostructures. *J. Am. Ceram. Soc.*, 91, 174-177, ISSN 0002-7820.

Choopun, S., Tabata, H. & Kawai, T. (2005). Self-assembly ZnO nanorods by pulsed laser deposition under argon atmosphere. *J. Crystal Growth*, 274, 167-172, ISSN 0022-0248.

Choopun, S., Tubtimtae, A., Santhaveesuk, T., Nilphai, S., Wongrat, E. & Hongsith, N. (2009). Zinc OxideNaonostructures for Applications as Ethanol Sensors and Dye-sensitized Solar Cells. *Appl. Surf. Sci.*, Article in press, ISSN 0169-4332.

Coleman, V.A. & Jagadish, C. (2006). Basic Properties and Applications of ZnO, In: *Zinc Oxide Bulk, Thin films and Nanostructures, Processing, Properties and Applications*, Jagadish, C. & Pearton, S. (Ed.), 1-20, Elsevier Limited, ISBN 0-08-044722-8, Radarweg 29 PO. Box 211, 1000 AE Amsterdam, The Netherlands.

Dai, Z. R., Pan, Z.W. & Wang, Z.L. (2003). Novel nanostructures of functional oxides synthesized by thermal evaporation. *Adv. Funct. Mater.*, 13, 9-24, ISSN 1616-301X.

Ellmer, K. & Klein, A. (2008). ZnO and Its Applications, In: *Transparent conductive ZnO: basics and applications in thin film solar cells*, Ellmer, K., Klein, A. & Rech, B., (Ed.), 1-33, Springer-Verlag Berlin Heidelberg, ISBN 3-540-73611-0, Springer Berlin Heidelberg Newyork.

Fan, H.J., Scholz, R., Kolb, F.M., & Zacharias, M. (2004). Two-dimensional dendritic ZnO nanowires from oxidation of Zn microcrystals. . *Appl. Phys. Lett.*, 85, 4142-4144, ISSN 0003-6951.

Hansen, B. J., Lu, G. & Chen, J. (2008). Direct Oxidation Growth of CuO Nanowires from Copper-Containing Substrates. *J. Nanomater.*, XXX, XXX-XXX, ISSN 1687-4110.

Henrich, V.E. & Cox, P.A. (1994). *The surface science of metal oxides*, The Press syndicate of the University of Cambridge, ISBN 0-521-44389-X, The Pitt Building, Trumpington Street, Cambridge.

Henrich, V.E. (2001). Metal oxide surfaces and interfaces: concepts and issues, In: *Oxide Surfaces*, Woodruff, D.P., (Ed.), 1-34, Elsevier Science B.V., ISBN 0-444-50745-0, Sara Burgerhartstraat 25 RO. Box 211, 1000 AE Amsterdam, The Netherlands.

Hirate, T., Sasaki, S., Li, W., Miyashita, H., Kimpara, T. & Satoh, T. (2005). Effects of laser-ablated impurity on aligned ZnO nanorods grown by chemical vapor deposition. *Thin Solid Films*, 487, 35-39, ISSN 0040-6090.

Hsueh, T.J. & Hsu, C.L. (2008). Fabrication of gas sensing devices with ZnO nanostructure by the low-temperature oxidation of zinc particles. *Sens. Actuators B.*, 131, 572-576, ISSN 0925-4005.

Jiang, X., Herricks, T. & Xia, Y. (2002). CuO Nanowires Can Be Synthesized by Heating Copper Substrates in Air. *Nano Lett.*, 2, 1333-1338, ISSN 1530-6984.

Jiang, X., Jia, C.L. & Szyszka, B. (2002). Manufacture of specific structure of aluminum-doped zinc oxide films by pattering the substrate surface. *Appl. Phys. Lett.*, 80, 3090-3092, ISSN 0003-6951.

Jie, J. S., Wang, G. Z., Han, X. H., Yu, Q., Liao, Y., Li, G. P. & Hou, J. G. (2004). Indium-Doped Zinc Oxide Nanobelts. *Chem. Phys. Lett.*, 387, 466-470, ISSN 0009-2614.

Kim, T.W., Kawazoe, T., Yamazaki, S., Ohtsu, M. & Sekiguchi, T. (2004). Low-temperature orientation-selective growth and ultraviolet emission of single-crystal ZnO nanowires. *Appl. Phys. Lett.*, 84, 3358-3360, ISSN 0003-6951.

Liang, H.Q., Pan, L.Z. & Liu, Z.J. (2008). Synthesis and photoluminescence properties of ZnO nanowires and nanorods by thermal oxidation of Zn precursors. *Mater. Lett.*, 62, 1797-1800, ISSN 0167-577X.

Lyu, S.C., Zhang, Y., Ruh, H., Lee, H.J., Shim, H.W., Suh, E.K., & Lee, C.J. (2002). Low temperature growth and photoluminescence of well-aligned zinc oxide nanowires. *Chem. Phys. Lett.*, 363, 134-138, ISSN 0009-2614.

Manmeet Kaur, Muthe, K.P., Despande, S.K., Shipra Choudhury, Singh, J.B., Neetika Verma, Gupta, S.K. & Yakhmi, J.V. (2006). Growth and branching of CuO nanowires by thermal oxidation of copper. *J. Crystal Growth*, 289, 670-675, ISSN 0022-0248.

Martin, M. & Fromm, E. (1997). Low-temperature oxidation of metal surfaces. *J. Alloy. Compd.*, 258, 7-16, ISSN 0925-8388.

Nguyen, D. H., Nguyen, V. Q. & Nguyen, V. H. (2009). Facile synthesis of p-type semiconducting cupric oxide nanowires and their gas sensing properties. *Physica E*, Article in Press, ISSN 1386-9477.

Raksa, P., Gardchareon, A., Chairuangsri, T., Mangkorntong, P., Mangkorntong, N. & Choopun, S. (2009). Ethanol sensing properties of CuO nanowires prepared by an oxidation reaction. *Ceram. Int.*, 35, 649-652, ISSN 0272-8842.

Raksa, P., Nilphai, S., Gardchareon, A., Choopun, S. (2009). Copper oxide thin film and nanowire as a barrier in ZnO dye-sensitized solar cells. *Thin Solid Films*, 517, 4741-4744, ISBN 0040-6090.

Ren, S., Bai, Y.F., Jun Chen, Deng, S.Z., Xu, N.S., Wu, Q.B. & Shihe Yang (2007). Catalyst-free synthesis of ZnO nanowire arrays on zinc substrate by low temperature thermal *oxidation*. *Mater. Lett.*, 61, 666-670, ISSN 0167-577X.

Ronning, C., Gao, P. X., Ding, Y., Wang, Z. L. & Schwen, D. (2004). Manganese-Doped ZnO Nanobelts for Spintronics. *Appl. Phys. Lett.*, 84, 783-795, ISSN 0003-6951.

Schroeder, P., Kast, M., Halwax, E., Edtmaier, C., Bethge, O. & Brückl1, H. (2009). Morphology alterations during postsynthesis oxidation of Zn nanowires. *J. Appl. Phys.*, 105, 104307, ISSN 0021-8979.

Schubert, L., Werner, P., Zakharov, N.D., Gerth, G., Kolb, F.M., Long, L. & Gösele, U. (2004). Silicon nanowhiskers grown on <111> Si substrates by molecular-beam epitaxy. *Appl. Phys. Lett.*, 84, 4968-4970, ISSN 0003-6951.

Sekar, A., Kim, S.H., Umar, A. & Hahn, Y.B. (2005). Catalyst-free synthesis of ZnO nanowires on Si by oxidation of Zn powders. *J. Crystal Growth,* 277, 471-478, ISSN 0022-0248.

Shewmon, P. (1989). *Diffusion in solids*, A Publication of The Minerals, Metal & Materials Society, ISBN 0-87339-105-5, 420, Commonwealth Drive Warrendale, Pennsylvania.

Smith, D.L. (1995). *Thin-Film Deposition: principles and Practice,* McGraw-Hill, Inc., ISBN 0-07-058502-4, New York, USA.

Stolen, S., Grande, T. & Allan, N.L. (2004). *Chemical thermodynamics of materials Macroscopic and microscopic aspects,* John Wiley & Sons Ltd., ISBN 0-4714-9230-2, The Atrium, Southern Gate, Chichester West Sussex PO19 8SQ, England.

Wagner, C. & Grunewald, K. (1938). *Z. Phys. Chem.* (B), 40, 455, ISSN xxxx-xxxx.

Wang, N., Cai, Y. & Zhang, R.Q. (2008). Growth of nanowires. *Mater. Sci. Eng. R-Rep.,* 60, 1-51, ISSN 0927-796X.

Wongrat, E., Pimpang, P. & Choopun, S. (2009). Comparative study of ethanol sensor based on gold nanoparticles: ZnO nanostructure and gold: ZnO nanostructure. *Appl. Surf. Sci.,* Article in press, ISSN 0169-4332.

Wu, Z.H., Mei, X.Y., Kim, D., Blumin, M. & Ruda, H.E. (2002). Growth of Au-catalyzed ordered GaAs nanowire arrays by molecular-beam epitaxy. *Appl. Phys. Lett.,* 81, 5177-, ISSN 0003-6951.

Yu, W., & Pan, C. (2009). Low temperature thermal oxidation synthesis of ZnO nanoneedles and the growth mechanism. *Mater. Chem. Phys.,* 115, 74-79, ISSN 0254-0584.

Zeng, J., Xu, J., Wang, S., Tao, P. & Hua, W. (2009). Ferromagnetic behavior of copper oxide-nanowire-covered carbon fibre synthesized by thermal oxidation. *Mater. Charact.,* 60, 1068-1070, ISSN 1044-5803.

Zhou, G. (2009). Nuleation thermodynamics of oxide during metal oxidation. Appl. *Phys. Lett.,* 94, 201905, ISSN 0003-6951.

6

p-type Phosphorus Doped ZnO Wires for Optoelectronic Applications

B. Q. Cao[2,1], M. Lorenz[1], G. Zimmermann[1], C. Czekalla[1], M. Brandt[1], H. von Wenckstern[1], and M. Grundmann[1]
[1]*Institut für Experimentelle Physik II, Universität Leipzig, Leipzig,*
[2]*School of Materials Science and Engineering, University of Jinan, Shandong,*
[1]*Germany*
[2]*China*

1. Introduction

Semiconductor nanowires are especially attractive building blocks for assembling active and integrated nanosystems since the individual nanostructures can function as both device elements and interconnects. Among wide band gap II-VI semiconductors, ZnO seems to be one of the most promising materials for optoelectronic applications. This is due to its stable excitons, having a large binding energy of 60 meV [Ellmer et al., 2008, Jagadish & Pearton, 2006], which is important for applications of UV light-emitting devices and laser diodes with high efficiency. Therefore, growth of p-type conductive ZnO material is a prerequisite step since ZnO is intrinsically n-type [Look et al., 2001]. The growth of semiconductor nanowires with reproducible electronic properties, including the controlled incorporation of n-type and/or p-type dopants, has been realized in silicon, indium phosphide, and gallium nitride [Lieber & Wang, 2007]. In comparison with the semiconductors mentioned above, doping of ZnO seems more difficult not only for wires but films and bulk crystals due to the low dopant solubility and the self-compensation effect of intrinsic defects [Park et al., 2002]. It is also desired to more deeply understand the underlying doping physics [Zhang et al., 2001] for the achievement of a high-quality compound-semiconductor p-n junction.

Until now, there are only few reports on doped ZnO nanowires for p-type conductivity, possibly due to the obvious difficulties in both growth and optical/electrical characterization. Liu et al. [2003] prepared boron-doped ZnO nanowires and ZnO:B/ZnO nanowire junction arrays by a two-step vapor transport method in pores of anodic aluminum oxide membrane. A p-n junction-like rectifying behavior was observed. Lin et al. [2005] also reported a p-n rectification behavior of nitrogen doped ZnO (ZnO:N) nanowires/ZnO film homojunctions, where the ZnO:N nanowires were prepared by a post-growth NH_3 plasma treatment. Both reports did not contain any optical or electrical characterization of single (probably p-type) doped ZnO nanowires. Lee et al. [2004] prepared arsenic-doped ZnO nanowires by post annealed ZnO nanowires grown on GaAs substrate and observed arsenic induced acceptor bound exciton emission at 3.358 eV. Shan et al. [2007] reported the preparation of arsenic and phosphorus (co)doped ZnO nanowires also by an annealing method using ZnSe nanowires and GaAs/InP substrates as precursor and dopants, respectively. Recently, Xiang et al. [2007]

reported the p-type ZnO nanowire growth with chemical vapor deposition using phosphorus as dopants. The p-type conductivity of the doped wires was illustrated with gate-voltage dependent conductance measurements of single ZnO:P nanowires with field effect transistor (FET) configuration. However, p-type conductivity was not stable and converted to n-type three months later. Similarly, Yuan et al. [2008] reported the growth by chemical vapor deposition (CVD) of nitrogen doped ZnO wires which showed p-type FET characteristics. Lu et al. [2009] reported the growth of p-type ZnO nanowires also with CVD using Zn_3P_2 as dopant. Moreover, energy conversion using the p-type ZnO NWs has also been demonstrated and the p-type ZnO NWs produce positive output voltage pulses when scanned by a conductive atomic force microscope (AFM) in contact mode while the n-type nanowires produced a negative voltage signal. Consequently, it seems that doped p-type ZnO wires come into reality. But, the doping induced optical fingerprints were not clearly identified and p-n junctions built from p-type ZnO nanowires are still not available.
In this Chapter, after the brief review of the research progress on p-type ZnO wires, some important results obtained in our group on p-type ZnO nanowires and microwires are summarized. In the second section, the high-pressure pulsed laser deposition (PLD) and carbothermal evaporation growth methods for doped wire growth will be described. Optical and electrical characterizations of the doped wires will be discussed in detail in the third section. This Chapter ends with a brief summary, which also includes our personal remarks on future research of p-type ZnO wires.

2. Growth of phosphorus doped ZnO wires

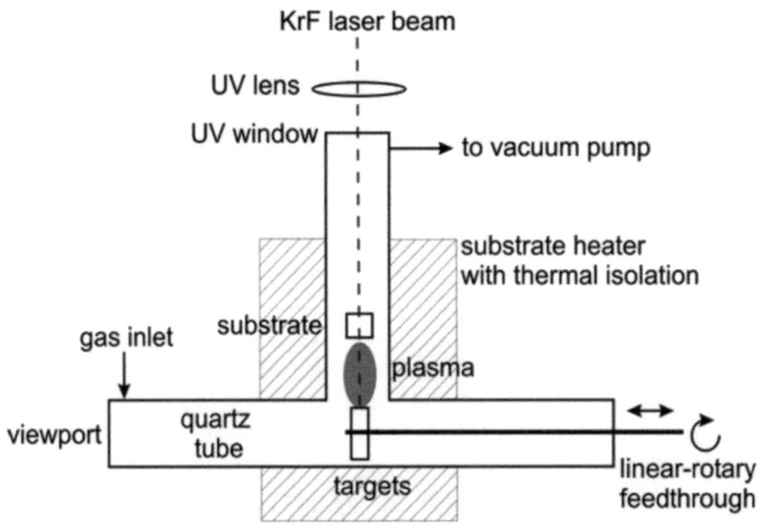

Fig. 1. Schematic illustration of the high-pressure PLD chamber for pure and doped ZnO nanowire growth with a T-shaped quartz tube. From Ref. [Lorenz et al., 2005].

Two methods, high-pressure pulsed laser deposition and carbothermal evaporation, were applied to grow phosphorus doped ZnO wires with diameter varied from 100 nm to 10μm.
PLD is a well established growth method for thin films by condensation of laser plasma ablated from a target, excited by the high-energy laser pulses far from equilibrium [Willmott

& Huber, 2000]. Figure 1 shows the scheme of the high-pressure PLD chamber specially designed for ZnO nanostructures using a T-shaped quartz tube with an outer diameter of 30 mm in the Leipzig semiconductor physics group [Lorenz et al., 2005]. A KrF excimer laser beam enters along the center of the tube and is focused on the cylindrical surface of the selected targets. The laser energy density on the target is about 2 J/cm^2, which is similar to conventional PLD film growth conditions. An encapsulated heater with an arrangement of KANTHAL wire in ceramic tubes and FIBROTHAL isolation material is built around the T-shape quartz tube. The growth temperature is usually chosen between 500 and 950 °C as measured by a thermocouple. Argon flow of 0.05 to 0.2 l/min was usually selected as carrier gas and the growth pressure ranged between 25 and 200 mbar. The a- or c-plane sapphire substrates (size 1×1 cm^2) were arranged of-axis, i.e. parallel to the expanding plasma plume.

PLD generally facilitates stoichiometric transfer of the chemical composition of a multielement source target into the grown samples [Chrisey & Hubler, 1994]. Therefore, if the ablated target is doped, the corresponding samples are expected to have the similar composition to the targets. For phosphorus doping of ZnO nanowires, phosphorus pentoxide (P_2O_5) was select as dopants for acceptors. Figure 2 shows a scanning electron microscopy (SEM) image of ZnO wires using doped ZnO targets with different P_2O_5 concentration. Figure 3 shows typical energy dispersive x-ray spectroscopy (EDX) results detected from a single doped nanowire. The phosphorus signal is clearly observed. The

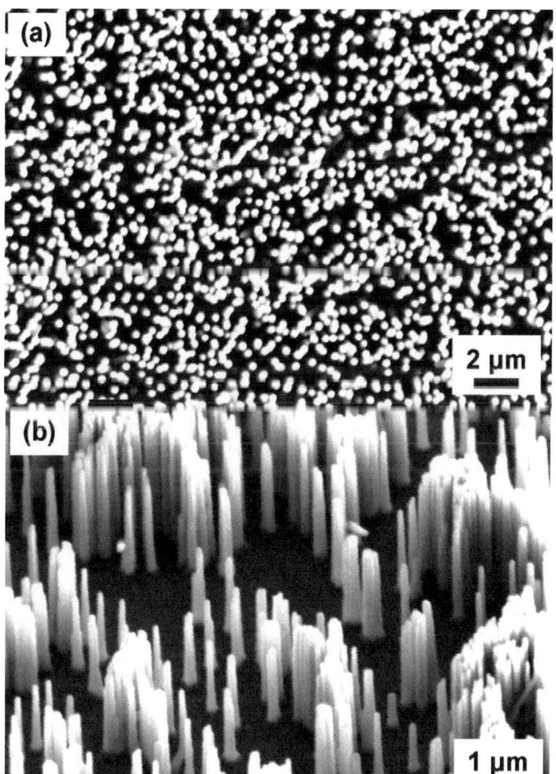

Fig. 2. SEM images of ZnO:P wire grown with PLD using (a) 1 wt% P_2O_5 and (b) 2 wt% P_2O_5 doped ZnO targets at 100 Torr and 900 °C.

weight percentage of phosphorous in the nanowire is about 1 wt%, which is smaller than the phosphorus concentration (1.7 wt%) in the target. This is probably caused by the different vapor pressures of zinc and phosphorus related species in the laser ablated plume and different sticking coefficients [Ohtomo et al., 1998].

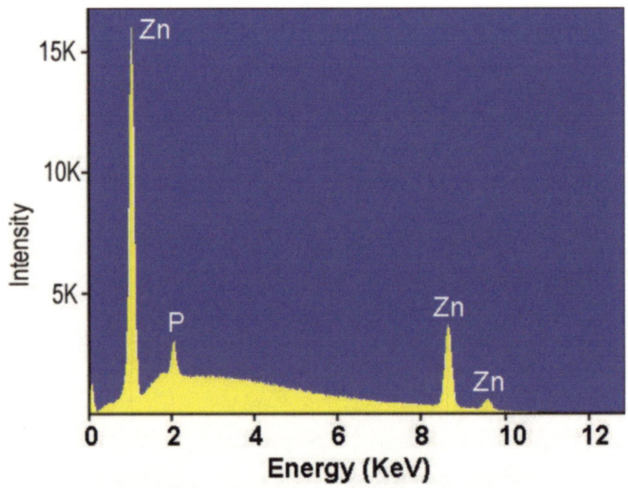

Fig. 3. Energy dispersive x-ray spectrum of a single doped ZnO:P wire with obvious phosphorus signal, excited by electron beam in a SEM.

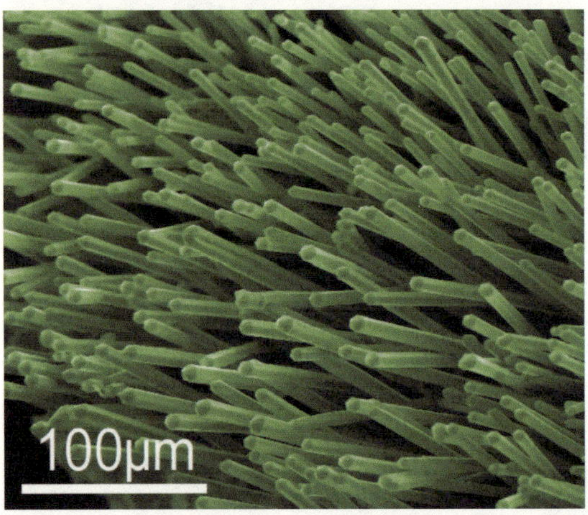

Fig. 4. A typical SEM image of doped ZnO microwires grown directly on pressed source material target. From Ref. [Cao et al., 2008].

In our group, vapor phase transport and deposition process was also employed for the growth of ZnO nanowires and microwires, as proposed by Huang and Yao [Huang et al., 2001; Yao et al., 2002]. This method, also called carbothermal evaporation, is widely used for growth of nanowires, and it has relations to other methods, such as chemical vapor deposition (CVD), metal-organic chemical vapor deposition (MOCVD), physical vapor deposition (PVD). Carbothermal evaporation has been conventionally used to synthesize one-dimensional and other complex oxide nanostructures, including ZnO as reviewed recently by Wang et al. [2009]. Here, the standard process was developed to synthesize doped ZnO microwires by direct thermal evaporation of a pressed ZnO/graphite (mass ratio 1:1) target doped with P_2O_5 (15 wt%) at ambient pressure. The growth temperature was as high as 1100 °C and the microwires grew directly on the pressed target. Figure 4 shows a typical SEM image of doped ZnO microwires produced by this method. Such microwires are easily visible with optical microscopy and, therefore, are more convenient for device fabrications.

3. Optical and electrical properties of ZnO:P wires

3.1 Optical properties of ZnO:P wires

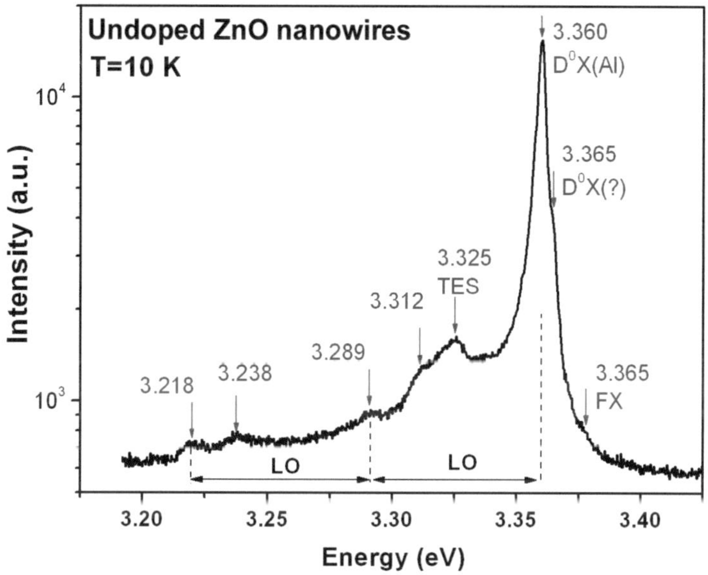

Fig. 5. CL spectrum of undoped ZnO nanowires measured at a temperature of 10 K.

To study the doping effect on the physical properties of ZnO nanowires and also identify the conductivity type of the doped nanowires, optical properties were first investigated by cathodoluminescence (CL) spectroscopy. More detailed description of the CL measurement system can be found in Ref. [Nobis et al., 2004].

A low temperature CL spectrum at T=10 K of undoped ZnO nanowires grown with PLD using pure ZnO target is shown in Figure 5. It is dominated by an emission peak at 3.360 eV (I_6) originating from the aluminum-related neutral donor-bound excitons (D^0, X) [Meyer et al., 2004]. On the high energy side of this peak, the minor peaks at 3.365 and 3.377 eV are the I_{3a} donor-bound exciton and the free-exciton (FX) recombination, respectively. The peak at 3.325 eV is due to two-electron satellite (TES) emission. The weak peaks in the lower energy range from 3.20 eV to 3.33 eV are the optical phonon replicas of the D^0X.

Figure 6(a) depicts the CL spectra of a single ZnO:P nanowire measured at three points as indicated, which show similar spectroscopic features. The spectra are clearly different from that of undoped ZnO nanowires. Three groups of new emission peaks at 3.356, 3.314, 3.234

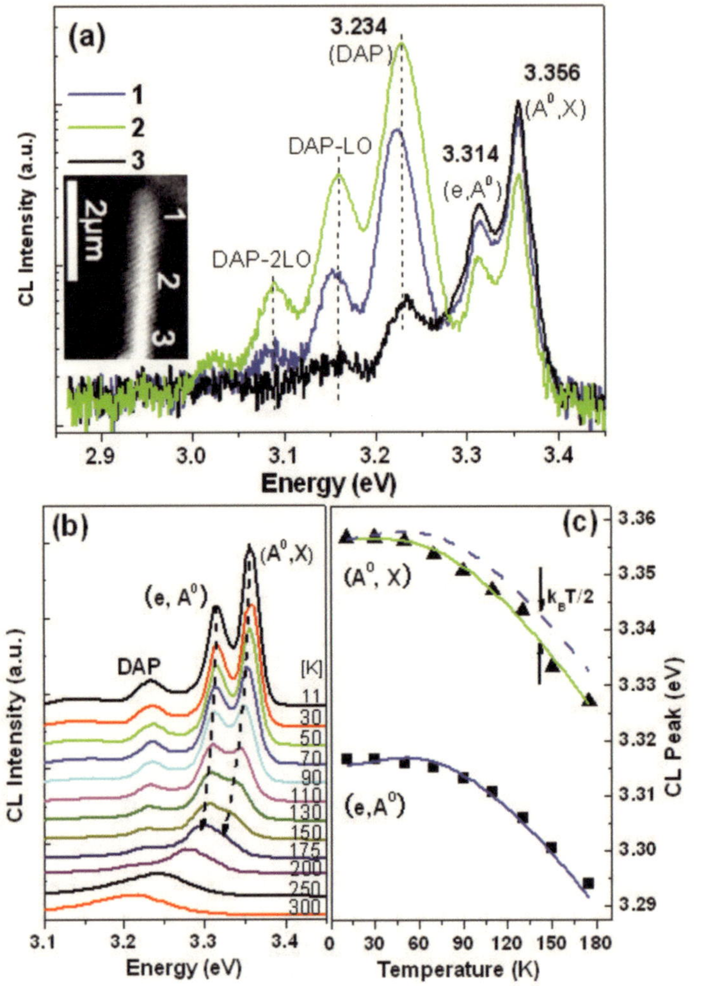

Fig. 6. (a) CL spectra at temperature of 10 K of a single ZnO:P nanowires measured at three parts, as indicated in the inset. (b) Temperature-dependent CL spectra of ZnO:P nanowires. (c) CL peak energies versus temperatures taken from (b). Solid symbols (■ and ▲) are experimental data and solid lines are fitted with equations [1] and [2], respectively. The dashed line is shifted to demonstrate the slope of the (e, A^0) fitted line (blue) near the (A^0, X) data points. From Ref. [Cao et al., 2007].

eV and their phonon replica were detected. It indicates that these emission peaks are due to the intentional phosphorous doping. The peak at 3.356 eV can be ascribed to the phosphorus-related neutral acceptor-bound exciton (A^0, X) emission [Hwang et al., 2005, Vaithianathan et al., 2005]. As for the peak at 3.314 eV, it is a prominent optical characteristic of most p-type ZnO material doped with group V elements. But, its origin is still a matter of debate [Look, 2005]. The origin of the dominant peak at 3.234 eV is also controversial. But, it shows strong longitudinal phonon replica that means that it comes from acceptor-related recombination process due to the phosphorus doping of ZnO. Because, for donor-related lines, only weekly coupling phonon lines can be observed, as shown in Figure 5. Therefore, the peaks at 3.314 and 3.234 eV could be tentatively ascribed respectively to free electron-to-acceptor (e, A^0) emission and donor-to-acceptor pair (DAP) emission in agreement with the discussion of nitrogen doped ZnO [Xiong, 2005].

Temperature-dependent luminescence behavior usually can prove the origin of the near-band-gap emissions of semiconductors. Figure 6(b) shows the temperature-dependent CL spectra of ZnO:P nanowires. The peak of (A^0, X) shows a continuous redshift due to the bandgap shrinkage with increasing temperature, which can be well fitted by the Bose-Einstein model [O'Donnell & Chen, 1991],

$$E_{A^0,X}(T) = E_{A^0,X}(0) - 2\alpha_B\Theta_B[\coth\left(\frac{\Theta_B}{2T}\right) - 1], \quad (1)$$

with $E_{A^0,X}(0)$=3.356 eV, a_B=1.2(±0.2)*10^{-4} eV/K, and Θ_B=306(±45) K. Its intensity decreases gradually, and eventually vanishes above 175 K due to the ionization of excitons from the acceptor. In this process, no emission of free exciton (FX) can be observed due to the dissociation of free excitons into free electrons and holes caused by the potential fluctuations of the band edges. It suggests that a high concentration of phosphorous dopants has been incorporated. The DAP peak intensity also decreases gradually with increasing temperature and finally disappears above 200 K. This temperature-dependent behavior is typical for the thermal ionization of shallow donors via the pathway of (D^0, A^0) → D^+ + A^0 + e. The ionized free electrons in the conduction band prefer to recombine with acceptors. As a result, the spectra are dominated by (e, A^0) at temperatures of 130-175 K. Figure 6 (c) also depicts the peak energy of (e, A^0) can be fitted with

$$E_{e,A^0}(T) = E_g(T) - E_A + \frac{1}{2}k_BT, \quad (2)$$

where E_{e,A^0}(T), E_g(T), and E_A are the peak position of (e, A^0), band gap, and binding energy of the acceptor, respectively. The fitting coefficients α_B and Θ_B were taken from the fitted values of (A^0, X) with Eq. (1). The fitted E_A of phosphorus acceptor is 122 (±1) meV. It is noted that the temperature dependence of (A^0, X) differs from that of (e, A^0) by $k_BT/2$, which is also the character of the free electron to acceptor transition [Ye et al., 2007]. In summary, the temperature-dependent CL results provide additional evidence by optical fingerprints of the phosphorus acceptor recombination in ZnO.

3.2 Electrical properties of ZnO:P wires

Homogeneous doped semiconductor nanowires represent key building blocks for a variety of electronic devices. After confirming the successful incorporation of phosphorus acceptors into the ZnO nanowires, the next step to identify the conductivity of the doped nanowires would be electrical measurements by device fabrication, which are always highly desirable. As direct Hall measurements on nanowires are difficult since the necessary two-dimensional configuration of the Hall contacts is hard to achieve, other practicable methods to obtain the conducting type of nanowires are adopted. In this section, p-n junction and field effect transistor (FET) structures were proposed and built with ZnO:P wires for electrical studies.

3.2.1 ZnO:P nanowires / ZnO:Ga film p-n homojunction

p-n junctions are of great importance both in modern electronic application and in understanding other semiconductor devices. A typical proposed two-terminal junction composed of doped nanowires and n-type ZnO:Ga film was shown in Figure 7(a). It was completely grown with PLD.

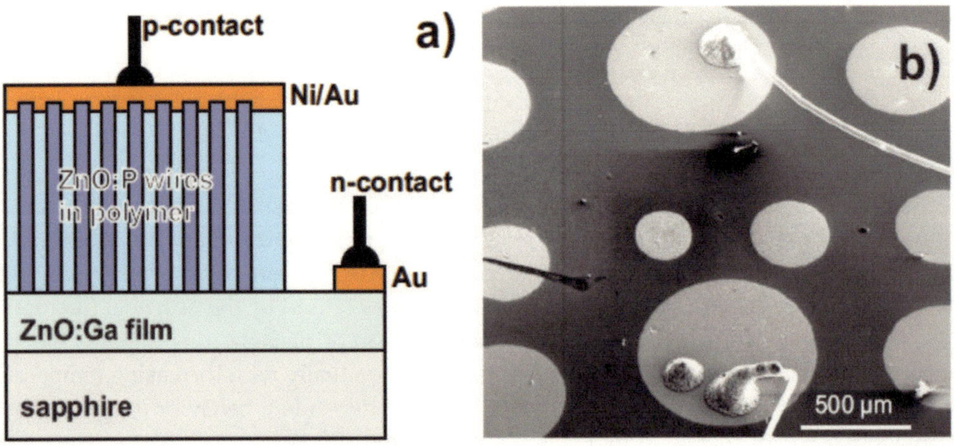

Fig. 7. (a) Scheme of the ZnO:P/ZnO:Ga p-n junctions including one p- and one n-contact. (b) A top view on the Ni–Au contact pads on top of the embedded ZnO:P nanowire array. The connection to the bond wires is made with silver glue. From Ref. [Lorenz et al., 2009].

Gallium-doped, n-type conducting ZnO thin films were grown with conventional low-pressure PLD using a ZnO:Ga (5 wt%) target. After deposition of the gold contact strip, these ZnO:Ga film templates were transferred into the high-pressure PLD chamber for growth of the ZnO:P nanowire arrays. The growth conditions were similar to that described in Section 2. After the nanowire growth, the samples were a spin-coated with polystyrene and then exposed to RF plasma at air pressure of 3×10^{-2} mbar to free the nanowires' tips from the polystyrene. Finally, Ni/Au top contacts were deposited with a mask on the top of the polystyrene and contacted with gold wire using conductive silver glue. Figure 7(b) shows a SEM image the top-view of the junction device.

Figure 8(a) shows the I-V curves measured from p_1-n and p_2-n, which are two parallel junction structures. The rectifying I-V behavior of the phosphorus doped wires on n-type ZnO film can be reproducibly observed from ten samples with over 50 p-n junctions. Moreover, the reverse parts of the I-V curves measured from p_1-n, and p_2-n agree reasonably well with the I-V curve of p_1-n-p_2, which is composed of two p-n junction diodes both under reverse bias. This is a strong indication for the p-type conductivity of the doped ZnO:P nanowires. The typical voltage for I=100 μA of the p-n junctions is up to 3.2 V and the forward to reverse current ratio is about 100 at ±3.5 V. The series resistance R_s of the p-n junction is about 23 kΩ. The low-voltage forward parts of the I-V curves were fitted with the equation $I=I_s\{\exp[e(V-R_sI)/nk_BT]-1\}$, i.e., the I-V characteristics of a real diode with negligible shunt resistance, as shown in Figure 8(b). The ideality factor n of the particular junction in Figure 8 was about 7.

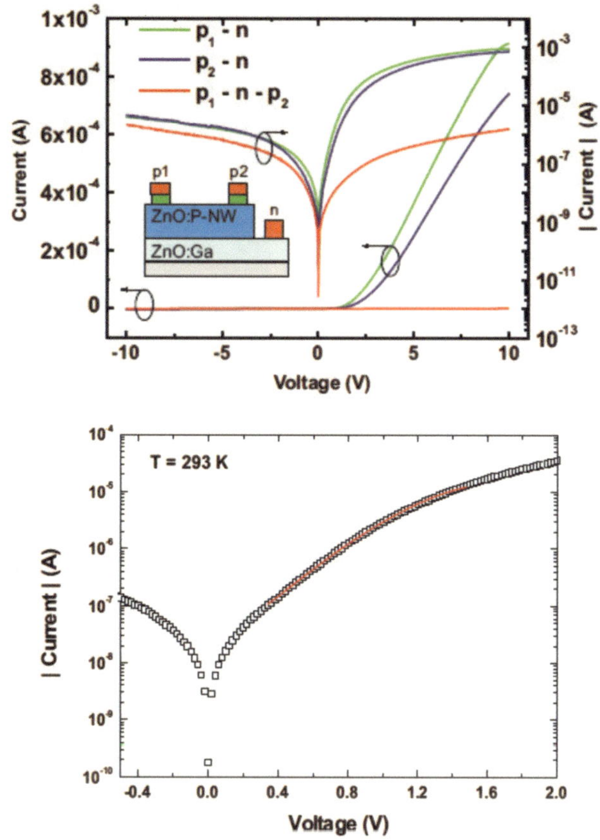

Fig. 8. (a) Rectifying I-V curves of two typical ZnO p-n junctions on the same substrate, denoted p_1-n and p_2-n, plotted in both linear and logarithmic current scale. The I-V curves of these two p-contacts is also shown, which corresponds to the p_1-n-p_2 configuration of two opposite p-n diodes. (b) I-V curve fitted with the equation, $I = I_s\{\exp[e(V-R_sI)/nk_BT]-1\}$ (red curve), to a typical p-n junction (open squares). The fit results are as follows: I_s=15.8 nA; R_s=23.0 kΩ; n=7.03. From Ref. [Lorenz et al., 2009].

3.2.2 p-type ZnO:P microwire FET

Another prototypical semiconductor device with broad applications is the nanowire FET. Nanowires configured as FETs have shown to operate at ultralow power below microwatts with enhanced operation speed. For example, studies of nanowire FETs fabricated from boron and phosphorus-doped Si nanowires have shown that the devices can exhibit performance comparable to the best reported value for planar devices made from the same materials [Cui et al., 2003]. Studies have also demonstrated the high electron mobility of epitaxial InAs nanowire FETs with a wrap-around cylindrical gate structure surrounding a nanowire [Thelander et al., 2008]. Moreover, the change in conductance of semiconductor wires as a function of gate voltage can be used to determine the conductive type of a given wire since the conductance will vary oppositely for increasing positive and negative gate voltages [Cui et al., 2003]. Therefore, the FET characteristics analyses were also used to study the doping effect of semiconductor nanowires.

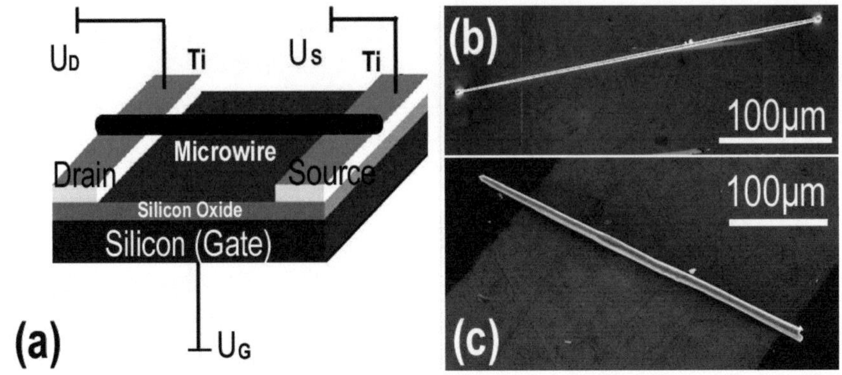

Fig. 9. (a) Schematic illustration of the back-gate FET configured with a microwire as conductive channel. (b) SEM image of a undoped ZnO microwire fixed on titanium electrodes with IBID. (c) SEM image of phosphorus doped ZnO microwire directly connecting with two titanium contacts.

Figure 9(a) schematically shows the FET structure assembled with a microwire. Silicon substrates (65 mΩcm) covered by a SiO_2 layer with thickness of 190 nm served as the global back-gate and dielectric gate oxide, respectively. Titanium strip contacts were then defined on such silicon substrates with photolithography. Then the microwires were transferred onto the predefined silicon substrate and laterally moved to connect them with titanium stripes, which formed the source and drain contacts. Figure 9(b) shows the SEM image of an undoped ZnO microwire FET, where the source and drain contact characteristics were further improved with tungsten contact pads deposited with ion-beam induced deposition (IBID). In this way, the contacts between ZnO wires and titanium strips are ohmic, as demonstrated in Figure 10(a) with the linear current-voltage dependence in a wide voltage range from -30 V to +30 V. The dependence of source-drain current (I_{SD}) through the ZnO wire on source-drain voltage (U_{SD}) was measured at various gate voltages, as shown in Figure 10(b). The I_{SD} increases with increasing U_{SD} and the slopes of the I_{SD} versus U_{SD} curves are dependent on the gate voltage (U_G). Figure 10(c) shows the corresponding transfer characteristics (I_{SD} - U_G) at a constant U_{SD} of 2 V. In general, the source-drain current increases with increasing gate voltage. This positive slope of the I_{SD} vs. U_G curve indicates that the undoped ZnO wire is n-type conducting. It has to be noted that usually the ZnO wire channel could not be fully depleted with such a back-gate configuration. A possible reason is the finite air gap between the gate oxide and the ZnO wire [Keem et al., 2007].

Because gallium is a donor in ZnO, the source and drain contacts of the doped ZnO:P microwire FETs were formed by connecting the ZnO:P wire directly to two separated titanium electrodes, as seen in Figure 9(c), without further FIB support [Weissenberger et al., 2007]. Therefore, the electrical characteristics of the ZnO:P wire-FETs are sensitively dependent on the particular contacts between the wire and electrodes. Figure 11 shows two groups of typical current-voltage (I_{SD} - U_{SD}) curves of ZnO:P wire-FETs under different back-gate voltages. All curves are slightly nonlinear, which indicates that the contacts between wire and titanium electrodes are not ideally ohmic. However, in both graphs of Figure 11 the I_{SD} - U_{SD} curves show an inverse arrangement concerning the gate voltage

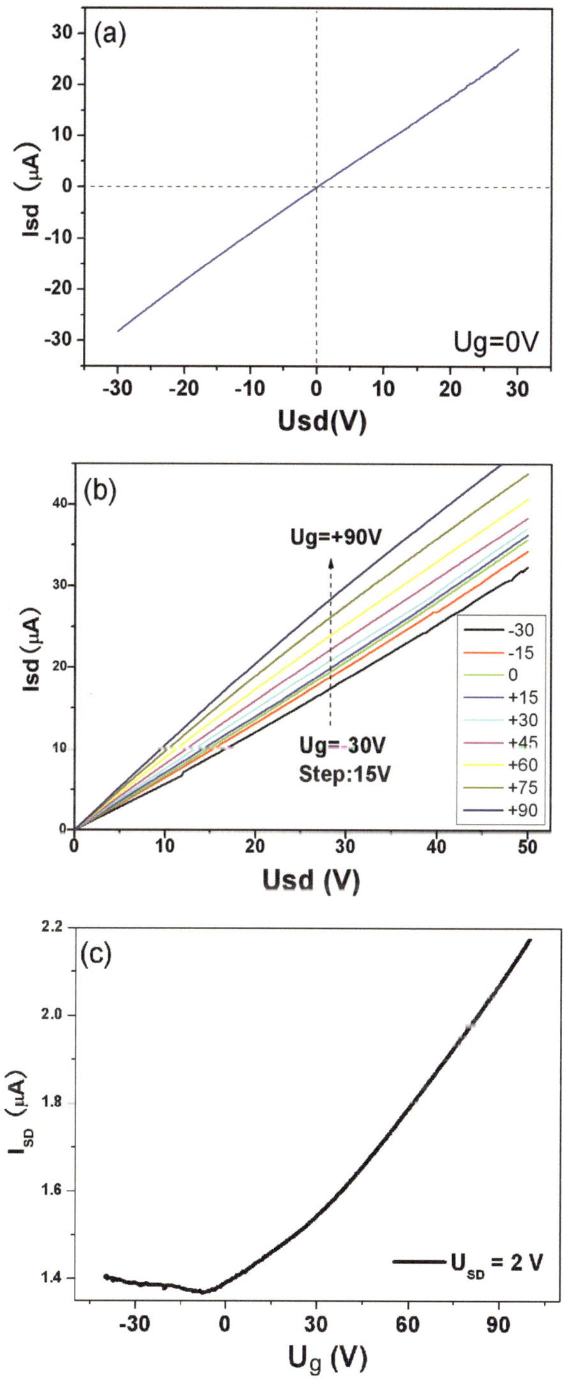

Fig. 10. (a) I_{SD} vs U_{SD} curve at U_G=0 V indicating good ohmic contacts between ZnO wires and titanium strips; (b) Output characteristics I_{SD} vs U_{SD} and (c) transfer characteristics I_{SD} vs U_g of the nominally undoped, n-type ZnO microwire FET. From Ref. [Cao et al., 2008].

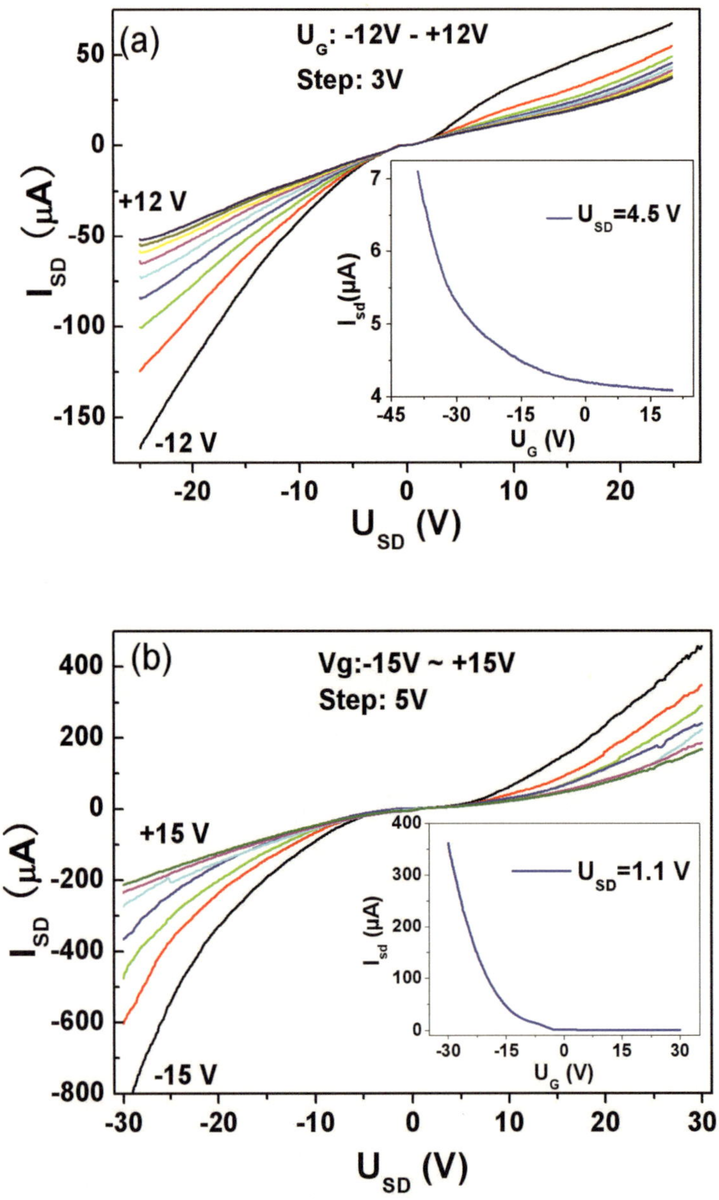

Fig. 11. Output characteristics (I_{SD} vs U_{SD}) and transfer characteristics (I_{SD} vs U_G, inset) of two different ZnO:P wire-FETs, indicating p-type conductivity of the doped ZnO microwires. From Ref. [Cao et al., 2008].

compared to that of the undoped n-type wire FET in Figure 10(b). This is further demonstrated by the insets of Figure 11, where the corresponding transfer characteristics of the ZnO:P wire FETs are shown. The source-drain current decreases with increasing gate voltage. This negative I_{SD}-U_G slope is opposite to that of the undoped n-type ZnO-FET. It indicates unambiguously that the conductivity of the phosphorus doped ZnO wires is p-type. In most cases, such ZnO:P wire-FETs could not be fully depleted as shown in Figure 11(a). However, Figure 11(b) is an example of a fully depleted ZnO:P microwire channel.

At last, the time stability of the p-type conductivity of such ZnO:P wires were investigated by repeated FET and p-n junction measurements. p-type conducting FET characteristics built with microwires grown by carbothermal evaporation were reproducibly observed within six months since their growth. No obvious change of the FET characteristics was found. Thus the time stability of p-type conductivity of the ZnO:P microwires is at least six months. As for the p-n junctions built with ZnO:P doped nanowires grown by PLD after one year, in total about 20 junction samples, about ten still showed the typical rectifying diode behavior as shown in Figure 8. Therefore, we can state a corresponding time stability of the ZnO junctions. Moreover, electroluminescence experiments with pulsed current excitation (6V/8mA peak current with 1 kHz and 10% duty cycle) on selected junctions showed a very weak, diffuse blue-green light emission. This light emission was visible to the naked eye in complete darkness.

4. Summary

In summary, we have reviewed the latest progress in the growth and optical/electrical characterizations of p-type doped ZnO wires, especially the work performed in the Leipzig group. Two growth methods, high-pressure PLD and carbothermal evaporation, have been adopted to grow phosphorus doped ZnO wires. Detailed optical and electrical characterizations of such doped wires indicated their p-type conductivity with one-year stability. In our opinion, more work is still needed to make further progress on this topic. First, the quality and stability of the p-type ZnO wires need to be further improved. This will require even better control the background n-type conductivity, development of new growth methods and search for new acceptor dopants. Second, high-quality p-n junctions built with single p-type ZnO nanowires showing good breakdown characteristics need to be demonstrated. Third, electrically pumped ZnO nanowire LED and laser diode showing strong band-edge emission are highly expected. Once these milestones are achieved, it would be the beginning of the ZnO applications for solid state lighting.

5. Acknowledgements

The work was financially supported by the European Union within STReP project NANDOS (Grant No. FP6-016924) and by the Deutsche Forschungsgemeinschaft within FOR 522. BQC also thanks the University of Jinan, for the start-up research fund for new faculty.

6. References

Cao B. Q.; Lorenz M.; Rahm A.; von Wenckstern H.; Czekalla C.; Lenzner J.; Benndorf G. & Grundmann M. (2007). Phosphorus acceptor doped ZnO nanowires prepared by pulsed-laser deposition, *Nanotechnology,* 18: 455707 (5 pages)

Cao B. Q.; Lorenz M.; Brandt M.; von Wenckstern H.; Lenzner J.; Biehne G. & Grundmann M. (2008) P-type conducting ZnO:P microwires prepared by direct carbothermal growth, *phys. stat. sol. (RRL)* 2(1): 37-39

Cao B. Q.; Lorenz M.; Brandt, von Wenckstern H.; Lenzner J.; Biehne G. & Grundmann M. (2008). p-type conducting ZnO:P microwires prepared by direct carbothermal growth, *phys. stat. sol. (RRL),* 2(1): 37-39.

Chrisey D. B. & Hubler G. K. (1994). *Pulsed laser deposition of thin films*, Wiley, ISBN: 978-0-471-59218-1, New York

Cui Y.; Zhong Z. H.; Wang D. L.; Wang W. U. & Lieber C. M. (2003). High performance silicon nanowire field effect transistors, *Nano Lett.* 3(2). 149-152.

Cui Y.; Duan X. F.; Hu J. T. & Lieber C. M. (2003). Doping and electrical transport in silicon nanowires, *J. Phys. Chem. B* 104(22). 5213-5216.

Ellmer K.; Klein A. & Rech B. (2008). *Transparent Conductive Zinc Oxide*, Springer, ISBN: 978-3-540-73611-0, Berlin, Heidelberg and New York

Huang M. H.; Wu Y. Y.; Feick H.; Tran N.; Weber E. & Yang P. D. (2001). Catalytic growth of zinc oxide nanowires by vapor transport, *Adv. Mater.* 13(2): 113-116

Hwang D. K.; Kim H. S.; Lim J. H.; Oh J.Y.; Yang J.H.; Park S. J.; Kim K. K.; Look D. C. & Park Y. S. (2005). Study of the photoluminescence of phosphorus-doped p-type ZnO thin films grown by radio-frequency magnetron sputtering, *Appl. Phys. Lett.* 86: 151917 (3 pages)

Jagadish C. & Pearton S. (2006). *Zinc Oxide Bulk, Thin films and Nanostructures: Processing, Properties and Applications,* Elsevier, ISBN: 978-0-08-044722-3, Amsterdam

Keem K. Y.; Kang J. M.; Yoon C. J.; Yeom D. H.; Jeong D. Y.; Moon B. M. & Kim S. S. (2007), A fabrication technique for top-gated ZnO nanowire field effect transistors by a photolithography process, *Microelectr. Eng.* 84: 1622-1626.

Lee W.; Jeong M. C. & Myoung J. M. (2004). Optical characteristics of arsenic-doped ZnO nanowires, *Appl. Phys. Lett.* 85(25): 6167 (3pages)

Lieber C. M & Wang Z. L. (2007). Functional Nanowires, *MRS Bulletin,* 32(Feb.): 99-108

Lin C. C.; Chen H. P. & Chen S. Y. (2005). Synthesis and optoelectrical properites of arrayed p-type ZnO nanorods grown on ZnO film/Si wafer in aqueous solutions, *Chem. Phys. Lett.* 404: 30-34

Liu C. H.; Yiu W. C.; Au. F. C. K. ; Ding J. X.; Lee C. S. & Lee S. T. (2003), Electrical properties of zinc oxide nanowires and intramolecular p-n junctions, *Appl. Phys. Lett.* 83(15): 3168 (3 pages)

Look D. C.; Claflin B.; Alivov Ya. I. & Park S. J. (2001). The future of ZnO light emitters, *phys. stat. sol. (a),* 201(10): 2203-2212

Look D. C. (2005), Electrical and optical properties of p-type ZnO, *Semicond. Sci. Technol.* 20: S55-S61

Lorenz M.; Kaidashev E. M.; Rahm A.; Nobis Th.; Lenzner J.; Wagner G.; Spemann D.; Hochmuth H. & Grundmann M. (2005). $Mg_xZn_{1-x}O$ ($0 \le x < 0.2$) nanowire arrays on sapphire grown by high-pressure pulsed-laser deposition, *Appl. Phys. Lett.* 86: 143113 (3 pages)

Lorenz M.; Cao B. Q.; Zimmermann G.; Biehne G.; Czekalla C.; Frenzel H.; Brandt M.; von Wenckstern H. & Grundmann M. (2009), Stable p-type ZnO:P nanowire/n-type

ZnO:Ga film junctions, reproducibly grown by two-step pulsed laser deposition, *J. Vac. Sci. Technol. B.* 27(3): 1693-1697.

Lu M. P.; Song J. H.; Lu M. Y. ; Chen M. T. ; Gao Y. F. ; Chen L. J. & Wang Z. L. (2009). Piezoelectric nanogenerator using p-type ZnO nanowire arrays, *Nano Lett.* 9(3): 1223-1227

Meyer B. K.; Alves H.; Hofmann D. M.; Kriegseis W.; Forster D.; Bertram F.; Christen J.; Hoffmann A.; Straßburg M.; Dworzak M.; Haboeck U. & Rodina A. V. (2004), Bound excitons and donor-acceptor pair recombinations in ZnO, *phys. stat. sol. (b)* 241(2): 231-260

Nobis T.; Kaidashev E. M.; Rahm A.; Lorenz M.; Lenzner J. & Grundmann M. (2004). Spatially inhomogenous impurity distribution in ZnO micropillars, *Nano Lett.* 4(5): 797-800.

O'Donnell K. P. & Chen X. (1991). Temperature dependence of semiconductor band gaps, *Appl. Phys. Lett.* 58: 2924 (3 pages)

Ohtomo A, Kawasaki M, Koida T, Masubuchi K, Koinuma H, Sakurai Y, Yoshida Y, Yasuda T and Segawa Y (1998). $Mg_xZn_{1-x}O$ as a II–VI widegap semiconductor alloy, *Appl.Phys. Lett.* 72: 2466 (3 pages)

Park C. H., Zhang S. B. & Wei S. H. (2002). Origin of p-type doping difficulty in ZnO: The impurity perspective, *Phys. Rev. B.* 66: 073202 (3 pages)

Shan C. X.; Liu Z.; Wong C. C. & Hark S. K. (2007). Doped ZnO nanowires obtained by thermal annealing, *J. Nanosci. Nanotech.* 7: 700-703

Thelander C.; Fröberg L. E.; Rehnstedt C.; Samuelson L. & Wernersson L. (2008). Vertical enhancement-mode InAs nanowire field-effect transistor with 50-nm wrap gate, *IEEE Electron Device Lett.* 29(3). 206-208.

Vaithianathan V.; Lee B. T. and Kim S. S. (2005). Pulsed-laser-deposited p type ZnO films with phosphorus doping, *J. Appl. Phys.* 98: 043519 (4 pages)

Wang Z. L. (2009). ZnO nanowire and nanobelt platform for nanotechnology, *Mater. Sci. Eng. R.* 64: 33-71.

Weissenberger D.; Mürrschnabel M.; Gerthsen D.; Pérez-Willard F.; Reiser A.; Prinz G. M.; Feneberg M.; Thonke K. & Sauer R. (2007). Conductivity of single ZnO nanorods after Ga implantation in a focused-ion-beam system, *Appl. Phys. Lett.* 91: 132110 (3 pages)

Willmott P. R. & Huber J. R. (2000). Pulsed laser vaporization and deposition, *Rev. Mod. Phys.* 72(1): 315-328

Xiang B.; Wang P. W.; Zhang X. Z.; Dayeh S. A.; Aplin D. P. R.; Soci C.; Yu D. P. & Wang D. L. (2007). Rational synthesis of p-type Zinc oxide nanowire arrays using simple chemical vapor deposition, *Nano Lett.* 7(2): 323-328

Xiong G.; Ucer K. B.; Williams R. T.; Lee J.; Bhattacharyya D.; Metson J. & Evans P. (2005). Donor-acceptor pair luminescence of nitrogen-implanted ZnO single crystal, *J. Appl. Phys.* 97: 043528 (4 pages).

Yao B. D.; Chan Y. F. & Wang N. (2002). Formation of ZnO nanostructures by a simple way of thermal evaporation, *Appl. Phys. Lett.* 81: 757 (3 pages).

Yuan G. D.; Zhang W. J.; Jie J. S.; Fan X.; Zapien J. A.; Leung Y. H.; Luo L. B.; Wang P. F.; Lee C. S. & Lee S. T. (2008). p-type ZnO nanowire arrays, *Nano Lett.* 8(8): 2591-2597

Ye J. D.; Gu S. L.; Li F.; Zhu S. M.; Zhang R.; Shi Y.; Zheng Y. D.; Sun X. W.; Lo G. Q. & Kwong D. L. (2007). Correlation between carrier recombination and p-type doping in P monodoped and In–P codoped ZnO epilayers, *Appl. Phys. Lett.* 90: 152108

Zhang S. B., Wei S. H. & Zunger A. (2001), Intrinsic n-type versus p-type doping asymmetry and the defect physics of ZnO, *Phys. Rev. B.* 63: 075205 (7 pages)

7

Lateral Surface Nanowires and Quantum Structures Based on ZnO

Hiroaki Matsui and Hitoshi Tabata
The University of Tokyo
Japan

1. Introduction

Zinc oxide (ZnO) has been important for the development of practical devices such as thin film transistors, magnetic semiconductors, transparent electrodes, and so on. ZnO has a large exciton energy of 60 meV, which raises the interesting possibility of utilizing excitonic effects at temperatures higher than 300 K (Thomas, 1960). Optically pumped UV stimulated emissions from ZnO layers have been demonstrated (Yu et al., 1997). Furthermore, $Mg_xZn_{1-x}O$ alloys are attracting a great deal of interest since they possess a higher band gap than ZnO (Sharma et al., 1999) and have been utilized for $Mg_xZn_{1-x}O$/ZnO multiple and single-quantum wells (Chen et al., 2000; Makino et al., 2000). These structures can form low-dimensional systems and produce interesting quantum phenomena such as an increased excitonic binding energy (Coli & Bajaj, 20001) and two-dimensional (2-D) electron transport aspects that contribute to both basic science and practical applications. (Tsukazaki et al., 2007).

A variety of nanostructures in semiconductor materials have been made and investigated. The number of papers concerning nanostructures in ZnO is increasing yearly. Self-organized techniques provide advantages for nanoscale engineering and have yielded many impressive results. Therefore, surface nanostructures in Si and GaAs have been fabricated using various growth mechanisms. Stranski-Krastanov (S-K) growth on lattice mismatched systems induces three-dimensional (3-D) nanodots on the 2-D wetting layers. Lateral surface nanowires have been fabricated by a step-faceting mode on vicinal surfaces (Schönher et al., 2001). These surface nanostructures have been developed for zero-dimensional (0-D) quantum dots and one-dimensional (1-D) quantum wires, respectively (Wang & Voliotis, 2006). Low-dimensional properties are currently receiving attention as advantages for optoelectronics with ZnO.

In epitaxial growth, lattice mismatch between an epilayer and a substrate plays a crucial role in epitaxy. Growth studies concerning ZnO epitaxy have been carried out using *c*- and *a*-sapphires (Vispute et al., 1997; Fons et al., 2000). Heteroepitaxial layers have a high dislocation density of 10^9 - 10^{10} cm^{-2} due to large mismatches in the lattice structure and thermal expansion (Viguè et al., 2001). The use of a ZnO substrate not only allows a reduction of the number of lattice defects involved in the epilayers, but also permits the selection of various growth directions without any lattice mismatch, which results in a direct understanding of growth dynamics. The growth polarity in ZnO is a primary factor. Zn (0001) and O (000-1) polarities have isotropic atom arrangements and possess spontaneous polarization along growth directions. On the other hand, the *M*-nonpolar (10-10) surface has an anisotropic atom structure, and the spontaneous polarization occurs parallel to a surface plane (Parker et al., 1998). For example, Zn-polar growth produces atomically flat surfaces due to a layer-by-layer mode (Kato et al., 2003; Matsui et al., 2004), whereas *M*-nonpolar ZnO layers result in anisotropic morphologies with a nanowires structure based on a step-edge barrier effect (Matsui & Tabata, 2005). Thus, the difference in growth directions influences the surface state, as well as optical and electrical properties in ZnO layers, which are more conspicuous through quantum structures (Matsui & Tabata, 2008). Quantum structures on various surface morphologies exhibit novel electronic and optical properties because quantized energy levels can be tailored by varying the geometric dimensions.

This chapter is organized as follows. In Section 2, we first give a description of polar and nonpolar growth on ZnO layers and outline a difference of surface nanowires between ZnO and GaAs systems. The surface nanowires on the *M*-nonpolar ZnO (10-10) layer surface are largely different from those on high-index GaAs layer surfaces, indicating that the growth origin of surface nanowires on ZnO results from a new bottom-up process. In Section 3, we introduce layer growth using a pulse laser ablation technique. In Section 4, we discuss the growth origin of surface nanowires on ZnO layer surfaces from various viewpoints. Sections 4 and 5 are devoted to reports of anisotropic optical and electrical properties that are remarkably modulated by the anisotropic surface morphology. Some concluding remarks and future research directions in this field are given in Section 5.

2. Difference in surface nanowires of ZnO and GaAs

2.1 Polar and nonpolar ZnO layer surfaces

ZnO has a hexagonal wurtzite structure (a = 0.325 nm, c = 0.5201 nm) in which each Zn^{2+} ion bonds by a tetrahedron of four O^{2-} ions, representing a structure that can be described as a number of alternating planes of Zn and O ions stacked along the *c*-axis [Fig. 1a]. Various surface-sensitive methods have been well used to investigate the polar surfaces in ZnO from fundamental and applied points of view. For example, the surface morphology was quite different for opposite polar surfaces when ZnO crystals were chemically etched (Mariano & Hanneman, 1963). Thus, epitaxy in ZnO with varying polarity should show different

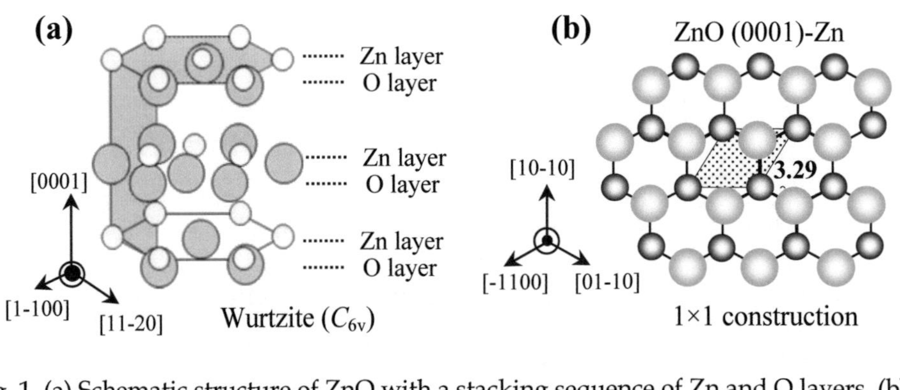

Fig. 1. (a) Schematic structure of ZnO with a stacking sequence of Zn and O layers. (b) Structural models showing the bulk-terminated Zn-polar (0001) surfaces of ZnO. The surface unit cells are indicated.

kinetics and material characteristics. Therefore, it is important to understand the uppermost surface structure and morphology in a Zn-polar surface. Figure 1(b) shows a structural model of the Zn-polar (0001) surfaces of ZnO. All O atoms on the borders have three nearest neighbours, i.e., only one bond is broken. The Zn-polar surface is unstable due to the existence of a non-zero dipole moment perpendicular to the surface, which raises a fundamental question regarding stabilization mechanisms.

Figure 2 shows the surface morphologies of Zn-polar, O-polar and *M*-nonpolar ZnO layers. The Zn-polar ZnO layer showed a very flat surface with a roughness of 0.21 nm [Fig. 2(a)]. The O-polar ZnO layer displayed a rough surface with a roughness of 4.2 nm and formed hexagonal islands [Fig. 2(b)]. The difference in surface morphology between Zn- and O-polar ZnO layers is similar to that for Ga- and N-polar GaN (0001) and Zn- and Se-polar ZnSe (111), which can be explained for in terms of the difference in the number of dangling bonds (Sumiya et al., 2000; Ohtake et al., 1998). In the case of Zn-polarity, the Zn atoms of ZnO molecules generated from the laser ablation are likely to be incorporated with less migration due to three dangling bonds. This suggests that Zn-polar growth should be dominated by a two-dimensional mode, resulting in very smooth surfaces. On the other hand, an O-polar surface has longer surface migration due to the single dangling bond and is adhered to the sites of the step edges with two dangling bonds. These result in hexagonal islands that originated from a spiral growth mode. The growth kinetics of O-polar growth has been suggested using molecular-dynamics crystal-growth simulations (Kubo et al., 2000).

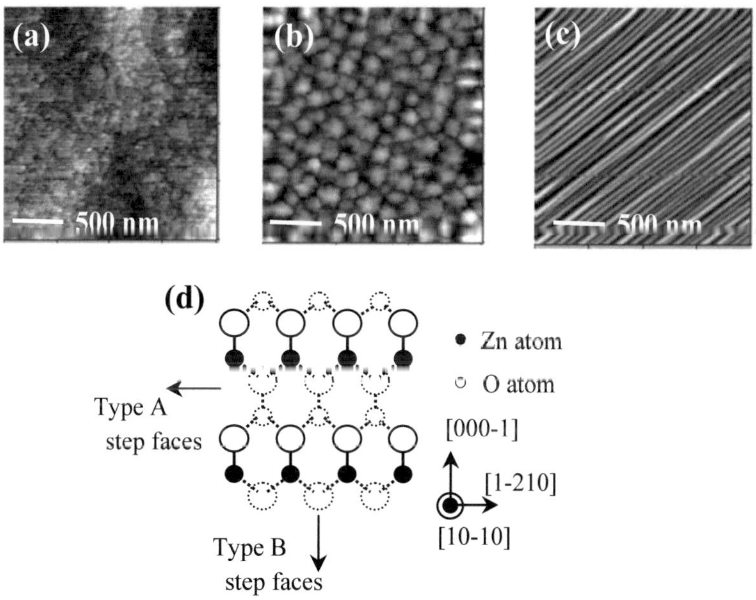

Fig. 2. AFM images of Zn-polar (a), O-polar (b) and *M*-nonpolar (c) ZnO layer surfaces. (d) Structural models showing the *M*-nonpolar (10-10) surfaces of ZnO.

On the other hand, the *M*-nonpolar ZnO layer surface showed a highly anisotropic surface morphology with self-organized surface nanowires elongated along the [0001] direction [Fig. 2(c)] (Matsui & Tabata, 2006). The stoichiometric ZnO (10-10) surface is auto-compensated since it contains an equal number of Zn and O ions per unit area. Zn and O atoms of the surface form dimer rows running along the [-12-10] direction, as shown in Fig.

2(d), which produces two types of A and B step edges consisting of stable low-index (-12-10) and (0001) planes, respectively (Dulub et al., 2002). The [-12-10] direction represents an auto-compensated nonpolar surface, while the [0001] direction consists of a polar surface with either Zn or O termination. This type of anisotropic surface structure has been utilized in scientific studies of heterogeneous catalytic processes involving the absorption of molecular and metallic atoms on nonpolar surfaces (Cassarin et al., 1999).

2.2 High-index GaAs layer surfaces

Similar surface nanostructures have been formed by a step-faceting growth mode on vicinal GaAs (775) *B* and (553) *B* substrates (Ohno et al., 2000; Yan et al., 2001). In GaAs, high-index or non-singular, planes are energetically unstable and tend to break up into low-index facets at normal epitaxial growth temperatures to minimize their surface energies. This process could produce periodic corrugations composed of nanometer-sized microscopic facets on originally flat surfaces. These ordered microscopic step arrays have been employed as the template for quantum wires. The most investigated high-index surfaces include (311) *A*, (331) *A* and (775) *B*. As an example, we present an AFM image of the GaAs (331) *A* layer surface in Fig. 3, which shows a highly anisotropic nanowire structure. The GaAs (331) *A* surface is thermally decomposed to a stable (111) *A* step and (110) terrace during homoepitaxial growth (Hong et al., 1988). This growth behaviour in GaAs indicates that the growth mechanism of the surface nanowires formed on non-vicinal ZnO (10-10) substrates differs from that of vicinal GaAs substrates. Thus, the surface nanowires on the low-index ZnO (10-10) surfaces are not fully related to the step-faceting process that was applied on the high-index GaAs surfaces.

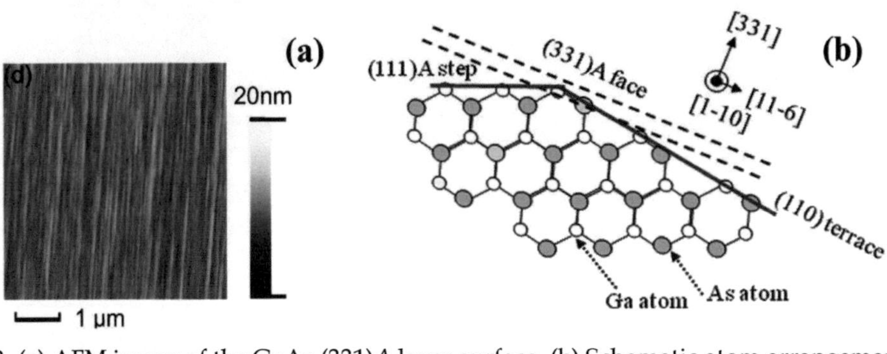

Fig. 3. (a) AFM image of the GaAs (331)*A* layer surface. (b) Schematic atom arrangement on cross-section with the [1-10] azimuth of the GaAs (331) *A* surface.

3. Experimental

ZnO layers and $Mg_{0.12}Zn_{0.88}O$/ZnO quantum wells (QWs) were grown at 400 - 600°C on *M*-plane ZnO (10-10) substrates (Crystec GmbH, Germany) using laser molecular beam epitaxy (laser-MBE). Figure 4 shows a schematic representation of the laser-MBE apparatus. The ZnO substrates were annealed *ex situ* at 1100°C in an oxygen atmosphere conducive to the formation of atomically flat surfaces, and then preannealed *in situ* at 600°C with an oxygen flow of 10^{-5} mbar for 20 min prior to laser-MBE growth. ArF excimer laser pulses (Compex 103: λ = 193 nm) were focused on ZnO and MgZnO targets located 4.5 cm from the substrates in an oxygen flow of 10^{-3} mbar. The number of Mg atoms in the MgZnO layer was

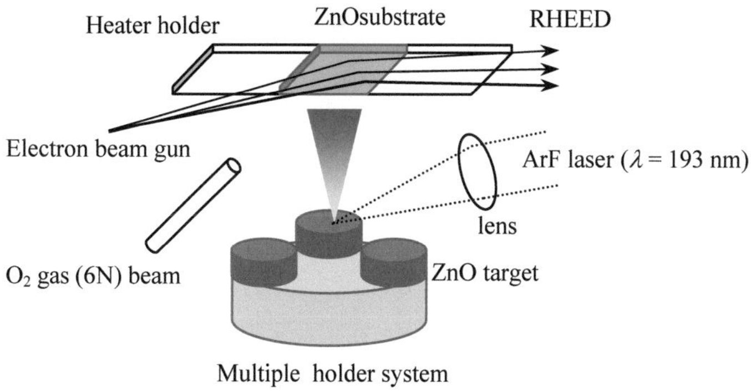

Fig. 4. A schematic figure of the laser-MBE apparatus.

estimated using an electron probe micro-analyzer (EPMA). The growth process was monitored using reflection high electron energy diffraction (RHEED).
Atomic force microscopy (AFM: Seiko SPI-3800) was used for observations of surface morphologies. Local structure analyses were conducted by means of high-resolution transmittance electron microscopy. Structural properties were characterized by high-resolution x-ray diffraction (HR-XRD: Philips X'pert) using a double-crystal monochromator. Micro-PL [(μ)-PL] spectroscopy was carried out at room temperature using a fourth harmonic generation of yttrium aluminium garnet laser (Nd^{3+}: YAG laser, 266 nm) excitation and a 0.85-m double monochromator (SPEX 1403) equipped with a nitrogen charge-coupled device camera. A reflective-type objective lens was used for this measurement to focus the laser to a diameter of 1 μm on the sample surface (Matsui et al., 2005). For polarized PL measurement, the sample was excited by a He-Cd laser (λ = 325 nm) in temperatures from 10 to 300 K. The luminescence was directed toward a Glan-Taylor prism to pick up polarization, and then passed through a depolarizer located behind the prism to eliminate the polarization. The spectrum was recorded using a 0.5-m single monochromator (SPEX 500M) equipped with a 1200 grooves/cm grating blazed at 500 nm. Electrical properties were measured using a four-probe Hall bar configuration with the perpendicular arms of the Hall bar aligned carefully in the [0001] and [11-20] directions. The Hall bars were fabricated by Ar ion milling of the samples through a photolithography-defined resist mask. The *ex situ* annealed ZnO substrate was treated as a semi-insulating substrate showing electrical resistivity in the order of 10^6 - 10^7 Ω cm.

3. Origin of surface nanowires

3.1 Growth evolution and structural quality

We describe the growth process and morphological evolution of the surface nanowires on the basis of RHEED and AFM investigations. The ZnO layers were grown at 550°C. At the very beginning of layer growth up to 8 nm in thickness, a 2D streak pattern appeared in place of sharp patterns of the ZnO substrates [Fig. 5(a) and 5(b)]. This is related to 2D nucleation at the initial growth stage, as evidenced by the smooth layer surface [Fig. 5(f)]. Continued growth of ZnO changed to a mixed pattern, which relates to the onset of the transition from 2D to 3D modes. This resulted from the appearance of a self-assembly of

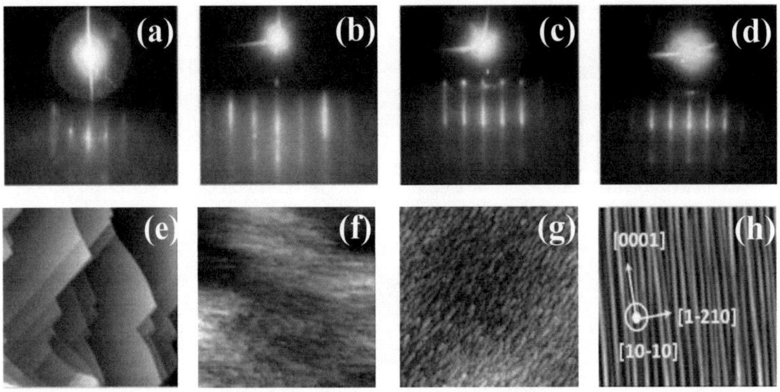

Fig. 5. RHEED patterns with the [0001] azimuth of the treated ZnO substrate (a) and ZnO layers with a thickness of (b) 8, (c) 20 and (d) 240 nm. AFM top view (2 x 2 μm^2) of the treated ZnO substrate (e) and ZnO layers with different thicknesses [(f)-(h)]. Layer thicknesses are (f) 8, (g) 20 and (h) 250 nm

anisotropic 3D islands [Fig. 5(c) and 5(g)]. Finally, the RHEED pattern showed 3D spots due to an island growth mode that originated from the formation of surface nanowires [Fig. 5(d) and 5(h)]. Surface nanowires with high density (10^5 cm^{-1}) that formed on the ZnO layers were homogeneously elongated along the [0001] direction above 5 μm with a few branches. Due to lattice strains at the heterointerface of a layer/substrate, S-K growth naturally induces 3D islands that are surrounded by high-index facets on 2D wetting layers. This has been observed in InGaAs/GaAs heteroepitaxy (Guha et al., 1990; Matsui et al., 2006). In an effort to examine the crystallinity in greater detail, plan-view and X-TEM observations were conducted to investigate the structural quality of the layer. Figure 6 (a) shows a low-

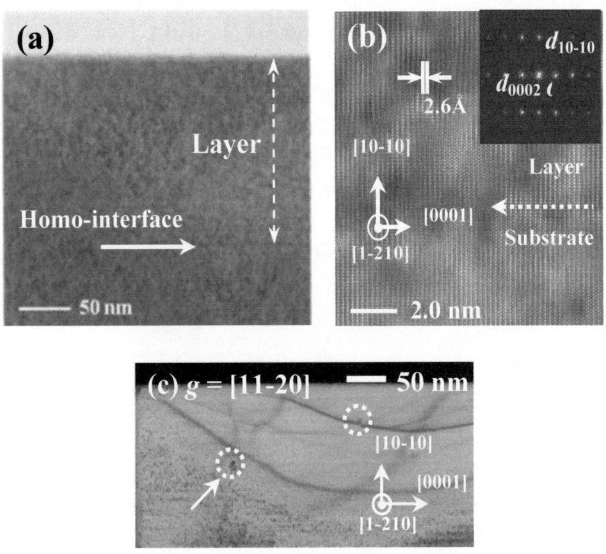

Fig. 6. (a) Low- and high-resolution X-TEM images of the ZnO layer taken with the [11-20] zone axis. Inset shows the RSD obtained by FFT analysis. (c) A bright field plan-view TEM image of the ZnO layer with g = [11-20] excitation under two-beam conditions.

resolution X-TEM image with the [11-20] zone axis. Threading dislocations induced by lattice relaxation between the layer and substrate were not observed. The high-resolution X-TEM image in Fig. 6(b) reveals a lattice arrangement between a smoothly connected layer and substrate. A 3 x 3 nm^2 space area selected from the layer region was utilized for a fast Fourier transform (FFT) analysis to examine local lattice parameters, and yielded a reciprocal space diffractogram (RSD) pattern [inset of Fig. 3(b)]. From the RSD pattern, the estimated strains (ε_{yy} and ε_{zz}) at the interface were approximately 0.10% and 0.18% with x, y and z being parallel to the [0001], [10-10] and [11-20] directions, respectively. Figure 6(c) shows a bright field plan-view TEM image with g = [11-20] excitation under two-beam conditions. Out-of-plane dislocations, marked by white open circles, were observed with a density of 3.2 x 10^7 cm^{-2}, and originated from threading dislocations running perpendicular to the layer surface. On the other hand, there were no in-plane dislocations propagated along the [0002] and [11-20] directions for different g vector excitations. These results indicate that the homoepitaxial interface was almost strain free. Thus, the elongated 3D islands that appeared on the 2D layers were formed under coherent homoepitaxy and had no correlation with S-K growth.

3.2 Characteristics of surface nanowires

Figure 7(a) and 7(b) show low-and high-resolution X-TEM images with the [0001] zone axis, respectively. A cross section of the surface nanowires displayed a triangular configuration with a periodicity of 84 nm. A high-resolution X-TEM image, marked by a white circle, revealed that the side facets did not consist of high-index facets, but instead had a step-like structure with a height of 0.27 nm that corresponded to half a unit of the m- axis. Side facets of the surface nanowires possessed uniform step spacing ranging from 0.1 to 0.2 nm, and were not surrouded by the high-index facets. A large number of surface nanowires showed flat tops with a (10-10) face and were separated laterally by deep grooves, as illustrated schematically in Fig. 7(e). A similar structure was also seen in the anisotropic 3D islands on the 20 nm thick layers, which indicated that the surface nanowires resulted from a coarsening of anisotropic 3-D islands formed at the initial growth stage.

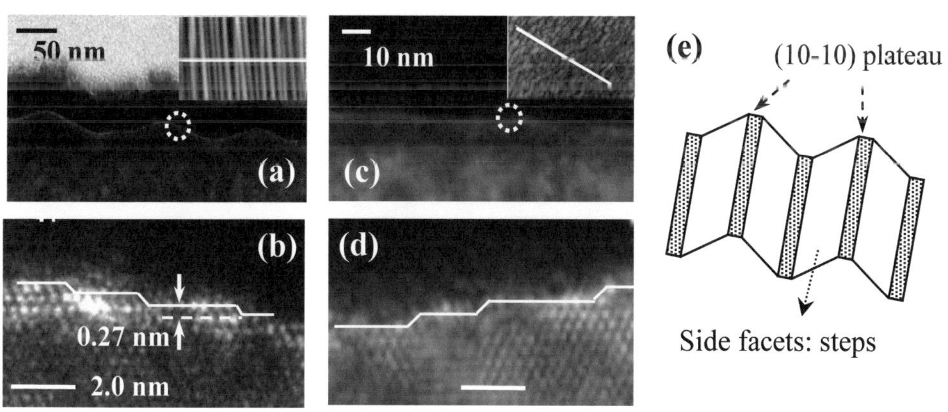

Fig. 7. (a) Low-and (b) high-resolution X-TEM images of the ZnO layer with a thickness of 240 nm. (c) Low-and (d) high-resolution X-TEM images of a 20 nm-thick ZnO layer. Insets (a) and (c) represent AFM images of the ZnO layers used for X-TEM observations. (e) Schematic representation of surface nanowires identified from X-TEM images.

The dependence of lateral periodicity of arrays on the thickness of the ZnO layers at a growth temperature (T_g) of 420°C was initially investigated, as shown in Fig. 8(a). The ZnO layers with layer thickness below 10 nm possessed flat surfaces. With increasing a layer thickness up to 15 nm, anisotropically small islands formed along the [0001] direction formed on the layer surface, and then developed to the surface nanowires. The lateral periodicity was 28 nm for the 15 nm-thick ZnO layer, and was almost saturated at 43 nm for ZnO layers with a thickness between 50 and 380 nm. The critical layer thickness, which completely developed to the surface nanowires, was approximately 0.1 μm. The dependence of the saturated lateral periodicity on growth temperature was also investigated. The lateral periodicity increased monotonically from 42 nm at T_g = 420°C to T_g = 600°C [Fig. 8(b)].
The saturation of the lateral periodicity with the layer thickness suggested that the surface migration of Zn-related ablation species, such as ZnO and Zn, supplied from the ablation targets is limited by the terrace width of the side facets (Ohtomo, et al., 1998). This was in agreement with the notion that the increase in periodicity with increasing growth temperature was due to prolonged surface diffusion of ablated species at high temperatures. The importance of surface diffusion was demonstrated using $Mg_xZn_{1-x}O$ alloys. An inhomogeneity in nanowire length with increasing Mg content was found in surface nanowires on layer surfaces of $Mg_xZn_{1-x}O$ (10-10) layers, although the lateral periodicity remained unchanged [Figs. 8(c) and 8(d)]. This may have been due to differences in surface migration and sticking probabilities of Zn- and Mg-related species. Thus, surface diffusion plays an important role in determining the size of nanowires, depending on the growth conditions and surface compositions. Moreover, highly anisotropic morphologies must be related because surface diffusion is much faster along the [0001] direction.

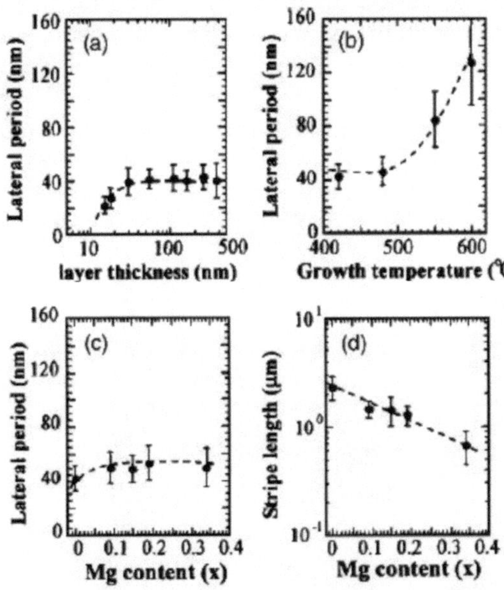

Fig. 8. (a) Dependence of lateral periodicity on layer thickness of ZnO layers at T_g = 420°C. (b) Dependence of saturated lateral periodicity on growth temperature of ZnO layers. (c) Dependence of lateral periodicity on Mg content of $Mg_xZn_{1-x}O$ layers at T_g = 420°C. (d) Dependence of nanowire length on Mg content (x) of $Mg_xZn_{1-x}O$ layers at T_g = 420°C.

3.3 Growth mechanism of surface nanowires

A multilayer morphology is determined not only by the transport of atoms within an atom layer (*intralayer transport*), but also by the transport of atoms between different atomic layers (*interlayer transport*). Thus, evolution of mound shapes is understood in terms of activation of atomic processes along the step edge. Therefore, a sequence of multilayer growth is governed by activation of atomic processes which enable exchange and hopping of atoms between different layers [Fig. 9(a)]. Schwoebel and Shipsey introduced the schematic potential energy landscape near a step that become the signature of what is often referred to as the Ehrlich-Schwoebel barrier (ESB) with a barrier energy of ΔE (Schwoebel, 1966). The mass transport of atoms between different layers is inhibited by a strong ESB effect, resulting in mound formation. This induces a nucleation of islands on the original surface together with inhibited interlayer transport. Once the islands are formed, atoms arriving on top of the islands will form second layer nuclei, and on top of this layer, a third layer will nucleate. This repetition leads to an increase in surface roughness with increasing layer thickness (Θ), resulting in the formation of mound shapes.

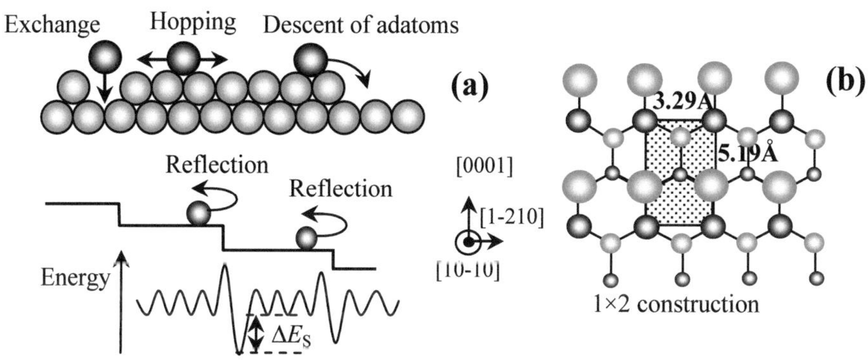

Fig. 9. (a) Upper part of the figure shows the descent of adatoms from an island by hopping and exchange. The lower part illustrates the energy landscape for hopping and the definition of ΔEs. (b) Structural models showing the M-nonpolar (10-10) surfaces of ZnO. The surface unit cells are indicated.

Mound formation is often observed on various systems such as semiconductors, metals, and organic materials. A mound structure possesses a small flat plateau at the top and a side facet with constant step spacing, and is separated from other mounds by deep grooves. This structure has been observed on dislocation-free metal homoepitaxial surfaces such as Pt/Pt (111) and Ag/Ag (100) systems, and is often referred to as a wedding cake (Michely & Krug, 2004). Here, mound formation emerging under reduced interlayer transport is described using the coarsening $\lambda \sim \Theta^n$, and the surface width $w \sim \Theta^\beta$. λ and w values are the height-height correlation between the nanowires and the surface roughening, respectively. As seen in Figs. 10(a) and 10(b), a coarsening exponent was indicated by $n = 0.23$, which was close to the n value of mound formation appearing during homoepitaxy. Moreover, the high β value of 0.60 was suitable for mound growth based on the ESB. This indicates that the surface nanowires formed during *M*-nonpolar ZnO homoepitaxy are due to the growth process originating at the ESB (Yu & Liu, 2008). The ESB was also seen for layer growth of O-polar ZnO with a hexagonal island surface. The appearance of an anisotropic morphology is related closely to a difference in surface diffusion and sticking probability as an important parameter. In *M*-nonpolar ZnO, the stoichiometric surface is auto-compensated since it contains an equal number of Zn and O ions per unit area. Surface Zn and O atoms form dimer rows running along the [1-210] direction, as shown in Fig. 6(b). This produces two types of A and B step edges consisting of stable low-index (1-210) and (0001) planes, respectively. The [1-210] direction represents an auto-compensated nonpolar surface, while the [0001] direction consists of a polar surface with either Zn or O termination. Thus, the origin of the surface nanowires is based not only on an ESB effect, but a difference in surface diffusion and sticking coefficient of atoms between the two types of step edges.

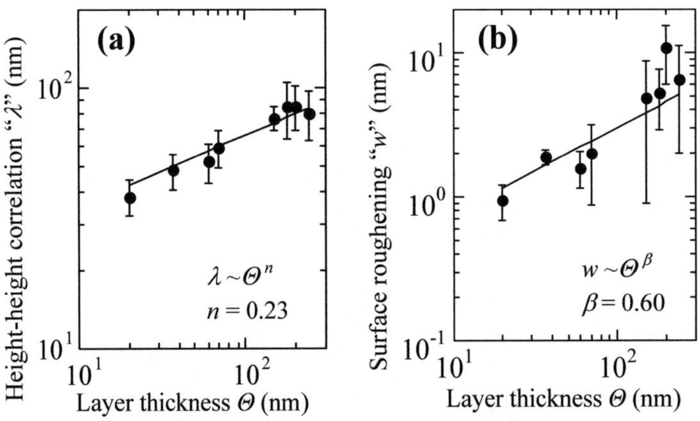

Fig. 10. (a) Height-height correlation (λ) and surface roughening (w) as a function of layer thickness.

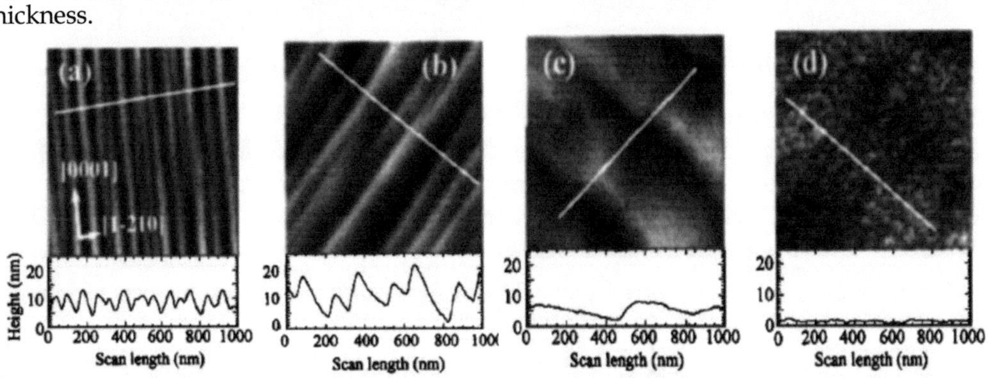

Fig. 11. AFM images (1 μm x 1μm) and cross-section profiles of ZnO layers on vicinal ZnO (10-10) substrates with off-angle of 0°-, 5°-, 10°-, and 15° from the (10-10)-oriented surface toward the (21-10)-oriented surface, indicated by (a), (b), (c), and (d), respectively. All nanowire arrays were elongated along the [0001] direction.

Anisotropic island growth on GaAs (001) based on the ESB effect is suppressed using vicinal GaAs (001) substrates with an off angle above a certain value (Johnson et al., 1994). AFM images of the ZnO layers on the vicinal substrates are shown in Fig. 11. With an increasing off angle, the cross-section profile of the AFM images gradually changed from symmetric to asymmetric shapes following the increase in lateral periodicity. The ZnO layer yielded a smooth surface with a roughness below 2 nm when the off angle of the substrate reached 15°. The off angle of 15° was close to that of the inclination of the surface nanowires. Figures 10 and 11 indicate that the surface nanowires originated from the ESB mechanism.

4. Band gap engineering and quantum wells (QWs)

4.1 $Mg_xZn_{1-x}O$ alloys

The discovery of tenability of a band gap energy based on ZnO has made the alloy system a promising material for use in the development of optoelectronic devices. Characterization of alloys such as (Mg,Zn)O or (Cd,Zn)O is important from the viewpoint of band gap engineering and the *p-n* junction. It was found that a $Mg_xZn_{1-x}O$ alloy was a suitable material for the barrier layers of ZnO/(Mg, Zn)O super-lattices due to its wider band gap. Since the ionic radius of Mg (0.56Å) is very close to that of Zn^{2+} (0.60Å), Mg-rich (Mg, Zn)O alloys with a wurtzite phase have been stably converted even when a rock salt-structured MgO is alloyed. A Mg content doped into a ZnO layer usually depends on the surface polarity, a growth technique, and type of substrate. Figure 12(a) shows the Mg content in $Mg_xZn_{1-x}O$ layers as a function of the target Mg content. Under growth conditions in this work, the Mg content in Zn-polar layers was always 1.6 times higher than the content in the ablation targets. This can be attributed to the low vapor pressure of Mg-related species compared to that of Zn. The incorporation efficiency of Mg atoms into the layers is more enhanced for O-polarity. On the other hand, the incorporation efficiency of Mg atoms in the case of a *M*-nonpolar surface is located between Zn- and O-polarities because the *M*-nonpolar surface has two dangling bonds, while Zn- and O-polarities have three and one-dangling bond, respectively. With increasing Mg content, the band gap systematically increased at 300 K as shown by the results of reflectance spectroscopy [Fig. 12(b)].

4.2 MgZnO/ZnO superlattices

A micro (μ)-PL technique is a powerful technique and can be employed to examine phase separation problems in MgZnO alloy layers. Figure 13(a) shows the variation in non-polarized μ-PL spectra at room temperature for $Mg_xZn_{1-x}O$ layers (x = 0 - 0.34). Band-edge emissions of all layers were systematically shifted to high energies with the Mg content alloys with different Mg contents at micro-scale. A driving force for the phase separation is closely related to lattice relaxation based on observations of Zn- and O-polar $Mg_xZn_{1-x}O$ homoepitaxial layers (Matsui et al., 2006). Therefore, a single phase of *M*-nonpolar $Mg_xZn_{1-x}O$ layers is only obtained below x = 0.12. 7-period $Mg_{0.12}Zn_{0.88}O$/ZnO MQWs were grown

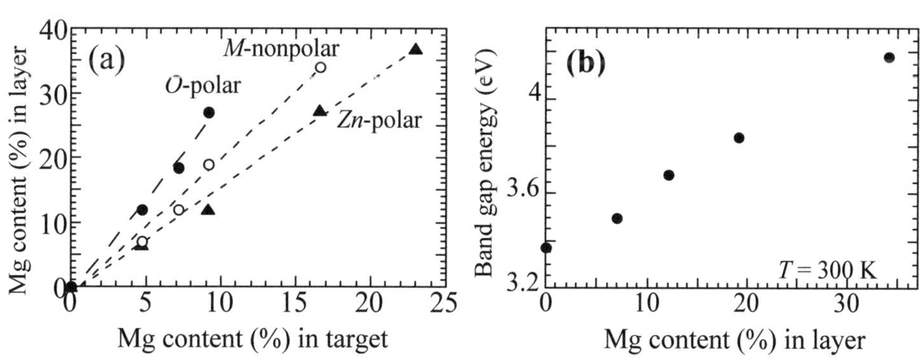

Fig. 12. (a) Mg contents in Zn-polar, O-polar and *M*-nonpolar $Mg_xZn_{1-x}O$ layers as a function of the target contents. (b) Dependence of band gap energy on Mg content for M-nonpolar $Mg_xZn_{1-x}O$ layers.

on ZnO layers with surface nanowires at T_g = 550°C and $p(O_2)$ = 10^{-3} mbar. The barrier thickness was set to 7 nm, and the well width was controlled from 1.4 to 4 nm. Figure 13(b) shows an AFM image of MQWs with a L_W of 2.8 nm. Surface nanowires elongated along the [0001] direction were homogeneously retained even after growing the MQWs. Figure 9(c) shows an X-TEM image of MQWs with the [0001] zone axis. The layer with a bright contrast represents MgZnO barriers, while the dark layers represent ZnO wells, which indicate that the MgZnO layers repeat the surface structure of the underlying ZnO layers.

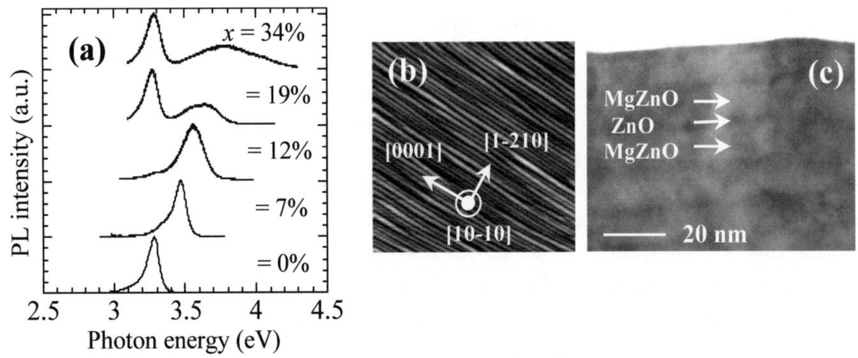

Fig. 13. (a) Non-polarized μ-PL spectra of $Mg_xZn_{1-x}O$ layers (x = 0, 0.07, 0.12, 0.19 and 0.34). The emission peak at 3.2 eV originates from the ZnO substrates. (b) Surface morphology and (c) X-TEM image of $Mg_{0.12}Zn_{0.88}O$/ZnO QWs with a L_W = 2.8 nm.

5. Anisotropic optical properties of MQWs

5.1 Linear polarized emissions

ZnO has attracted great interest for new fields of optical applications. Interesting characteristics of wurzite structure include the presence of polarization-induced electric fields along the c-axis. However, the optical quality of a quantum-well structure grown along the c-axis suffers from undesirable spontaneous and piezoelectric polarizations in well layers, which lower quantum efficiency. The use of nonpolar ZnO avoids this problem due to an equal number of cations and anions in the layer surface. Nonpolar ZnO surfaces have in-plane anisotropy of structural, optical, acoustic, and electric properties, which is useful for novel device applications. In this session, we discuss polarized PL of *M*-nonpolar ZnO layers and $Mg_{0.12}Zn_{0.88}O$/ZnO QWs.

Figure 14 shows splitting of the valence band (VB) in ZnO under the influence of crystal-field splitting and spin-orbit coupling. The VB of ZnO is composed of *p*-like orbitals. Spin-orbit coupling leads to a partial lifting of the VB degeneracy, and the former six-fold degenerate VB is split into a four-fold (j = 3/2) and two-fold (j = 1/2) band. The spin-orbit coupling is negative. The j = 1/2 band is at a higher energy than the j = 3/2 band. On the other hand, the crystal field in ZnO results in further lifting of the VB degeneracy due to the lower symmetry of wurtzite compared to zinc blende. The crystal field causes a splitting of *p* states into Γ_5 and Γ_1 states. Crystal-field splitting Δ_{cf} and spin-orbit coupling Δ_{so} together give rise to three types of two-fold degenerate valence bands, which are denoted as A (Γ_9-symmetry), B (Γ_7) and C (Γ_7) (Reynolds et al., 1999). These energies are formulated as follows:

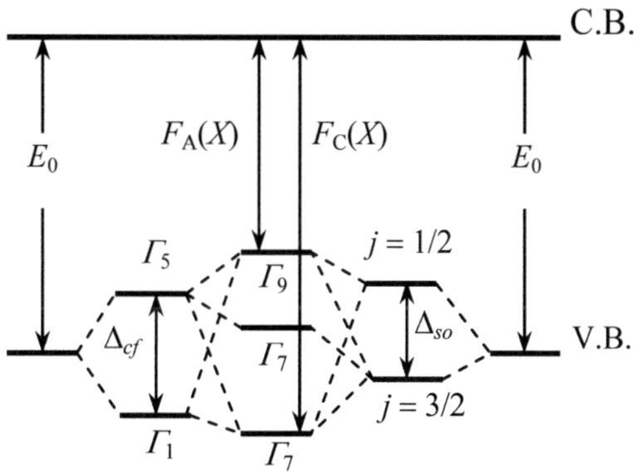

Fig. 14. Schematic energy level of band splitting by the crystal-field (Δ_{cf}) and spin-orbit (Δ_{so}) interactions in a wurtzite structure. $F_A(X)$ and $F_C(X)$ correspond to *A* and *C*-excitons, respectively, which are indicated in the middle. In the electronic energy levels proposed by Park *et al.* and Reynolds *et al.*, the uppermost Γ_9 and Γ_7 levels are interchanged.

$$E_A(\Gamma_9) - E_B(\Gamma_7) = -\frac{\Delta_{SO} + \Delta_{cf}}{2} + \frac{\sqrt{(\Delta_{SO} + \Delta_{cf})^2 - \frac{8}{3}\Delta_{SO}\Delta_{cf}}}{2} \tag{1}$$

$$E_A(\Gamma_9) - E_C(\Gamma_7) = \sqrt{(\Delta_{SO} + \Delta_{cf})^2 - \frac{8}{3}\Delta_{SO}\Delta_{cf}} \tag{2}$$

For ZnO, the experiment gave E_A-E_B = 0.0024 eV and E_C = 0.0404 eV (Fan et al., 2006). Solving the above two equations, we obtain Δ_{cf} (0.0391 eV) and Δ_{so} (-0.0035 eV). *A*- and *B*-excitons are referred to as heavy (HH) and light hole (LH) bands, respectively, and the crystal-field split-off hole (CH) was related to the C-exciton. The detection of $E \perp c$ and $E // c$ points to *A*-exciton (X_A) and C-exciton (X_c), respectively, where *E* represents the electric field vector (Reynolds et al., 1999).

Figure 15(a) shows the $E \perp c$ and $E // c$ components of the normalized PL spectra of strain-free ZnO layers (Matsui &Tabata, 2009). Polarization direction is referenced in Fig. 16 (a). The peak energies of X_A and X_C were located at 3.377 and 3.419 eV, respectively. These energies coincided with the X_A (3.377 eV) and X_C (3.4215 eV) peaks in ZnO crystals, respectively. The dependence of peak intensities on temperature could be fitted using the Bose-Einstein relation with a characteristic temperature of 315 and 324 K from the X_A and X_C peaks, respectively. The bound exciton (D^0X) peak disappeared at 120 K due to the activation energy of 16 meV. The polarization degree (*P*) is defined as $(I_\perp - I_{//}) / (I_\perp + I_{//})$, where $I_\perp$ and $I_{//}$ are the peak intensities for $E \perp c$ and $E // c$, respectively. Figure 15 (b) shows the polarization-dependent PL spectra at 300 K. The layer strongly emitted polarized light. The *P* value was calculated as 0.49. Significant spectral shifts in PL were detected when altering the polarization angle. This is attributed to a difference in carrier distribution in the VB between the HH and CH levels at 300 K. Figure 15(c) shows the dependence of polarization angle on PL intensity. Experimental data (triangle dots) were in agreement with the cos $(\theta)^2$ fit line (solid) obeyed by Malus' law.

5.2 In-plane anisotropy of MQW emissions

The polarization PL character in *M*-nonpolar MQWs is now discussed. ZnO wells are strain-free in the case of pseudomorphically grown MgZnO/ZnO MQWs. The PL spectra for $E{\perp}c$ and $E//c$ in MQWs with a well thickness (L_W) of 2.8 nm are shown in Fig. 16(a).

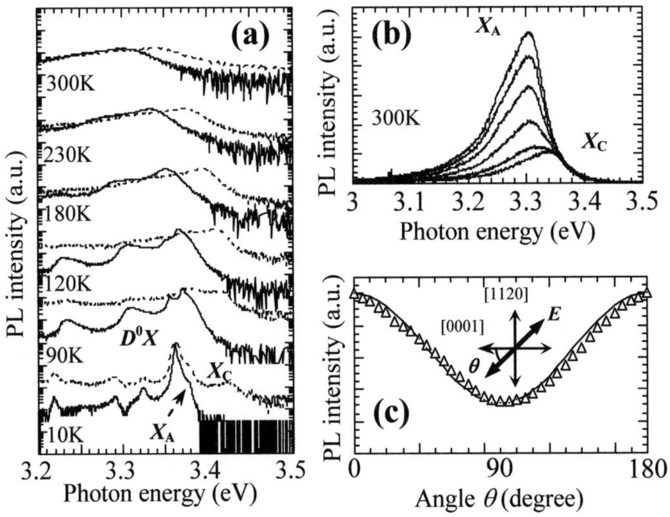

Fig. 15. (a) Temperature dependence of PL spectra on strain-free ZnO layers for $E \perp c$ (solid lines) and $E//c$ (dotted lines). (b) Polarization-dependent PL spectra at 300 K taken in steps of $\Delta\theta = 15^{o}$. (c) PL intensity as a function of polarization angle θ. Inset shows a schematic representation of the measurement geometry and sample orientation

The emission peaks around 3.6 eV correspond to 7 nm-thick $Mg_{0.12}Zn_{0.88}O$ barriers. At 300 K, an energy separation (ΔE) of 37 meV was found between the MQWs emissions of 3.372 eV ($E{\perp}c$) and 3.409 eV ($E//c$). The emission peak for $E{\perp}c$ appeared under conditions of $E//c$ below 120 K since the thermal distribution of carriers in the high-energy level for $E//c$ is negligible at 10 K. A polarization degree close to unity was found with a high P of 0.92 at 10 K [Fig. 16(c)]. In contrast, excited carriers at 300 K were sufficiently distributed in the high-energy level, resulting in a low *P* of 0.43. Furthermore, ΔE between the emission peaks for $E{\perp}c$ and $E//c$ was retained at around 40 meV even at 60 K [Fig. 16(d)]. This ΔE was close to the theoretical ΔE between the X_A and X_C states (Mang et al., 1999). For unstrained bulk ZnO, a polarization magnitude of zero and unity in the *C*-exciton is detected along the normal direction and along the *c*-axis, respectively [Fig. 15(a)]. However, the confinement of *M*-nonpolar MQWs takes place perpendicular to the quantization of the [10-10] direction. This result generates weak mutual mixing of the different *p* orbitals. Therefore, a π polarization component is expected for the A-excitonic state in these MQWs. In the case of *M*-nonpolar MQWs, it is predicted that a 10% p_z orbital component is involved with the *A*-excitonic states (Niwa et al., 1996), which is in agreement with the experimentally obtained *P* value of 0.92. MQWs with a L_W of 1.4 nm showed that the polarized PL spectra of $E{\perp}c$ and $E//c$ were separated by a small ΔE of 27 meV at 300 K [Fig. 16(b)]. ΔE decreased with temperature, and then completely disappeared at 60 K. The *P* value also dropped for all of the temperature regions. These behaviours are due to a large admixture of p_x to p_z orbitals for E//c, originating from an inhomogeneous roughening between the well and barrier layers. The interface roughness increased a potential fluctuation of quantized levels in the MQWs, being reflected by the broadened PL lines (Waag et al., 1991).

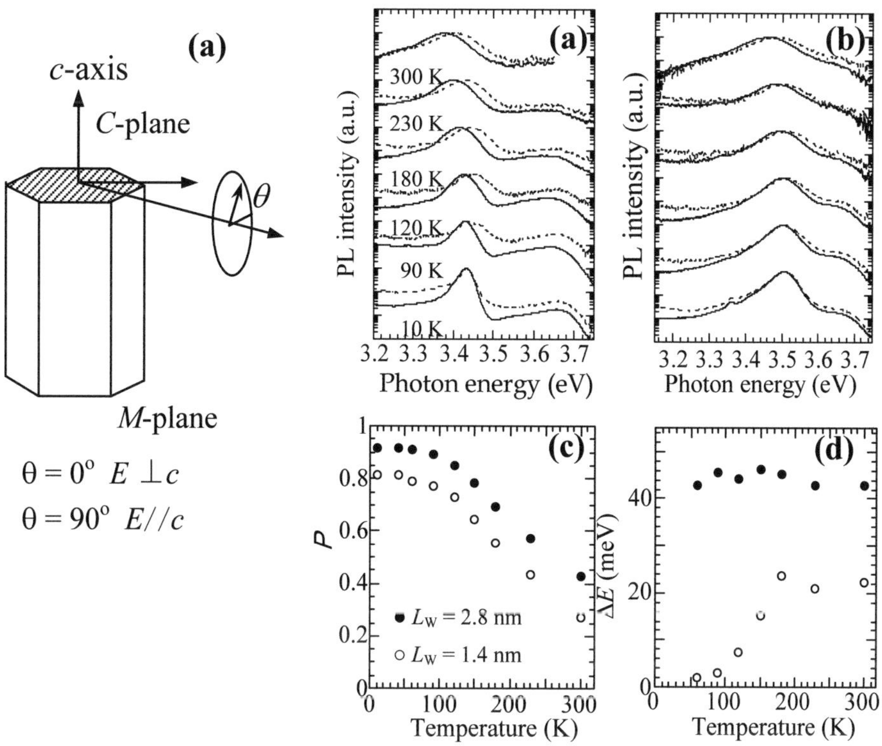

Fig. 16. PL spectra under $E\perp c$ (solid lines) and $E//c$ (dotted lines) on MQWs with L_W = 2.8 nm (a) and 1.4 nm (b). (c) and (d) show the relationship of temperature with polarization degree (P) and energy separation (ΔE) on MQWs with different L_W.

6. In-pane anisotropy of conductivity on MQWs for quantum wires

Transport properties were determined using a double Hall bar configuration with the [0001] and [1-210] directions [Fig. 17(a) and 17(b)]. Figure 17 (b) shows the temperature-dependent Hall mobility parallel ($\mu_{H[0001]}$) and perpendicular ($\mu_{H[1\text{-}210]}$) to the nanowires. $\mu_{H[0001]}$ gradually increased with decreasing temperature and was almost retained below 150 K due to a suppression of ionized impurity scattering. The electron concentration of MQWs also saturated below 70 K, suggesting that the whole electric current flows as 2D-like transport through the ZnO wells. On the other hand, $\mu_{H[1\text{-}210]}$ was much lower and resulted in large anisotropy of electron transport. The nature of the large transport anisotropy was evaluated by examining the temperature dependence of the conductivity (σ) for both directions [Fig. 17 (c)]. $\sigma_{[0001]}$ parallel to the nanowire arrays was almost independent of temperature below 70 K. Furthermore, the carrier concentration (n_e) was also constant below 70 K, suggesting that $\sigma_{[0001]}$ possessed metallic conductivity. In contrast, $\sigma_{[1\text{-}210]}$ perpendicular to the nanowire increased exponentially with T^{-1} in the region from 70 to 10 K. The activation energy E_a was

12 meV, indicating that an effective potential barrier height of 12 meV is formed for electron motion perpendicular to the nanowire arrays.

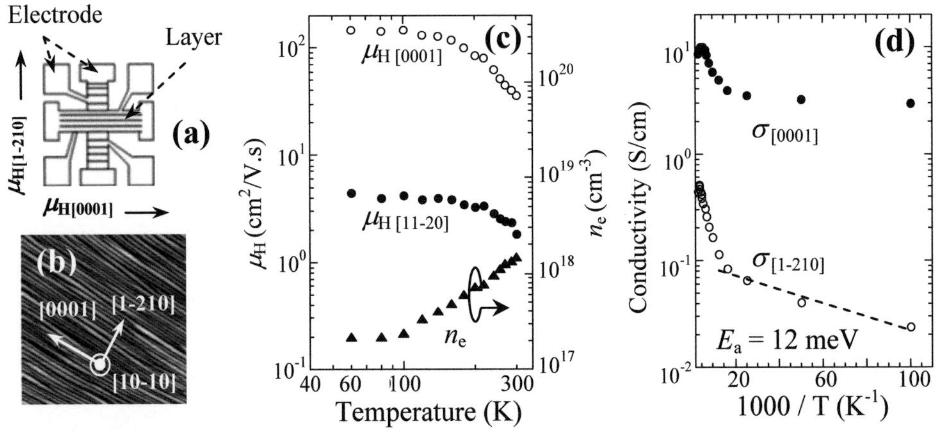

Fig. 17. (a) Hall bar patterned used to investigate anisotropic transport. (b) An AFM image of the $Mg_{0.12}Zn_{0.88}O/ZnO$ MQWs. (c) Temperature dependence of $\mu_{H[0001]}$ and $\mu_{H[1\text{-}210]}$ for MQWs with a L_W of 2.8 nm. (d) Temperature dependence of $\sigma_{[0001]}$ and $\sigma_{[1\text{-}210]}$ for MQWs with a L_W of 2.8 nm.

Figure 18 (a) shows the ratio of $\mu_{H[0001]}$ and $\mu_{H[1\text{-}210]}$ as a function of temperature. The curves correpond to different L_W of 2.2, 2.8 and 4 nm. For MQWs with a L_W of 4 nm, we observed no anisotropic behavior. However, the anisotropy of the Hall mobiliyu increased to 52 for MQWs with a L_W of 2.8 nm at low temperatures. The transport properties indicate that an electron can move almost freely along the nanowires, but are blocked from moving perpendicular to the nanowires. We discuss a possible mechansim for this type of activation barrier. The large anisotropy of electron transport disappeared when a flat surface was realzied using Zn-polar MQWs, as shown in Fig. 18(b). Zn-polar MQWs shows an isotropic surface morphology and has sharp MgZnO/ZnO heterointerfaces, as confirmed using high-resolution X-TEM image [inset of Fig. 18(b)]. Interface-roughness scattering dominates low-temperature mobility in MQWs (Sakai et al., 1987). A slight roughness of the heterointerfaces induces a large fluctuation in quantization energy of confined electrons. This acts as a scattering potential barrier for electron motion and reduces mobility. Therefore, electrons may readily undergo frequent scattering in a direction perpendicular to the nanowires by potential barriers produced between nanowires, and, consequently, may become extremely immobile.

On the other hand, parallel conductance along the nanowires involves a lower scattering probality than perpendicular transport due to a weak heterointerface modulation. However, the P value for MQWs with a L_W of 2.2 nm decreased with a decrease in $\mu_{H[0001]}$. Inspection of polarized PL spectra showed that the energy fluctuations in the quantum well gradually increaed with decreasing L_W [Fig. 18]. A decreased Hall moblity with a narrowing of LW has been observed on very thin InAs/GaSb MQWs since energy fluctuations in a quantum well are caused by an increase in interface roughness (Tsujino et al., 2004; Szmulowicz et al., 2007). It is concluded, therefore, that the large transport anisotropy was obtained through both a quantum size effect and small energy fluctuations in the quantum well, i.e., when L_W was in the vicinity of 3 nm.

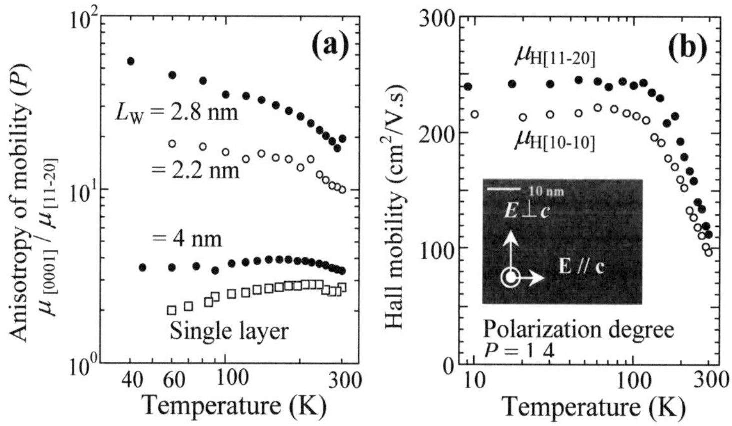

Fig. 18. (a) Temperature dependence of anisotropy of mobility (*P*) for M-nonpolar $Mg_{0.12}Zn_{0.88}O$/ZnO MQWs with different well thicknesses (L_W) of 2.2, 2.8 and 4 nm. (b) Temperature depednence of Hall mobiliy ($\mu_{H[11\text{-}20]}$) and ($\mu_{H[10\text{-}10]}$) for Zn-polar $Mg_{0.27}Zn_{0.73}O$/ZnO MQWs with a L_W of 2.4 nm.

Finally, the whole $\mu_{H[0001]}$ for MQWs was higher than that for *M*-nonpolar ZnO single layers as a result of the MQWs structure. However, it was lower than the mobility of ZnO single crystals, which is associated with thermal diffusions of extrinsic impurities from the hydrothermal ZnO substrates as a mobility-limited mechanism. Al atoms of the order of 1016 cm^{-3} were incorporated thermally into the samples from the substrates, as confirmed by secondary ion mass spectroscopy. It is noted that the presence of a background impurity results in a decrease of the whole Hall mobility for the MQWs.

6. Conclusion

We introduced a growth process for the anisotropic morphology that formed naturally on *M*-nonpolar ZnO (10-10) layer surfaces. Surface nanowires of high density elongated over 2 μm in length were observed during homoepitaxial growth. The surface nanowires gradually developed from elongated 3D mounds that originated from an ESB effect. Fabrication of *M* nonpolar MQWs allowed examination of the relationship between electron transport and surface morphology. The observed transport anisotropy correlated strongly with the surface morphology, and was dependent on the crystal quality of the MQWs as determined by low-temperature polarized PL spectroscopy. The MQWs showed strong polarization light at 300 K with a *P* value of 0.43 for $E \perp c$. Polarization anisotropy was based on the selection rule and is attributed to emissions from *A*- and *C*-excitonic states. Deviation of the *P* value from unity at 10 K in MQWs was associated with the confinement-induced admixture of the P_z orbit to the *A*-excitonic states, being a characteristic of the *M*-plane quantum well. As a possible mechanism of the large conductance anisotropy, we proposed that electron motion perpendicular to the nanowire arrays was restricted by the potential barrier caused by an interface roughness between the surface nanowires.

In this chapter, we reported the formation of electrical channels with a 1-D conductance as in quantum wires using *M*-nonpolar MQWs on the lateral surface of a nanowires structrue. Semiconductor quantum wires have been the subject of extensive theoretical and experimental studies over the past two to three decades. These are motivated by various unique quantum effects predicted in 1-D electronic systems, such as strong Coulomb correlation, suppression of electron scattering and an increase of quantum confinement. Q uantum wires have many potential applications to optoelectronic devices, such as high-mobility field-effect transistors. These surface nanowire structures contribute additional degrees of freedom for future studies of electron transport in field-effect transistors and magnetoelectric devices.

7. Acknowledgments

This work was supported in part by a Grant-in-Aid for Young Scientists (No. 18760231) from the Japan Society for the Promotion of Science, and a research grant from the Iketani Science and Technology Foundation (No. 081085-A)

8. References

Cassarin, M.; Maccato, M. & Vittadini, A. (1999). A comparative study of the NH_3 chemisorption on ZnO (10-10) and Cu_2O (111) nonpolar surfaces. *Chemical Physics Letters*, 300, 403-408.

Chen, Y.; Ko, H.-J.; Hong, S.-K.; Sekiuchi, T.; Yao, T. & Segawa, Y. (2000). Plasma-assisted molecular beam epitaxy for ZnO based II-V semiconductor oxides of their heterostructures. *J. Vacuum Science & Technology B*, 18, 1514-1522.

Coli, G & Bajaj, K.K. (2001). Excitonic transitions in ZnO/MgZnO quantum well heterostructures. *Applied Physics Letters*, 78, 2861-2863.

Dulub, O.; Boatner, L.A. & Diebold, U. (2002). STM study of the geometric and electronic structure of ZnO (0001)-Zn, (000-1)-O, (10-10) and (11-20) surfaces, *Surface Science*, 519, 201-217.

Fan, W.J.; Xia, J.B.; Agus, P.A.; Tan, S.T.; Yu, S.F. & Sur, X.W. (2006). Band paramters and electronic structures of wurtzite ZnO and ZnO/MgZnO quantum wells. *Journal of Applied Physics*, 013702(1)-013702(4).

Fons, P.; Iwata, K.; Niki, S.; Yamada, A.; Matsubara, K. & Watanabe, M. (2000). Uniaxial locked growth of high-quality epitaxied ZnO films on (11-20) α-Al_2O_3. *Journal of Crystal Growth*, 209, 532-537.

Guha, S.; Madhukar, A. & Rajkumar, C. (1990). Onset of incoherency and defect introduction in the initial stages of molecular beam epitaxial growth of highly strained $In_xGa_{1-x}As$ on GaAs (100). *Applied Physics Letters*, 57, 2110-2112.

Hong, S.; Young, K. & Kahn, A. (1988). Structural studies od (331) GaAs surface. *Jounral of Vacuum Science & Technology A*, 7, 2039-2043.

Johnson, M.D.; Orme, C.; Hunt, W.; Graff, D.; Sudijono, J.; Sander, L.M. & Orr, B.G. (1994). Stable and unstable growth in molecular beam epitaxy. *Physical Review Letters*, 72, 116-119.

Kato, H.; Sano, M.; Miyamoto, K. & Yao. T. (2003). Effect of O/Zn flux ratio on crystalline quality of ZnO films grown by plasma-assisted molecular beam epitaxy. *Japanense Journal of Applied Physics*, 42, 2241-2244.

Kubo, M.; Oumi, Y.; Takaba, H.; Chatterjee, A.; Miyamoto, A.; Kawasaki, M.; Yoshimoto, M. & Koinuma, H. Homoepitaxial growth mechanism of ZnO (0001): Molecular-dynamics simulations. *Physical Review B*, 61, 16087-16092.

Maag, A.; Reimann, K. & Rübenacke, St. (1999). Band gaps, crystal-field splitting, spin-orbit coupling, and exciton binding energies in ZnO under hydrostatic pressure. *Solid State Communications*, 94, 251-254.

Makino, T.; Chia, C.H.; Tuan, N.T.; Sun, H.D.; Segawa, Y.; Kawasaki, M.; Ohtomo, A.; Tamura, K. & Koinuma, H. (2000). Room-temperature luminescence of wxcitons in ZnO/(Mg, Zn)O multiple quantum wells on lattice-matched substrates. *Applied Physics Letters*, 77, 4250-4252.

Mariano, A.M. & Hanneman, R.E. (1963). Crystallographic polarity of ZnO crystals. *Journal of Applied Physics,* 34, 384-388.

Matsui, H.; Saeki, H.; Kawai, T.; Sasaki, A.; Yoshimoto, M.; Tsubaki, M & Tabata, H. (2004). Characteristics of polarity-controlled ZnO films fabricated using the homoepitaxy technique. *Journal of Vacuum Science & Technology B*, 22, 2454-2461.

Matsui, H. & Tabata, H. (2005). Self-organized nanostripe arrays on ZnO (10-10) surfaces formed during laser molecular beam epitaxy growth. *Applied Physics Letters*, 87, 143109(1)-143109(3).

Matsui, H.; Tabata, H.; Hasuike, N.; Harima, H. & Mizobuchi, B. (2005). Epitaxial growth and characteristics of N-doped anatase TiO_2 films grown using a free-radical nitorgen oxide source. *Journal of Applied Physics,* 97, 123511(1)-123511(8).

Matsui, H. & Tabata, H. (2006). Correlation of self-organized surface nanostructures and anisotropic electron transport in nonpolar ZnO (10-10) homoepitaxy. *Journal of Applied Physics*, 99, 124307 (1) – 124307 (8).

Matsui, H.; Hasuike, N.; Harima, H. & Tabata, H. (2006). Critical thickness and lattice relaxation of Mg-rich strained $Mg_{0.37}Zn_{0.63}O$ (0001) layers towards multiple-quantum wells. *Journal of Applied Physics*, 99, 024902(1)-024902(7).

Matsui, H.; Hasuike, N.; Harima, H.; Tanaka, T. & Tabata, H. (2006). Stranski-Krastanov growth in $Mg_{0.37}Zn_{0.63}O$/ZnO heteroepitaxy: Self-organized nanodots and local composition separation. *Applied Physics Letters*, 89, 091909(1)-091909(3).

Matsui, H.; Hasuike, N.; Harima, H. & Tabata, H. (2008). Growth evolution of surface nanowires and large anisotropy of conductivity on MgZnO/ZnO quantum wells based on *M*-nonpolar (10-10) ZnO. *Jounral of Applied Physics,* 104, 094309(1)-094309(6).

Matsui, H. & Tabata, H. (2009). In-plane anisotropy of polarized photoluminescence in *M*-plane (10-10) ZnO and MgZnO/ZnO multiple quantum wells. *Applied Physics Letters*, 94, 161907(1)-161907(3).

Michely, T. & Krug, J. (2004). Patterns and processes in crystal growth far from equilibrium. *Islands, Mounds, Atoms*. Springer Series in Surface Sicence 42, 1-305. (Springer, New York).

Niwa, A.; Ohtoshi, T. & Kuroda, T. (1996). Tight binding analysis of the optical matrix element in wurtzite- and zincblende- GaN quantum wells. *Japanense Jounral of Applied Physics*, 35, L599-L601.

Ohno, Y.; Higashiwaki, H.; Shimomura, S. & Hiyamizu, S. (2000). Laser operation at room temperature of self-organized $In_{0.1}Ga_{0.9}As/(GaAs)_6(AlAs)_1$ quantum wires grown on (775) B-oriented GaAS substrates by molecular beam epitaxy. *Journal of Vacuum Science & Tecnology B*, 18, 1672-1674.

Ohtake, A.; Miwa, S.; Kuo, L.H.; Kinura, K.; Yasuda, T.; Jin, C. & Yao, T. Polar surface dependence of epitaxy processes: ZnSe on GaAs {111}A, B-(2 x 2). *Applied Surface Science*, 130-132, 398-402.

Ohtomo, A.; Kawasaki, M.; Koida, T.; Masubuchi, K.; Koinuma, H.; Sakurai, Y.; Yoshida, Y.; Yasuda, T. & Segawa, Y. (1998). $Mg_xZn_{1-x}O$ as II-IV wide gap semicondcutor alloy. *Applied Physics Letters*, 72, 2466-2468.

Parker, T.M.; Condon, N.G.; Lindsay, R.; Leibsle, F.M. & Thornton, G. (1998). Imaging the polar (000-1) and nonpolar (10-10) surfaces of ZnO with STEM. *Surface Science,* 415, L1046-L1050.

Reynolds, D.C.; Look, D.C.; Jogai, B.; Litton, C.W.; Cantwell, G. & Harsch, W.C. (1999). Valence band ordering in ZnO. *Physcial Reviewv B*, 60, 2340-2344.

Reynolds, D.C.; Look. D.C.; Jogai, B.; Litton, C.W.; Collins, T.C.; Harris, T.; Callahan, M.J.; Bailey, J.S. (1999). Strain splitting of the Γ_5 anf Γ_6 free excitons in ZnO. *Journal of Applied Physics*, 86, 5598-5600.

Sakai, H.; Noda, T.; Hirakawa, K.; Tanaka, M. & Matsusue, T. (1987). Interface roughnes scattering in GaAs/AlAs quantum wells. *Applied Physics Letters*, 51, 1934-1936.

Schönher, H.P.; Fricke, J.; Niu, Z.; Friedland, K.J.; Nötzel, R. & Ploog, K.H. (2001). Uniform multiatomic step arrays formed by atomic hidrogen assisted molecular beam epitaxy on GaAs (331) substrates. *Applied Physics Letters*, 72, 566-568.

Schwoebel, R.L. (1966). Step motion on crystal surfaces. *Journal of Applied Physics*, 37, 3862-3686.

Sumiya, M.; Yoshimoto, K.; Ito, T.; Ohtsuka, K.; Fuke, S.; Mizuno, K.; Yoshimoto, M.; Koinuma, H.; Ohtomo, A. & Kawasaki, M. (2000). Growth mode and surface morphology of a GaN film deposited along the N-face polar direction on c-plane sapphire substrate. *Jounral of Applied Physics*, 88, 1158-1165.

Szmulowicz, F.; Elhanri, S.; Haugan, H.J.; Brown, G.J. & Mitchel, W.C. (2007). Demonstration of interface-scattering-limitted electron mobilities in InAs/GaSb superlattices. *Journal of Applied Physics*, 043706(1)-043706(5).

Thomas, D.G. (1960). The exciton spectrum of zinc oxide. *Journal of Physical & Chemcial Solids*, 15, 86-89.

Tsujino, S.; Falub, C.N.; Müller, E.; Scheinert, M.; Drehl, L.; Gennser, D.; Fromherz, T.; Borak, A.; Sigg, H.; Grützmacher, D.; Campidelli, Y.; Kermarrec, O. & Bensahel, D. (2004). Hall mobility of narrow $Si_{0.2}Ge_{0.8}$-Si quantum wells on $Si_{0.5}Ge_{0.5}$ relaxed buffer substrates. *Applied Physics Letters*, 84, 2829-2831.

Tsukazaki, A.; Ohtomo, A.; Kita, T.; Ohno, Y.; Ohno, H. & Kawasaki, M. Quantum Hall effect in polar oxide heterostructures. *Science*, 78, 2861-2863.

Sharma, A.K.; Narayan, J.; Muth, J.F.; Teng, C.W.; Jin, C.; Krit, A.; Kobas, R.M. & Holland, O.W. (1999). Optical and structural properties of epitaxial $Mg_xZn_{1-x}O$ alloys. *Applied Physics Letters*, 75, 3327-3329.

Viguè, F.; Vennégnès, P.; Vèzian, S.; Laùgt, M. &Fourie, J.-P. (2001). Defect characterization in ZnO layers grown by plasma-enhanced molecular beam epitaxy on (0001) sapphire substrates. *Applied Physics Letters*, 79, 194-196.

Vipute, R.D.; Talyansky, V.; Tranjanovic, Z.; Choopun, S.; Downes, M.; Sharma, R.P.; Venkatesan, T.; Wood, M.C.; Lareau, R.T.; Jones, K.A. & Ilidais, A.A. (1997). High quality crystalline ZnO buffer layers on sapphire (0001) by pulse laser deposition for III-V nitrides. *Applied Physics Letters*, 70, 12735-12737.

Waag, A.; Schmeusser, S.; Bicknell-Tassius, R.N.; Yakovlev, D.R.; Ossau, W.; Landwehr, G.; Uraltsev, I.N. (1991). Molecular beam epitaxial growth of ultra-thin CdTe-CdMnTe quantum wells and their characterization. *Applied Physics Letters*, 59, 2995-2997.

Wang, K.L. & Voliotis, V. (2006). Epitaxial growth and optical properties of semiconductor quantum wires. *Journal of Applied Physics*, 99, 121301(1)-121302(38).

Yan, F.W.; Zhang, W.J.; Zhang, R.G.; Cui, L.Q.; Liang, C.G. & Liu, S.Y. (2001). Formation and characterization of (553) *B* $In_{0.15}Ga_{0.85}As$/GaAs quantum wire structure. *Journal of Applied Physics*, 90, 1403-1406.

Yu, P.; Tang, Z.K.; Wong, G.K.L.; Kawasaki, M.; Ohtomo, A.; Koinima, H. & Segawa, Y. (1997). Ultraviolet spontaneous and stimulated emissions from ZnO microcrystalline thin films at room temperature. *Solid State Communications*, 103, 459-464.

Yu, Y.M. & Liu, B.G. (2008). Contrasting morphologies of O-rich ZnO epitaxy on Zn- and O-polar thin film surfaces: Phase-field model. *Physical Review B*, 195327(1)-195327(6).

8

Characterization of Room-Temperature Ferromagnetic $Zn_{1-x}Co_xO$ Nanowires

Yi-Ching Ou[1], Zhong-Yi Wu[2], Fu-Rong Chen[2],
Ji-Jung Kai[2] and Wen-Bin Jian[1]
[1]*National Chiao Tung University,*
[2]*National Tsing Hua University,*
Taiwan

1. Introduction

The manipulation and detection of an electron's charge and, simultaneously, its spin orientation in electronic devices have been developed to be a new emerging field of spintronics (or magnetoelectronics) (Prinz, 98; Wolf et al., 2001). At present, the most notable spintronic applications could be the hard disk read heads and the magnetic random access memory which are based on metal magnetic materials and are assorted into *metallic* spintronic devices. The establishment of metallic spintronics might be ascribed to a discovery of giant magnetoresistance (Baibich et al., 1988; Binasch et al., 1989) and, subsequently, understanding and exercise of a spin-valve scheme (Moodera et al., 1995), and tunnelling magnetoresistance (Dieny et al., 1991) in ferromagnetic multilayers. On the other hand, in order to integrate with the modern industrial technology, new semiconductor materials such as diluted magnetic semiconductors (DMSs) (Furdyna, 1988), also known as ferromagnetic semiconductors (Ohno, 1998), have been searched for a supply of a spin-polarized carrier source. Those devices building on a transport of spin current in semiconductors are categorized into *semiconductor* spintronics. Spin injection, maintenance of a spin coherence, spin detection, and a spin carrier source in semiconductors are all important issues for semiconductor spintronics.

The DMSs, based on host materials of II-VI and IV-VI semiconductors, have been studied for several decades. Although the indirect exchange mechanisms between *3d* transition metal dopants in these semiconductors have been inspected experimentally and discussed theoretically (Story et al., 1986; Sawicki et al., 1986; Furdyna, 1988) for a long time, the Curie temperature (T_C), below which a spontaneous magnetization and a spin-polarized current in the DMSs arise, was too low to be capable of employment. Until recent advance in III-V DMSs of (In,Mn)As and (Ga,Mn)As (Ohno et al., 1996), T_C's of some new DMSs such as (Ga,Mn)As have been raised up to ~110 K. These new III-V DMS materials were exploited to demonstrate tunneling magnetoresistance in (Ga,Mn)As ultrathin heterostructures (Hayashi et al., 1999), electrical spin injection in a ferromagnetic semiconductor heterostructure (Ohno et al., 1999), electric-field control of ferromagnetism (Ohno et al., 2000}, electrical manipulation of magnetization reversal (Chiba et al., 2003), and current-induced domain-wall switching (Yamanouchi et al., 2004). On the other hand, the other approach of spin-current injection into semiconductors from ferromagnetic metals has recently been achieved, so as to realize semiconductor spintronics at room temperature.

By using Zener model description, Dietl et al. (Dietl et al., 2000) have theoretically sustained the fact of a 110-K high T_C for *p*-type (Ga,Mn)As with a manganese concentration of just 5%. In addition, they argued the presence of a T_C above room temperature in Mn doped ZnO or GaN with hole carriers of 3.5×10^{20} cm^{-3}. These theoretical arguments drew much attention on search for room-temperature ferromagnetism (RTFM) in new DMS materials. For example, Matsumoto et al. (Matsumoto et al., 2001) discovered RTFM in Co doped TiO_2 with a magnetic moment of 0.32 Bohr magneton (μ_B) per Co atom and Toyosaki et al. (Toyosaki et al., 2004) observed anomalous Hall effect in this particular material. Else, Ueda et al. descried ferromagnetism and a T_C above 280 K in pulse laser deposited $Zn_{1-x}Co_xO$ films (Ueda et al., 2001). Cho et al. found ferromagnetic and antiferromagnetic ordering in $(Zn_{1-x}Mn_x)GeP_2$ at temperatures up to 312 K and below 47 K (Cho et al., 2002), respectively. Among all new as-proposed DMS materials, Co-doped metal oxides, such as $Ti_{1-x}Co_xO_2$ and $Zn_{1-x}Co_xO$, seem to be an appropriate candidate for a spin-polarized carrier source at room temperature (Janisch et al., 2005).

ZnO is recently a hot material and it is proposed to be valuable in many application fields such as blue/ultraviolet optoelectronics (Klingshirn, 2007; Pearton et al., 2004). It is a direct and wide band gap semiconductor and can be easily over-doped to form conductive and transparent films. ZnO is natively *n*-type doped to show relatively lower resistivity due to difficulties in control of point defects during the growth process. In addition, it shows ultraviolet (near band edge) and green (or blue) defect emission at ~3.2 and ~2.5 eV, respectively, in photoluminescence (PL) spectra. It is proposed that oxygen vacancies (Lanny & Zunger, 2005), zinc interstitials (Look et al., 1999), Zn_I-N_O complexes (Look et al., 2005), metastable conductive states (Lany & Zunger, 2007), or hydrogens (Van de Walle, 2000) can lead to the native *n*-type doping, the coloration, and the green emission. The point defects not only result in an increase of conductivity but also modulate magnetic ordering after ZnO is doped with paramagnetic *3d* transition metal dopants.

2. Important

There are so many experimental and theoretical reports on claiming that Co-doped ZnO is an intrinsically DMS (Schwartz & Gamelin, 2004; Coey et al., 2005; Neal et al., 2006; Zhang et al., 2009). Very recently, different magnetic mechanisms are uncovered in insulating and magnetic regimes (Behan et al., 2008). In particular, magnetic resistance has been observed in a magnetic tunnel junction fabricated by using Co-doped ZnO as one ferromagnetic electrode (Xu et al., 2008). There are, however, other contradictory reports exposing antiferromagnetism (Risbud et al., 2003), secondary phases in crystalline structure, clustering of Co metals or ions (Sati et al., 2007), or absence of ferromagnetism in this material. On the other hand, even for similar conclusions of ferromagnetism, the Curie temperature either above or below the room temperature is another issue. As we have emphasized, it is difficult to control point defects in ZnO during growth. Moreover, electrical resistivity and PL emission of a pure ZnO, and a magnetic ordering of a Co-doped ZnO can all be altered by thermal annealing after growth. Here we propose the employment of nanowires (NWs) for an exploration into magnetism because, after converted to a nanophase, the nanomaterials have a large surface to volume ratio, feasible for thermal treatments, and they are handy for a structural characterization by using transmission electron microscopy.

In our previous reports, we have observed the structure and annealing effect on ferromagnetic ordering (Jian et al., 2006; Wu et al., 2006), and have explored the size dependent behavior (Jian et al., 2007) in $Zn_{1-x}Co_xO$ NWs. In addition, we have discovered RTFM in high-vacuum annealed $Zn_{1-x}Co_xO$ NWs (Chen et al., 2008) and the effect of cross-sectional shape modulation (Wu et al., 2008). In this chapter, we present complete characterizations, including structure, optical, and magnetic property measurements on pure ZnO NWs, as-implanted $Zn_{1-x}Co_xO$ NWs, annealed $Zn_{1-x}Co_xO$ NWs, and a comparative sample of ZnO NWs sheathed in amorphous carbon with Co clusters, so as to explore the mechanism of RTFM in the DMS of $Zn_{1-x}Co_xO$ NWs.

3. Experiment

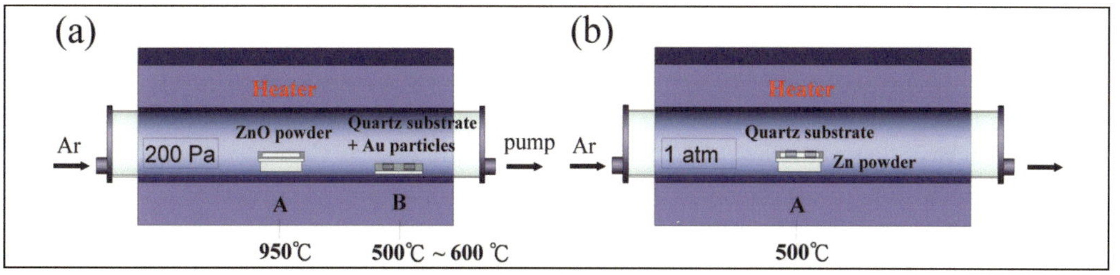

Fig. 1. Schematic illustration of the growth of pure ZnO NWs with (a) circular and (b) hexagonal cross sections.

Cylindrical and hexagonal ZnO NWs were grown by using a vapor-phase transport process. The growth of cylindrical ZnO NWs is schematically illustrated in Fig. 1(a). A quartz tube treated as a growth chamber was inserted in a furnace. ZnO powders were placed in a crucible in the growth chamber and heated to 950°C. The chamber was maintained at 200 Pa with a constant flow of argon and a pumping system. For a purpose of controlling NW diameter, gold nanoparticles as catalysts with specified average diameters of 5, 10, 20, 40, 70, and 100 nm were dispersed on quartz substrates. The substrates were positioned at the downstream end of the growth chamber and were maintained at 500-600°C. Cylindrical ZnO NWs with a controllable diameter were formed on substrates after a growth period of 8 h.

The growth of hexagonal ZnO NWs, as schematically illustrated in Fig. 1(b), is different from that of cylindrical NWs. Diameters of hexagonal NWs cannot be well regulated through the use of catalysts. During the synthesis process, Zn powders were placed in an alumina boat in the quartz tube chamber and heated to 500°C. The substrates were put on top of the alumina boat and the chamber was maintained at 1 atm with a constant flow of argon. Like the growth of cylindrical NWs, a 8-h synthesis period was retained for growth of hexagonal ZnO NWs. The crystalline structure and morphology of both cylindrical and hexagonal ZnO NWs were analyzed by using field-emission scanning electron microscope (SEM, JEOL JSM 7000F) and transmission electron microscope (TEM, JEOL JEM-2010F).

The as-grown ZnO NWs were implanted by Co ions with doses of (1-6) × 10^{16} cm^{-2}. By using a tandem accelerator (NEC 9SDH-2), the implantation was performed at room temperature. An accelerating energy of 72 keV was used for NWs with average diameters larger than ~70 nm. Thinner NWs were implanted by Co ions with an acceleration energy of 40 keV. A beam current of either 150 or 600 nA/cm^2 was used to make $Zn_{1-x}Co_xO$ NWs. The high bean current of 600 nA/cm^2 could somewhat turn out to be thermal treatment due to the high energy ion bombardment in a high vacuum. The fabrication process is schematically drawn in Fig. 2(a) and its corresponding side-view SEM image of the $Zn_{1-x}Co_xO$ and ZnO NWs is

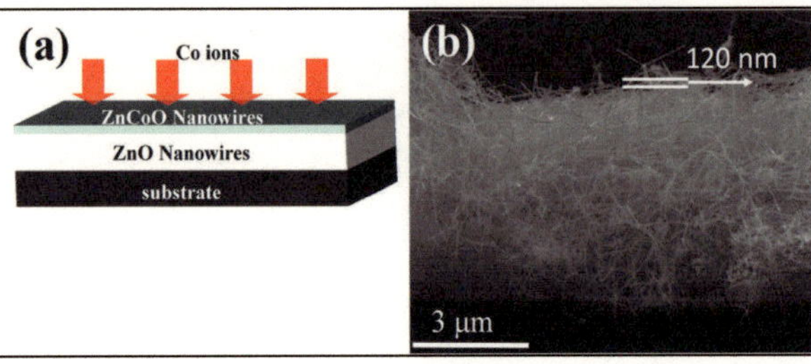

Fig. 2. (a) Schematic illustration of Co ion implantation. (b) Side-view SEM image of cylindrical ZnO NWs on a quartz substrate.

displayed in Fig. 2(b). As indicated in the figure, only a ~120-nm thin layer of $Zn_{1-x}Co_xO$ NWs could be formed on ~3-μm thick layer of pure ZnO NWs (Wu et al., 2006). The chemical composition as well as Co element distribution in $Zn_{1-x}Co_xO$ NWs were inspected through energy dispersive x-ray (EDX) and electron energy loss spectroscopy (EELS) mapping. In order to study the origins of ferromagnetism in $Zn_{1-x}Co_xO$ NWs, some specimens were post-annealed in argon, in a high vacuum of 5×10^{-5} torr, or in oxygen at 600°C (or 450°C) for several hours. In particular, multiple steps of thermal annealing in a high vacuum were carried out to produce a gradual transition of magnetic states of this DMS material. PL spectra of some specimens were measured at room temperature by using a 325-nm He-Cd laser as UV fluorescent light excitation.

In addition to DMS $Zn_{1-x}Co_xO$ NWs, ZnO NWs sheathed in amorphous carbon with Co clusters were produced for comparison. These purposely fabricated samples were treated with the same thermal annealing process as that for DMS NWs. Co metal clusters in carbon-coated ZnO NWs were intriguingly formed after a high-vacuum annealing. The morphology, crystalline structure and chemical composition of these comparative samples were analyzed in a similar way.

Magnetic properties of DMS $Zn_{1-x}Co_xO$ NWs and comparative samples (Co clusters on ZnO NWs) were measured by employing a SQUID magnetometer (Quantum Design MPMS-XL7) with the reciprocating sample option mode. Field cool (FC) and zero-field cool (ZFC) processes were conducted to obtain temperature dependent magnetization during the rising temperature sequence under an external magnetic field of 500 Oe. That is, the samples were subjected to oscillating with decreasing fields and were cooled from 300 K down to 2 K in a zero field. The samples were then warmed up to obtain ZFC magnetization as a function of temperatures in 500 Oe. They were cooled down in the same field and warmed up again to record the FC magnetization. Before NW growth, the magnetic susceptibility of a quartz substrate was estimated to be $\sim -1.1 \times 10^{-6}$ emu/cm^3 so that diamagnetic contribution of the substrate can be subtracted from the total magnetization. Magnetic data were presented in unit of μ_B per Co where the amount of Co ions was evaluated by multiplying the ion dose per cm^2 with the substrate area.

4. Growth, morphology, crystalline structure, and photoluminescence

In this section, the growth behavior of pure ZnO NWs is discussed. The morphology, crystalline structure, and optical properties of as-grown ZnO, as-implanted $Zn_{1-x}Co_xO$, high-vacuum annealed $Zn_{1-x}Co_xO$, and ZnO NWs sheathed in amorphous carbon with Co clusters are inspected by using electron microscopy and PL spectra analyses.

4.1 Growth behavior

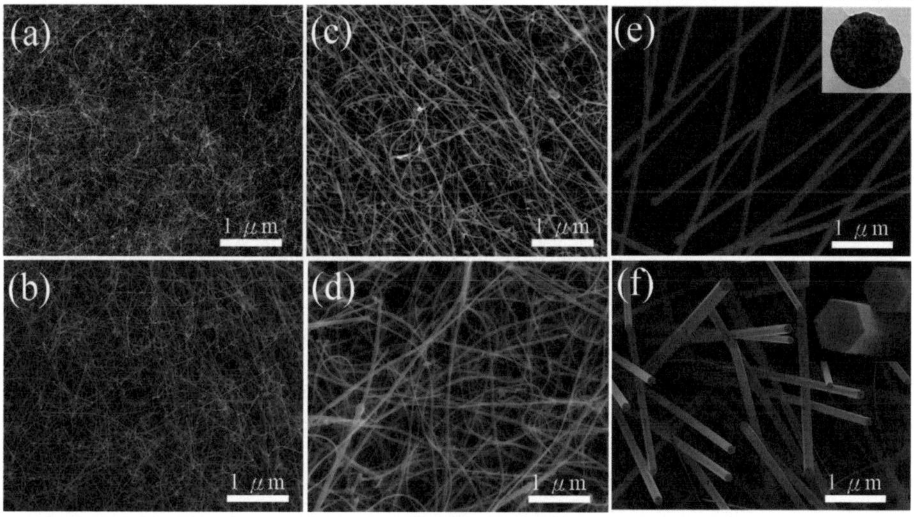

Fig. 3. SEM images of as-grown cylindrical ZnO NWs with average diameters of (a) 7 nm, (b) 12 nm, (c) 19 nm, (d) 38 nm, and (e) 113 nm. (f) SEM image of as-grown hexagonal ZnO NWs with an average diameter of 134 nm. The insets of Figs. 3(e) and (f) display cross-sections of cylindrical and hexagonal ZnO NWs, respectively.

Figures 3(a)-(e) display SEM images of cylindrical ZnO NWs with increasing average diameters and Fig. 3(f) displays a SEM image of hexagonal ZnO NWs. The cylindrical NWs with average diameters of 7, 12, 19, 38, and 113 nm were grown by using gold-nanoparticle catalysts with average sizes of 5, 10, 20, 40, and 100 nm, respectively. The NWs displaying in the same magnification SEM images demonstrate obviously distinct dimensions, implying a very well control of the NW diameter through the size of gold nanoparticles. In addition, the surface morphology of cylindrical NWs appearing in the inset of Fig. 3(e) indicates that the cross-section of ZnO NWs certainly conforms to the circular shape of gold nanoparticles. A different synthesis method resulting in an either circular or hexagonal cross-section could be discerned in the insets of Figs. 3(e) and (f). Besides, we have noticed a more and more curved feature for cylindrical ZnO NWs as compared with hexagonal ones, and for thinner NWs as compared with thicker ones. The same growth period of 8 h is kept and a considerably high density of small diameter NWs could be observed unambiguously in SEM images.

Figure 4 displays statistical information of diameters of our as-grown ZnO NWs. In Fig. 4(a), we demonstrate a representable diameter distribution of cylindrical ZnO NWs with a 12-nm average diameter. The standard deviation of the 12-nm diameter NWs is evaluated to be 2.7 nm (23%). This somewhat large deviation in NW diameter may come from a broad size distribution of our catalysts, gold nanoparticles, which is not investigated in this experiment yet. In contrast, Fig. 4(b) reveals a flat diameter distribution, indicating a large diameter deviation of hexagonal ZnO NWs, due to a disparate growth behaviour. The average diameter and standard deviation of hexagonal NWs are estimated to be about 134 nm and 74 nm (55%), respectively. Figure 4(c) reveals the average diameters and standard deviations of cylindrical ZnO NWs as a function of sizes of gold nanoparticles. A highly linear correlation between the nanoparticle and the NW diameters firmly corroborate again a well control of the NW diameter.

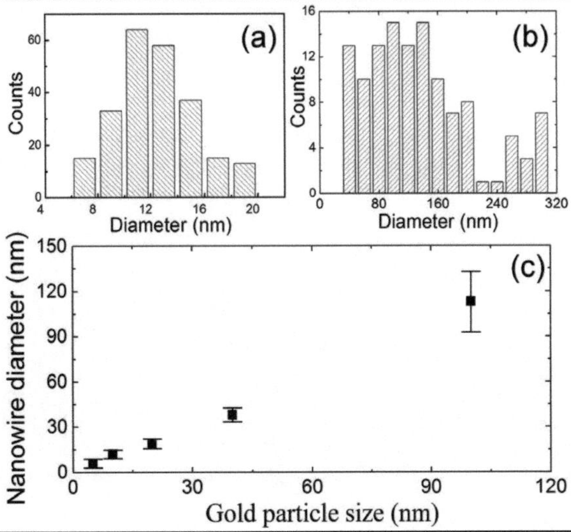

Fig. 4. (a) A typical statistical distribution of NW diameters for cylindrical ZnO NWs having an average diameter of 12 nm. (b) A statistical distribution of NW diameters for hexagonal ZnO NWs. (c) The average diameters with standard deviations of cylindrical NWs as a function of the diameter of the gold nanoparticles (catalysts).

4.2 Morphology and crystalline structure

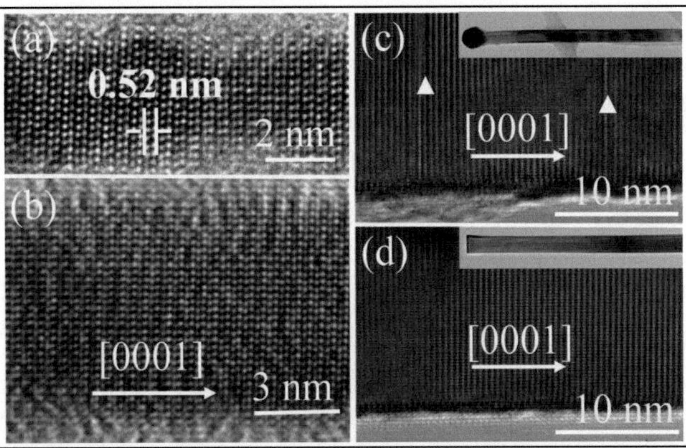

Fig. 5. High-resolution TEM images of as-grown cylindrical ZnO NWs with average diameters of (a) 7 nm, (b) 12 nm, and (c) 38 nm. (d) High-resolution TEM image of hexagonal ZnO NWs with an average diameter of 134 nm. The upper right insets in Figs. 5(c) and (d) show TEM images of as-grown ZnO NWs at low magnification. The upper triangles marked in Fig. 5(c) point to the planar defects of stacking faults.

Four representative high-resolution TEM images of as-grown cylindrical ZnO NWs with average diameters of 7, 12, and 38 nm are presented in Figs. 5(a), (b), and (c), respectively. A double layer spacing of 0.52 nm agrees well with the *c* lattice constant of a ZnO wurtzite crystal structure that also denotes the [0001] growth direction. A single crystalline structure in different average diameters of either cylindrical or hexagonal ZnO NWs has been inspected and verified. The insets in Figs. 5(c) and (d) demonstrate TEM images of cylindrical and hexagonal ZnO NWs, respectively, at low magnification. A gold nanoparticle sitting on one end of the cylindrical ZnO NW is observed in Fig. 5(c) as well. Further, as exhibited in Fig. 5(a), a nanoscale bumper edge surface is more evidently observed on thiner and cylindrical NWs than on thicker and hexagonal NWs. Moreover, many stacking faults, designating by upper triangles in Fig. 5(c), are identified in cylindrical NWs.

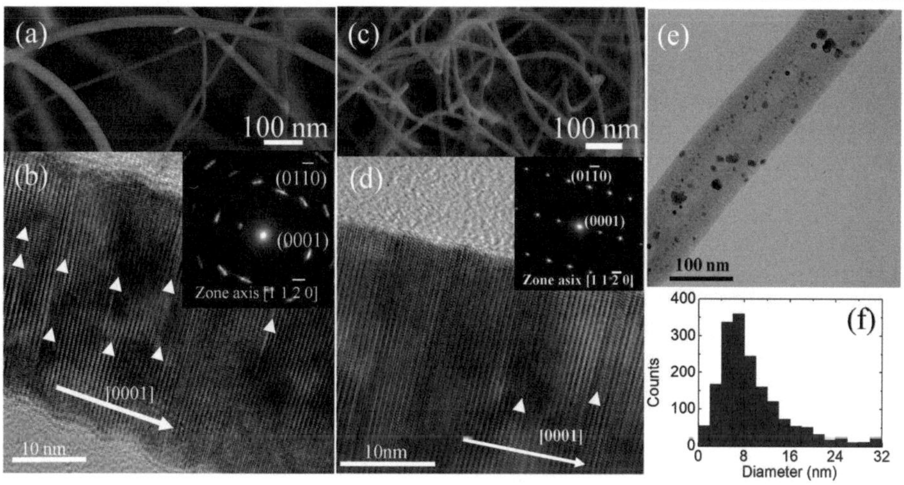

Fig. 6. SEM images of (a) as-implanted and (c) high-vacuum annealed $Zn_{1-x}Co_xO$ NWs with an average diameter of 38 nm and a Co ion does of 6×10^{16} cm^{-2}. High resolution TEM images of (c) as-implanted and (d) high-vacuum annealed $Zn_{1-x}Co_xO$ NWs. The insets in Figs. 6(b) and (d) display corresponding electron diffraction patterns. (e) TEM image of a ZnO NW sheathed in carbon amorphous with Co clusters after high-vacuum annealing. (f) Statistical distribution of Co-cluster diameters estimated from TEM images. The average diameter and standard deviation are 9.4 and 6.0 nm, respectively.

As-grown and pure ZnO NWs were doped by high energy Co ions to form $Zn_{1-x}Co_xO$ NWs. SEM and TEM images of $Zn_{1-x}Co_xO$ NWs with an average diameter of 38 nm and a Co ion dose of 6×10^{16} cm^{-2} are demonstrated in Figs. 6(a) and (b). In Fig. 6(a), the bending feature of as-implanted $Zn_{1-x}Co_xO$ NWs is appreciable. In addition, the as-implanted $Zn_{1-x}Co_xO$ NWs consist of lots of stacking faults, as designated by triangles in Fig. 6(b), and they exhibit a streaking of an electron diffraction pattern (see the inset). Although the bending feature can be detected in as-grown ZnO NWs, more and more stacking faults and an obvious streaking in an electron diffraction pattern are discovered in as-implanted $Zn_{1-x}Co_xO$ NWs. It is proposed that these structure defects could mainly come from a high-energy Co ion bombardment during the ion implantation process. After a high-vacuum annealing, SEM and TEM images of the same sample are shown in Figs. 6(c) and (d). We can see that the stacking faults (indicated as triangles in Fig. 6(d)) and streaking are removed after a thermal treatment. We notice that an annealing at 600°C could help to recover structure disorders and defects. We also found that annealing at a higher temperature may cause a meltdown of

ZnO NWs. A lower annealing temperature of 450°C was thereafter applied to subsequent experiments and no noticeable changes in morphology were observed after the thermal treatment.

In order to study the magnetic mechanism in $Zn_{1-x}Co_xO$ NWs, we have made a comparative sample, ZnO NWs sheathed in amorphous carbon by Co ion implantation. There are neither perceptible clusters nor nanocrystals before any thermal treatments (not shown in figures). After a high-vacuum annealing, Co clusters, having a broad size distribution, could be discovered in TEM images. Figures 6(e) and (f) show a typical TEM image of the Co clusters and a statistical distribution of diameters. The 40-nm diameter ZnO NW is embedded in a shell of carbon amorphous with a diameter of ~100 nm. The average diameter and standard deviation of the Co clusters are about 9.4 and 6.0 nm, respectively. This sample was fabricated by Co ion implantation with a dose of 4×10^{16} cm^{-2} and post-annealed in a high vacuum at 600°C. In contrast to the high-vacuum annealed $Zn_{1-x}Co_xO$ NWs, in which Co cluster have never been detected in TEM images (Fig. 6(d)), the sample of ZnO sheathed in amorphous carbon with Co-ion implantation exhibits obviously many Co clusters after a high-vacuum annealing. The result suggests that Co ions may have a longer diffusion length in amorphous carbon than that in ZnO.

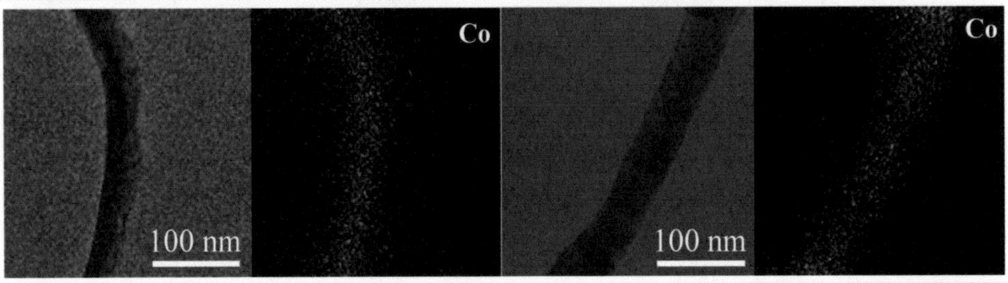

Fig. 7. TEM images of (a) as-implanted and (c) high-vacuum annealed $Zn_{1-x}Co_xO$ NWs with a dose of 6×10^{16} cm^{-2} and an average diameter of 38 nm. EDX mapping images of the Co element in (b) as-implanted and (d) high-vacuum annealed $Zn_{1-x}Co_xO$ NWs.

To identify Co ion distribution in $Zn_{1-x}Co_xO$ NWs, a EDX mapping of the Co element is employed. Figures 7(a) and (b) present TEM and EDX mapping images of as-implanted $Zn_{1-x}Co_xO$ NWs. For comparison, TEM and EDX mapping images of high-vacuum annealed $Zn_{1-x}Co_xO$ NWs are given in Figs. 7(c) and (d). Under the spatial resolution of the EDX chemical mapping, no perceptible aggregation of Co ions has ever been detected in all high-vacuum annealed $Zn_{1-x}Co_xO$ NWs. The results are in line with the TEM measurements shown in Fig. 6. We concluded, therefore, that a low temperature thermal treatment (below 600°C) can induce a recovery of a structure disorder but not a diffusion and aggregation of Co ions. Moreover, a high resolution technique, the compositional mapping of electron energy loss spectroscopy, a confirmation of a non-aggregated distribution of Co element in both as-implanted and high-vacuum annealed $Zn_{1-x}Co_xO$ NWs (Chen et al., 2008). Chemical compositions of $Zn_{1-x}Co_xO$ NWs were determined by using EDX spectra (Jian et al., 2007). Average Co-concentrations of $Zn_{1-x}Co_xO$ NWs were decided to be 2, 4, 6, 8, 10, and 11% for ZnO NWs with Co ion doses of 1, 2, 3, 4, 5, and 6×10^{16} cm^{-2}, respectively. The variation of NW diameters does not affect the average Co-concentration but causes a large standard deviation of Co-concentration in thinner $Zn_{1-x}Co_xO$ NWs.

4.3 Photoluminescence spectra

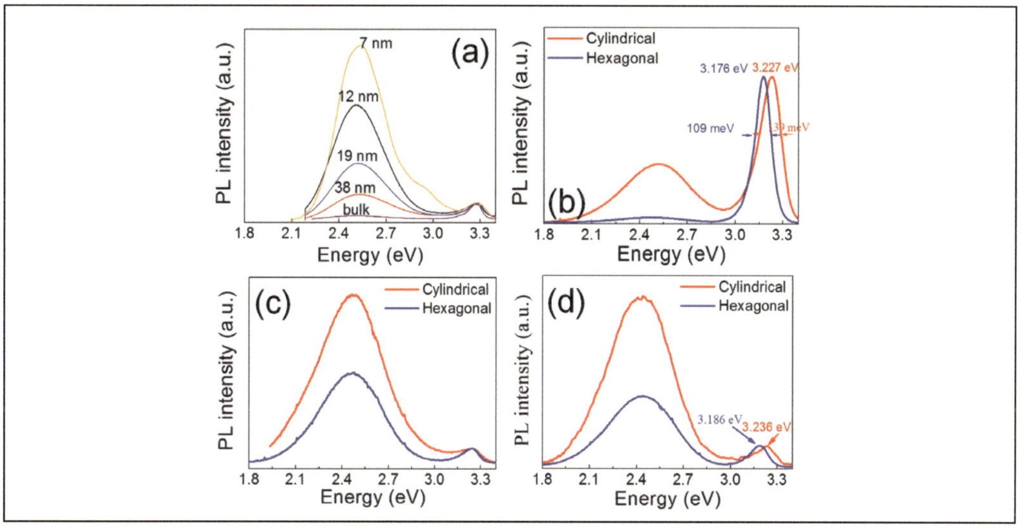

Fig. 8. (a) Room temperature PL spectra of as-grown (pure) ZnO bulk and NWs with average diameters as indicated on graph. Room temperature PL spectra of (b) as-grown, (c) as-implanted, and (d) high-vacuum annealed $Zn_{0.89}Co_{0.11}O$ NWs with average diameters of 113 and 134 nm for cylindrical and hexagonal cross sections.

The PL spectra of as-grown ZnO, as-implanted $Zn_{1-x}Co_xO$, and high-vacuum annealed $Zn_{1-x}Co_xO$ NWs are presented in Fig. 8. PL attributes show a green defect emission at ~2.5 eV and a near band edge emission at ~3.2 eV. The peak of the near band edge emission is normalized to be of the same height for a easy comparison of the defect emission. Figure 8(a) exhibits PL spectra of NWs with several average diameters. It shows that the intensity of the green emission (near band emission) is relatively higher (lower) for thinner ZnO NWs. As mentioned above, thinner NWs were examined to have a high density of structure defects (stacking faults), a bumper surface, and a curved feature. The green defect emission might be in connection with structure defects such as point defects of oxygen vacancies and zinc interstitials since an increase of point defects may cause a generation of more planar defects (stacking faults).

Figures 8(b), (c), and (d) present PL spectra of as-grown, as-implanted, and high-vacuum annealed $Zn_{0.89}Co_{0.11}O$ NWs with average diameters of 113 and 134 nm for cylindrical and hexagonal cross sections. The as-implanted $Zn_{0.89}Co_{0.11}O$ NWs exhibit a much higher intensity of a green emission (Fig. 8(c)) while the high-vacuum annealed NWs show a lower green emission peak (Fig. 8(d)). This result marks a correlation between the structure defects and the green emission as well. In addition to an intensity change of the defect emission, Fig. 8(b) points to a shift of the near band edge emission from 3.178 eV (cylindrical) to 3.227 eV (hexagonal) for as-grown ZnO NWs. Both cylindrical and hexagonal, as-implanted NWs reveal a blue shift in the near band edge emission peak (see Fig. 8(c)). After annealing in a high vacuum, the blue shift disappears and the band emission peaks move back to 3.186 and 3.236 eV for cylindrical and hexagonal $Zn_{0.89}Co_{0.11}O$ NWs, respectively. Though structure defects, strains, and surface effects (surface roughness) could all be the rationales, we believe that the anomalous blue shift could predominantly come from the surface effects.

5. Magnetic properties

The morphology and structure analyses indicate that Co-ions are randomly disributed without aggregation in as-implanted and high-vacuum annealed $Zn_{1-x}Co_xO$ NWs. Meanwhile, the high-vacuum annealing will induce an aggregation of Co ions in amorphous carbon coated on ZnO NWs and result in Co clusters. In this section, a SQUID magnetometer is employed to study temperature and field dependent behaviours of as-implanted $Zn_{1-x}Co_xO$, high-vacuum annealed $Zn_{1-x}Co_xO$, and ZnO NWs sheathed in amorphous carbon with Co clusters.

5.1 Temperature dependent magnetization

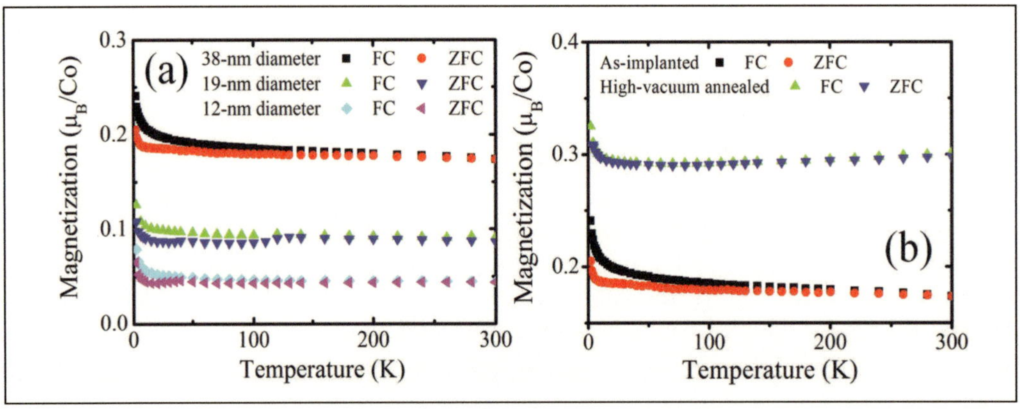

Fig. 9. (a) FC and ZFC behaviors of temperature dependent magnetization of as-implanted $Zn_{0.92}Co_{0.08}O$ NWs with average diameters of 12, 19, and 38 nm. (b) FC and ZFC magnetizations of as-implanted and high-vacuum annealed $Zn_{0.92}Co_{0.08}O$ NWs with an average diameter of 38 nm. The annealing time is 12 h.

In a magnetic field of 500 Oe, temperature dependent magnetizations of various average diameters of $Zn_{0.92}Co_{0.08}O$ NWs are shown in Fig. 9(a). The magnetization per Co ion of as-implanted $Zn_{0.92}Co_{0.08}O$ NWs depends strongly on the NW diameter. Thicker NWs exhibit higher magnetization. In addition, magnetizations under field cooled (FC) and zero-field cooled (ZFC) procedures show a division into two separate curves with decreasing temperature. In a similar way, the transition temperature, at which the FC and ZFC magnetization curves bifurcate, is higher for the thicker NWs. The difference in FC and ZFC magnetization suggests an existence of small magnetic domains in $Zn_{0.92}Co_{0.08}O$ NWs. In addition, the high transition temperature implies larger magnetic domains existing in thicker NWs.

In a previous report (Chen et al., 2008), we argued that either oxygen vacancies or zinc interstitials could result in a ferromagnetic coupling between the Co ions. It is conjectured that the as-implanted $Zn_{0.92}Co_{0.08}O$ NWs consist of the same concentration of oxygen vacancies (zinc interstitials) so the size of magnetic domains of non-aggregated Co ions may be larger in thicker NWs. On the other hand, the size dependent magnetization and hysteresis loop could be owing to the generation of planar defects, stacking faults and streaking, during the ion bombardment process (Jian et al., 2006). Moreover, planar defects could hinder an oxygen-vacancy mediated ferromagnetic ordering so as to abate magnetization and coercivity of thinner, as-implanted $Zn_{1-x}Co_xO$ NWs.

Figure 9(a) displays a non-vanishing and non-decreasing magnetization up to a room temperature, signifying a ferromagnetic ordering as well as RTFM. After annealing in a high vacuum, the temperature behavior of $Zn_{0.92}Co_{0.08}O$ NWs with an average diameter of 38 nm is displayed in Fig. 9(b), including reproduced data of as-implanted NWs for comparison. The temperature behavior demonstrates a much higher magnetization (a strong ferromagnetic state) and a coincidence and overlapping of FC and ZFC magnetization. This result indicates a growth and development of large magnetic domains, formed by non-aggregated Co ions in high-vacuum annealed $Zn_{0.92}Co_{0.08}O$ NWs. This phenomena can be observed in all $Zn_{1-x}Co_xO$ NWs having different diameters and Co-concentrations. It implies that a high-vacuum annealing produces oxygen vacancies (zinc interstitials) to enhance a ferromagnetic interaction between Co ions and to intensify a magnetic state.

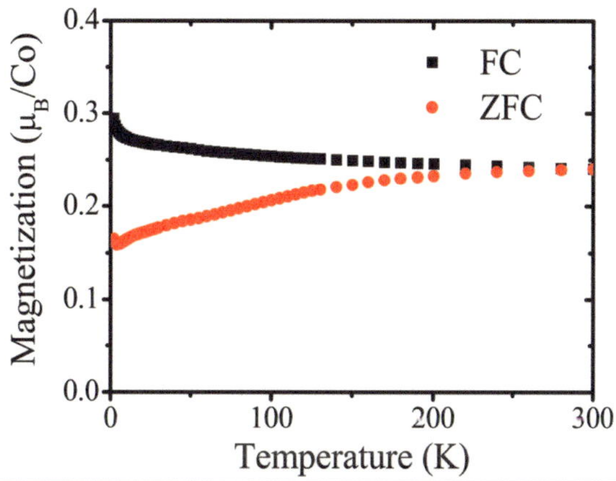

Fig. 10. FC and ZFC magnetization of ZnO sheathed in amorphous carbon with Co clusters. The Co ion dose and average diameter of ZnO are 4×10^{16} cm^{-2} and 38 nm, respectively, for this sample.

The temperature dependent behavior of ZnO sheathed in amorphous carbon with Co clusters is presented in Fig. 10. FC and ZFC magnetizations are separated into two parts with a decrease of temperature. The undeniable bifurcation of temperature dependent magnetization in FC and ZFC procedures stands for a superparamagnetic feature of ferromagnetic collloids of Co clusters (Bean & Livingston, 1959). This feature will be evident if the Co clusters are monodispersed and uniform in size. As we have shown in Fig. 6(f), the Co clusters have a wide distribution and a standard deviation of ~6.0 nm in diameter that causes a relatively small deviation in FC and ZFC magnetization at low temperatures in comparison with ideal ferromagnetic colloids. The magnitude of several tenths of μ_B in magnetization is in the same order of magnitude as that of DMS $Zn_{1-x}Co_xO$ NWs (see Fig. 9). This finding indicates that magnetic moments of cluster samples and DMS $Zn_{1-x}Co_xO$ NWs do originate from aggregated Co nanoparticles and non-aggregated Co ions, respectively.

5.2 Field dependent magnetization

In addition to a temperature dependent behavior, data of field dependent magnetizations as well as hysteresis loops were taken at several different temperatures. Figure 11 exhibits hysteresis loops of as-implanted $Zn_{1-x}Co_xO$ NWs. Having an equal Co-concentration of 8%,

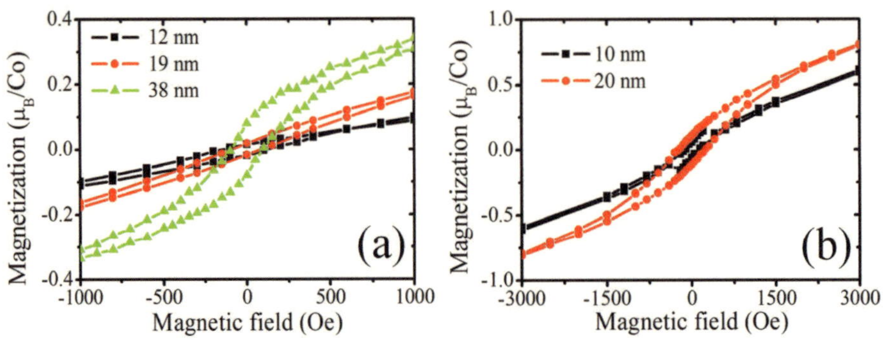

Fig. 11. (a) Hysteresis loops of as-implanted $Zn_{0.92}Co_{0.08}O$ NWs with three different average diameters marked on graph. The data were taken at 5 K. (b) Hysteresis loops of as-implanted $Zn_{0.96}Co_{0.04}O$ NWs with two different average diameters marked on graph. The data were taken at 2 K.

thick NWs reveal a high magnetization and a larger hysteresis loop (see Fig. 11(a)). Figure 11(b) presents a similar manner of a size dependence to convince us this general phenomena observed in as-implanted $Zn_{1-x}Co_xO$ NWs. The consequence of a high magnetization in thick NWs agrees with the temperature dependence delineated in Fig. 9(a).

We have argued that the implantation of a high beam current of 600 nA/cm^2 could somewhat introduce a high-vacuum annealing and create oxygen vacancies (zinc interstitials) in ZnO NWs so as to turn on an exchange interaction between non-aggregated Co ions. The Co ions occupying in a certain volume of a ZnO form a magnetic domain. If the ZnO is cut into smaller pieces such as NWs, the magnetic domain and magnetization (moment) will be abated and reduced. This splitting and diminishing of magnetic domains lead to the size effect observed in as-implanted $Zn_{1-x}Co_xO$ NWs. Moreover, the small hysteresis loop indicating a low coercive field (force) in thin $Zn_{1-x}Co_xO$ NWs may be due to a weak interaction between size-reduced magnetic domains or to planar defects (stacking faults and streaking) induced a reduction of ferromagnetic interactions.

We have observed an increase in magnetization from temperature dependent studies after a high-vacuum annealing (Section 5.1). To learn the annealing effect, multiple steps of high-vacuum annealing for hours are employed and the field dependent magnetizations are investigated after each step of annealing. Figure 12(a) and (b) demonstrate a change in hysteresis loops of $Zn_{0.92}Co_{0.08}O$ NWs with average diameters of 38 and 19 nm after each step of a high-vacuum annealing. The magnetization as well as the loop becomes higher and larger after several steps of high-vacuum annealing. The results of multiple-step annealing implies a diffusion of composing elements of the $Zn_{1-x}Co_xO$ material. It has been confirmed from EDX, EELS mapping, and high-resolution TEM inspections that the annealing will not induce detectable diffusion and clustering of Co ions in the DMS NWs. We argued, therefore, that the annealing effect produces oxygen vacancies (zinc interstitials) to enhance an exchange interaction between Co ions.

In addition to the dependence of annealing time, different surface ratios of thin and thick NWs may give rise to dissimilar responses to annealing time. Figure 12(b) reveals a larger increase and expansion in magnetization and field-dependent loops for thinner (19-nm average diameter) NWs. A decrease of annealing time and steps for thinner $Zn_{1-x}Co_xO$ NWs is due to a

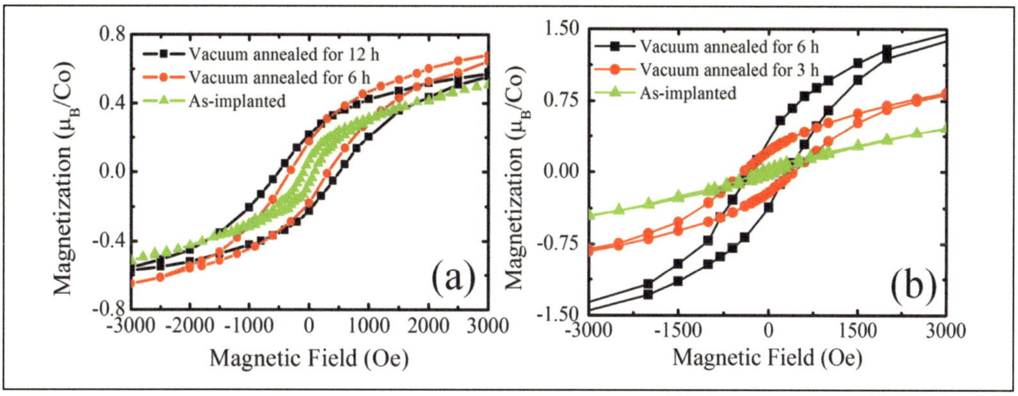

Fig. 12. (a) Hysteresis loops, taken at 10 K, of as-implanted, 6-h vacuum annealed, and 12-h vacuum annealed $Zn_{0.92}Co_{0.08}O$ NWs with a 38-nm average diameter. (b) Hysteresis loops, taken at 2 K, of as-implanted, 3-h vacuum annealed, and 6-h vacuum annealed $Zn_{0.92}Co_{0.08}O$ NWs with a 19-nm average diameter.

large surface-to-volume ratio for oxygen diffusion and a large increase in magnetization could be related to the above-mentioned reduction of magnetization in thinner NWs. To confirm the creation of oxygen vacancies during the high-vacuum annealing process, the sample is annealed in oxygen to exhibit a weak magnetic state of a low magnetization and small a hysteresis loop (not shown here), and they are subsequently annealed in a high vacuum to recover a strong magnetic state in high-vacuum annealed $Zn_{1-x}Co_xO$ NWs.

To learn more about the high vacuum annealing enhancement of ferromagnetic ordering, temperature dependence of hysteresis loops of $Zn_{0.92}Co_{0.08}O$ NWs with 70-nm average diameter are displayed in Fig. 13(a). Unlike a bulk magnet which shows a weak temperature dependence of hysteresis loops, the DMS NWs display a strong temperature dependence as presented in Fig. 13(a). They exhibit the largest hysteresis loop at 2 K and a shrinkage of the loop like a paramagnetic linear response above room temperature. We infer that a splitting and dividing of magnetic domains (from bulk) into a small volume in NWs leads to a strong temperature dependence of hysteresis loops. This phenomena is similar to a

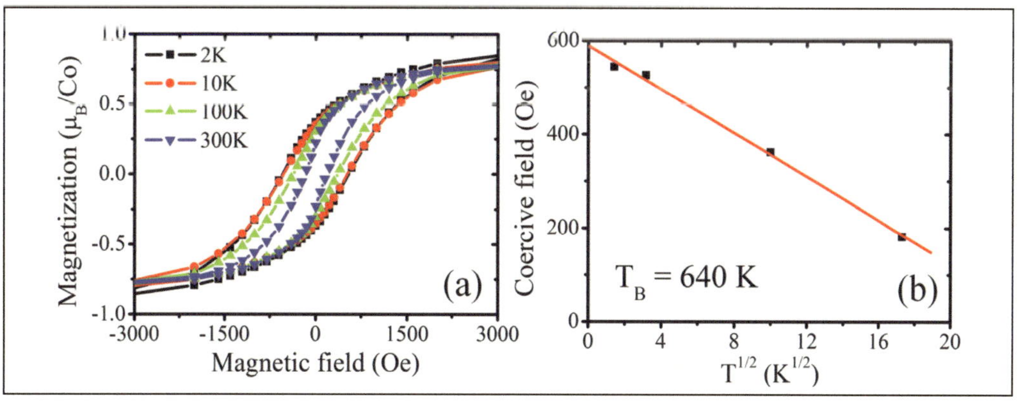

Fig. 13. (a) Hysteresis loops of $Zn_{0.92}Co_{0.08}O$ NWs with 70-nm average diameter at several temperatures after annealing in a high vacuum for 12 h. (b) The coercive field, estimated from Fig. 13(a), as a function of square root of temperature.

superparamagnetic effect on ferromagnetic colloids or magnetic clusters. On the other hand, if the exchange interaction is mediated by oxygen vacancies (zinc interstitials), a random distribution of these vacancies in $Zn_{1-x}Co_xO$ NWs could give a oxygen-vacancy depleted and non-ferromagnetic regime. This non-ferromagnetic regime separates and splits magnetic domains of non-aggregated Co ions into smaller ones. An analysis method similar to that used in a study of superparamagnetism is employed and the coercive fields evaluated from Fig. 13(a) are presented as a function of square root of temperature in Fig. 13(b). A linear dependence can be derived appreciably.

On the contrary, if we assume that the temperature dependent coercivity is originated from Co clusters, we may estimate the cluster diameter according to the equation (McHenry et al., 1994):

$$T_B = \frac{K\langle V\rangle}{30k_B}, \tag{1}$$

where T_B is the blocking temperature, $K \approx 5 \times 10^6$ erg/cm^3 is the anisotropy energy of Co metal, k_B is the Boltzmann constant, and $<V>$ is the average volume of Co clusters. The T_B can be estimated to be 640 K from the x-axis intercept of the red fitting line in Fig. 13(b). Assume a spherical geometry for Co clusters, an average diameter of ~9 nm is derived. Such a large cluster of ~9 nm in diameter, if any exist, should be detectable in electron microscopy analyses. The Co clusters are nevertheless invisible in electron microscopy images. The temperature dependence of coercivity is therefore owing to magnetic domains formed by non-aggregated Co ions in ZnO NWs. Moreover, Fig. 13(a) demonstrates a temperature independence of magnetization saturation that is consistent with the result shown in Fig. 9(b). The ferromagnetic ordering remains up to room temperature so the RTFM in the high-vacuum annealed $Zn_{1-x}Co_xO$ NWs is confirmed.

The field dependent magnetization of the Co clustering sample at various temperatures is displayed in Fig. 14 for a comparative study. The superparamagnetic attribute of a shrinkage of hysteresis loops as well as a decrease in coercive fields with increasing temperature is perceived. The non-vanishing magnetization and coercive field at 300 K implicate that both the Curie and blocking temperatures, T_C and T_B, are above room temperature. The coercive field as a function of square root of temperature is delineated in the inset of Fig. 14. The data are fitted according to the equation (McHenry et al., 1994):

$$H_C(T) = H_C(0)\left(1 - \sqrt{T / T_B}\right), \tag{2}$$

where $H_C(T)$ and $H_C(0)$ are coercive fields at temperatures T and 0 K, respectively. The blocking temperature is determined to be ~420 K via a least square fitting, shown as a red line in the inset of Fig. 14. The average diameter of ~9 nm can be estimated by using Eq. 1 with T_B = 420 K. The average diameter of ~9 nm agrees very well with that calculated from a statistical distribution of cluster diameters from TEM measurements (9.4 nm in Fig. 6(f)). This result sustains the analyses and deductions used in this work. It is noted that all of the three characteristics of superparamagnetic Co clusters as well as ferromagnetic colloids have been observed. These features are a bifurcation of FC and ZFC magnetication (Fig. 10), a temperature dependent coercive field (Fig. 14), and the same average diameter evaluated from both TEM measurements (Fig. 6(f)) and T_B estimations (Eq. 2). It is emphasized that a wide distribution in cluster diameters (standard deviation of ~6.0 nm in this case) could smooth out superparamagnetic characteristics.

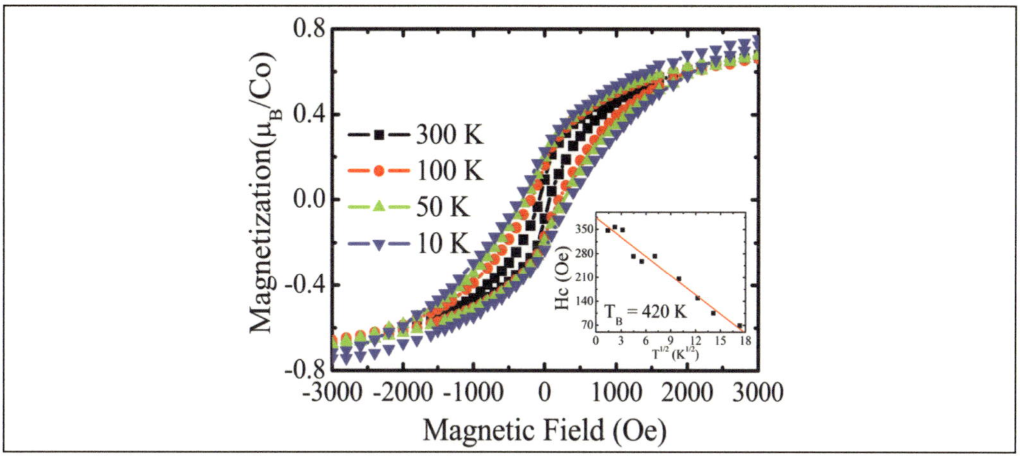

Fig. 14. Hysteresis loops of ZnO sheathed in amorphous carbon with Co clusters. The Co ion dose and average diameter of ZnO are 4×10^{16} cm^{-2} and 38 nm, respectively. The inset shows the coercive field as a function of square root of temperature.

In contrast to Co clustering samples, as-implanted (DMS) $Zn_{1-x}Co_xO$ NWs display a slight distinction between FC and ZFC magnetization in Fig. 9. After a high-vacuum annealing, FC and ZFC magnetization of DMs NWs cannot be separated from each other. High-vacuum annealed $Zn_{1-x}Co_xO$ NWs present a temperature dependence of hysteresis loops and coercive fields as shown in Fig. 13. Such a large cluster diameter of ~9 nm is estimated from the temperature dependent coercivity, but no perceptible Co clusters have ever been detected in $Zn_{1-x}Co_xO$ NWs (see Fig. 6). The Co ions in the NWs show non-aggregated random distribution (see Fig. 7) before and even after annealing in a high vacuum. It is conjectured, therefore, that a magnetic domain could be composed of non-aggregated Co ions in $Zn_{1-x}Co_xO$ NWs. Field dependent magnetization shown in Fig. 12 endorses our conjecture. Further, a vacuum annealing can help to generate oxygen vacancies (zinc interstitials) and induce a ferromagnetic interaction between Co ions. Moreover, the size effect shown in Figs. 9 and 11 implicates that $Zn_{1-x}Co_xO$ NWs, like a large magnetic domain in bulk being cut into small pieces, may reveal a relatively low magnetization as well as a weak magnetic state. That is why the $Zn_{1-x}Co_xO$ NWs exhibit superparamagnetic features. Through a systematic analysis and a comparative study with Co clustering samples, we come to a conclusion that $Zn_{1-x}Co_xO$ NWs are a DMS material. Moreover, a ferromagnetic order in $Zn_{1-x}Co_xO$ NWs remains up to room temperature, implying the RTFM in this particular material.

6. Conclusion

Various average diameters of single-crystalline, either cylindrical or hexagonal ZnO NWs with a [0001] growth direction are synthesized by using the vapor transport method. The diameters of cylindrical ZnO NWs can be well regulated by using gold nanoparticles as catalysts while the diameters of hexagonal NWs have a wide statistical distribution. The hexagonal NWs show straight in the growth direction whereas the cylindrical NWs show a bending feature, structure defects of stacking faults and point defects, and bumpy surfaces. The thinner the NWs are the higher the structure defect density and the more obvious the bending manner exist. In addition, the optical properties of thin ZnO NWs show a stronger green defect emission.

The as-grown ZnO NWs are implanted with different doses of Co ions to form $Zn_{1-x}Co_xO$ NWs ($x < 0.12$). The as-implanted $Zn_{1-x}Co_xO$ NWs possess a high density of bombardment-induced structure defects and exhibit either a paramagnetic or a weak ferromagnetic state. Thinner as-implanted $Zn_{1-x}Co_xO$ NWs exhibit a larger hysteresis loop and a higher magnetization. This NW-diameter dependence indicates that a bulk magnet with a large magnetic domain is divided into many pieces of NWs with lots of small-size domains. Annealing in a high vacuum reduces structural, planar defects of stacking faults and streaking, and creates point defects of oxygen vacancies (zinc interstitials) in $Zn_{1-x}Co_xO$ NWs to induce a strong ferromagnetic state. By using EDX mapping, it is observed that Co ions are randomly distributed without any aggregations in both as-implanted and annealed $Zn_{1-x}Co_xO$ NWs. The annealing effect further supports the idea of oxygen vacancies (zinc interstitials) induced ferromagnetic interactions between Co ions. Moreover, an oxygen-vacancy depleted, non-magnetic regime and a NW-divided small volume give rise to a separation and partition of a ferromagnetic domain, leading to a superparamagnetic feature. After high-vacuum annealing, more oxygen vacancies are generated and the magnetic domains grow up. The superparamagnetic features gradually disappear in the high-vacuum annealed $Zn_{1-x}Co_xO$ NWs. The ferromagnetic properties are observed at room temperature to assure the RTFM in the high-vacuum annealed $Zn_{1-x}Co_xO$ NWs, and to confirm the T_C above room temperature.
In particular, ZnO NWs sheathed in amorphous carbon with Co clusters have been produced after annealing in a high vacuum. The clustering sample show superparamagnetic features of FC and ZFC magnetization separation, temperature dependent coercivities, and a blocking temperature with which average diameters of Co clusters have been evaluated. The result of a comparative study with Co clustering samples corroborates our measurements and analyses of DMS $Zn_{1-x}Co_xO$ NWs.

7. Acknowledgments

This work was supported by the Taiwan National Science Council under Grant No. NSC 95-2112-M-009-045-MY3 and by the MOE ATU Program. The magnetization measurements were performed on a SQUID magnetometer (MPMS XL-7) at the National Chiao Tung University.

8. References

Baibich, M. N.; Broto, J. M.; Fert, A.; Nguyen Van Dau, F.; Petroff, F.; Eitenne, P.; Creuzet, G.; Friederich, A. & Chazelas, J. (1988). Giant Magnetoresistance of (001)Fe/(001)Cr Magnetic Superlattices. *Phys. Rev. Lett.*, Vol. 61, pp. 2472-2475.

Bean, C. P. & Livingston, J. D. (1959). Superparamagnetism. *J. Appl. Phys.*, Vol. 30, pp. 120S-129S.

Behan, A. J.; Mokhtari, A.; Blythe, H. J.; Score, D.; Xu, X. H.; Neal, J. R.; Fox, A. M. & Gehring, G. A. (2008). Two Magnetic Regimes in Doped ZnO Corresponding to a Diluted Magnetic Semiconductor and a Diluted Magnetic Insulator. *Phys. Rev. Lett.*, Vol. 100, p. 047206.

Binasch, G.; Grünberg, P.; Saurenbach F. & Zinn, W. (1989). Enhanced magnetoresistance in layered magnetic structures with antiferromagnetic interlayer exchange. *Phys. Rev. B*, Vol. 39, pp. 4828-4830.

Chen, I. J.; Ou, Y. C.; Wu, Z. Y.; Chen, F. R.; Kai, J. J.; Lin, J. J. & Jian, W. B. (2008). Size Effects on Thermal Treatments and Room-Temperature Ferromagnetism in High-Vacuum Annealed ZnCoO Nanowires. *J. Phys. Chem. C*, Vol. 112, pp. 9168-9171.

Chiba, D.; Yamanouchi, M.; Matsukura, F. & Ohno, H. (2003). Electrical Manipulation of Magnetization Reversal in a Ferromagnetic Semiconductor. *Science*, Vol. 301, pp. 943-945.

Cho, S.; Choi, S.; Cha, G. B.; Hong, S. C.; Kim, Y.; Zhao, Y. J.; Freeman, A. J.; Ketterson, J. B.; Kim, B. J.; Kim, Y. C.; Choi, B. C. (2002). Room-Temperature Ferromagnetism in $(Zn_{1-x}Mn_x)GeP_2$ Semiconductors. *Phys. Rev. Lett.*, Vol. 88, p. 257203.

Coey, J. M. D.; Venkatesan, M. & Fitzgerald, C. B. (2005). Donor impurity band exchange in dilute ferromagnetic oxides. *Nat. Mater.*, Vol. 4, pp. 173-179.

Dieny, B.; Speriosu, V. S.; Parkin, S. S. P.; Gurney, B. A.; Wilhoit, D. R. & Mauri, D. (1991). Giant Magnetoresistance in soft ferromagnetic multilayers. *Phys. Rev. B*, Vol. 43, pp. 1297-1300.

Dietl, T.; Ohno, H.; Matsukura, F.; Cibert, J. & Ferrand, D. (2000). Zener Model Description of Ferromagnetism in Zinc-Blende Magnetic Semiconductors. *Science*, Vol. 287, pp. 1019-1022.

Furdyna, J. K. (1988). Diluted magnetic semiconductors. *J. Appl. Phys.*, Vol. 64, pp. R29-R64.

Hayashi, T.; Shimada, H.; Shimizu, H. & Tanaka, M. (1999). Tunneling spectroscopy and tunneling magnetoresistance in (GaMn)As ultrathin heterostructures. *J. Cryst. Growth*, Vol. 201/202, pp. 689-692.

Janisch, R.; Gopal, P. & Spaldin, N. A. (2005). Transition metal-doped TiO2 and ZnO--present status of the field. *J. Phys. Condens. Mater.*, Vol. 17, pp. R657-R689.

Jian, W. B.; Wu, Z. Y.; Huang, R. T.; Chen, F. R.; Kai, J. J.; Wu, C. Y.; Chiang, S. J.; Lan, M. D. & Lin, J. J. (2006). Direct observation of structure effect on ferromagnetism in $Zn_{1-x}Co_xO$ nanowires. *Phys. Rev. B*, Vol. 73, p. 233308.

Jian, W. B.; Chen, I. J.; Liao, T. C.; Ou, Y. C.; Nien, C. H.; Wu, Z. Y.; Chen, F. R.; Kai, J. J. & Lin, J. J. (2007). Size Dependent Magnetization and High Vacuum Annealing Enhanced Ferromagnetism in $Zn_{1-x}Co_xO$ Nanowires. *J. Nanosci. Nanotechnol.*, Vol. 8, pp. 202-211.

Klingshirn, C. (2007). ZnO : Material, Physics and Applications. *ChemPhysChem*, Vol. 8, pp. 782-803.

Lany, S. & Zunger, A. (2005). Anion vacancies as a source of persistent photoconductivity in II-VI and chalcopyrite semiconductors. *Phys. Rev. B*, Vol. 72, p. 035215.

Lany, S. & Zunger, A. (2007). Dopability, Intrinsic Conductivity, and Nonstoichiometry of Transparent Conducting Oxides. *Phys. Rev. Lett.*, Vol. 98, p. 045501.

Look, D. C.; Hemsky, J. W. & Sizelove, J. R. (1999). Residual Native Shallow Donor in ZnO. *Phys. Rev. Lett.*, Vol. 82, pp. 2552-2555.

Look, D. C.; Farlow, G. C.; Reunchan, P.; Limpijumnong, S.; Zhang, S. B. & Nordlund, K. (2005). Evidence for Native-Defect Donors in *n*-Type ZnO. *Phys. Rev. Lett.*, Vol. 95, p. 225502.

McHenry, M. E.; Majetich, S. A.; Artman, J. O.; DeGraef, M. & Staley, S. W. (1994). Superparamagnetism in carbon-coated Co particles produced by the Kratschmer carbon arc process. *Phys. Rev. B*, Vol. 49, pp. 11358-11363.

Matsumoto, Y.; Murakami, M.; Shono, T.; Hasegawa, T.; Fukumura, T.; Kawasaki, M.; Ahmet, P.; Chikyow, T.; Koshihara, S. & Koinuma, H. (2001). Room-Temperature Ferromagnetism in Transparent Transition Metal-Doped Titanium Dioxide. *Science*, Vol. 291, pp. 854-856.

Moodera, J. S.; Kinder, L. R.; Wong, T. M. & Meservey, R. (1995). Large Magnetoresistance at Room Temperature in Ferromagnetic Thin Film Tunnel Junctions. *Phys. Rev. Lett.*, Vol. 74, pp. 3273-3276.

Neal, J. R.; Behan, A. J.; Ibrahim, R. M.; Blythe, H. J.; Ziese, M.; Fox, A. M. & Gehring, G. A. (2006). Room-Temperature Magneto-Optics of Ferromagnetic Transition-Metal-Doped ZnO Thin Films. *Phys. Rev. Lett.*, Vol. 96, p. 197208.

Ohno, H.; Shen, A.; Matsukura, F.; Oiwa, A.; Endo, A.; Katsumoto, S. & Iye, Y. (1996). (Ga,Mn)As : A new diluted magnetic semiconductor based on GaAs. *Appl. Phys. Lett.*, Vol. 69, pp. 363-365.
Ohno, H. (1998). Making Nonmagnetic Semiconductors Ferromagnetic. *Science*, Vol. 281, pp. 951-956.
Ohno, H.; Chiba, D.; Matsukura, F.; Omiya, T.; Abe, E.; Dietl, T., Ohno, Y. & Ohtani, K. (2000). Electric-field control of ferromagnetism. *Nature*, Vol. 408, pp. 944-946.
Ohno, Y.; Young, D. K.; Beschoten, B.; Matsukura, F.; Ohno, H. & Awschalom, D. D. (1999). Electrical spin injection in a ferromagnetic semiconductor heterostructure. *Nature*, Vol. 402, pp. 790-792.
Pearton, S. J.; Norton, D. P.; Ip, K.; Heo, Y. W. & Steiner, T. (2004). Recent advances in processing of ZnO. *J. Vac. Sci. Technol. B*, Vol. 22, pp. 932-948.
Prinz, G. A. (1998). Magnetoelectronics. *Science*, Vol. 282, pp. 1660-1663.
Risbud, A. S.; Spaldin, N. A.; Chen, Z. Q.; Stemmer, S. & Seshadri, R. (2003). Magnetism in polycrystalline cobalt-substituted zinc oxide. *Phys. Rev. B*, Vol. 68, p. 205202.
Sati, P.; Deparis, C.; Morhain, C.; Schäfer, S. & Stepanov, A. (2007). Antiferromagnetic Interactions in Single Crystalline $Zn_{1-x}Co_xO$ Thin Films. *Phys. Rev. Lett.*, Vol. 98, p. 137204.
Schwartz, D. A. & Gamelin, D. R. (2004). Reversible 300 K Ferromagnetic Ordering in a Diluted Magnetic Semiconductor. Adv. Mater., Vol. 16, pp. 2115-2119.
Story, T.; Galazka, R. R.; Frankel, R. B. & Wolff, P. A. (1986). Carrier-Concentration-Induced Ferromagnetism in PbSnMnTe. *Phys. Rev. Lett.*, Vol. 56, pp. 777-779.
Sawicki, M.; Dietl, T.; Kossut, J.; Igalson, J.; Wojtowicz, T. & Plesiewicz, W. (1986). Influence of s-d Exchange Interaction on the Conductivity of $Cd_{1-x}Mn_xSe$:In in the Weakly Localized Regime. *Phys. Rev. Lett.*, Vol. 56, pp. 508-511.
Toyosaki, T.; Fukumura, T.; Yamada, Y.; Nakajima, K.; Chikyow, T.; Hasegawa, T.; Koinuma, H. & Kawasaki, M. (2004). Anomalous Hall effect governed by electron doping in a room-temperature transparent ferromagnetic semiconductor. *Nat. Mater.*, Vol. 3, pp. 221-224.
Ueda, K.; Tabata, H. & Kawaib, T. (2001). Magnetic and electric properties of transition-metal-doped ZnO films. *Appl. Phys. Lett.*, Vol. 79, pp. 988-990.
Van de Walle, C. G. (2000). Hydrogen as a Cause of Doping in Zinc Oxide. *Phys. Rev. Lett.*, Vol. 85, pp. 1012-1015.
Wolf, S. A.; Awschalom, D. D.; Buhrman, R. A.; Daughton, J. M.; S. von Molnár, Roukes, M. L.; Chtchelkanova, A. Y. & Treger, D. M. (2001). Spintronics: A Spin-Based Electronics Vision for the Future. *Science*, Vol. 294, pp. 1488-1495.
Wu, Z. Y.; Chen, F. R.; Kai, J. J.; Jian W. B. & Lin, J. J. (2006). Fabrication, characterization and studies of annealing effects on ferromagnetism in $Zn_{1-x}Co_xO$ nanowires. *Nanotechnology*, Vol. 17, pp. 5511-5518.
Wu, Z. Y.; Chen, I. J.; Lin, Y. F.; Chiu, S. P.; Chen, F. R.; Kai, J. J.; Lin, J. J. & Jian W. B. (2008). Cross-sectional shape modulation of physical properties in ZnO and $Zn_{1-x}Co_xO$ nanowies. *New J. Phys.*, Vol. 10, p. 033017.
Xu, Q.; Hartmann, L.; Zhou, S.; Mcklich, A.; Helm, M.; Biehne, G.; Hochmuth, H.; Lorenz, M.; Grundmann, M. & Schmidt, H. (2008). Spin Manipulation in Co-doped ZnO. *Phys. Rev. Lett.*, Vol. 101, p. 076601.
Yamanouchi, M.; Chiba, D.; Matsukura, F. & Ohno, H. (2004). Current-induced domain-wall switching in a ferromagnetic semiconductor structure. *Nature*, Vol. 428, pp. 539-542.
Zhang, Z. H.; Wang, X.; Xu, J. B.; Muller, S.; Ronning, C. & Li, Q. (2009). Evidence of intrinsic ferromagnetism in individual dilute magnetic semiconducting nanostructures. *Nat. Nanotechnol.*, Vol. 4, pp. 523-527.

9

On-Chip Tungsten Oxide Nanowires Based Electrodes for Charge Injection

Anıl Ağıral and J. G. E. (Han) Gardeniers
Mesoscale Chemical Systems, MESA+ Institute for Nanotechnology, University of Twente Enschede, The Netherlands

1. Introduction

Transition metal oxides are fundamental to the development of many potential applications in nanoelectronics, optoelectronics and sensor devices (Wang, 2003 and Simon et al., 2001). Nanostructured metal oxides are widely investigated for scientific and technological applications. For example, binary semiconducting oxides have distinctive properties as transparent conducting oxide materials (Pan et al., 2001). Nanowires, which are stimulated by carbon nanoubes, have attracted wide interest due to their potential for bringing basic issues like dimensionality and transport phenomena in nanoscale dimensions. Among the transition metal oxide nanowires, tungsten oxide nanorods/nanowires show good sensing and field emission (FE) properties. FE from tungsten oxide nanowires shows good stability and high emission current density due to large aspect ratio, low turn-on field and stability at high pressures of 10^{-6}-10^{-3} Torr (Kim et al., 2005 and Seelaboyina et al., 2006). Synthesis methods for tungsten oxide nanowires include heating and oxidation of tungsten filaments (Liu et al., 2005), foils or films in vacuum at temperatures above 1000°C (Liu et al, 2003 and Cho et al., 2004). To the best of our knowledge, there are no studies atmospheric pressure FE performances of tungsten oxide nanowires. Synthesis of uniform and crystalline $W_{18}O_{49}$ nanowires on tungsten thin films by thermal annealing in ethane and nitrogen will be described in the first part of this chapter. Growth mechanism and atmospheric pressure FE analysis of a diode device based on nanowires will be discussed. The oxides have mixed cation valences and an adjustable oxygen deficiency which form the bases for tuning the electrical, chemical, optical and magnetic properties.

In the second part of this chapter, application of tungsten oxide nanowires in a miniaturized plasma device will be demonstrated. Miniaturized plasma sources have generated wide interest, recently, due to a number of important applications, light sources and chemical reactors (Becker et al., 2006). Performing an atmospheric pressure plasma process in a microreactor leads to precise control of process parameters such as residence time and heat transfer, and also extreme quenching conditions, enabling control over the reactants to selectively produce desirable products (Nozaki et al., 2004). $W_{18}O_{49}$ nanowires gave outstanding field emission characteristics with low threshold voltages due to local field enhancement and high aspect ratio (Guillorn et al., 2001 and Zhou et al., 2005). Process of field emission and field ionization can supply free electrons and ions and it can contribute to a pre-breakdown current during the initiation of streamer of discharge. It was shown previously that multiplication of electrons in a series of impact ionizations can lower the breakdown voltage in microplasma devices (Park & Eden, 2004). Eden et al. incorporated multiwall carbon nanotubes (CNTs) into the cathode and it resulted in lowering the ignition voltage and increasing the radiative efficiency. It was also reported that field emitted electrons can vibrationally excite molecules near a surface and lower the activation energy for dissociative adsorption (Tas, 1995). In order to test the contribution of FE from tungsten oxide nanowires during atmospheric pressure microplasma production, $W_{18}O_{49}$ nanowires on a silicon chip electrode placed in a glass microreactor channel will be described. The characteristics of dielectric barrier discharge with $W_{18}O_{49}$ nanowires will be reported and compared with plane-to-plane electrodes without nanostructures.

2. Synthesis of tungsten oxide nanowires

Tungsten oxide nanowires were grown in a quartz reactor with a glass plate to support n-type Si (100) (1- 10 Ohm cm) samples, on which a 300 nm thick tungsten film was deposited by DC plasma sputtering from a 99.999 % pure tungsten target. The temperature was monitored with a single thermocouple positioned just above the samples. Gases were mixed at atmospheric pressure, flushed through the tube furnace, collected and analyzed with an online gas chromatography. The whole chemical vapour deposition (CVD) apparatus is shown in Fig. 1.

(a) (b) (c) (d)

Fig. 1. CVD setup for tungsten oxide growth apparatus with *(a)* the sample holder with two samples, *(b)* gas chromatography, *(c)* the tube furnace and *(d)* the gas supply with flow controllers and mixing board.

In a typical growth procedure, samples were loaded at room temperature. Then the temperature was increased with a ramp of 6°C/min. When the desired growth temperature was reached, the carbon (gas) source was opened while the ouput gas composition was monitored. No pre-treatment procedure was needed. When the hyrogen production was detected in the chromatogram, it is evidence that hydrocarbon gas was dissociated on the tungsten film. The sample was heated in N_2 (80 sccm) and ethene was introduced in order to start nanowire growth for 20 mins. After nanowire growth, ended by switching off the carbon source, the temperature was ramped down with 10°C/min until the temperature was 400°C. At this temperature, the heater was switched off resulting in passive cooling process.

Chemical characterization of nanowires was carried out by using scanning electron microscopy (SEM) (LEO 1550 FEG) and high resolution tranmission electron microscopy (HRTEM) equipped with EDX analysis (Philips CM300ST-FEG microscope) and selected area diffraction (SAD). X-ray Photoelectron spectroscopy (XPS) analysis was also carried out in a PHI Quantera Scanning ESCA Microprobe system.
Atmospheric pressure FE measurements were performed in a microfabricated diode device in a laminar air flow cabinet at room tempereture. n-type silicon substrates were oxidized to form SiO_2 as dielectric spacer and the oxide was etched with HF solution isotropically to create a rectangular window, with an area of 0.7 cm^2, which was defined by photolithography. 2 μm thermal oxide was grown at 1150 °C in a wet (H_2O) ambient oxidation process for 10 hours. The oxide thickness was verified by ellipsometry using two wavelengths (632.8 and 1552 nm), and turned out to be around 1945 nm. When a large voltage is applied at an electrode, a leakage current may flow between cathode and anode through the oxide layer. In order to check the oxide quality, the leakage current was measured using special test structures. W dots were sputtered on oxidized n^{++} substrates (2 μm oxide) using a shadow mask and the back of wafer was sputter coated with 200 nm platinium (Pt). Wafers were HF dipped (1%, 1 min) prior to the platinium deposition, tungsten dots were deposited after platinium sputtering. These samples were not subjected to high temperature processes. The leakage current was measured at one stop with the back of the wafer as second electrode using an Agilent 4145 IV meter in combination with a Microtech Cascade probe station. A typical result is shown in Fig. 2:

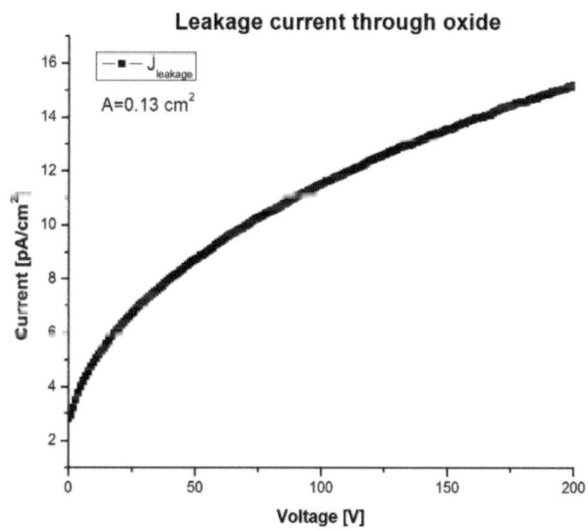

Fig. 2. Leakage current through oxide (2 μm) for a typical tungsten spot (4 mm Ø).

This figure indicates that the gate leakage current is very low; in the range of pA/cm.
SiO_2 layer was etched with HF solution in order to form dielectric spacer between nanowires and anode electrode in the diode device. After wet etching of SiO_2, a tungsten film was sputtered into the window. After nanowires growth on the cathode, the silicon substrate with a 10 nm gold layer as extractor electrode was clamped to the cathode to ensure a 2-μm electrode distance in a diode structure. Voltage current data were obtained with a Keithley 237 source meter unit.

Fig. 3. Cross section SEM image of tungsten oxide nanowires grown on tungsten thin films exposed to 27 sccm of C_2H_4 and 80 sccm of N_2 at 700°C at atmospheric pressure.

Fig. 3 shows a SEM image of tungsten oxide nanowires which were grown on tungsten thin films exposed to 27 sccm of of C_2H_4 and 80 sccm of N_2 at 700°C at atmospheric pressure. Nanowires were grown uniformly with high yield on the tungsten thin film with diameters and average length in the range of 15 - 20 nm and 50 - 250 nm, respectively.

To further illuminate the microstructure of as-deposited films, nanowires were also grown on both sides of freestanding tungsten thin films fabricated based on gated FE structure. Fig. 4 shows the etch profile of the silicon dioxide with freestanding tungsten thin film. In order to create the device, three main steps in the process flow were developed. First step is the photolithography, making a large array of holes of 2 μm in diameter with acceptable uniformity requires a high level of control of the lithographic process. In the single mask process one layer of photoresist was used as etch mask for both gate and oxide etch steps. Final device is compatible with high temperature nanowire growth step and survive thermal stress problems (e.g. adhesion, stability). For the substrate selection for nanowire growth, silicon substrates were chosen. As the charge carriers in field emitters are electrons, the doping of the silicon substrate is taken into consideration. Highly doped wafers were selected as they minimize the substrate resistance. In p-type silicon the number of electrons (minority charge carriers) is limited by thermal generation which might give rise to current saturation at high fields during field emission. By choosing for a highly doped n-type substrate, these effects are avoided. Therefore highly doped n-type wafers were selected as substrates. These one side polished 100 mm silicon wafers have arsenic (As) doping, (100)-orientation, a thickness of 525 ± 25 μm and a resistivity of 0.001-005 Ωcm. Prior to processing, wafers were cleaned in the standard cleaning. Before oxidation, wafers were were dipped in 1 % HF to remove native oxide.

Thermal oxidation resulted in a 2 μm high quality oxide as shown in Fig. 4 and cross section SEM image of tungsten oxide nanowires growth on both sides of freestanding thin films. Thin films have parallel ridges, which is a characteristic feature of sputtered tungsten films (Gosnet et al., 1989), covered with a dense nodular nanostructure at the tungsten/nanowire interface.

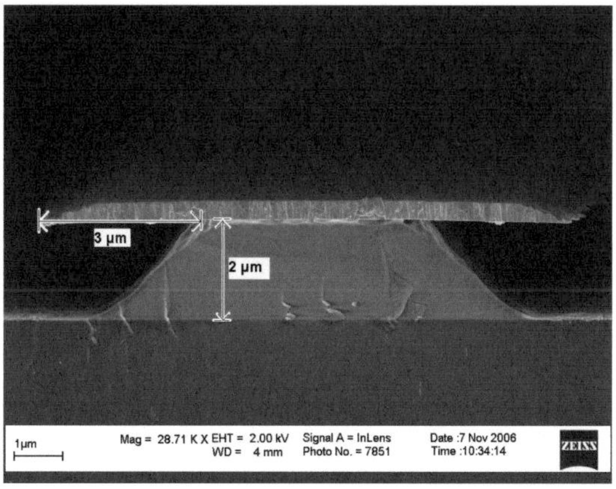

Fig. 4. Cross section SEM image of tungsten oxide nanowires grown on tungsten thin films exposed to 27 sccm of C_2H_4 and 80 sccm of N_2 at 700°C at atmospheric pressure.

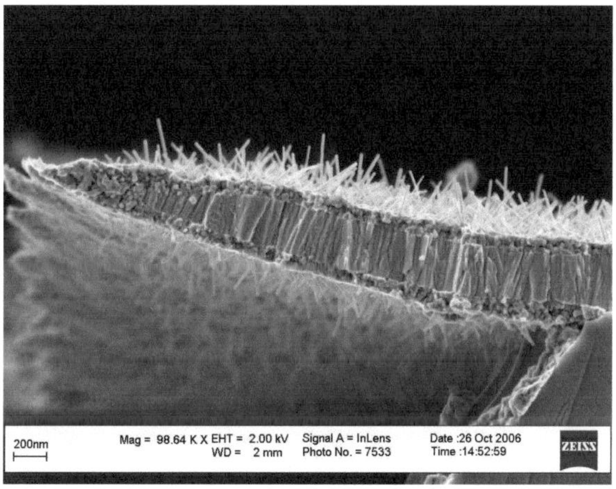

Fig. 5. Cross section SEM image of tungsten oxide nanowires grown on tungsten thin films exposed to 27 sccm of C_2H_4 and 80 sccm of N_2 at 700°C at atmospheric pressure nanowire growth on both sides of freestanding thin films.

Fig. 6 shows the samples before and after nanowires growth. Sample was covered with black substance which was formed during ethylene dissociation into carbon and flow pattern of the gases can be seen around the sample during the growth. Nanowires can be seen grown on the gate electrode.

X-ray diffraction analysis on the bulk sample indicated characteristic peaks for WO_{3-x}, α-W_2C, β-W, α-W and Si. WO_3 and $W_{18}O_{49}$ show similar peaks which are difficult to distinguish in diffractogram, therefore SAD and HRTEM analyses were used to determine chemical structure and composition. From the HRTEM images in Fig. 7, it is derived that the d-spacing along the long side of the nanowire is 3.78 Å, which is in agreement with the SAD analysis and corresponds to the (010) plane of monoclinic $W_{18}O_{49}$ (JCPDS no. 36-0101). (Agiral & Gardeniers, 2008). EDX analysis was performed on the central part of the area of nanowires which confirmed that oxygen is in the order of 2.73 times the amount of tungsten, supporting the evidences found in HRTEM and SAD measurements. Interfacial phase

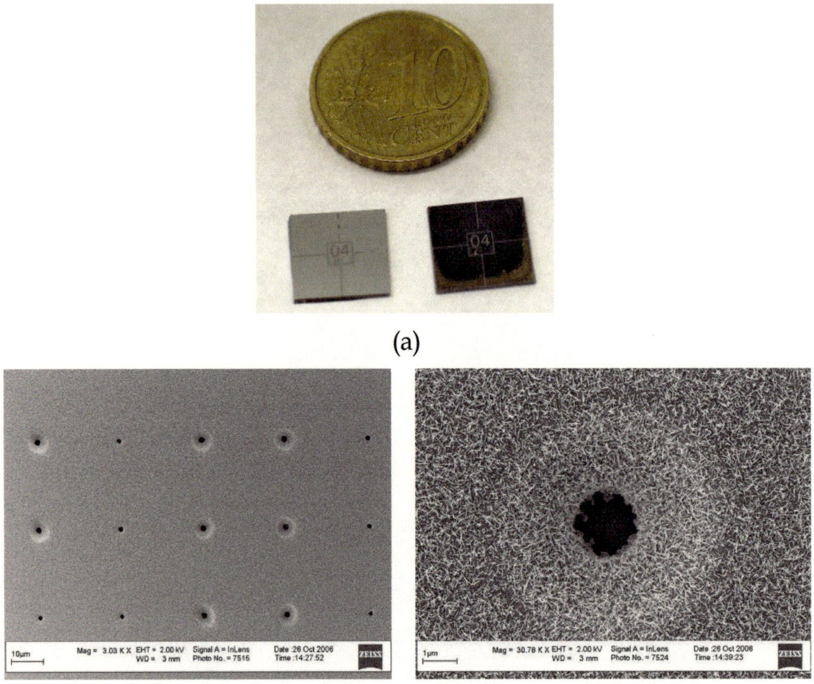

(a) (b) (c)

Fig. 6. Samples before (left) and after (right) the nanowires growth process. (b, c) SEM images of growth experiments on gated field emitters.

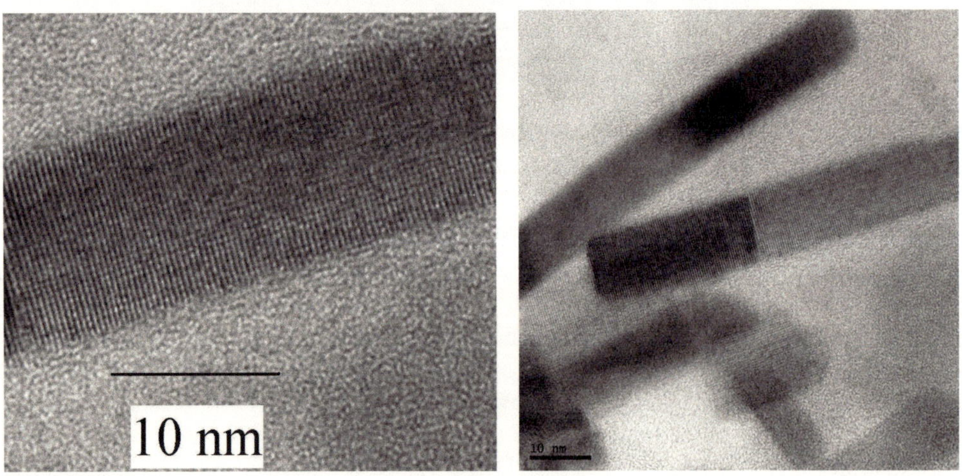

Fig. 7. HRTEM image of $W_{18}O_{49}$ nanowires [Agiral].

between nanowires and tungsten thin film was analyzed by SAD and it was found out that interfacial phase has d-spacings 2.36 Å and 1.36 Å corresponds to the phases α-W_2C(002) and α-W_2C(103), respectively (Wang et al., 2004).

Chemical bonding states on the surface of the tungsten thin film and the nanowires were investigated by XPS (not shown). W 4f spectra were deconvoluted into Lorentzian-Gaussian peaks which were assigned to ~31.4 (W $4f_{7/2}$ of W) (Mueller & Shih, 1988), ~33.5 (W $4f_{5/2}$ of W) (Rogers et al., 1982), ~33.2 (W $4f_{7/2}$ of W^{4+}, WO_2) (Salvati et al., 1981), ~35.3 (W $4f_{5/2}$ of W^{4+}, WO_2), ~34.5 (W $4f_{7/2}$ of W^{5+}, WO_x) (Jeon & Yong, 2007), ~36.6 (W $4f_{5/2}$ of W^{5+}, WO_x), ~35.7 (W $4f_{7/2}$ of W^{6+}, WO_3) (Fleisch & Mains, 1982), ~37.8 (W $4f_{5/2}$ of W^{6+}, WO_3). It was observed that after nanowire growth, the intensity ratio of binding states of oxidized tungsten to metallic tungsten increased, which is consistent with $W_{18}O_{49}$ formation. A low intense peak at about 31.7 eV was assigned to tungsten carbide W $4f_{7/2}$ which is possibly bound in the nanowires/film interface (Katrib et al., 1994).

Oxygen on the tungsten thin film originally comes from the native oxide layer that was formed when the sample exposed to the atmosphere. It was reported that molecular oxygen in the air can dissociatively chemisorb with oxygen atoms at triply-coordinated sites (Feydt et al., 1999). In order to determine the effect of chemisorbed oxygen present on the tungsten surface, thin film was reduced by introducing hydrogen at 700 °C for 20 min before nanowire growth. After ethene exposure, a dramatic decrease in the number of nanowires was observed. These results show the fact that tungsten oxide nanowire growth was caused by the oxygen bound on the surface. Carbon deposition rate, ethene dissociation and hydrogen evolution during nanowire growth were followed by online gas chromatography. It was found out that (Agiral & Gardeniers, 2008) dissociation of ethane into carbon and hydrogen took place according to the reaction:

$$C_2H_{4(g)} \leftrightarrow 2C_{(s)} + 2H_{2(g)} \quad (1)$$

A solid-state reaction took place between tungsten and deposited carbon in order to form a tungsten carbide phase during nanowire growth:

$$2W_{(s)} + C_{(s)} \leftrightarrow W_2C_{(s)} \quad (2)$$

Examination of the role of tungsten carbide phase for nanowire formation in the presence of only pure nitrogen indicated the fact that without ethene gas the nanowires could only grow with very low yield. Since tungsten carbide was not incorporated in the nanowires, it is assumed that it promoted nucleation and the growth. Vapor-solid and Vapor-Liquid-Solid growth models could not be explanations since the nanowires were formed with most of the oxygen on the tungsten surface and there were no nanoparticles at the end of the nanowires (Wang et al., 2005 & Wagner & Ellis, 1964). Here, tungsten carbide formation at the W/WO_x interface enhances the strain which drives whisker formation for nanowire growth. In -situ hydrogen formation during ethene dissociation can contribute to enhance oxygen diffusion especially through the grain boundaries. $W_{18}O_{49}$ nanowires can grow on dislocation sources which originate from the defects along the grain boundaries. Nanowires growth is enhanced by the interfacial strain created by tungsten carbide formation (Agiral & Gardeniers, 2008).

3. Atmospheric pressure field emission properties of $W_{18}O_{49}$ nanowires

Atmospheric pressure field emission measurements of tungsten oxide nanowires were done in air for a contact gap close to 2 µm. The mean free path of an electron in air is close to 2 µm (Hopwood & Iza, 2004) and probability of ionization from collisions of gas particles with FE electrons is very limited in this diode type device. However, only few collisions which develop positive ions in front of nanowires can even increase the current density (Schwirzke et al., 1993).

Fig. 8 illustrates the field emission measurements. The macroscopic current density J_M (mA/cm^2) was calculated to be the emission current (I) divided by the surface area of the diode window (A_M = macroscopic film area) on the cathode ($0.7\ cm^2$). The macroscopic electric field F_M (V/µm) was defined as the applied voltage divided by the electrode gap (~2 µm) between the cathode and the gold coated anode. The FE turn-on field is ca. 3.3 V/µm and defined as the electric field required to extract a current density of $10\ \mu A/cm^2$. The J_M-F_M plotted in Fig. 8 shows the FE diode characteristics for 3 consecutive sweeps (plots a-b-c) of negative cathode voltage. The exponential dependence of current density on electric field suggests a FE process taking place. A similar experiment performed using a parallel plate diode structure (tungsten surface as cathode) without nanowires did not show rectification behavior (plot d).

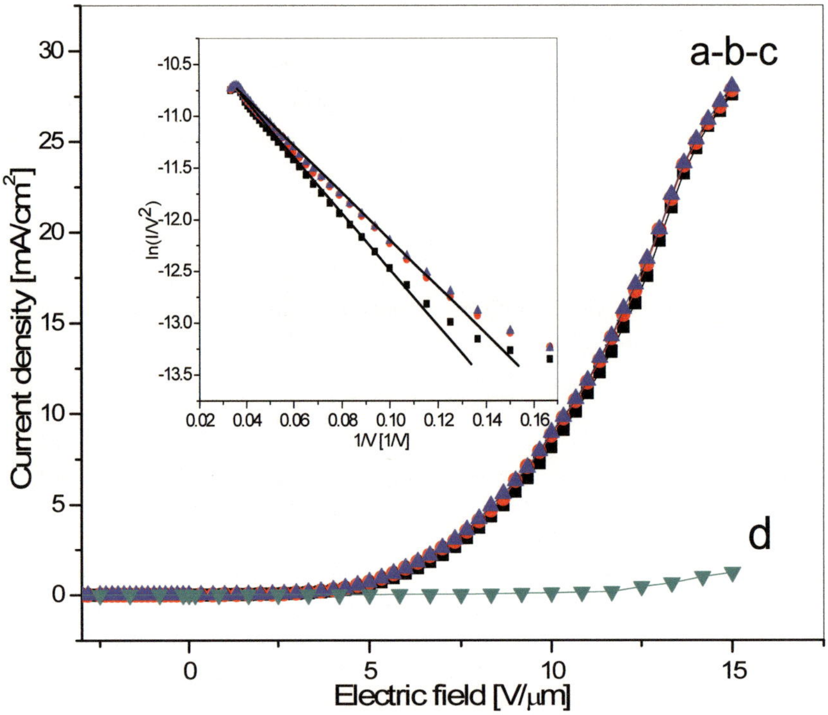

Fig. 8. Atmospheric pressure FE properties of tungsten oxide nanowires measured in air for a contact gap ~ 2µm. The macroscopic current density vs. macroscopic electric field (J_M-F_M) plots a-b-c correspond to 3 consecutive sweeps of negative cathode voltage with nanowires, while plot d shows the results of a parallel-plate diode without nanowires (insets are the corresponding Fowler Nordheim plots) (Agiral and Gardeniers, 2008).

The emission characteristics were analyzed by standard Fowler-Nordheim (FN) tunneling theory using the current-density equation (Forbes et al., 2007):

$$J = [\tau_F^{-2} a \varphi^{-1} F^2] exp[-\upsilon_F b \varphi^{3/2} / F] \quad (3)$$

where the 'first Fowler-Nordheim constant' $a = e^3/8\pi h_p = 1.541434 \times 10^{-6}$ AeVV^{-2}, 'second Fowler-Nordheim constant' $b = (8\pi/3)(2m_e)^{1/2}/eh_p = 6.830890 \times 10^9\ eV^{-3/2}Vm^{-1}$, e is the elementary positive charge, m_e is the electron mass, h_p is Planck's constant, J is local current density, F is local field, φ is local work function of emitting surface, τ_F and υ_F are the values of τ (decay rate correction factor) and υ (tunneling exponent correction factor) which apply to a barrier of unreduced height h equal to the local work function φ. For the elementary triangular barrier, the correction factors υ and τ are set equal to unity. Assuming that the local work function is uniform, the emission current I can be written as $I = A_N J$, where A_N is notional emission area. This gives:

$$I = A_N J = A_N \tau_F^{-2} a \varphi^{-1} F^2 exp[-\upsilon_F b \varphi^{3/2}/F] \quad (4)$$

Geometric field enhancement may lead to the local field F to be greater than F_M by a factor (γ) which is called the field enhancement factor:

$$\gamma = F/F_M \quad (5)$$

In the literature, FN plots from tungsten oxide nanowires are analyzed using FN theory for metals (Huang et al., 2007). If the macroscopic electric field F_M is used as FN variable in Eq. (4) to replace F, the formula can be written as:

$$ln[I/F_M^2] = ln[A_N \tau_F^{-2} a \varphi^{-1}\ \gamma^2] - [\upsilon_F b \varphi^{3/2}/\ \gamma]/F_M \quad (6)$$

The theoretical model for a FN plot is the tangent to Eq. (6), taken at F_{M1} of macroscopic field (Forbes, 2001):

$$ln[I/F_M^2] = ln[A_N r_1 a \varphi^{-1}\ \gamma^2] - [s_1 b \varphi^{3/2}/\ \gamma]/F_M \quad (7)$$

where r_1 and s_1 are the generalized intercept correction factor and the generalized slope correction factor, respectively, taken at field F_{M1} (Forbes, 1999). s_1 is approximately 1 and therefore omitted in the interpretation of the slope of the FN plot. Assuming that the emission current comes mainly from one band, and φ is 5.7 eV, similar to WO_3, the field enhancement factor found from the FN plot is 7228.

Estimation of the apex field-enhancement factor γ of a protrusion on a flat planar surface has been described with 'hemisphere on a post' and 'hemi-ellipsoid on a plane' models (Forbes et al., 2003). Both these geometrical models can be applied to tungsten oxide nanowires, considering their geometry. The 'hemisphere on a post' model gives the simple formula:

$$\gamma = 0.7(L/\rho) \quad (8)$$

where L and ρ are protrusion length and base radius, respectively. In the 'hemi-ellipsoid on a plane' model γ can be written as:

$$\gamma = \zeta^3/[(\ \upsilon ln(\upsilon + \zeta)) - \zeta] \quad (9)$$

where L is the semi-major axis length, ρ is the semi-minor axis length, $\upsilon\ (=L/\rho)$ is the ratio of L to ρ and ζ equals $(\upsilon^2-1)^{1/2}$. If the formula of the 'hemisphere on a post' model for a nanowire of 15 nm diameter and 250 nm length is used, γ is calculated to be around 23 which is far below the experimental field enhancement factor. However, SEM and HRTEM analysis revealed that there are few nanowires on the sample with length and diameter of ~1 µm and ~12 nm, respectively. Applying the 'hemi-ellipsoid on a plane' formula, γ can be calculated to be 5776, which is of the order of the experimental value. A small part of the sample surface may have protruding nanowires with longer length and smaller diameter than the average dimensions.

At high electric fields (> 13 V/µm), J_M diverged from the Fowler-Nordheim equation where saturation affects were observed and eventually breakdown occurred around 15.2 V/µm.

Emission stability of nanowires was also measured at 13.6 V/µm, however no significant degradation in emission current was observed over 1 hour of testing period. Current fluctuations can be attributed to surface chemical interactions with adsorbents which can affect the tunneling state of nanowires or residual ions impacting on the surface. In order to demonstrate the viability of tungsten oxide nanowires as a pulsed electron source, a 100 Hz square wave voltage was applied to the cathode and pulsed electron emission with good repetition rate was observed.

As the F_M reached the value of 15.2 V/µm, breakdown occurred. SEM analysis (not shown) on the surface is suggestive of an explosive FE process due to the presence of craters on the cathode surface and melted film traces on the anode surface. Possible mechanisms for explosion of emitter can be high density FE currents, resistive heating and the Nottingham mechanism. Nottingham effect can arise when temperature inside the emitters increases faster than that of the surface, due to the energy difference between emitted electrons and their replacement in the lattice through the electrical circuit. Under these conditions, temperature can rise to a critical level and a phase transition of the field emitters into a dense plasma and loss of cathode material is observed.

4. Incorporation of nanostructures into microplasma reactor

As an application example, tungsten oxide nanowires were incorporated in a glass microchannel to generate barrier discharge between the nanowires and a copper foil electrode protected by a glass dielectric layer. 250 nm titanium-tungsten alloy and 50 nm tungsten metal layers were deposited on silicon chips by plasma sputtering. Silicon chips with W layers were placed in the CVD setup to grow nanowires as mentioned in the previous sections. Silicon chips with nanowires were placed in glass microreactors with rectangular cross section of 50 mm in length and 20 mm in width. To finish the fabrication of whole device, microchannels, inlet and outlet holes for gas flow were created by powder blasting and the glass layers were thermally bonded to seal the microchannel hermetically. Fig. 9a and 9b show the final devices. Fig. 9c shows the general diagram of device and operating plasma device is shown in Fig. 9d.

A combination of a high voltage amplifier (Trek 610E) and a function generator (Agilent 3322A) was used to generate a barrier discharge at 1 kHz in the microreactor. Voltage drop across a resistor and a capacitor connected in series with the electrodes were used to calculate the transferred current and charge, respectively. Light emission from the discharge was collected by an optical fiber inserted into microchannel and optical fiber was connected to an optical emission spectrometer (HR 4000, Ocean optics). Lissajous figures measured with an oscilloscope were evaluated in order to calculate the absorbed power by discharge.

The amount of current that is transferred across the gap is also much higher in the case of microreactor with nanowires electrode. The reduction of breakdown voltage and increase in transferred current is suggesting electric field amplification and emission of electrons on the nanowires surface. Electric field in the gas gap of microreactor (5 V/µm) is enough to generate field emitted electrons before breakdown. Work functions of planar and tungsten

oxide nanowires surfaces are similar since there is a native oxide layer a few nanometer thickness on the tungsten film.
Comparison of power absorbed by the discharge as a function of measured voltage shows that at the same measured voltage, a higher energy density microdischarge was generated with the nanowires due to generation of higher number of streamers in the gas gap. Local electric field enhancement on the surface helps to increase the number of streamers and microplasma became more homogeneous.

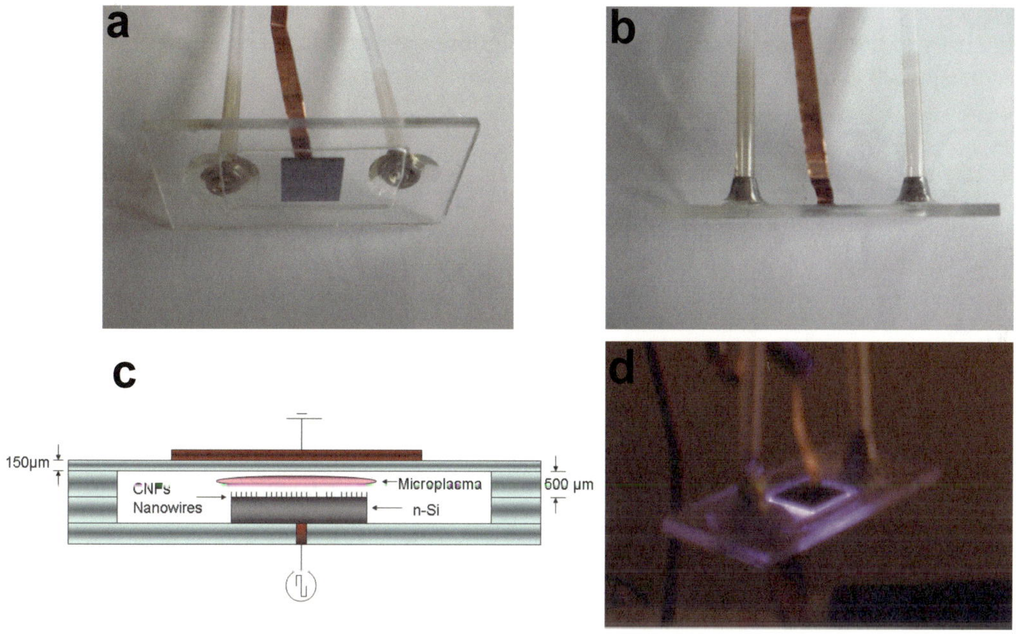

Fig. 9. Microplasma reactors with integrated tungsten oxide nanowires electrode (a) and (b). General diagram of microreactor (c) and the plasma device in operation (Agiral et al., 2008).

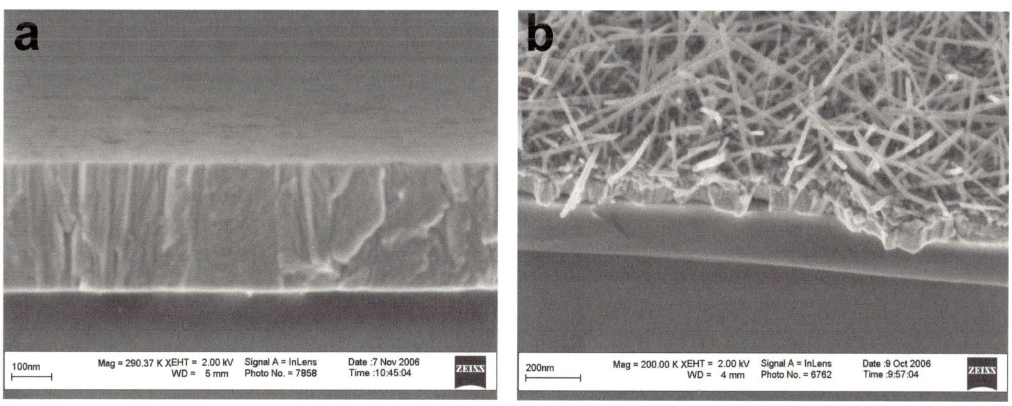

Fig. 10. SEM images of planar tungsten thin film electrode (a) and electrode covered with tungsten oxide nanowires (b).

In order to analyze light emission from microplasma, an optical fiber coupled to the emission spectrometer, inserted into microchannel. In case of nanowires, the intensity of the emission lines is found to be much higher than planar electrodes. Jani et al. (Jani et al., 1999) analyzed the variation of electron temperature by spectroscopic measurement using the second positive system of N_2 (337.1 nm) and the first negative (FNS) of N_2^+ (391.4 nm). It was shown that the ratio of the SPS to the FNS changes with electron temperature in the discharge, and the intensity ratio of the lines increase with increasing average electron energy. In our case, intensity ratio was calculated to be 0.061 and 0.058 for nanowires and planar electrodes, respectively, which shows no remarkable difference although the intensity of the emission lines is much higher for the spectrum for nanowire electrode at the same applied voltage.

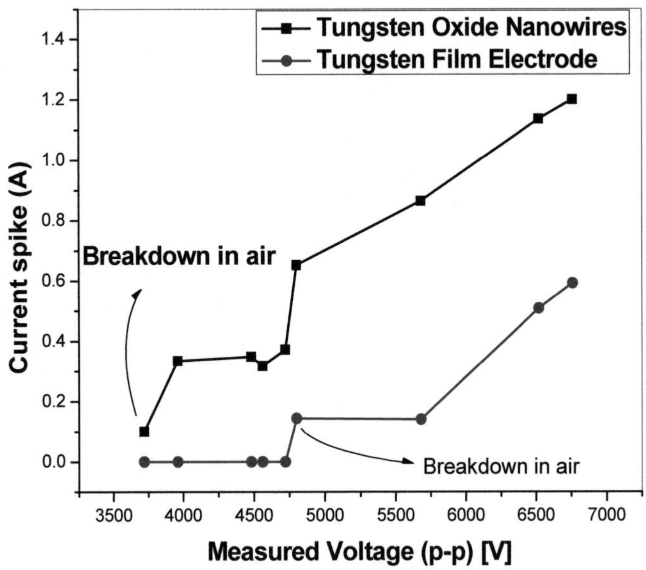

Fig. 11. Comparison of discharge current spikes measured with tungsten oxide nanowires and with a planar tungsten film electrode for varying applied peak-to-peak voltage; f = 1 kHz. Points show the current measured with oscilloscope at corresponding measured voltage values. Breakdown voltage is defined as voltage level at which light emission was detected by optical emission spectrometer (Agiral et al., 2008).

5. Conclusion

In summary, uniform and highly crystalline $W_{18}O_{49}$ nanowires were successfully grown with high yield on sputter deposited tungsten films by thermal annealing at 700°C in ethene and nitrogen at atmospheric pressure. It has been shown that tungsten carbide (W_2C) formation at the surface of the tungsten film increased the nucleation and growth of nanowires by generating dislocation sources and interfacial strain. Field emission measurements at atmospheric pressure indicated that stable atmospheric pressure field emission and pulsed field emission took place in air at a current density up to 28 mA/cm^2. The field emission current obeyed the Fowler-Nordheim equation.

A microplasma reactor having tungsten oxide nanowires as electrodes were fabricated in a barrier discharge configuration. Field enhancement at the apex of the tips of nanostructures resulted in electron emission at atmospheric pressure in air. For nanowires, a decrease in breakdown voltage during barrier discharge generation resulted in a higher number of microdischarges and a higher power deposition at the same measured potential compared to planar cathodes.

6. References

Wang, Z. L. (2003) Nanobelts, Nanowires, and Nanodiskettes of Semiconducting Oxides-From Materials to Nanodevices. *Adv. Mater.*, 15, 432.

Simon, I.; Bârsan, N.; Bauer, M. & Weimar, U. (2001). Micromachined metal oxide gas sensor: opportunities to improve sensor performance. *Sens. Actuators B*, 73, 1.

Pan, Z. W.; Dai, Z. R. & Wang, Z. L. (2001). Nanobelts of semiconducting oxides. *Science* , 291, 1947.

Kim, Y. S.; Ha, S.-C.; Kim, K.; Yang, H.; Choi, S.-Y.; Kim,Y. T.; Park, J. T.; Lee, C. H.;

Choi, J.; Paek, J. & Lee, K. (2005). Room temperature semiconductor gas sensor base don stoichiometric tungsten oxide nanorod film. *Appl. Phys. Lett.*, 86, 213105.

Seelaboyina, R.; Huang, J.; Park, J.; Kang, D. H. & Choi, W. B. (2006). Multistage field enhancement of tungsten oxide nanowires and its field emission in various vacuum conditions. *Nanotechnology*, 17, 4840.

Liu, K.; Foord, D. T. & Scipioni, L. (2005). Easy growth of undoped and doped tungsten oxide nanowires with high purity and orientation, *Nanotechnology*, 16, 10.

Liu, Z.; Bando, Y. & Tang, C. (2003). Synthesis of tungsten oxide nanowires, *Chem. Phys. Lett.*, 372, 179.

Cho, M. H. et al. (2004). Evolution of tungsten oxide whiskers synthesized by a rapid thermal annealing treatment. *J. Vac. Sci. Technol. B*, 22, 1084.

Becker K. H., Schoenbach K. H. & Eden J. G., (2006). Microplasmas and applications *J. Phys. D: Appl. Phys.*, 39, R55.

Nozaki T., Hattori A. & Okazaki K., (2004). Partial oxidation of methane using a microscale non-equilibrium plasma reactor. *Catal. Today*, 98, 607.

Guillorn M. A., Melechko A. V., Merkulov V. I., Ellis E. D., Britton C. L., Simpson M. L., Lowndes D. H. & Baylor L. R., (2001). Operation of a gated field emitter using and individual carbon nanofiber cathode, *Appl. Phys. Lett.*, 79, 3506.

Zhou J., Gong L., Deng Z., Chen J., She J. C., Xu N. S., Yang R. & Wang Z. L. (2005). Growth and field emission property of tungsten oxide nanotip arrays. *Appl. Phys. Lett.*, 87, 223108.

Park S. J. & Eden J. G. (2004). Carbon nanotube enhanced performance of microplasma devices. *Appl. Phys. Lett.*, 84, 4481.

Tas, M. A. (1995) Plasma Induced Catalysis: PhD Thesis (Technical University of Eindhoven) p 14.

Gosnet, A. M. H.; Ladan, F. R.; Mayeux, C.; Launois, H. & Jancour, M. C. (1989). Stress and microstructure in tungsten sputtered thin films. *J. Vac. Sci.Technol. A*, 7, 2663.

Agiral, A. & Gardniers, J.G.E. (2008). Synthesis and atmospheric pressure field emission operation of W18O40 nanorods. *Journal of Physical Chemistry C.*, 112, 39.

Wang, S. J.; Chen, C. H.; Chang, S. C.; Uang, K. M.; Juan, C. P. & Cheng, H. C. (2004). Growth and characterization of tungsten carbide nanowires by thermal annealing of sputter deposited WCx films. *Appl.Phys. Lett.*, 85, 12, 2358.

Mueller, D. & Shih, A. (1988). A Synchotron radiation study of BaO Films on W(001) and their interaction with H_2O, CO_2 and O_2. *J. Vac. Sci. Technol. A*, 6, 1067.

Rogers, J. D.; Sundaram, V. S.; Kleiman, G. G.; Castro, C. G. C.; Douglas, R. A. & Peterlevitz, A. C. (1982). High resolution study of M45N67N67 and M45N45N67 Auger transitions in the 5D series. *J. Phys. F.*, 12, 2097.

Salvati, L.; Makovsky, L. E.; Stencel, J. M.; Brown, F. R. & Hercules, D. M. (1981). Surface spectroscopic study of tungsten alumina catalysts using x-ray photoelectron, ion scattering, and raman spectroscopies. *J. Phys. Chem.*, 85, 3700.

Jeon, S. & Yong, K. (2007). Direct synthesis pf W18O49 nanorods from W2N film by thermal annealing. *Nanotechnology*, 18, 245602.

Fleisch, T. H. & Mains, G. J. (1982). An XPS study of the UV reduction and photochromism of MoO3 and WO3. *J. Chem. Phys.*, 76, 780.

Katrib, A.; Hemming, F.; Hilaire, L.; Wehrer, P. & Maire, G. (1994). XPS studies of supported tungsten carbides. *J. Electron Spectrosc. Relat. Phenom.*, 68, 589-595.

Feydt, J.; Elbe, A.; Egelhard, H.; Meister, G. & Goldmann, A. (1999). Normal emission photoelectron studies of the W(110)(1 x 1)O surface. *Surf. Sci.*, 440, 213.

Wang, S. J.; Chen, C. H.; Ko, R. M.; Kuo, Y. C.; Wong, C. H.; Wu, C. H.; Uang, K. M.;

Chen, T. M. & Liou, B. W. (2005). Preparation of tungsten oxide nanowires from sputter deposited WCx films using an annealing/oxidation process. *Appl. Phys. Lett.*, 86, 263103.

Wagner, R. S. & Ellis, W. C. (1964). Vapour-liquid-solid mechanism of single crystal growth. *Appl. Phys. Lett.*, 4, 89.

Hopwood, J.; Iza, F. (2004). Ultrahigh frequency microplasmas from 1 pascal to 1 atmosphere. *J. Anal. At. Spectrom.*, 19, 1145-1150.

Schwirzke, F.; Hallal, M. P. & Maruyama, X. K. (1993). Onset of breakdown and formation of cathode spots. *IEEE Trans. Plasma Sci.*, 21, 410-415.

Forbes, R. G. & Deane, J. H. B. (2007). Reformulation of the standard theory of Fowler-Nordheim tunnelling and cold field electron emission. *Proc. R. Soc. A*, 463, 2907-2927.

K. Huang, et al. (2007). Synthesis and field emission properties of the tungsten oxide nanowire arrays. *Physica E*, doi:10.1016/j.physe.2007.04.007.

Forbes, R. G. (1999). Field emission:New theory for the derivation of emission area from a Fowler-Nordheim plot. *J. Vac. Sci. Technol. B*, 17(2), 526-533.

Forbes, R. G. (2001). Low macroscopic field electron emission from carbon films and other electrically nanostructured heterogeneous materials:hypotheses about emission mechanism. *Solid-State Electronics*, 45, 779-808.

Forbes, R. G.; Edgcombe, C. J. & Valdré, U. (2003). Some comments on models for field enhancement. *Ultramicroscopy*, 95, 57-65.

Agiral, A.; Groenland, A. W.; Chinthaginjala, J. K.; Seshan, K.; Lefferts, L. & Gardeniers, J. G. E. (2008). On-chip microplasma reactors using carbon nanofibers and tungsten oxide nanowires as electrodes. *Journal of Physics D: Applied Physics*, 41, 194009.

Jani M. A.; Takaki K. & Fujiwara T., (1999). Streamer polarity dependence of NOx removal by dielectric barrier discharge with a multipoint to plane geometry. *J. Phys. D: Appl. Phys.*, 32, 2560.

10

Advanced Electron Microscopy Techniques on Semiconductor Nanowires: from Atomic Density of States Analysis to 3D Reconstruction Models

Sònia Conesa-Boj[1], Sònia Estradé[1], Josep M. Rebled[1], Joan D. Prades[1], A. Cirera[1], Joan R. Morante[1,2], Francesca Peiró[1], and Jordi Arbiol[1,3,*]
[1]*Dept. d'Electrònica, Universitat de Barcelona*
[2]*IREC, Catalonia Institute for Energy Research*
[3]*ICREA Research Professor at Institut de Ciència de Materials de Barcelona, CSIC*
Spain

1. Introduction

Technology at the nanoscale has become one of the main scientific world challenges as new quantum physical effects appear and can be modulated at will (Kastner, 1992). Superconductors, materials for spintronics, electronics, optoelectronics, chemical sensing, and new generations of functionalized materials are taking advantage of the low dimensionality, improving their properties and opening a new range of applications (De Franceschi et al., 2003; Samuelson et al., 2004; Fischer et al., 2006, Pettersson et al., 2006; Wang et al., 2006; Hernández-Ramírez et al., 2007; Qin et al., 2007; Appenzeller et al., 2008; Boukai et al., 2008; Hochbaum et al., 2008; Wang et al., 2008; Colombo et al., 2009; Thunick et al., 2009). These new materials for future applications are being synthesized at the nanoscale (ultrathin layers, nanoparticles, nanowires or nanotubes functionalized). Among all these new materials, one dimensional (1D) nanostructures such as nanowires, are one of the most used and promising morphologies (Lieber, 2003, Yang, 2005, Thelander et al., 2006; Lieber & Wang, 2007). Groups all around the world in the area of materials science, physics, chemistry and biology work in close collaboration with nanoscopy tools as there is a critical need for the structural, chemical and morphological characterization of the synthesized nanostructures at atomic scale in order to correlate these results with the physical and chemical properties and functionalities they present. In order to obtain an accurate control and understanding of these new materials properties, it is essential to access their structure and chemistry at atomic scale. Electron Microscopy and more precisely (scanning) transmission electron microscopy ((S)TEM) and electron spectroscopy related techniques (also known as electron nanoscopies) have thus a preeminent role in advanced materials science. Recent developments in electron microscopy, such as aberration correctors and monochromators are allowing us to reach sub-angstrom and sub-eV, spatial and energy resolutions, respectively. In addition to these advances, the possibility to obtain 3D models of our nanostructures by means of electron tomography, have shown that Electron Microscopy related techniques are the most promising to fully characterize complex nanostructures. In the present chapter we will show how advanced electron microscopy techniques can be applied to obtain a deeper characterization of complex structures in semiconductor nanowires. The chapter will be distributed in two parts: in the first one, we will show the advantages of using aberration-corrected STEM and monochromated electron energy loss spectroscopy (EELS) in order to deeply characterize 1-D nanostructures at atomic scale. As an advanced example of how these techniques can improve Nanowire characterization, we will show how they can allow the local analysis of the atomic scale modifications in the local density of states that occur in a complex system in which high Mg-doping induces formation of transversal twin defects (perpendicular to the growth axis) on GaN Nanowires (NWs). Formation of these defects generates local structural phase changes, thus implying a different optoelectronic behavior of the nanowires, induced by the local change in the material bandgap. The considered techniques can allow measuring the bandgap exactly in every atomic column, thus leading to a complete characterization. In the second part of the chapter, we will focus on how 3D reconstruction of 1D nanostructures by means of electron tomography can improve the morphological characterization at the nanoscale. As advanced examples of this part, we will show how to obtain 3D models of complex coaxial nanowire heterostructures.

2. Aberration-corrected scanning transmission electron microscopy and monochromated electron energy loss spectroscopy

2.1 State of the art

Nowadays, given the eventual reduction in scale of the challenges at hand and the progressive instrumental advances that have taken place in microscopy in general and in EELS in particular, EELS has become one of the most crucial tool in materials science and even the life sciences (Calvert et al., 2005; Pan et al., 2009). As EELS is performed in the Transmission Electron Microscope (TEM), it benefits from the very high spatial resolution that can be achieved with electron optics, which can focus the electron beam to form a subnanometric probe. In particular, if a field emission gun is used, sufficient current can be obtained for a 1 nm probe. Within aberration-corrected instruments, this figure can be reduced to 0.1 nm. In addition, EELS can be easily combined with structural information as obtained from the TEM imaging and diffraction modes, and even with complementary X-ray energy-dispersive spectroscopy (EDXS) if needed. There is a fundamental limit to the minimum lateral resolution that can be achieved by EELS, irrespective of the electron optics. This limit is given by the delocalisation produced in inelastic scattering, and depends on the energy loss (the lower the loss, the greater the delocalisation) (Egerton, 1996). Yet, fortunately, this limit does not prevent from getting EELS signal from single atom columns at core-loss (Allen, 2003) or subnanometric resolution in low-loss experiments (Grogger, 2005). With the recent advances in instrumentation (spherical aberration correctors, electron monochromators, new energy filters and CCD detectors) EELS experiments can now be performed with a spatial resolution well below 0.1 nm and an energy resolution better than 0.1 eV. One of the instrumental highlights in the history of TEM is the recent introduction of systems to compensate for spherical and even chromatic aberrations (Haider et al., 1998; Batson et al., 2002). Using aberration corrected microscopy, an electron probe smaller than 1 Å can be achieved, which allows imaging of single atoms, clusters of a few atoms, and atomic columns. A multipole corrector built into the illumination system of a STEM increases the image resolution and allows more current to be focused in a given probe. This is of great importance for spectroscopy, as both lateral resolution and signal-to-noise ratio are enhanced. If EELS presents a lower energy resolution when compared to other spectroscopies as XAS, the limitation does not lie in the capabilities of the spectrometers, but in the energy width of the electron source. This energy dispersion is typically 1–2 eV for a thermionic source, 0.5–1 eV for a Schottky or hot field-emission tip and around 0.3-0.35 eV for a cold field-emission tip. For comparison, synchrotron X-ray sources and beam-line spectrometers commonly provide a resolution below 0.1 eV for absorption spectroscopy, and even below 1 meV in certain cases (Egerton, 2003). In order to reduce the source energy spread, monochromators have been recently introduced. Nowadays, the monocromators yield a beam current in the order of several 100 pA (Tsuno, 2000; Egerton, 2003; Sigle, 2005). In particular, and due to all these achievements, HREELS associated to STEM in HRHAADF mode, can be used to solve the local electronic properties in nanostructured materials. In this way, bulk plasmon peak position can be used as an indirect compositional measure, and has been extensively used as a local chemical characterization tool, especially in the case of semiconductors. It can be primarily used as an identification tag for determining which compound is there at a given region of the studied specimen (Topuria et al., 2003; Irrera et al., 2005). As the plasmon peak position depends on the lattice parameter (as well as the bandgap energy and the dielectric constant) it can also give an indirect measure of structural properties (Shen et al., 2000; Sanchez et al., 2006). The need for characterization techniques that provide precise information regarding the bandgap and general optical properties at high spatial resolution seems to be out of question, given the scaling down that has taken place in the field of materials science and the rapidly widening use of nanostructures. In this sense, standard optical techniques such as vacuum ultra-violet spectroscopy do not provide the spatial resolution required to probe a material on the nanometer scale. Low-loss EELS seems to be a most fitting technique for the local characterization of optoelectronic properties at the nanoscale. For insulators or semiconductors with a sufficiently wide bandgap (that can be less than 1 eV using a monochromated STEM (Erni & Browning, 2005)), interband transitions can be observed in the EELS spectrum. It is possible to identify through EELS the bandgap energy of given nanostructures (Kuykendall et al., 2007; Arenal et al., 2008; Iakoubovskii et al., 2008). It is also possible to assess the existence of localized states within the bandpgap, which may be due to the presence of dislocations or other kinds of defects, for instance (Batson et al., 1986; Xin et al., 2000), which create new energy levels in the local DOS.

2.2 Particular case: Local electronic properties of Mg doped GaN NWs

In recent years much attention has been focused on the growth of quasi-one-dimensional (1D) nanostructures for the controlled fabrication of nanodevices (Arbiol et al., 2002; Hernandez-Ramirez et al., 2007; Hochbaum et al., 2008; Nesbitt, 2007; Stern et al., 2007). In particular, III-nitrides (InN, GaN, AlN and their alloys) have shown promising properties. Their direct band gap can be controlled from 0.7 eV (InN) to 3.4 eV (GaN) and to the deep UV spectral range of 6.0 eV (AlN). This makes them excellent candidates for the fabrication of heterostructures for optoelectronic applications, such as light emitting diodes, laser diodes, or quantum well infrared photodetectors as well as high electron mobility transistors. The synthesis of p-type 1D nanostructures based on these III-nitrides is still a challenging topic. To exploit these material properties also in nanoscaled devices, catalyst-induced processes like the vapour liquid solid (VLS) mechanism (Wagner & Ellis, 1964;

Morral et al., 2007), using metal droplets like Au, Ni (Chen et al., 2001; Kim et al., 2003), or Fe (Chen et al., 2001) as catalysts have been developed for the growth of GaN nanorods (NRs) or nanowires (NWs) with vertical orientation with respect to the substrate, with horizontal orientation, or as free structures. GaN NWs grown by laser assisted catalytic growth have been reported and heterodiodes have been realized by deposition of n-type GaN NWs on p-type Si substrates (Huang et al., 2002). However, for the integration of more complex structures such as p–n junctions or quantum wells, and in order to avoid the negative effects of catalysts in group III–V NRs or NWs, molecular beam epitaxy (MBE) is the growth method of choice (Calleja et al., 2000; Colombo et al. 2008; Morral et al., 2008b). As mentioned above, for the realization of nanoelectronic devices doping is one of the most important issues. In the case of Mg doped (p-type) GaN NRs and NWs, only a few studies have been devoted to analyze their optoelectronic (Furtmayr et al., 2008a; Furtmayr et al., 2008b; Park et al., 2006a; Zhong et al., 2003; Lai et al., 2006; Pal et al., 2006), transport, and electronic properties (Cheng et al., 2003; Zhong et al., 2003; Park et al., 2006b). Some of these works have provided limited information on the morphology (Park et al., 2006a) and crystal structure (Cheng et al., 2003; Zhong et al., 2003; Cimpoiasu et al., 2006) of the Mg doped GaN NRs and NWs. In addition, the analysis of GaN NWs as a model system for single crystals with a very low defect density can also contribute to solving some of the still existing technological problems of p-type doping of GaN with Mg in general. To this end, in the following we will show how combination of atomically resolved high resolution electron energy loss (EEL) measurements with specific *ab-initio* calculations can be a useful way to demonstrate that the presence of Mg atoms during the growth of GaN nanowires results in direct structural and morphological modifications with significant effects on the local electronic structure.

2.3 Experimental details: Growth method

Gallium nitride nanowires were grown using plasma assisted molecular beam epitaxy (PAMBE). Nitrogen radicals were supplied by an Oxford Applied Research RF-plasma source whereas for Ga and Mg thermal effusion cells were used. Low resistivity n-type Si(111) substrates were etched in 5% hydrofluoric acid for 10 s to remove the surface oxide layer prior to transfer into the load lock chamber. For all samples a substrate temperature of 790 °C, a nitrogen pressure of $2.7\times10{-}5$ mbar and a Ga beam equivalent pressure (BEP) of 3.3×10^{-7} mbar was used. The substrates were exposed to the nitrogen plasma at a temperature of 790 °C for 2 min directly before GaN growth. For the magnesium doped samples the Mg effusion cell temperature (TMg) was varied between 205 °C and 355 °C, corresponding to a BEP of 1×10^{-11} mbar to 2×10^{-8} mbar, respectively. Growth duration was 90 min unless mentioned otherwise. Samples were grown by Florian Furtmayr, Christoph Stark, Martin Stutzmann and Martin Eickhoff (Walter-Schottky-Institut, Technische Universität München) and Andreas Laufer (I Physikalisches Institut, Justus-Liebig-Universität).

2.4 Transmission electron microscopy characterization

In order to analyze the influence of the Mg doping on the atomic structure of the Mg NWs, selected area electron diffraction (SAED) patterns were obtained on several samples, prepared in XTEM geometry (Fig. 1. SAED analysis). Fig. 1.(a) shows a bright field STEM (BFSTEM) general view of the undoped GaN NWs. Its corresponding SAED pattern is displayed in Fig. 1.(b).

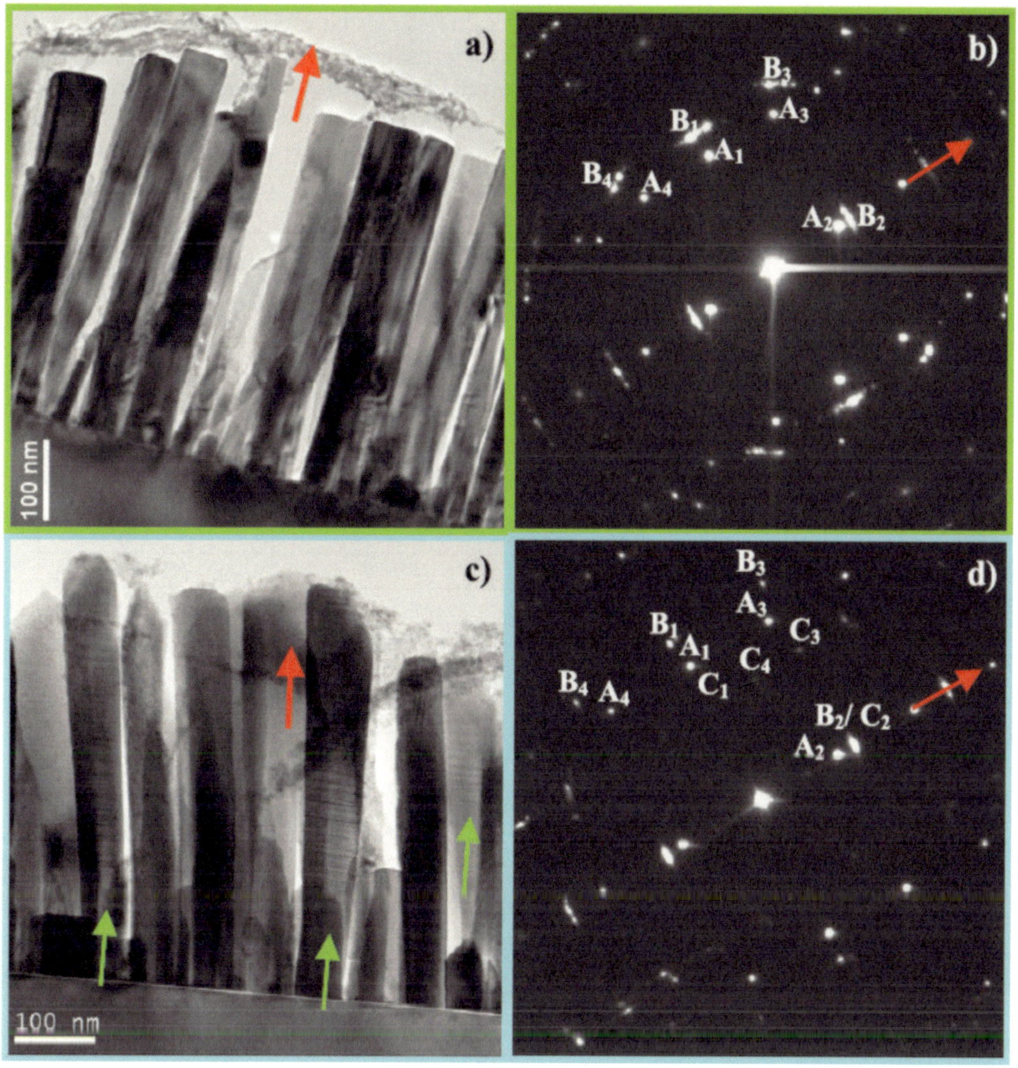

Fig. 1. SAED analysis

On the other hand, Fig. 1.(c) shows a BFSTEM general view of the highly Mg doped (TMg = 355 ºC) GaN NWs. Some of the NWs present twin defects (marked with green arrows). In this case, Fig. 1.(d) displays its SAED pattern. A new crystal orientation appears in this case, that has been related to the twinned NWs (spots labelled as C). Top arrows (red) are pointing to the same equivalent growth direction in SAED patterns and BFSTEM micrographs. After this analysis (see the corresponding indexation in Table 1. SAED patterns indexation) it was observed that GaN NWs follow a preferential epitaxial relationship versus the Si substrate, namely: (0001)[0-110] GaN // (111)[-1-12] Si (Arbiol et al., 2009). Undoped GaN NWs appear to be defect-free, whilst Mg doped samples present some NWs with twin (T) defects along the growth axis, as shown in Fig. 1.(a) and (b) -some twinned NWs have been marked with green arrows in Fig. 1.(c). When a SAED pattern of a NW region populated with few twinned NWs is obtained, a new orientation relationship superimposed to the one found for undoped samples can be observed. As shown in Fig. 1.(d), the new spots (indexed in Table 1. SAED patterns indexation), labeled as C_n, are rather weak. The presence of these weak spots can be attributed to the presence of Ts in the NWs. The new epitaxial relationship which is found is: (0001)[1-210]GaN // (111)[-1-12]Si. The twinned NWs still grow along the [0001] direction; however, they are rotated 30º along the growth axis with respect to the non-twinned NWs.

Spot #	**d (nm)**	**(deg) versus spot A_1**	**Indexation**
A1	1.92	–	(2-20)
A2	3.13	90	(111)
A3	1.64	31	(3-11)
A4	1.64	31	(1-3-1)
Zone axis: [-1-12] Si			

Spot #	**d (nm)**	**(deg) versus spot B_1**	**Indexation**
B1	1.55	–	(-2110)
B2	2.59	90	(0002)
B3	1.30	32	(-2112)
B4	1.30	32	(-211-2)
Zone axis: [0-110] GaN			

Spot #	**d (nm)**	**(deg) versus spot C_1**	**Indexation**
C1	2.76	–	(-1010)
C2	2.59	90	(0002)
C3	1.89	43	(-1012)
C4	2.44	62	(-1011)
Zone axis: [1-210] GaN			

Table 1. SAED patterns indexation

The density of twins dramatically increases with increasing Mg concentration and, thus, the formation of twins can be directly correlated to the increasing presence of Mg. Twin defects along the growth axis have been commonly observed in some other one-dimensional semiconductor nanostructures such as in III–V NWs (Huang et al., 2002; Mikkelsen et al., 2004; Johansson et al., 2006; Davidson et al., 2007; Karlsson et al., 2007; Arbiol et al., 2009), or even in Si NWs (Arbiol et al., 2007; Arbiol et al., 2008b).

High resolution TEM (HRTEM) analysis (Fig. 2. HRTEM analysis of the highly Mg-doped GaN NWs) confirms that NWs with Ts grow in the (0002)[1-210]GaN // (111)[-1-12]Si orientation and are rotated 30º from the defect-free NWs. As shown in Fig. 2.(e), red arrows are pointing towards GaN(0001) wurtzite planes, while the cyan ones are pointing to the GaN twinned planes (in particular to a triple-twin region). As observed, the stacking of the planes is different in the twinned regions. On the other hand, an amorphous layer between the substrate and the GaN NW is apparent in Fig. 2.(a). Core-loss EELS quantification along the interface was carried out to find out the chemical nature of this amorphous layer. As Si

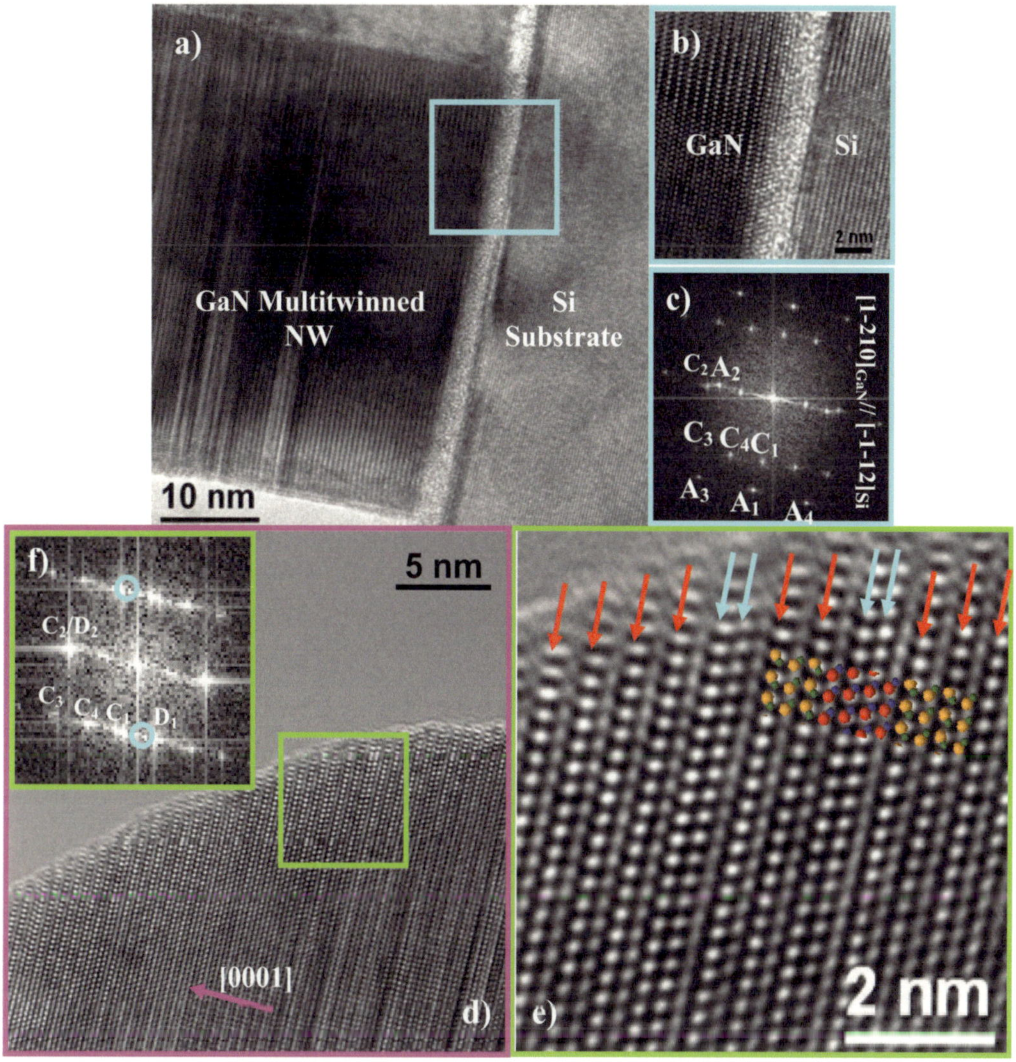

Fig. 2. HRTEM analysis of the highly Mg-doped GaN NWs

$L_{2,3}$ and Ga $M_{2,3}$ edges overlap, quantification was not carried out in the usual manner (computing background subtracted integrated signals for each element, sigma correcting them and dividing them by each other), but rather treating the overlapping region as a linear combination of the Si and Ga edges. This way, it was found that the amorphous layer contained 50 ± 5 % of Si, 50 ± 5 % of N and no Ga (Fig. 3. GaN/Si interface EELS analysis). This Si nitride layer was very likely created during the nitridation process that occurred during the first growth steps. A scheme of the twin formation is also shown (Fig. 4. Twin formation scheme). Fig. 4.(a) represents the conventional GaN wurtzite (WZ) structure, while in Fig. 4.(c) the formation of a single twin, equivalent to a 180° rotation in the WZ structure around the [0001] growth axis, is shown. It is important to point out that the consecutive stacking of [0001] axial twins leads to the formation of the GaN zinc-blende (ZB) structure, which is cubic instead of hexagonal (Fig. 4.(b) for a scheme of the GaN ZB structure).

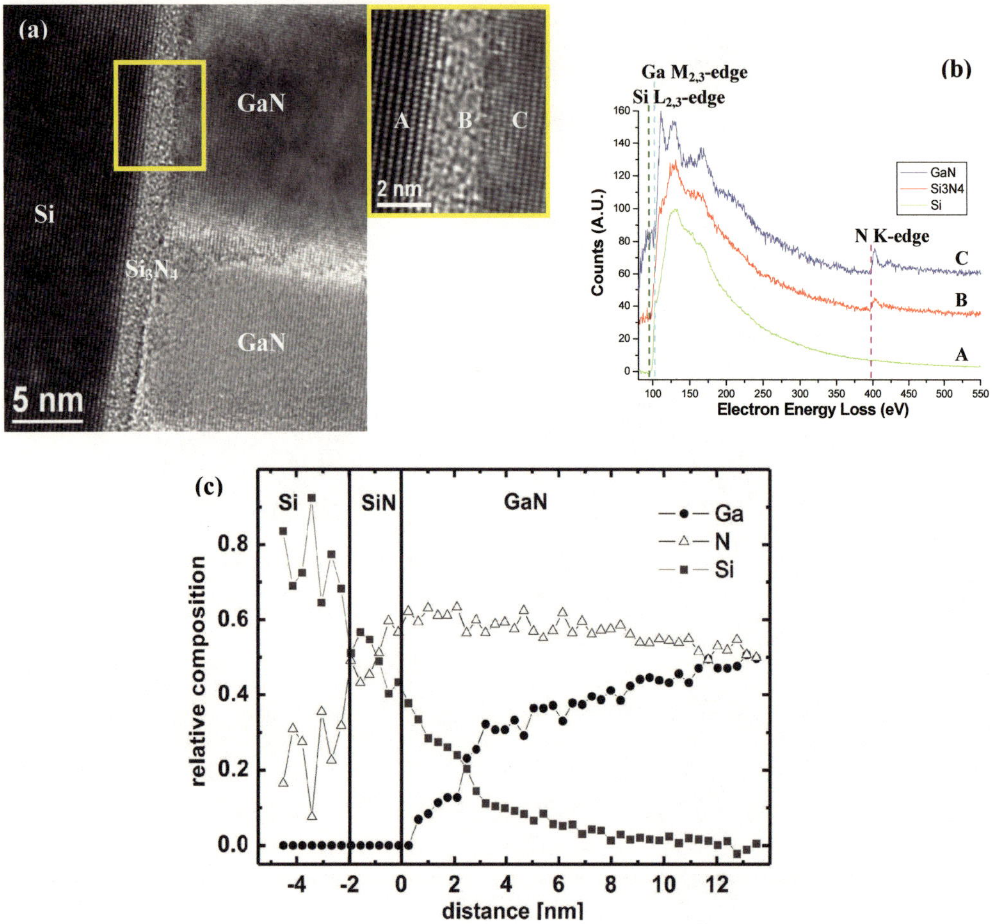

Fig. 3. GaN/Si interface EELS analysis. For Fig. 3.(c) see (Arbiol et al., 2008b).

In particular, it is displayed how a triple-twin creates a three-cell ZB domain (Figs. 2.(e) & 4.(e)). In this way, in the power spectrum shown in Fig. 2.(f), the D1 spot corresponds to the (-11-1) GaN ZB plane, while the D2 spot is the (111) plane. The appearance of alternating wurtzite and zinc-blende structures due to the presence of twins has been widely reported for III–V NWs (Banerjee et al., 2006; Ihn et al., 2006; Arbiol et al., 2008b) and the change of crystal phases in 1D nanostructures due to twinning is a widespread phenomenon in several materials (Arbiol et al., 2008a). In addition, the formation of wurtzite and zinc-blende heterostructures in a chemically homogeneous nanowire material is a hot scientific topic as it is opening up new possibilities for band-structure engineering (Algra et al., 2008; Arbiol et al., 2008b; Bao et al., 2008). These heterostructures can have dramatic implications on the electronic properties of the NWs, as a change in the crystal structure also implies a variation in the density of states and thus on the carrier transition energy.

In the present case, due to the different band gap, the resulting inclusion of ZB GaN in the WZ GaN NW creates quantum well regions along the NW axis. In a recent work (Bao et al., 2008), it was shown how the presence of twinned planes between WZ quantum domains in InP NWs (they attributed it to WZ–ZB alternance) could influence their photoluminescence properties.

In their case, the excitation power dependent blueshift of the observed photoluminescence could be explained in terms of the predicted staggered band alignment of the rotationally twinned ZB/WZ InP heterostructure and of the concomitant diagonal transitions between localized electron and hole states responsible for radiative recombination.

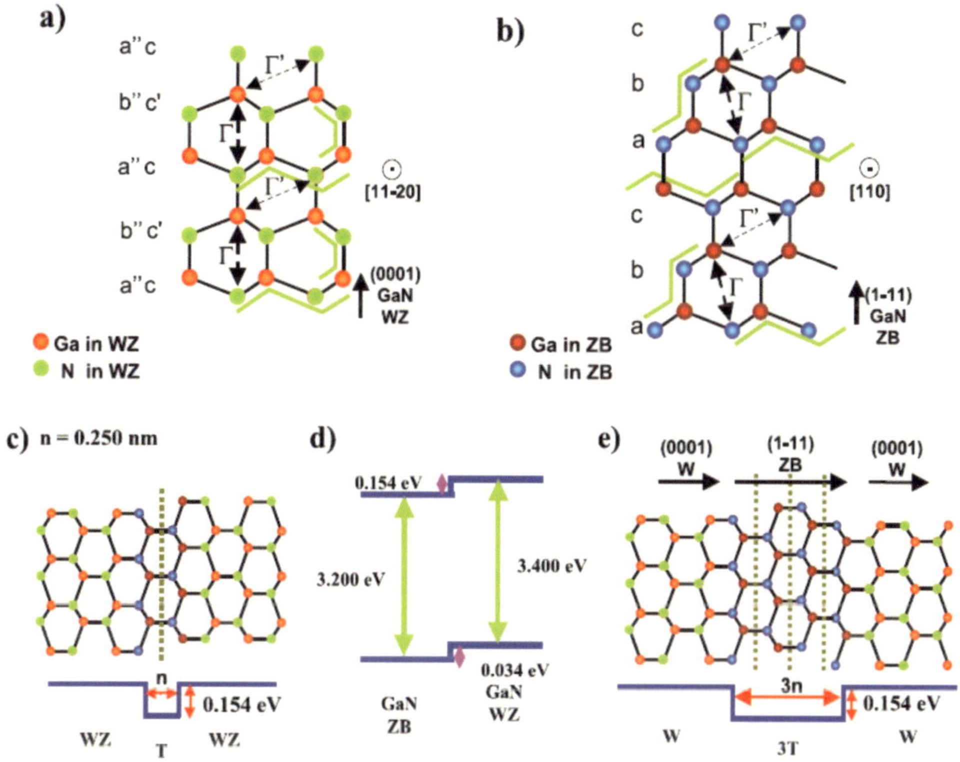

Fig. 4. Twin formation scheme

2.5 Local optical properties addressed through EELS: Plasmon position and structural considerations

Low-loss spectra were acquired along the doped NWs using a GIF2001 spectrometer coupled to a JEM2010F microscope. The plasmon position was found to remain constant along the NW. Now, the plasmon position depends on bandgap, but it also depends on other factors that may cancel out. In particular, it is possible to assume that plasmon position depends on the composition and the distance to the first neighboring atom, which are the same for WZ and ZB regions in the NWs, so that even if the bandgap changes, the plasmon energy remains the same. In other words, the unchanging plasmon position does not need to correspond to unchanging bandgap energy. This affirmation will be further sustained by ab-initio simulations in the following section. On the other hand, it is clear that EELS spectra with a better energy resolution are needed to access the bandgap energy.

Bandgap estimation

Aberration-corrected STEM, and EELS spectra with a 0.3 energy resolution were used to move one step further and locally analyze the local density of states at the atomic scale that occur in the interface between the triple-twin (3T) planes (ZB) and the WZ GaN heterostructure. Experiments were performed on a dedicated VG HB 501 STEM retrofitted with a Nion quadrupole-octupole corrector (SuperSTEM 1) at Daresbury (Arbiol et al., 2009). Cs-corrected HRHAADF STEM micrographs, where the displayed bright spots directly correspond to the actual atomic positions, corroborate the proposed structural model for the 3T domains (Fig. 5. High-resolution HAADF twin domain and local HREELS Bandgap analysis). EEL spectra maps of the NWs were obtained with a zero loss peak (ZLP) energy FWHM of 0.3 eV and a probe size of about 0.1 nm. The much reduced probe size allowed us to obtain several EEL spectra in the direction perpendicular to the NW growth direction for wurtzite and 3T planes (see square blue and red marks in Fig. 5.(a) for the EELS selected areas). Then, the spectra obtained at the exact atomic positions with respect to the wurtzite and the 3T regions were processed and analyzed. These spectra are displayed in Fig. 5.(b). It is generally accepted that the ZLP shape is Gaussian (Egerton, 1996), and thus that it contributes to the low-loss signal as A $\exp(-rx^2)$, where x is the energy loss. In the present work, it was subtracted from the low-loss region by fitting an A $\exp(-rx^2)$ function to the positive tail of the ZLP. The resulting spectra (Fig. 5.(b)) show several interesting features -before analyzing the low-loss spectra, it should yet be noticed that the peak observed at about 24 eV is in fact the Ga 3d transition. The obtained band gap is found to be higher for the WZ region (measured to be 3.4 eV) than for the 3T region (measured to be 3.2 eV). The values are comparable to the band gap measurements for WZ and ZB phases obtained by EELS in literature (Bangert et al., 1998; Lazar et al., 2003) but, interestingly, lower than the values obtained using other techniques. On the other hand, features B and C in Fig. 5.(b) are found at higher energies for zinc-blende GaN than for wurtzite GaN. These peaks were expected, from first principles calculations (Gavrilenko & Wu, 2000), to arise from transitions from the three upper valence bands to the third and fourth conduction bands (B), and from the same bands to the fifth and upper conduction bands (C), occurring at higher energies for ZB GaN than for WZ GaN. Finally, the higher intensity of the EELS spectrum in the region <15 eV and the broadening of the plasmon peak for the faulty region can be related to the occurrence of surface modes (Egerton, 1996; Erni & Browning, 2005) corresponding to the WZ-ZB-WZ interface (Fig. 5.(b)). Specific ab initio simulations of the EELS spectra of pristine bulk wurtzite and defective bulk wurtzite (triply-twinned inclusion) were carried out (Figs. 5.(c) & (e)) in order to corroborate the origin of these experimental features. Ab initio calculations were carried out using the SIESTA (Soler et al., 2002) code, which combines density functional theory (DFT), normconserving pseudopotentials, and local basis set functions.

We used the generalized gradient approximations (GGAs) with the Perdew, Burke, and Ernzerhof (PBE) parameterization (Perdew et al., 1996). For Ga and N atoms, the double ζ local basis set was used with polarization. Additional d-electrons were included in the valence electron set of Ga. Well converged spectra were obtained with a real space mesh cut-off of 250 Ryd and Monkhorst-Pack sets larger than $34 \times 34 \times 18$ for wurtzite structures. Experimental HRTEM lattice parameters were used to build all crystal models. Atomic positions were determined by performing structural relaxations using conjugate gradient minimization of the energy, until the forces on all the atoms were smaller than 0.04 eV $Å^{-2}$. In the relaxation of the models, lattice dimensions were kept constant (in accordance with the experimental values) and no constraints were imposed on the atomic positions within the supercell. To model the planar defects, defective inclusions were embedded in pristine bulk wurtzite. Eight [0001] layers of Ga-N dimers were used to separate the periodic images

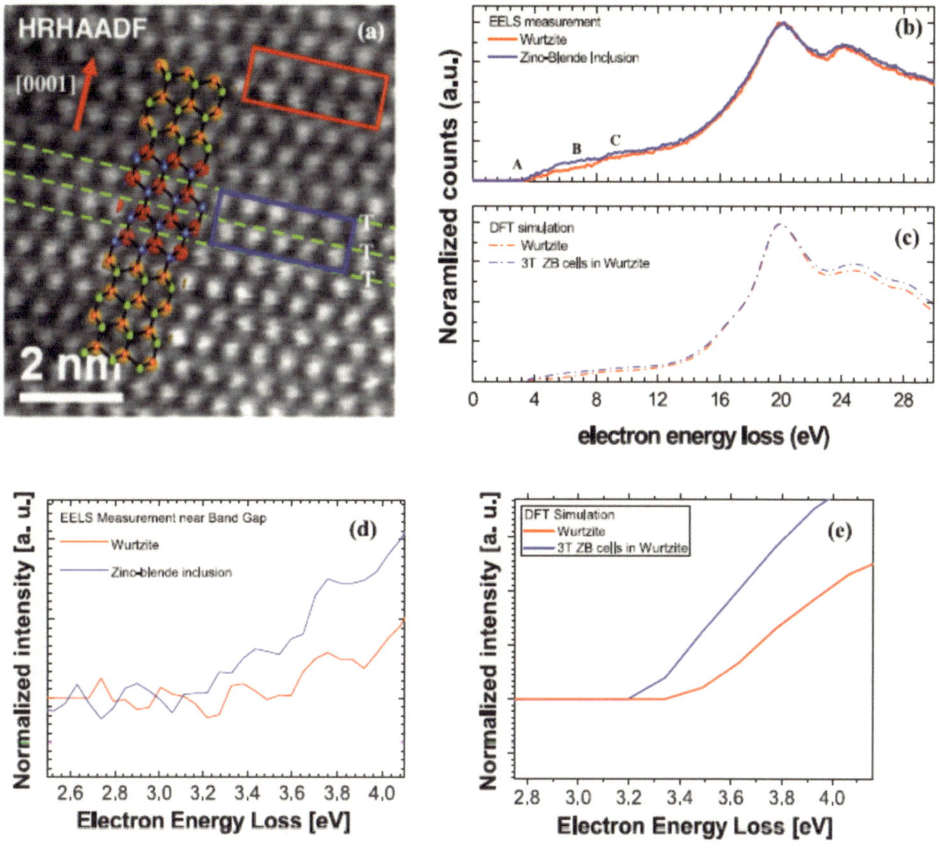

Fig. 5. High-resolution HAADF twin domain and local HREELS Bandgap analysis

of the defects. Variations in the electronic structure of less than 0.2% were observed for 16 spacing layers. The loss functions were obtained using first-order time-dependent perturbation theory to calculate the dipolar transition matrix elements between occupied and unoccupied single-electron eigenstates, as implemented in SIESTA 2.0. The optical matrix elements were calculated including the corrections due to the nonlocality of the pseudopotentials (Read & Needs, 1991), which were then used to obtain the dielectric function $\varepsilon(\omega)$ and the loss function $\text{Im}\{-1/\varepsilon(\omega)\}$ that is directly comparable with the EELS spectra. It is well known that electronic structure calculations within DFT-GGA generally underestimate the band gap of semiconductors (Jones & Gunnarsson, 1989). In this case, the calculated band gap for the pristine wurtzite phase was EW g(GGA) = 2.42 eV, clearly lower than the experimental value EW g(exp) = (3.4 ± 0.1) eV. We followed the most common procedure in the literature to circumvent this difficulty that is to apply scissor operators (Levine & Allan, 1989; Hughes & Sipe, 1996) that rigidly shift the conduction band in order to match the position of the main peak of the calculated and experimental EELS spectra (located at 20.1 eV in Figs. 5.(b) & (c)). Figs. 5.(c) & (e) show the simulated spectrum of the pristine wurtzite bulk and the local spectrum of the 3T cells embedded in wurtzite. These calculations not only reproduce the main features (A, B, C in Fig. 5.(b)) of both situations but also predict the experimental band gap change (Figs. 5.(d) & (e)). Notice that all spectra have been normalized to the maximum of the plasmon peak in Figs. 5.(b) & (c).

3. 3D reconstruction of 1D complex nanostructures by means of electron tomography

3.1 State of the art

Tomography

Tomography is a technique used to reconstruct the 3D morphology of an object from its projecting images. The mathematical principles which allow these techniques were established since the theorem made by Radon in 1917 (Radon, 1917). However, the necessary computer tools were not developed enough in order to perform the complex calculations. The necessity to obtain higher dimension structures from data series of fewer proportions is present in many scientific fields. The first real application was made by Bracewell in the area of astronomy in 1956 (Bracewell, 1956). He proposed a method to recreate a 2D map related to the microwaves emission of sun from 1D fan beam series profiles measured with a microwave telescope. The usefulness of 3D reconstructions was confirmed by 2 Nobel Prizes. The first, in 1979, was given to A. Cormack (Cormack, 1980) and G.N Hounsfield (Hounsfield, 1980). They developed the Computerized Axial Tomography, a very well known diagnosis technique. The second Nobel Prize was given to Aaron Klug in 1982 (Klug, 1982). His work was pioneer in 3D reconstructions of molecular structures through projecting images obtained with electron microscopy. In nanoscience and nanotechnology it is very well known that the shape, size, and morphology of a nano-object are very important factors, sometimes with similar relevance than the material composition in the final physical and chemical behavior. Nanostructured materials like nanowires, for example, are becoming materials of a huge importance since its physical properties dissociate in a significant way from the behavior that the same material would have in bulk. In general, the effects due to the nanostructuration start from 1 to few hundred nanometers. Several experimental characterization techniques are able to obtain 3D information at nanometric scale. One example is the atom-probe-field-ion microscopy which can be applied to conductive samples used in mass spectrometry (Humphreys, 2007). Another example would be the serial sectioning approach in which a tridimensional model is reconstructed through a series of slices. From this, we can extract images trough cryo TEM or even with an Atomic Force Mircroscopy (AFM). The major problem of these cross-sectional techniques is that they are destructive. Moreover, image formation techniques based on magnetic resonance or X-ray absorption remind restricted outside the nanoscale. It is for this reason that Electron Tomography becomes a very important tool for the structural modeling of nanometric objects. Computerized tomography is directly related with Radon theorem. The Radon transform is defined as a mapping into the so-called Radon space of a function describing a real space object, by the projection, or line integral, through that function along all possible lines (Fig. 6. Radon transform scheme). The Radon transform R can be visualized as the integration through a body D in real space f(x,y) along all possible line integrals L, with its normal at an angle to the horizontal.

$$Rf = \int_L f(x,y)ds \quad (1)$$

Thus, given a sufficient number of projections, an inverse Radon transform of this space should reconstruct the object (Midgley & Weyland, 2003). A discrete sampling of the Radon transform is geometrically equivalent to the sampling of an experimental object by some form of transmitted signal or projection. The consequence of such equivalency is that the

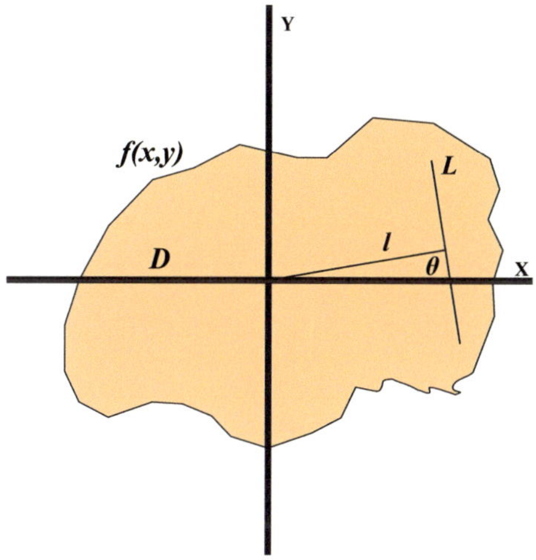

Fig. 6. Radon transform scheme

reconstruction of the structure of an object *f(x,y)* from projections *Rf* can be achieved by implementation of the inverse Radon Transform (Midgley & Weyland, 2003). In practice, the reconstruction that comes from projections is based on the knowledge between the relationship of the projections in the real space and Fourier space. The theorem of the central slice establishes that a projection of a given angle is a central section trough the Fourier transform of this object (Fig. 7. Sampling in Fourier space for tilting with equal increments). Thus, if a series of projections are acquired at different tilt angles, each projection will equate to part of an object's Fourier transform, sampling the object over the full range of frequencies in a central section.

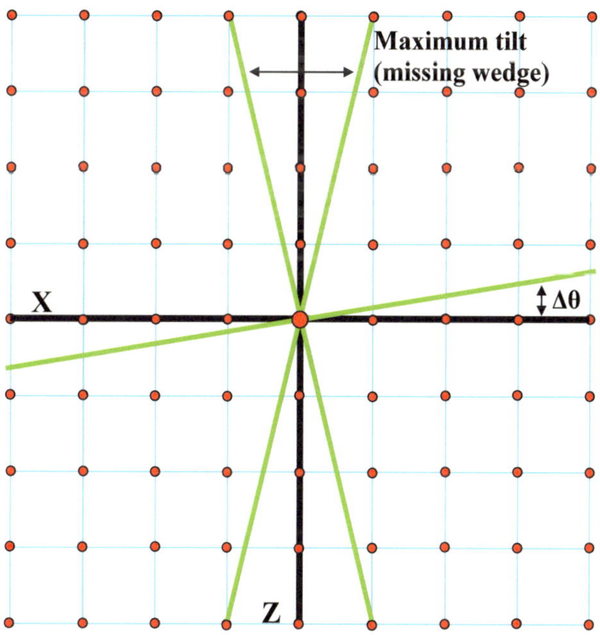

Fig. 7. Sampling in Fourier space for tilting with equal increments

Therefore tomographic reconstruction is possible from an inverse Fourier transform of the superposition of a set of Fourier transformed projections: an approach known as direct Fourier reconstruction. However, if projections are missing from an angular range, brought about by a limit on the maximum tilt angle, then Fourier space is under-sampled in those directions and as a consequence the back transform of the object will be degraded in the direction of this missing information. In this way, different sampling geometries, which are associated with the technical characteristics of the microscope sample holder and its capacity to fill the Fourier space can be used: single-axis, double-axis and conical (Penczek et al., 1995; Frank, 2006). Moreover, two problems appear in this formulation. Firstly, the projection data is always sampled at discrete angles leaving regular gaps in Fourier space. As the inverse transform requires a continuous function, radial interpolation is required to fill the gaps in Fourier space. Secondly, there is a non-uniform sampling of the data, which results in that the central zone of Fourier space has more information than the zones that are more remote. Thus, direct reconstruction methods have been replaced by the retroprojection method, which requires less compute power and allows to solve the problem of the inhomogeneities in the sampling in the Fourier space via application of some kind of filters, in general ramp filters: this method is know as Weighted BackProjection (WBP). The method of backprojection is based on inverting the set of recorded images, projecting each image back into an object space at the angle at which the original image was recorded. Using a sufficient number of backprojections, from different angles, the superposition of all the backprojected "slices" will reconstruct the original object. A schematic diagram of this approach is shown (Fig. 8. Scheme of 3D reconstruction via backprojection).

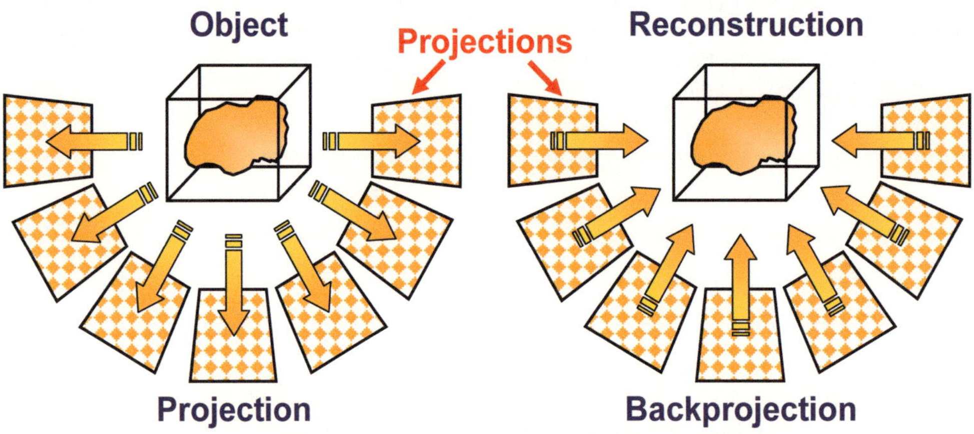

Fig. 8. Scheme of 3D reconstruction via backprojection. An object is sampled by projection from a range of angles and then reconstructed by backprojecting these projections at the original sampling angle into the object space

To provide maximum 3D information as many projections as possible should be acquired over as wide a tilt range as possible. Once we have introduced tomography fundamentals, we will focus our efforts on the electron tomography (Midgley & Weyland, 2003).

Electron tomography

The transmission electron microscope (TEM) is a very important tool in the structural and analytical characterization of objects at the nanoscale since it can offer us different sources of information depending on the detectors that are used. However, most of the techniques that are associated with TEM are based on simple 2D projections obtained after the electron beam is transmitted through a 3D object. As it has been said before, in order to recover the tridimensional information lost with the projections we can apply some tomographic reconstruction techniques (Fig. 9. TEM tomography geometries). These techniques are widely used in the areas of medicine. Nevertheless, electronic tomography has been very little used in materials science basically for two major reasons: a) in general, inorganic materials have few 3D structure and they can be very well described by 2D projections; b) In contrast to the biology area or organic composites, tomography based on bright field (BF) images does not give accurate results as far as the crystalline samples are concerned (Weyland, 2002; Midgley et al., 2007). The explanation to this is that in order that a reconstruction can be faithful to the real object, the intensity of the projecting image must be a monotonous function of some characteristics of the real object. The amount of material projected in a parallel direction to electron beam could be a good example of this. This is known as a projection requirement. In crystalline samples, the contrast in BF conditions is ruled by the diffraction, which is related to the Bragg conditions and not necessary to the thickness of the material projected. Even all this, in the last years there has been a great effort to solve this problem by using the TEM operative mode High Angle Annular Dark Field (HAADF) (Koguchi et al., 2001; Midgley et al., 2001). Electrons scattered to low angles are predominantly coherent in nature and therefore, conventional BF and DF images exhibit sudden contrast changes depending on specimen thickness, orientation or defocus. This observation mode feature does not obey the projection requirement.

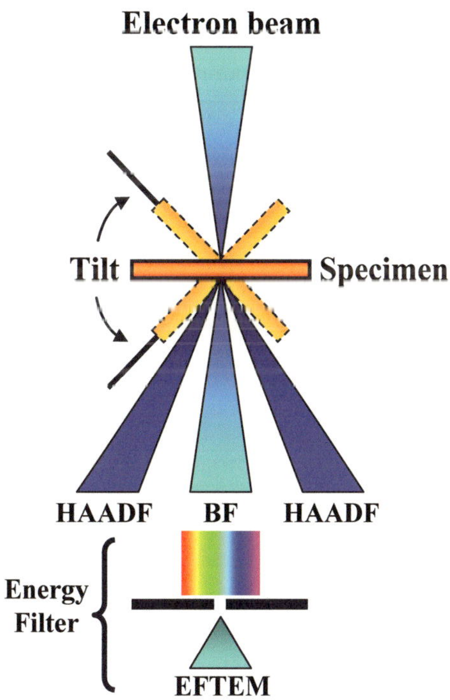

Fig. 9. Summary of the geometries of the various signals that may be used for tomographic reconstruction in TEM

However, electrons scattered to high angles are predominantly incoherent, and images formed using HAADF detector do not show the contrast changes associated with coherent scattering. Such high angle scattering is associated with electron interaction close to the nucleus of the atom and thus the cross-section for HAADF scattering approaches the unscreened Rutherford cross-section, which is strongly dependent on the atomic number Z. In practice the unscreened limit is never reached and the exact dependence depends on many factors. Nevertheless, medium-resolution STEM images formed with a HAADF detector are very sensitive to changes in specimen composition with the intensity varying monotonically with composition and specimen thickness, thus satisfying the projection requirement and giving an extra information on sample composition as it is possible to discern between different materials composing the studied reconstructed structure. For a comparison between the results obtained in crystalline samples by BF-TEM and HAADF-TEM see (Fiedrich et al., 2005). On one hand, it must be said that if HAADF gives the chance to obtain 3D structural reconstructions of crystalline nanostructures, while the Energy Filtered TEM (Möbus et al., 2003) and Electron Energy Loss Spectroscopy (van den Broek et al., 2006) have demonstrated to be a valid tool in order to find 3D compositional maps of nanostructured materials. On the other hand, the combination of HAADF with the Scanning TEM mode shows a great advantage since in scanning mode the electron dose on the surface area is time limited and then the sample damage is reduced. In the last years electron tomography has been applied to obtain solve 3D morphological and compositional problems related to nanowire synthesis. Up to our knowledge the first time that electron tomography was applied on NWs, was in 2004 to discern the growth of Metal Nanowires (Pt) inside hard template mesoporous silica (Arbiol et al., 2004a; Arbiol et al., 2004b). Lately, and thanks to the development of the HAADF STEM tomography technique, application of electron tomography to crystalline nanostructures such as nanowires has been extended. In this way, it has been applied for example to analyze the 3D morphology of core-shell GaP-GaAs NWs (Verheijen et al., 2007), the 3D structure of helical and zigzagged nanowires (Kim et al., 2008), the microsctructure of magnetic $CoFe_2O_4$ nanowires inside carbon nanotubes (Ersen et al., 2008), the 3D surface defects in core-shell nanowires (Arslan et al., 2008), the homogeneity of prismatic heterostructures on the facets of catalyst-free GaAs nanowires (Heigoldt et al., 2009), the 3D chemical arrangement on Ge-Si Nanostructures (Montoro et al., 2009) or the 3D line edge roughness in Cu NWs (Ercius et al., 2009).

3.2 Particular case: HAADF STEM tomography of coaxial multi-quantum wells in semiconductor nanowires

In order to demonstrate the capabilities of the Electron Tomography as characterization tool in the field of inorganic nanostructured crystalline materials, we have applied STEM-HAADF Tomography to coaxial nanowire heterostructures with variable quantum well thickness. In recent times the interest in III-V semiconductor nanowires has received renewed attention due to their applications in electronics and optoelectronics. The improvements of new methods of synthesis and characterization, has turned nanowires (and other nanostructures) into testers of quantum-mechanical effects (Hu et al., 2007; Shorubalko et al., 2008; Heigoldt et al., 2009). More complex structures have been obtained by combining materials coaxially and axially along the growth direction of the nanowires. Up to now, coaxial heterostructures have been considered to improve the performance of nanowire devices to confine the carriers at the core, leading to the reduction of surface scattering (Wang et al., 2005). Core-shell structures have also been used for engineering the optoelectronical properties of the core, for example for the fabrication of multi-color light emitting diodes or lasers (Quian et al., 2008). Little attention has been paid to the geometry of the deposited shell, as the main functionality continued to be reserved to the core of the nanowire. Recently, we have shown that is possible to uniformly coat the nanowires with successive epitaxial layers resulting into multiple quantum heterostructures defining for example prismatic quantum wells (p-QW) (Morral et al., 2008b). Growth of p-QWs constitutes an additional functionality to the nanowire and, accordingly, an increased freedom design for nanostructures and devices. As a result, new architectures based in coaxial heterostructures (Fig. 10. HRTEM transversal section of AlAs-GaAs MQW nanowires grown on (001) and (111)B GaAs substrates) are being synthesized and investigated, due to their optoelectronic properties and consequent applications in the fabrication of light-emitting diodes (LEDs). Compared to bulk samples, where electrons and holes can propagate in all three dimensions, there are new effects emerging when this is no longer the case. By introducing potential barriers for these carriers one can confine them in one, two or three spatial directions. If the dimension of the confinement is of the order of the Fermi wavelength, this results into quantum confinement, ending up with carriers which can only propagate freely in less than three dimensions, speaking then of quantum wells (2D), wires (1D) or dots (0D) respectively (Wegscheider et al., 1993; Schedelbeck et al., 1997). In the example selected to illustrate the electron tomography potentiality, we used a layer sequence of $Al_{0.35}Ga_{0.55}As$, GaAs, $Al_{0.35}Ga_{0.55}As$ for building the quantum well and a final layer of GaAs to prevent oxidation of $Al_{0.35}Ga_{0.55}As$.

Fig. 10. HRTEM transversal section of AlAs-GaAs MQW nanowires grown on (001) and (111)B GaAs substrates, left and right, respectively

For this material combination, a type I quantum well is formed in the GaAs layer sandwiched between the $Al_{0.35}Ga_{0.55}As$ barrier layers. By varying the thickness of the QW, it is possible to control the confinement energy of the carriers. In this way the wavelength of the emitted light can be tuned, adding new functionality to the heterostructured NWs. Nanowires have been grown on a (001) and (111)B GaAs substrates by molecular beam epitaxy (MBE). The substrate was coated with a SiO_2 layer of about 10nm thick. A recent work (Morral et al., 2008a) showed that when the thickness of SiO_2 is less than 30nm an epitaxial growth of the nanowire with respect to the substrate exists, thanks to the penetration of the Ga catalyst through nanometric pinholes allowing the contact between substrate and wire. By changing the growth conditions, it is possible to switch between nanowire axial growth and typical MBE planar growth. This results in the possibility of depositing layers in an epitaxial way on the nanowire facets. Then, due to the directionality of the molecular beam, the thickness of the layers will depend if the nanowire is standing perpendicular to the substrate or with an angle (Heigoldt et al., 2009). A scheme of the epitaxial growth on the nanowire facets and the respective band alignement of quantum wells grown on the facets, depending on substrate orientation is shown in Fig. 11. As [111]B growth axis is the preferentially followed by the nanowires (Morral et al., 2008a), the growth of the epitaxed nanowire over the (001) GaAs substrate leads to a non perpendicular orientation of the nanowire (35° leaning with respect to the substrate), implying a inhomogeneous configuration of the coating over the different faces of the hexagonal nucleus with quantum wells of distinct thickness in each lateral side. Oppositely, those nanowires grown on [111]B substrates, would be oriented perpendicularly to the substrate, thus implying a homogeneous shell covering on all the lateral facets.

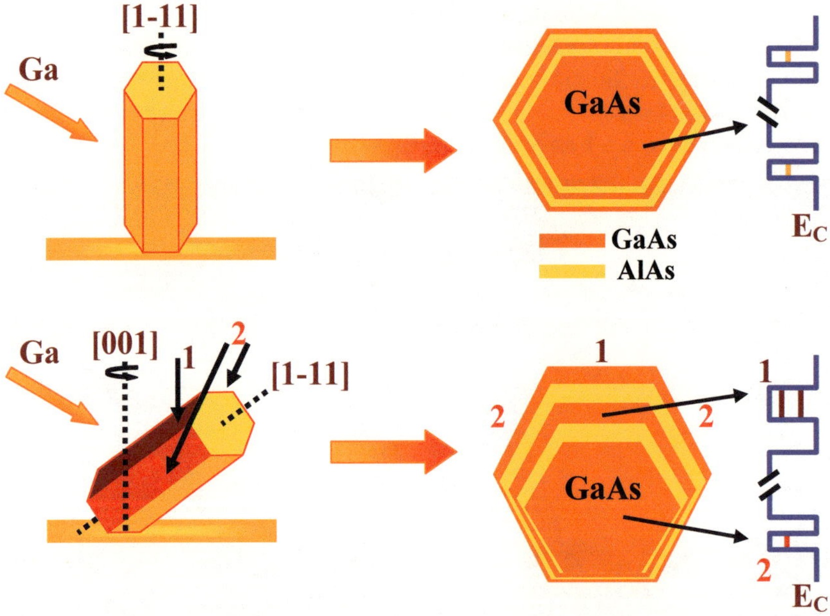

Fig. 11. Scheme of Nanowires growth and band alignment of quantum wells grown epitaxially on the facets depending on substrate orientation

As a consequence, several emission peaks in photoluminescence (PL) are observed, depending on the nanowire configuration on the substrate. This is in good agreement with the proposed scheme. (Morral et al., 2008b). In the case where six identical quantum well-like states exists, one would expect other confinement states at the six corners where the QWs cross. Due to the effectively increased distance between the barriers in such a corner, i.e. the wavefunctions of electrons and holes should be less confined than in a QW, forming a one dimensional quantum wire like state, as shown in previous calculations (Heigoldt et al., 2009). Such states have already been demonstrated experimentally at the intersection of two quantum wells, using for example the technique of cleaved edge overgrowth or by overgrowth of so call V-grooves (Kaufman et al., 1999; Merano et al., 2006). The distance in energy of such a one dimensional state to the ground state in the QWs is small, usually in the order of meV, demanding a very good homogeneity of the QW width, in order to be resolved experimentally by photoluminescence (Wegscheider et al., 1994; Wegscheider et al., 1999).

3.3 HAADF STEM tomography

In order to carry out this experimental approach we have used a TEM JEOL 2100 with a HAADF detector. To obtain the data we used the GATAN 3D Tomography-Acquisition Software as well as the 3D Reconstruction and 3D Visualization PlugIns integrated in the well-known Digital Micrograph package. The microscope holder used was an EM-21010/21311 HTR-High Tilt Holder, specially designed to allow a high tilt performance up to ± 80°. This software is able to account for the acquisition, image processing and electron energy loss spectroscopy (EELS) tools to quantitatively extract structural and chemical information of the sample. Since the samples object of the present research were of crystalline nature, we used STEM-HAADF mode for the development of tomography. In all the reconstructions made in the present work we have followed 3D Simultaneous Iterative Reconstruction Technique (SIRT) by plane, which consists in multiple iteration algorithms to try to minimize errors in the reconstruction that arise from the reconstruction process. These techniques have been useful for minimizing the effect of incomplete tilt range coverage commonly referred to at the "missing wedge". The number of iterations has been 15 since the results we obtained were successful. Complete tomographic reconstruction movies related to the samples analyzed will be available in the following web site: http://www.icmab.cat/gaen

3D reconstruction model of the MQW NWs grown on (001) substrates

A series of AlAs and GaAs layers of thicknesses ranging from 8 to 30 nm was grown on inclined wires, as grown on (001) GaAs substrates. As schematized in Fig. 11., in this case where the NWs are oriented with an angle of 35° with respect to the surface, the flux of adatoms is different for each of the six nanowire sidewalls. The top facets are facing the flux and therefore the thickness of their grown layers should be the thickest.

As it can be observed in Fig. 10. (left), this is also what it is observed from the cross-section HAADF (S)TEM analysis. The growth rate on the other facets is smaller, with the facets facing the substrate surface showing almost no growth. A 3D HAADF Tomography reconstruction of a p-QW NW grown on (001) GaAs substrate has been also obtained in order to verify the 3D geometrical configuration. For this purpose we selected a NW grown on a (001) GaAs substrate and obtained the corresponding tilt series on HAADF. The 3D tomographic reconstruction is shown (Fig. 12. 3D tomographic reconstruction of the multiwalled nanowire nature and the inhomogeneous layer thickness), where we can appreciate that the difference in thickness of the various facets of the NW. The tomographic reconstruction has been made from the series of images taken at the top end of the nanowire which at -28° has a diameter of 235 nm, and with projections from -72° to +56°, eventhough the last 10 projections were removed because of pollution problems. In order to remove the shift 23 iterations have been made. The core of the nanowire (GaAs) has a diameter of 128

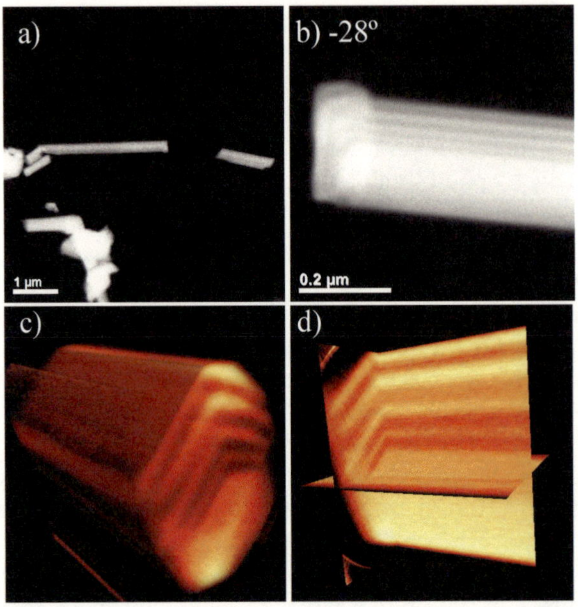

Fig. 12. a) general view of the characterized nanowire. b) HAADF image at -28° tilt (the presence of multiple shells is observed). c) Volumetric reconstruction of the nanowire. d) Ortho slices in the tomographic reconstruction that show the multiwalled nanowire nature and the inhomogeneous layer thickness.

nm and the nanowire presents four coatings of successive layers of AlAs and GaAs, respectively. The AlAs layers increase their width between 10 nm to 15 nm and the GaAs between 11 nm to 19 nm, in radial direction. Finally, the 3D reconstruction allows obtaining additional information which was not possible in the case of a cross section. Indeed, we observe that the QWs on individual facets are of equal thickness along the nanowire facet, indicating a homogeneous epitaxial growth along the whole nanowire. This is in agreement with the homogeneity of the optical properties along the nanowire axis, as reported elsewhere (Heigoldt et al., 2009).

3D reconstruction model of the MQW NWs grown on (111)B substrates

In the case of those NWs grown on (111)B GaAs substrates, we expect a homogeneous coaxial structure in all of the facets, as schematized in Fig. 11 and observed in Fig. 10. (right), from the cross-section HAADF (S)TEM analysis.

As in the previous case, the complete tomographic reconstruction of one of these NWs was obtained in order to assure the perfect homogeneity of the p-QW within the 6 facets and along the NW. For this purpose we obtained the tilt series on the same area shown in Fig. 10. (right). In Fig. 13. the 3D reconstruction of the sectioned NWs shown in Fig. 10. (right) has been obtained. From these measurements we find that the quantum wells are flat and homogeneous in thickness along the whole nanowire in good agreement with the previous HRTEM results. Thicknesses measured on the 3D reconstruction are in good agreement with those obtained earlier by cross-sectional HAADF STEM. However, thanks to the 3D reconstruction, we can assure that the measured thicknesses are kept along the whole NW growth axis. Not surprisingly, it is found again that the scaling factor, e.g. the quotient of the measured and the nominal thickness is approximately constant for the AlAs layers. This means that diffusion from the substrate to the facets of the NW plays a minor role on the deposition mechanisms on the facets.

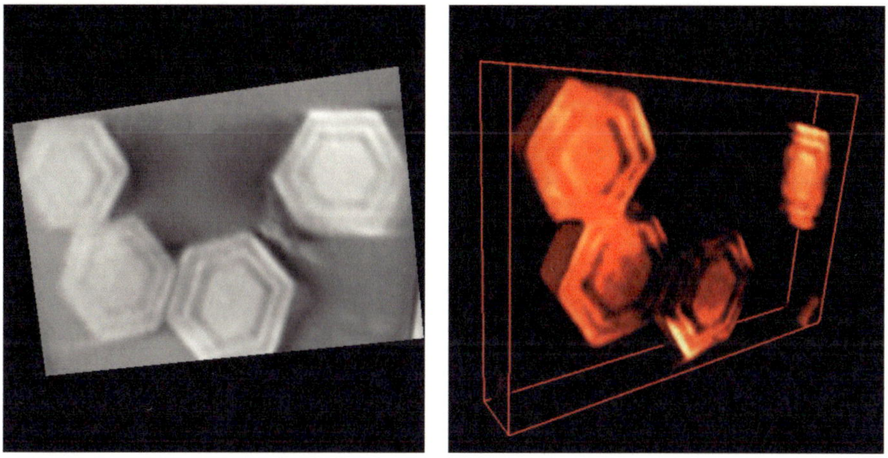

Fig. 13. Tomography slice obtained from the reconstructed 3D volume (left). 3D reconstructed model of several NWs with homogeneous MQWs.

4. Conclusion

In the present chapter we have shown how advanced tools in electron nanoscopy can help analyze complex nanostructured nanowires. On one hand, EELS combined with STEM has been applied to characterize the features produced by the Mg doping in GaN nanowires grown by plasma assisted molecular beam epitaxy to obtain p-type nanostructures and their effects on the nanowire. As observed by (S)TEM, the presence of Mg results in the formation of triple-twin (3T) defects, which increase with increasing Mg concentration. The high concentration of misplaced atoms gives rise to local changes in the crystal structure equivalent to three non-relaxed atomic cells of zinc-blende (ZB), which define quantum wells (QW) along the wurtzite (WZ) nanowire growth axis. Local EEL spectra obtained on the 3T and wurtzite planes show modifications in the local density of states, in which the band to band electronic transition energies change from 3.4 eV (for the wurtzite band gap) to 3.2 eV in the 3T lattice regions, the latter being in good agreement with the band gap measured in GaN ZB structures by EELS. These results are confirmed by specific ab initio atomistic simulations of these two situations. EELS assessment of the optoelectronic properties of the given nanostructures at a subnanometric scale has been successfully demonstrated. On the other hand, we have shown how Z-contrast or HAADF electron tomography is a perfect tool in order to characterize the morphology of complex heterostructures in nanowires. We have presented the implementation and optimization of the tomographic methods in electron microscopy and its best performance for NW applications. These methods have been tested to carry out tomographic reconstruction of several coaxial core-shell heterostructures in semiconductor nanowires, in order to obtain their corresponding 3D information. This last promising 3D technique allowed a better understanding of the structure of our coaxial nanowires and is an excellent complement to the high resolution cross-sectional imaging modes.

5. Acknowledgements

We kindly acknowledge Prof. Anna Fontcuberta i Morral groups from EPFL (Lausanne, Switzerland) and WSI-TUM (Garching, Germany) for the synthesis of GaAs multiquantum-well nanowires used in electron tomography analyses examples, as well as for her contribution and discussions on the present chapter. We also want to acknowledge Prof. Martin Eickhoff groups in WSI-TUM (Garching, Germany) and Justus-Liebig-Universität (Giessen, Germany) for the synthesis of Mg doped GaN nanowires used to illustrate advances in EEL spectroscopy applied to Nanowires. We also want to thank Serveis Cientificotècnics of Universitat de Barcelona for the use of their TEM facilities (Barcelona, Spain).

6. References

Algra, R. E.; Verheijen, M. A.; Borgstrom, M. T.; Feiner, L-F.; Immink, G.; van Enckevort, W. J. P.; Vlieg, E. & Bakkers, E. P. A. M. (2008). Twinning superlattices in indium phosphide nanowires. *Nature*, 456, 7220, (November 2008) 369-372, ISSN: 0028-0836

Allen, J. L.; Findlay, S. D.; Lupini, A. R.; Oxley, M. P. & Pennycook, S. J. (2003). Atomic-resolution electron energy loss spectroscopy imaging in aberration corrected scanning transmission electron microscopy. *Physical Review Letters*, 91, 10, (September 2003) 105503, ISSN: 0031-9007

Appenzeller, J.; Knoch, J.; Bjork, M. T.; Riel, H.; Schmid, H. & Riess, W. (2008). Toward nanowire electronics. *IEEE Transactions on electron devices*, 55, 11, (November 2008) 2827-2845, ISSN: 0018-9383

Arbiol, J.; Cirera, A.; Peiró, F.; Cornet, A.; Morante, J. R.; Delgado, J. J. & Calvino, J. J. (2002). Optimization of tin dioxide nanosticks faceting for the improvement of palladium nanocluster epitaxy. *Applied Physics Letters*, 80, 2, (January 2002) 329-331, ISSN: 0003-6951

Arbiol, J.; Rossinyol, E.; Cabot, A.; Peiro, F.; Cornet, A.; Morante, J. R.; Chen, F. L. & Liu, M. L. (2004). Noble metal nanostructures synthesized inside mesoporous nanotemplate pores. *Electrochemical and Solid State Letters*, 7, 7, (January 2004) J17-J19, ISSN: 1099-0062

Arbiol, J.; Rossinyol, E.; Cabot, A.; Peiro, F.; Cornet, A. & Morante, J. R. (2004). TEM 3D tomography of noble metal nanowires growth inside SiO2 mesoporous aggregates. *Materials Research Society Symposium Proceedings*, 818, (April 2004) 61-65, ISSN: 0272-9172

Arbiol, J.; Kalache, B.; Roca i Cabarrocas, P.; Morante, J. R. & Morral, A. F. I. (2007). Influence of Cu as a catalyst on the properties of silicon nanowires synthesized by the vapour-solid-solid mechanism. *Nanotechnology*, 18, 30, (August 2007) 305606, ISSN: 0957-4484

Arbiol, J.; Comini, E.; Faglia, G.; Sberveglieri, G. & Morante, J. R. (2008). Orthorhombic Pbcn SnO2 nanowires for gas sensing applications. *Journal of Crystal Growth*, 310, 1, (January 2008) 253-260, ISSN: 0022-0248

Arbiol, J.; Morral, A. F. I.; Estradé, S.; Peiró, F.; Kalache, B.; Roca i Cabarrocas, P. & Morante, J. R. (2008). Influence of the (111) twinning on the formation of diamond cubic/diamond hexagonal heterostructures in Cu-catalyzed Si nanowires. *Journal of Applied Physics*, 104, 6, (September 2008) 064312, ISSN: 0021-8979

Arbiol, J. ; Estrade, S. ; Prades, J. D. ; Cirera, A. ; Furtmayr, F. ; Stark, C. ; Laufer, A. ; Stutzmann, M. ; Eickhoff, M. ; Gass, M. H. ; Bleloch, A. L. ; Peiro, F. & Morante. J. R. (2009). Triple-twin domains in Mg doped GaN wurtzite nanowires: structural and electronic properties of this zinc-blende-like stacking. *Nanotechnology*, 20, 14, (April 2009) 145704, ISSN : 0957-4484; Fig. 3.(c) Reprinted with permission from this paper. Copyright [2008], American Institute of Physics.

Arenal, R.; Stéphan, O.; Kociak, M.; Taverna, D.; Loiseau, A. & Colliex, C. (2008). Optical gap measurements on individual boron nitride nanotubes by electron energy loss spectroscopy. *Microscopy and Microanalysis*, 14, 3, (June 2008) 274-282, ISSN: 1431-9276

Arslan, I.; Talin, A. A. & Wang, G. T. (2008). Three-dimensional visualization of surface defects in core-shell nanowires. *Journal of Physical Chemistry C*, 112, 30, (July 2008) 11093-11097, ISSN: 1932-7447

Banerjee, R.; Bhattacharya, A.; Genc, A. & Arora, B. M. (2006). Structure of twins in GaAs nanowires grown by the vapour-liquid-solid process. *Philosophical Magazine Letters*, 86, 12, (December 2006) 807-816, ISSN: 0950-0839

Bangert, U.; Harvey, A.; Davidson, J.; Keyse, R. & Dieker, C. (1998). Correlation between microstructure and localized band gap of GaN grown on SiC. *Journal of Applied Physics*, 83, 12, (June 1998), 7726-7729, ISSN: 0021-8979

Bao, J.; Bell, D. C.; Capasso, F.; Wagner, J. B.; Martensson, T.; Trägardh, J. & Samuelson, L. (2008). Optical properties of rotationally twinned InP nanowire heterostructures. *Nanoletters*, 8, 3, (March 2008) 836-841, ISSN: 1530-6984

Batson, P. E.; Kavanah, K. L.; Woodall, J. M. & Mayer, J. M. (1986). Electron-energy-loss scattering near a single misfit dislocation at the gaas/gainas interface. *Physical Review Letters*, 57, 21, (November 1986) 2729-2732, ISSN: 0031-9007

Batson, P. E.; Dellby, N. & Krivanek, O. L. (2002). Sub-angstrom resolution using aberration corrected electron optics. *Nature*, 418, 6898, (August 2002) 617-620, ISSN: 0028-0836

Boukai, A. I.; Bunimovich, Y.; Tahir Kheli, J.; Yu, J. K.; Goddard, W. A. & Heath, J. R. (2008). Silicon nanowires as efficient thermoelectric materials. *Nature*, 451, 7175, (January 2008) 168-171, ISSN: 0028-0836

Bracewell, R. N. (1956). Two dimensional aerial smoothing in radio astronomy. *Australian Journal of Physics*, 9, 3, (January 1956) 297-314, ISSN: 0004-9506

Calleja, E.; Sánchez-García, M. A.; Sánchez, F. J.; Calle, F.; Naranjo, F. B.; Muñoz, E.; Jahn, U. & Ploog, K. (2000). Luminescence properties and defects in GaN nanocolumns grown by molecular beam epitaxy. *Physical Review B*, 62, 24, (December 2000) 16826-16834, ISSN : 1098-0121

Calvert, C. C.; Brown, A. & Brydson, R. (2005). Determination of the local chemistry of iron in inorganic and organic materials. *Journal of Electron Spectroscopy and Related Phenomena*, 143, 2-3, (May 2005) 173-187, ISSN: 0368-2048

Chen, C-C.; Yeh, C-C.; Chen, C-H.; Yu, M-Y.; Liu, H-L.; Wu. J-J.; Chen, K-H.; Chen, L-C.; Peng, J-Y. & Chen, Y-F. (2001). Catalytic growth and characterization of gallium nitride nanowires. *Journal of the American Chemical Society*, 123, 12, (March 2001) 2791-2798, ISSN: 0002-7863

Cheng, G. S.; Kolmakov, A.; Zhang, Y.; Moskovits, M.; Munden, R.; Reed, M. A.; Wang, G.; Moses, D. & Zhang, J. (2003). Current rectification in a single GaN nanowire with a well-defined p-n junction. *Applied Physics Letters*, 83, 8, (August 2003) 1578-1580, ISSN: 0003-6951

Cimpoiasu, E.; Stern, E.; Klie, R.; Munden, R. A.; Cheng, G. & Reed, M. A. (2006). The effect of Mg doping on GaN nanowires. *Nanotechnology*, 17, 23, (December 2006) 5735-5739, ISSN: 0957-4484

Colliex, C. (1991). The impact of EELS in materials science. *Microscopy Microanalysis and Microstructures*, 2, 2-3, (April-June 1991) 403-411, ISSN: 1154-2799

Colombo, C.; Spirkoska, D.; Frimmer, M.; Abstreiter, G. & Morral, A. F. I. (2008). Ga-assisted catalyst-free growth mechanism of GaAs nanowires by molecular beam epitaxy. *Physical Review B*, 77, 15, (April 2008) 155326, ISSN: 1098-0121

Colombo, C.; Heiβ, M.; Gratzel, M. & Morral, A. F. I. (2009). Gallium arsenide p-i-n radial structures for photovoltaic applications. *Applied Physics Letterrs*, 94, 17, (April 2009) 173108, ISSN: 0003-6951

Cormack, A. M. (1980). Early two-dimensional reconstruction and recent topics stemming from it. *Science*, 209, 4464, (September 1980) 1482-1486, ISSN: 0036-8075

Davidson, F. M.; Lee, D. C.; Fanfair, D. D. & Korgel, B. A. (2007). Lamellar twinning in semiconductor nanowires. *Journal of Physical Chemistry C*, 111, 7, (February 2007) 2929-2935, ISSN: 1932-7447

De Franceschi, S.; van Dam, J. A.; Bakkers, E. P. A. M.; Feiner, L. F.; Gurevich, L. & Kouwenhoven, L. P. (2003). Single-electron tunneling in InP nanowires. *Applied Physics Letterrs*, 83, 2, (July 2003) 344-346, ISSN: 0003-6951

Egerton, R. F. (1996). *Electron Energy Loss in the Electron Microscope (2nd ed.)*, Plenum Press, ISBN-10: 0-306-45223-5 , New York

Egerton, R. F. (2003). New techniques in electron energy-loss spectroscopy and energy-filtered imaging. *Micron*, 34, 3-5, (August 2002) 127-139, ISSN: 0968-4328

Ercius, P.; Gignac, L. M.; Hu, C. K. & Muller, D. A. (2009). Three-Dimensional Measurement of Line Edge Roughness in Copper Wires Using Electron Tomography. *Microscopy and Microanalysis*, 15, 3, (June 2009) 244-250, ISSN: 1431-9276

Erni, R. & Browning, N. D. (2005). Valence electron energy-loss spectroscopy in monochromated scanning transmission electron microscopy. *Ultramicroscopy*, 104, 3-4, (October 2005) 176-192, ISSN: 0304-3991

Ersen, O.; Begin, S.; Houlle, M.; Amadou, J.; Janowska, I.; Greneche, J. M.; Crucifix, C. & Pham-Huu, C. (2008). Microstructural investigation of magnetic CoFe2O4 nanowires inside carbon nanotubes by electron tomography. *Nano Letters*, 8, 4, (April 2008) 1033-1040, ISSN: 1530-6984

Fischer, S. F.; Apetrii, G.; Kunze, U.; Schuh, D. & Abstreiter, G. (2006). Energy spectroscopy of controlled coupled quantum-wire states. *Nature Physics*, 2, 2, (February 2006) 91-96, ISSN: 1745-2473

Friedrich, H.; McCartney, M. R. & Buseck, P. R. (2005). Comparision of intensity distributions in tomograms from BF TEM, ADF STEM, HAADF STEM, and calculated tilt series. *Ultramicroscopy*, 106, 1, (December 2005) 18-27, ISSN: 0304-3991

Furtmayr, F.; Vielemeyer, M.; Stutzmann, M.; Arbiol, J.; Estradé, S.; Peirò, F.; Morante, J. R. & Eickhoff, M. (2008). Nucleation and growth of GaN nanorods on Si „111… surfaces by plasma-assisted molecular beam epitaxy - The influence of Si- and Mg-doping. *Journal of Applied Physics*, 104, 3, (August 2008) 034309, ISSN: 0021-8979

Furtmayr, F.; Vielemeyer, M.; Stutzmann, M.; Laufer, A.; Meyer, B. K. & Eickhoff, M. (2008). Optical properties of Si- and Mg-doped gallium nitride nanowires grown by plasma-assisted molecular beam epitaxy. *Journal of Applied Physics*, 104, 7, (October 2008) 074309, ISSN: 0021-8979

Gavrilenko, V. I. & Wu, R. Q. (2000). Energy loss spectra of group III nitrides. *Applied Physics Letters*, 77, 19, (November 2000) 3042-3044, ISSN: 0003-6951

Grogger, W.; Varela, M.; Ristau, R.; Schaffer, B.; Hoffer, F. & Krishnan, K. M.. (2005). Energy-filtering transmission electron microscopy on the nanometer length scale. *Journal of Electron Spectroscopy and Related Phenomena*, 143, 2-3, (May 2005) 139-147, ISSN: 0368-2048

Haider, M.; Uhlemann, S.; Schwan, E.; Rose, H.; Kabius, B. & Urban, K. (1998). Electron microscopy image enhanced. *Nature*, 392, 6678, (April 1998) 768-769, ISSN: 0028-0836

Heigoldt, M.; Arbiol, J.; Spirkoska, D.; Rebled, J. M.; Conesa-Boj, S.; Abstreiter, G.; Peiro, F.; Morante, J. R. & Morral, A. F. I. (2009). Long range epitaxial growth of prismatic heterostructures on the facets of catalyst-free GaAs nanowires. *Journal of Materials Chemistry*, 19, 7, (January 2009) 840-848, ISSN: 0959-9428

Hernandez-Ramirez, F.; Tarancon, A.; Casals, O.; Arbiol, J.; Romano-Rodriguez, A. & Morante, J. R. (2007). High response and stability in CO and humidity measures using a single SnO2 nanowire. *Sensors and Actuators B*, 121, 1, (January 2007) 3-17, ISSN: 0925-4005

Hochbaum, A. I.; Chen, R.; Diaz Delgado, R.; Liang, W.; Garnett, E. C.; Najarian, M.; Majumdar, A. & Yang , P. (2008). Enhanced thermoelectric performance of rough silicon nanowires. *Nature*, 451, 7175, (January 2008) 163-167, ISSN: 0028-0836

Hounsfield, G. N. (1980). Computed Medical Imaging. *Science*, 210, 4465, (October 1980) 22-28, ISSN: 0036-8075

Hu, Y.; Churchill, H. O. H.; Reilly, D. J.; Xiang, J.; Lieber, C. M. & Markus, C. M. (2007). A Ge/Si heterostructure nanowire-based double quantum dot with integrated charge sensor. *Nature Nanotechnology*, 2, 10, (October 2007) 622-625, ISSN: 1748-3387

Huang, Y.; Duan, X.; Cui, Y. & Lieber, C. M. (2002). Gallium nitride nanowire nanodevices. *Nano Letters*, 2, 2, (February 2002) 101-104, ISSN: 1530-6984

Hughes, J. L. P. & Sipe, J. E. (1996). Calculation of second-order optical response in semiconductors. *Physical Review B*, 53, 16, (April 1996) 10751-10763, ISSN: 0163-1829

Humphreys, C. J.; Galtrey, M. J.; van der Laak, N.; Oliver, R. A.; Kappers, M. J.; Barnard, J. S.; Graham, D. M. & Dawson, P. (2007). The Puzzle of Exciton Localisation in GaN-Based Structures: TEM, AFM and 3D APFIM Hold the Key. *Microscopy of semiconducting materials 2007, Springer Proceedings in Physics*, 120, (April 2008) 3-12, ISSN: 0930-8989

Iakoubovskii, K.; Mitsuishi, K. & Furuya, K. (2008). High-resolution electron microscopy of detonation nanodiamond. *Nanotechnology*, 19, 15, (April 2008) 155705, ISSN: 0957-4484

Ihn, S-G.; Song, J-I.; Kim, Y-H. & Lee, J. Y. (2006). GaAs nanowires on Si substrates grown by a solid source molecular beam epitaxy. *Applied Physics Letters*, 89, 5, (July 2006) 053106, ISSN: 0003-6951

Irrera, A.; Iacona, F.; Franzò, G.; Boninelli, S.; Pacifici, D.; Miritello, M.; Spinella, C.; Sanfilippo, D.; Di Stefano, G.; Fallica P. G. & Priolo, F. (2005). Correlation between electroluminescence and structural properties of Si nanoclusters. *Optical Materials*, 27, 5, (February 2005) 1031-1040, ISSN: 0925-3467

Johansson, J.; Karlsson, L. S.; Svensson, C. P. T.; Martensson, T.; Wacaser, B. A.; Deppert, K.; Samuelson, L. & Seifert, W. (2006). Structural properties of (111)B-oriented III-V nanowires. *Nature Materials*, 5, 7, (July 2006) 574-580, ISSN: 1476-1122

Jones, R. O. & Gunnarsson, O. (1989). The density functional formalism, its applications and prospects. *Review of Modern Physics*, 61, 3, (July 1989) 689-746, ISSN: 0034-6861

Karlsson, L. S.; Dick, K. A.; Wagner, J. B.; Malm, J. O.; Deppert, K.; Samuelson, L. & Wallenberg, L. R. (2007). Understanding the 3D structure of GaAs < 111 > B nanowires. *Nanotechnology*, 18, 48, (December 2007) 485717, ISSN: 0957-4484

Kastner, M. A. (1992). The single-electron transistor. *Reviews of Modern Physics*, 64, 3, (July 1992) 849-858, ISSN: 0034-6861

Kaufman, D.; Berk, Y.; Dwir, B.; Rudra, A.; Palevski, A. & Kapon, E. (1999). Conductance quantization in V-groove quantum wires. *Physical Review B*, 59, 16, (April 1999) 10433-10436, ISSN: 1098-0121

Kim, H. S.; Hwang, S. O.; Myung, Y.; Park, J.; Bae, S. Y. & Ahn, J. P. (2008). Three-dimensional structure of helical and zigzagged nanowires using electron tomography. *Nano Letters*, 8, 2, (February 2008) 551-557, ISSN: 1530-6984

Kim, T. Y.; Lee, S. H.; Mo, Y. H.; Shim, H. W.; Nahm, K. S.; Suh, E-K.;Yang, J. W.; Lim, K. Y. & Park, G. S. (2003). Growth of GaN nanowires on Si substrate using Ni catalyst in vertical chemical vapor deposition reactor. *Journal of Crystal Growth*, 257, 1-2, (September 2003) 97-103, ISSN: 0022-0248

Klug, A. (1982). From macromolecules to biological assemblies. *Nobel lecture*, Stockholm, December 1982.

Koguchi, M.; Kakibayashi, H.; Tsuneta, R.; Yamaoka, M.; Niino, T.; Tanaka, N.; Kase, K. & Iwaki, M. (2001). Three-dimensional STEM for observing nanostructures. *Journal of Electron Microscopy*, 50, 3, (January 2001) 235-241, ISSN: 0022-0744

Kuykendall, T.; Ulrich, P.; Aloni, S. & Yang, P. (2007). Complete composition tunability of InGaN nanowires using a combinatorial approach. *Nature Materials*, 6, 12, (December 2007) 951-956, ISSN: 1476-1122

Lai, F-I.; Kuo, S. Y.; Chang, Y. H.; Huang, H. W.; Chang, C. W.; Yu, C. C.; Lin, C. F.; Kuo, H. C. & Wang, S. C. (2006). Fabrication of magnesium-doped gallium nitride nanorods and microphotoluminescence characteristics. *Journal of Vacuum Science and Technology B*, 24, 3, (May-June 2006) 1123-1126, ISSN: 1071-1023

Lazar, S.; Botton, G. A.; Wu, M-Y.; Tichelaar, F. D. & Zandbergen, H. W. (2003). Materials science applications of HREELS in near edge structure analysis and low-energy loss spectroscopy. *Ultramicroscopy*, 96, 3-4, (September 2003) 535-546, ISSN: 0304-3991

Levine, Z. H. & Allan, D. C. (1989). Linear optical-response in silicon and germanium including self-energy effects. *Physical Review Letters*, 63, 16, (October 1989) 1719-1722, ISSN: 0031-9007

Lieber, C. M. (2003). Nanoscale science and technology: Building a big future from small things. *MRS Bulletin*, 28, 7, (July 2003) 486-491, ISSN: 0883-7694

Lieber, C. M. & Wang, Z. L. (2007). Functional nanowires. *MRS Bulletin*, 32, 2, (February 2007) 99-108, ISSN: 0883-7694

Merano, M.; Sonderegger, S.; Crottini, A.; Collin, S.; Pelucchi, E.; Renucci, P.; Malko, A.; Baier, M. H.; Kapon, E.; Ganiere, J. D. & Deveaud, B. (2006). Time-resolved cathodoluminescence of InGaAs/AlGaAs tetrahedral pyramidal quantum structures. *Applied Physics B*, 84, 1-2, (July 2006) 343-350, ISSN: 0946-2171

Midgley, P. A.; Weyland, M.; Thomas, J. M. & Johnson, B. F. G. (2001). Z-Contrast tomography: a technique in three-dimensional nanostructured analysis based on Rutherford scattering. *Chemical Communications*, 10, (February 2001), 907-908, ISSN: 1359-7345

Midgley, P. A. & Weyland, M. (2003). 3D electron microscopy in the physical science: the development of Z-contrast and EFTEM tomography. *Ultramicroscopy*, 96, 3-4, (September 2002), 413-431, ISSN: 0304-3991

Midgley, P. A.; Ward, E. P. W.; Hungría, A. B. & Thomas, J. M. (2007). Nanotomography in the chemical, biological and materials sciences. *Chemical Society Reviews*, 36, 9, (September 2007) 1477-1494, ISSN: 0306-0012

Mikkelsen, A.; Sköld, N.; Ouattara, L.; Borgström, M.; Andersen, J. N.; Samuelson, L.; Seifert, W. and Lundgren, E. (2004). Direct imaging of the atomic structure inside a nanowire by scanning tunnelling microscopy. *Nature Materials*, 3, 8, (August 2004) 519-523, ISSN: 1476-1122

Möbus, G.; Doole, R. C. & Inkson, B. J. (2003). Spectroscopic electron tomography. *Ultramicroscopy*, 96, 3-4, (September 2003) 433-451, ISSN: 0304-3991

Montoro, L. A.; Leite, M. S.; Biggemann, D.; Peternella, F. G.; Batenburg, K. J.; Medeiros-Ribeiro, G. & Ramirez, A. J. (2009). Revealing Quantitative 3D Chemical Arrangement on Ge-Si Nanostructures. *Journal of Physical Chemistry C*, 113, 21, (May 2009) 9018-9022, ISSN: 1932-7447

Morral, A. F. I.; Arbiol, J.; Prades, J. D.; Cirera, A. & Morante, J. R. (2007). Synthesis of silicon nanowires with wurtzite crystalline structure by using standard chemical vapor deposition. *Advanced Materials*, 19, 10, (May 2007) 1347-1349, ISSN: 0935-9648

Morral, A. F. I.; Colombo, C.; Abstreiter, G.; Arbiol, J. & Morante, J. R. (2008). Nucleation mechanism of gallium-assisted molecular beam epitaxy growth of gallium arsenide nanowires. *Applied Physics Letters*, 92, 6, (February 2008) 063112, ISSN: 0003-6951

Morral, A. F. I.; Spirkoska, D.; Arbiol, J.; Morante, J. R. & Abstreiter, G. (2008). Prismatic quantum heterostructures synthesized on molecular-beam epitaxy GaAs nanowires. *Small*, 4, 7, (July 2008) 899-903, ISSN: 1613-6810Nesbitt, D. J. (2007). Photonics - Charge of the light brigade. *Nature*, 450, 7173, (December 2007) 1172-1773, ISSN: 0028-0836

Pal, S.; Ingale, A.; Dixit, V. K.; Sharma, T. K.; Porwal, S.; Tiwari, P. & Nath, A. K. (2007). A comparative study on nanotextured high density Mg-doped and undoped GaN. *Journal of Applied Physics*, 101, 4, (February 2007) 044311, ISSN: 0021-8979

Pan, Y. H.; Sader, K.; Powell, J. J.; Bleloch, A.; Gass, M.; Trinick, J.; Warley, A. , Li, A.; Brydson, R. & Brown, A. (2009). 3D morphology of the human hepatic ferritin mineral core: New evidence for a subunit structure revealed by single particle analysis of HAADF-STEM images. *Journal of Structural Biology*, 166, 1, (April 2009) 22-31, ISSN: 1047-8477

Park, Y. S.; Na, J. H.; Taylor, R. A.; Park, C. M.; Lee, K. H. & Kang, T. W. (2006). The recombination mechanism of Mg-doped GaN nanorods grown by plasma-assisted molecular-beam epitaxy. *Nanotechnology*, 17, 3, (February 2006) 913-916, ISSN: 0957-4484

Park, Y. S.; Park, C. M.; Park, C. J.; Cho, H. Y.; Lee, S. J.; Kang, T. W.; Lee, S. H.; Oh, J-E.; Yoo, K-H. & Son, M.-S. (2006). Electron trap level in a GaN nanorod p-n junction grown by molecular-beam epitaxy. *Applied Physics Letters*, 88, 19, (May 2006) 192104, ISSN: 0003-6951

Penczek, P.; Marko, M.; Buttle, K. & Frank, J. (1995). Double-tilt electron tomography. *Ultramicroscopy*, 60, 3, (October 1995), 393-410, ISSN: 0304-3991

Perdew, J. P.; Burke, K. & Ernzerhof, M. (1996). Generalized gradient approximation made simple. *Physical Review Letters*, 77, 18, (October 1996) 3865-3868, ISSN: 0031-3844

Pettersson, H.; Tragardh, J.; Persson, A. I.; Landin, L.; Hessman, D. & Samuelson, L. (2006). Infrared photodetectors in heterostructure nanowires. *Nano Letters*, 6, 2, (February 2006) 229-232, ISSN: 1530-6984

Qian, F.; Li, Y.; Gradeak, S.; Park, H.-G.; Dong, Y.; Ding, Y.; Wang, Z. L. & Lieber, C. M. (2008). Multi-quantum-well nanowire heterostructures for wavelength-controlled lasers. *Nature Materials*, 7, 9, (September 2008) 701-706, ISSN: 1476-1122

Qin, Y.; Wang, X. D. & Wang, Z. L. (2008). Microfibre-nanowire hybrid structure for energy scavenging. *Nature*, 451, 7180, (February 2008) 809-U5, ISSN: 0028-0836

Radon, J. (1917). Über die Bestimmung von Funktionen durch ihre Integralwerte längs gewisser Mannigfaltigkeiten. *Berichte Sachsische Akademie der Wissenschaften Leipzig*, 69, (1917) 262-277, ISSN: 1867-7061

Read, A. J. & Needs, R. J. (1991). Calculation of optical matrix-elements with nonlocal pseudopotentials. *Physical Review B*, 44, 23, (December 1991), 13071-13073, ISSN: 0163-3844

Samuelson, L.; Thelander, C.; Bjork, M. T.; Borgstrom, M.; Deppert, K.; Dick, K. A.; Hansen, A. E.; Martensson, T.; Panev, N.; Persson, A. I.; Seifert, W.; Skold, N.; Larsson, M. W. & Wallenberg, L. R. (2004). Semiconductor nanowires for 0D and 1D physics and applications. *Physica E*, 25, 2-3, (November 2004) 313-318, ISSN: 1386-9477

Sanchez, A. M.; Beanland, R.; Papworth, A. J.; Goodhew, P. J. & Gass, M. H. (2006). Nanometer-scale strain measurements in semiconductors. *Applied Physics Letters*, 88, 5, (January 2006) 051917, ISSN: 0003-6951

Schedelbeck, G.; Wegscheider, W.; Bichler, M. & Abstreiter, G. (1997). Coupled quantum dots fabricated by cleaved edge overgrowth: From artificial atoms to molecules. *Science*, 278, 5344, (December 1997) 1792-1795, ISSN: 0036-8075

Shen, Y. G.; Mai, Y. W.; McKenzie, D. R. , Zhang, Q. C.; McFall, W. D. & McBride, W. E. (2000). Composition, residual stress, and structural properties of thin tungsten nitride films deposited by reactive magnetron sputtering. *Journal of Applied Physics*, 88, 3, (August 2000) 1380-1388, ISSN: 0021-8979

Shorubalko, I.; Leturcq, R.; Pfund, A.; Tyndall, D.; Krischek, R.; Schön, S. & Ensslin, K. (2008). *Nanoletters*, 8, 2, (February 2008) 382-385, ISSN: 1530-6984

Sigle, W. (2005). Analytical transmission electron microscopy. *Annual Review of Materials Research*, 35, (December 2005) 239-314, ISSN: 1531-7331

Soler, J. M.; Artacho, E.; Gale, J. D.; Garcia, A.; Junquera, J.; Ordejon. P. & Sanchez-Portal, D. (2002). The SIESTA method for ab initio order-N materials simulation. *Journal of Physics: Condensed Matter*, 14, 11, (March 2002) 2745-2778, ISSN: 0953-8984

Stern, E.; Klemic, J. F.; Routenberg, D. A.; Wyrembak, P. N.; Turner-Evans, D. B.; Hamilton, A. D.; LaVan, D. A.; Fahmy, T. M. & Reed, M. A. (2007). Label-free immunodetection with CMOS-compatible semiconducting nanowires. *Nature*, 445, 7127, (February 2007) 519-522, ISSN: 0028-0836

Thelander, C.; Agarwal, P.; Brongersma, S.; Eymery, J.; Feiner, L. F.; Forchel, A.; Scheffler, M.; Riess, W.; Ohlsson, B. J.; Gosele, U. & Samuelson, L. (2006). Nanowire-based one-dimensional electronics. *Materials Today*, 9, 10, (October 2006) 28-35, ISSN: 1369-7021

Thunich, S.; Prechtel, L.; Spirkoska, D.; Abstreiter, G.; Morral, A. F. I. & Holleitner, A. W. (2009). Photocurrent and photoconductance properties of a GaAs nanowire. *Applied Physics Letters*, 95, 8, (August 2009) 083111, ISSN: 0003-6951

Topuria, T.; Browning, N. D. & Ma, Z. (2003). Characterization of ultrathin dopant segregation layers in nanoscale metal-oxide-semiconductor field effect transistors using scanning transmission electron microscopy. *Applied Physics Letters*, 83, 21, (November 2003) 4432-4434, ISSN: 0003-6951

van den Broek, W.; Verbeeck, J.; De Backer, S.; Scheunders, P. & Schryvers, D. (2006). Acquisition of the EELS data cube by tomographic reconstruction. *Ultramicroscopy*, 106, 4-5, (March 2006) 269-276, ISSN: 0304-3991

Verheijen, M. A.; Algra, R. E.; Borgstrom, M. T.; Immink, G.; Sourty, E.; van Enckevort, W. J. P.; Vlieg, E. & Bakkers, E. P. A. M. (2007). Three-dimensional morphology of GaP-GaAs nanowires revealed by transmission electron microscopy tomography. *Nano Letters*, 7, 10, (October 2007) 3051-3055, ISSN: 1530-6984

Wang, X. D.; Song, J.; Liu, J. & Wang, Z. L. (2007). Direct-current nanogenerator driven by ultrasonic waves. *Science*, 316, 5821, (April 2006) 102-105, ISSN: 0036-8075

Wagner, R. S. & Ellis, W. C. (1964). Vapor-liquid-solid mechanism of single crystal growth (new method growth catalysis from impurity whisker epitaxial + large crystals Si e). *Applied Physics Letters*, 4, 5, (March 1964) 89-90, ISSN: 0003-6951

Wang, J. F.; Gudiksen, M. S.; Duan, X. F.; Cui, Y. & Lieber, C. M. (2001). Highly polarized photoluminescence and photodetection from single indium phosphide nanowires. *Science*, 293, 5531, (August 2001) 1455-1457, ISSN: 0036-8075

Wang, W. U.; Chen, C.; Lin, K. H.; Fang, Y. & Lieber, C. M. (2005). Label-free detection of small-molecule-protein interactions by using nanowire nanosensors. *Proceedings Of The National Academy of Sciences of The United States of America*, 102, 9, (March 2005) 3208-3212, ISSN: 0027-8424

Wegscheider, W.; Pfeiffer, L. N.; Dignam, M. M.; Pinczuk, A.; West, K. W.; McCall, S. L. & Hull, R. (1993). Lasing from excitons in quantum wires. *Physical Review Letters*, 71, 24, (December 1993) 4071-4074, ISSN: 0031-9007

Wegscheider, W.; Pfeiffer, L.; Dignam, M.; Pinzcuk, A.; West, K. & Hull, R. (1994). Lasing in lower-dimensional structures formed by cleaved edge overgrowth. *Semiconductor Science and Technolology*, 9, 11, (November 1994) 1933-1938, ISSN: 0268-1242

Wegscheider, W.; Rother, M.; Schedelbeck, G.; Bichler, M. & Abstreiter, G. (1999). Optical and transport properties of low-dimensional structures fabricated by cleaved edge overgrowth. *Microelectronic Engineering*, 47, 1-4, (June 1999) 215-219, ISSN: 0167-9317

Weyland, M. (2002). Electron tomography of catalysts. *Topics in Catalysis*, 21, 4, (December 2002) 175-183, ISSN: 1022-5528

Xin, Y.; James, E. M.; Arslan, I.; Sivananthan, S.; Browninga, N. D.; Pennycook, S. J.; Omnes, F.; Beaumont, B.; Faurie, J. & Gibart, P. (2000). Direct experimental observation of the local electronic structure at threading dislocations in metalorganic vapor phase epitaxy grown wurtzite GaN thin films. *Applied Physics Letters*, 76, 4, (January 2000) 466-468, ISSN: 0003-6951

Yang, P. D. (2005). The chemistry and physics of semiconductor nanowires. *MRS Bulletin*, 30, 2, (February 2005) 85-91, ISSN: 0883-7694

Zhong, Z. H.; Qian, F.; Wang & D. Lieber, C. M. (2003). Synthesis of p-type gallium nitride nanowires for electronic and photonic nanodevices. *Nanoletters*, 3, 3, (March 2003) 343-346, ISSN: 1530-6984

11

Low Temperature Phase Separation in Nanowires

Sheng Yun Wu
Department of Physics, National Dong Hwa University
Taiwan, R. O. C.

1. Introduction

The ways to develop one-dimensional (1D) nanostructures, such as nanowires, nanorods, nanobelts and nanotubes, are being studied intensively, due to their unique applications in mesoscopic physics and nanoscale electronic devices [1-3]. Structural phase transition between the wurtzite (WZ) and zinc-blend (ZB) GaN induced by the deposition conditions [4], temperature-mediated phase selection during the growth of GaN [5], and substrate control [6] by the crystallographic alignment of GaN have all been observed. It is known that x-ray scattering technique plays an important role in investigating the lattice excitations and structural transformation associated with thermal strain in 1D nanowires [7]. For example, Dahara and co-workers [8] reported a phase transformation from hexagonal to cubic in Ga^{+} implaned GaN nanowires (GaNWs). The SC16 phase of GaAs appears at high pressure can be transformed to the hexagonal WZ phase by reducing the pressure to the ambient one. WZ GaAs is stable in resisting a transformation to the ZB phase at temperatures up to 473 K at ambient pressure [9]. Currently, most of the studies on the crystalline structure of GaNWs are focused on the stable hexagonal α-GaN and metal-stable cubic β-GaN. In this work, we study the crystalline structure of GaNWs by using *in situ* low-temperature x-ray diffraction and Rietveld analysis [10]. Our findings show that the ZB phase starts to develop below 260 K. A finite size model wherein the random phase distribution is utilized to describe the development of short range atomic ordering. The phase separation was found to be reversible upon temperature cycling, and occurred through the exchange and interaction of the characteristic size of the ordered domain of the GaN nanowires.

2. Important

In situ low temperature x-ray diffraction was employed to investigate the phase separation of GaN nanowires. Observations showed that a distinct phase separation developed below 260 K, the Zinc-Blend phase, which was related to short range ordering. Surprisingly, the correlation lengths of the Zinc-Blend phase reached their maximum at 140 K but correlation length was still revealed at around 23 nm. Our results may be understood using the short range correlation model, and support the conclusion that the phase separation was reversible and occurred through the interaction of the characteristic size of the ordered domain of the GaN nanowires.

3. Experimental details

GaN is a direct wide band-gap semiconductor at room temperature. It is a prominent candidate for optoelectronic devices at blue and near ultra-violet wavelengths [11-14]. In addition, it exhibits high thermal conductivity and little radiation damage, suitable for high temperature and high power microelectronic devices[15]. GaN nanowires have been synthesized by several groups using different methods[16-22]. The randomly oriented GaNWs used in this study were synthesized by a low pressure thermal chemical vapor deposition (LPTCVD) technique. The samples were grown at 950 °C on Si [001] substrates precoated with a 5 nm Au catalyst layer by an E-Gun evaporator. Molten gallium was used as the source material and NH_3 (30 sccm) as the reactant gas in a horizontal tubular furnace. Details of the growth process may be found elsewhere [23]. A low temperature *in situ* X-ray diffractometer (Scintag 2000) was utilized to investigate the crystalline structure of the GaNWs produced at various temperatures. The specimens were mounted on background-free sample holders, which were then attached to a cold-head placed in a high vacuum (< 10^{-6} Torr) environment. The chamber was equipped with a beryllium hemisphere, and evacuated to reduce air scattering and absorption of the x-ray. No obvious differences were found in the x-ray diffraction patterns taken on different portions of the sample.

4. Results and discussion

Fig. 1. SEM micrograph of GaNWs homogeneously grown on the substrate.

4.1 SEM results

The morphology of sample was characterized by a field emission scanning electron microscope (FE-SEM, JEOL JSM-6500F) equipped with an energy dispersive x-ray spectroscope (EDS, Oxford Instrument INCA x-sight 7557). Atomic-resolution transmission electron microscopic (TEM) analysis and high-resolution transmission electron microscopy (HRTEM) images were taken with the CCD-camera of an electron microscope (JEOL JEM-2100) at 200 kV. Analysis software (Digital Micrograph) was employed to digitalize and analyze the obtained images. Figure 1 displays a portion of the SEM image showing the morphology of the GaNWs. The diameters of the GaNWs assembly ranged from 20 to 50 nm, with a length of several tens of microns. The diameter distribution of the GaNWs assembly, as shown in the Fig. 2, is quite asymmetric and can be described using a log-normal distribution function (solid line). The log-normal distribution is defined as follows:

$f(d)=\frac{1}{\sqrt{2\pi}d\sigma}\exp\left(-\frac{(\ln d-\ln\bar{d})^2}{2\sigma^2}\right)$, where $\bar{d}$ is the mean value and σ is the standard deviation of the function. The mean diameter obtained from the fit is <d>=40(3) nm. The small standard deviation (σ< 0.5) of the function indicates that the distribution is confined to a limited range. The broadening of the width of the distribution profile is due to crystalline and nanoparticle aggregation effects.

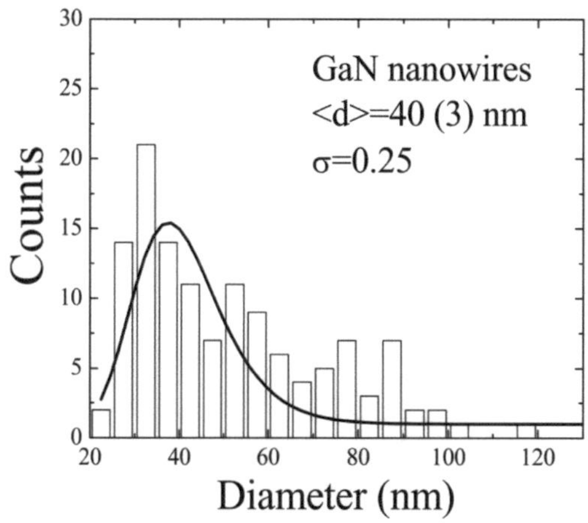

Fig. 2. The diameter distribution of the GaNWs obtained from SEM images.

4.2 TEM and HRTEM results

Figure 3 shows the TEM morphology of a typical nanowire. TEM image reveals that most of the nanowires are straight, and the diameter along the growth direction is uniform, with a mean diameter of 40(3) nm. Figure 4 shows the selected area electron diffraction (SAED) pattern taken on a region close to the surface of a single nanowire. It clearly reveals a single crystalline nature for the sample studied. The Bragg spots correspond to the [001] reflection of the wurtzite structure of the GaNW. The pattern of the main spots can easily be seen as hexagonal cells with lattice parameters of a=3.195 Å and c=5.193 Å, which indicates a predominantly polycrystalline hexagonal wurtzite GaN, shown in Fig. 5. In wurtzite

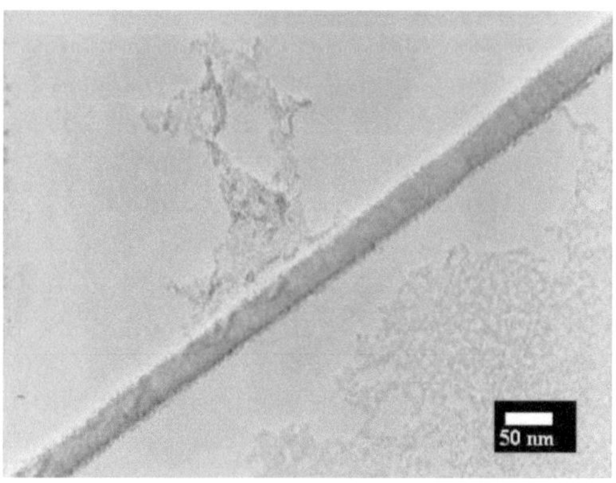

Fig. 3. TEM image of the GaNWs revealing a uniform diameter of ~40 nm.

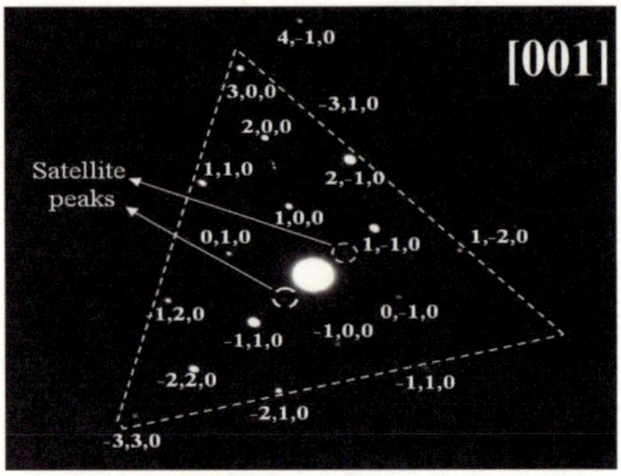

Fig. 4. SAED pattern of the GaNWs confirming the [001] growth direction.

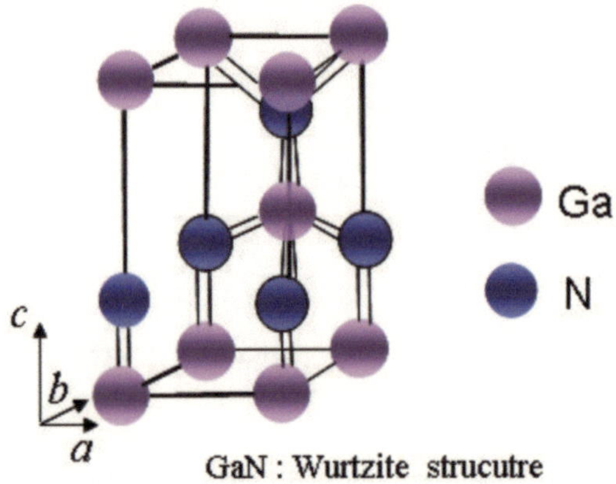

Fig. 5. Crystal structure of WZ-GaN.

structure of GaNWs, on the surface of [001], each Ga atom has three complete bonds to the underlying nitrogen atomic plane. Details of the description of crystal structure may be found with the earlier finding [24].

4.3 X-ray diffraction

X-ray diffraction patterns are known as the fingerprints of crystalline materials. They reveal details of the crystalline structure and their formation during synthesis, and even the crystalline phase transitions or separation at various temperatures. The x-ray and Rietveld refined diffraction patterns of the GaNWs, taken at 320 K and 80 K, are shown in Fig. 6 and 7, respectively. Diffraction patterns were utilized to characterize the crystalline structure in the prepared samples. The diffraction peaks appeared to be much broader than the instrumental resolution, reflecting the nano-size effects. The analysis was performed using the program package of the General Structure Analysis System (GSAS) [25] following the Rietveld method [10]. Several models with different symmetries were assumed during the preliminary analysis. In our structural analysis we then pay special attention to searching for the possible symmetries that can describe the observed diffraction pattern well. All the structural and lattice parameters were allowed to vary simultaneously, and refining processes were carried out until R_p, the weighted R_{wp} factor, differed by less than one part in a thousand within two successive cycles. Figure 6 shows the diffraction pattern (black cross) taken at room temperature, where the solid curve (red curve) indicates the fitted pattern and the differences (blue curve) between the observed and the fitted patterns are plotted at the bottom of Fig. 6. The refined lattice parameters at 320 K are $a=b=3.195(2)$ Å and $c=5.193(1)$ Å. This $c/a=1.625$ that we obtained for the WZ structure agrees very well with that obtained in a separated study [26], but is ~0.5% smaller than the theoretically expected value [27] of 1.633. The reasons for this are not completely clear, but could be due to the nanowires are expected to grow in the *c*-direction that resulted in a smaller length-to-width ratio.

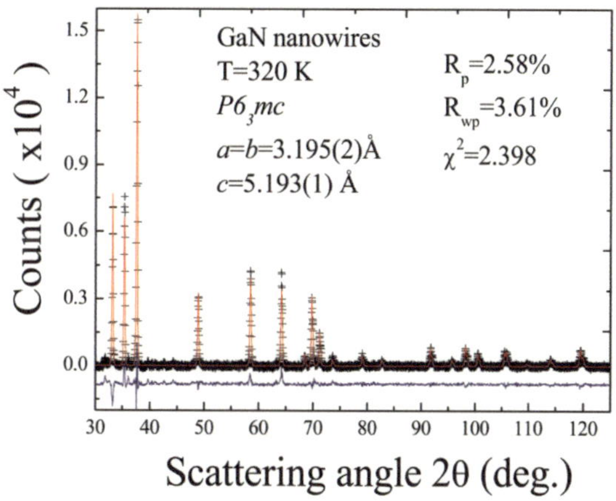

Fig. 6. The observed (crosses) and Rietveld refined (solid lines) x-ray diffraction patterns of GaNWs taken at 320 K.

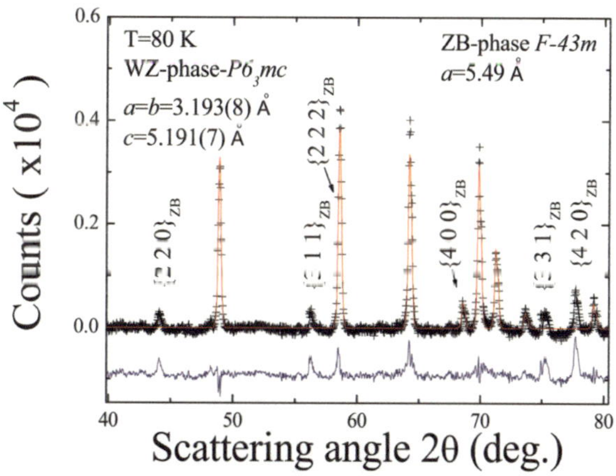

Fig. 7. The observed (crosses) and Rietveld refined (solid lines) x-ray diffraction patterns of GaNWs taken at 80 K. A new set of diffraction peaks that is associated with the zinc-blend phase appears in the pattern taken at 80 K.

A series of new peaks, at scattering angles of 44.08°, 56.22°, 58.2° 68.2°, and 75.3°, becomes visible in the diffraction patterns taken at 80 K, as can be seen in the Fig. 7. These peaks were not observed at 320 K and cannot be associated to the α-GaNW. They, however, may be indexed as the $\{220\}_{ZB}$, $\{311\}_{ZB}$, $\{222\}_{ZB}$, $\{400\}_{ZB}$, $\{331\}_{ZB}$, and $\{420\}_{ZB}$ reflections of the ZB phase, shown in Fig.8. All these new peaks may be identified to belong a cubic *F-43m* GaN structure of lattice constant a=5.49 Å. All the x-ray diffraction patterns taken on the sample holder, on the silicon substrate, and on the empty chamber reveal no such signals.

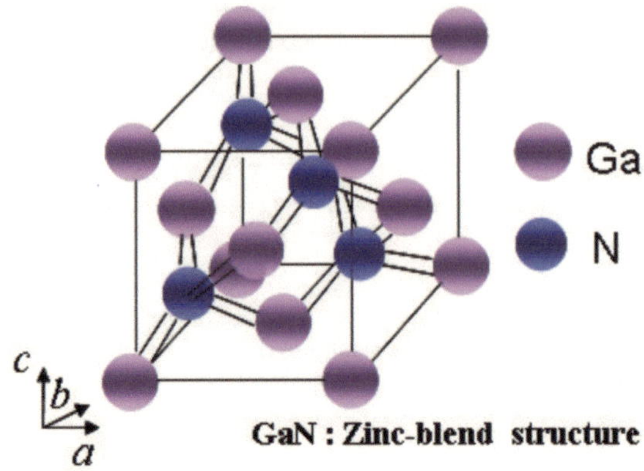

Fig. 8. Crystal structure of ZB-GaNWs.

4.4 *In situ* low temperature X-ray diffraction

Figure 9 shows the temperature dependency of the *in situ* x-ray diffraction patterns, where the color bars represent the diffraction intensity. The $\{112\}_{WZ}$, $\{201\}_{WZ}$, $\{004\}_{WZ}$, and $\{202\}_{WZ}$ reflections are clearly revealed at high temperatures, while the $\{331\}_{ZB}$ and $\{420\}_{ZB}$ reflections develop below 260 K. No obvious changes in the width of the diffraction peaks that belong to the WZ-phase may be identified in the temperature regime studied, as can be seen in the Fig. 10 where FWHM represents the full width at half maximum of the diffraction peak. Figure 11 and 12 show the temperature dependency of the integrated intensity and the FWHM of the $\{420\}_{ZB}$ reflection, respectively. Below 260 K, the integrated intensity of the

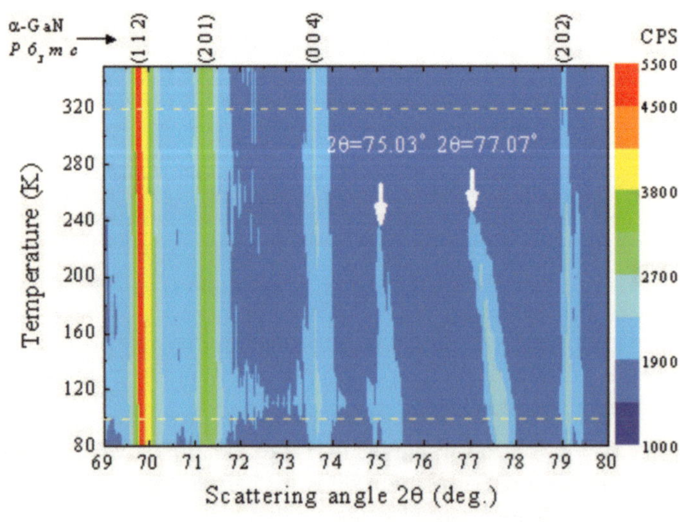

Fig. 9. Plots of the temperature dependence of the *in situ* low-temperature x-ray diffraction patterns.

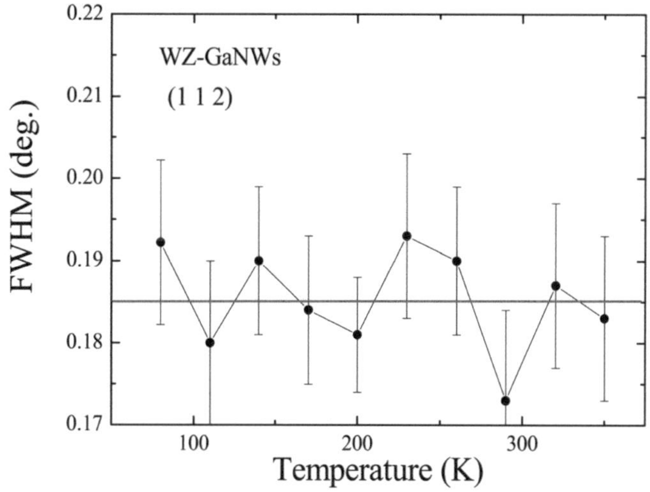

Fig. 10. The FWHM of the $\{112\}_{WZ}$ reflection taken at various temperatures, revealing a monotonic change of the FWHM is related to the fluctuation in temperature or to the fit of the error bar.

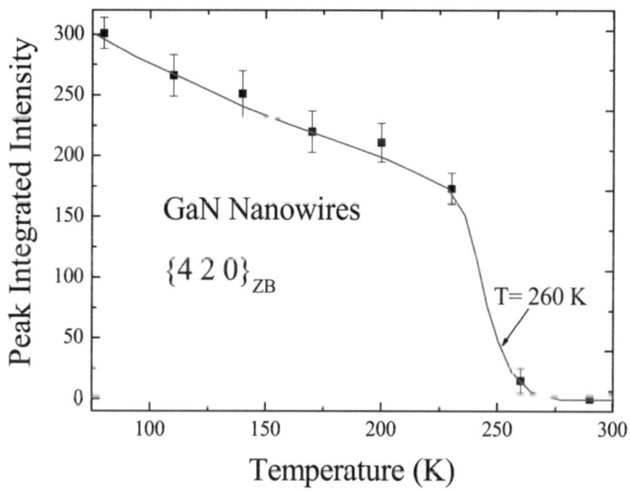

Fig. 11. Temperature dependence of the integrated intensity of the $\{420\}_{ZB}$ reflection, where the solid curve is guide to the eye only. A distinct structural transformation may be clearly seen to occur at around 260 K.

$\{420\}_{ZB}$ reflection increases rapidly, which is accompanied by a reduction in the peak width. Clearly, these behaviors signal the development of the ZB-phase GaNWs below 260 K.

It is known that the reduction in the peak width with decreasing temperature indicates the growth of the crystalline domain. The observed peak profiles for the ZB-phase are much broader then the instrument resolution function show that the crystalline domains are finite sized, which can be described by the finite lattice model [28]. It follows the instrumental resolution function, which can be well approximated by a Gaussian function. We propose that the intensity of the Bragg reflection from finite size systems can be described [29] as

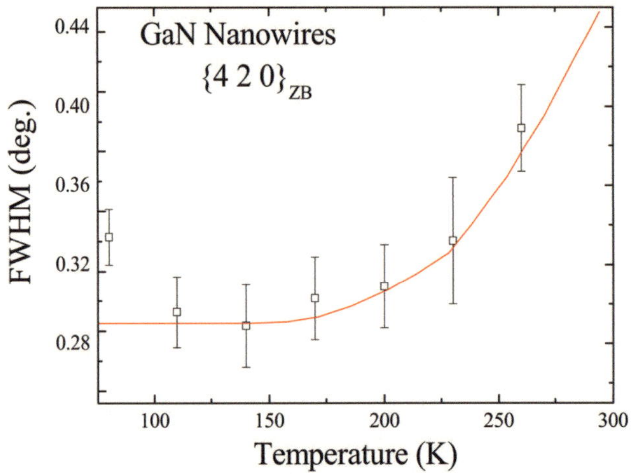

Fig. 12. Temperature dependence of the FWHM of the $\{420\}_{ZB}$ reflection, where the solid curves is guide to the eye only. The temperature dependency of the FWHM of the selected peak of $\{4\ 2\ 0\}_{ZB}$ indicates the structure of the ordering parameter with temperature.

$$I_{hkl}(\theta) = C\left(\frac{1}{v^2}\right)\left[\left|F_{hkl}\right|^2 p\left(\frac{1+\cos^2 2\theta}{\sin\theta}\right)\right]\frac{e^{-2M}}{2\mu}S(\theta) \quad (1)$$

where 2θ is the scattering angle, C is the instrumental constant, e^{-2M} is the Debye temperature factor, μ is the linear absorption coefficient, M is the multiplicity of the $\{h\ k\ l\}$ reflection, F_{hkl} is the structure factor, and the phase factor $S(\theta)$ reads

$$S(\theta) = \frac{1}{2\pi}\int_0^{\frac{\pi}{2}}\int_0^{\pi} \exp\left\{\frac{-8\pi^2}{\lambda^2}\zeta^2\left[\sin\theta(\sin\alpha\cos\beta + \sin\alpha\sin\beta + \cos\alpha) - \sin\theta_B\right]^2\right\} d\alpha d\beta \ . \quad (2)$$

Here λ is the wavelength of the incident x-ray, θ_B is the Bragg angle of the $\{h\ k\ l\}$ reflection, and ξ is the correlation length of the Bragg scattering that indicates the characteristic size of the crystalline domains. In Fig. 13 we show the development of the $\{420\}_{ZB}$ reflection with temperature. No significant ZB-phase crystallinity may be identified at above 260 K. At 230 K a broad peak at the $\{420\}_{ZB}$ position becomes evident, as shown in Fig. 13(f). The diffraction patterns taken at different temperatures show that this peak starts to develop at T~260 K, and becomes saturated in intensity at T=140 K. The solid curves shown in Figs.13(a)-(f) indicate the fits of the data to the above expression convoluted with the Gaussian instrumental resolutions function. This reflection originates from the development of finite size atomic crystalline domains that belong to the ZB-GaNWs phase. Fig. 13(i) shows a portion of the diffraction pattern taken in a subsequent warm up to 320 K. It shows that the occurrence of phase separation in temperature cycling is reversible.

This critical scattering originates from the short range ordered domains that can be indexed by the ZB-GaNWs, as observed by the *in situ* x-ray diffraction method. The correlation length ξ of the Bragg scattering that represents the characteristic size of the ordered domain can be used to investigate the growth of the GaNWs. Figure 14 shows the obtained correlation lengths of GaNWs versus temperatures. The results show that the self organization process is characterized by a rapid initial growth rate that slows down and self-terminates. This solid curve shown in Fig. 14 describes an exponential growth function [30], namely $\xi = \xi_o - \beta e^{\frac{T}{\tau}}$, where ξ_o =23.8 nm, τ=75.3 K, and β=0.776 nm represent the initial constants and the fitted parameters, respectively. Furthermore, the nanowire growth rate, defined by G=|dξ/dT|, can be used to probe the growth rate of short range domain. Thus, at T=230 to 80 K, we have a growth rate of 0.0103 Å/K and a self-terminated length of ξ_o=23.8 nm.

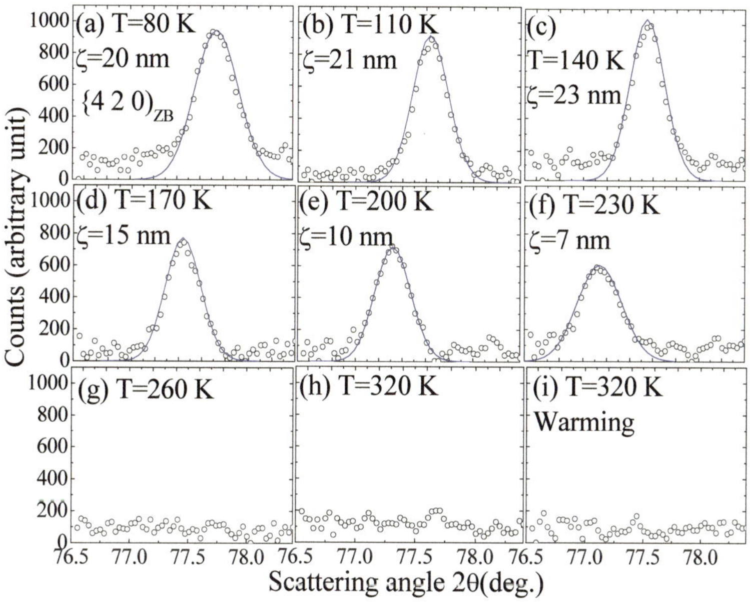

Fig. 13. Variations of the $\{420\}_{ZB}$ reflection with temperature. The solid curves indicate the fitted of the data to the diffraction profile for finite size structure.

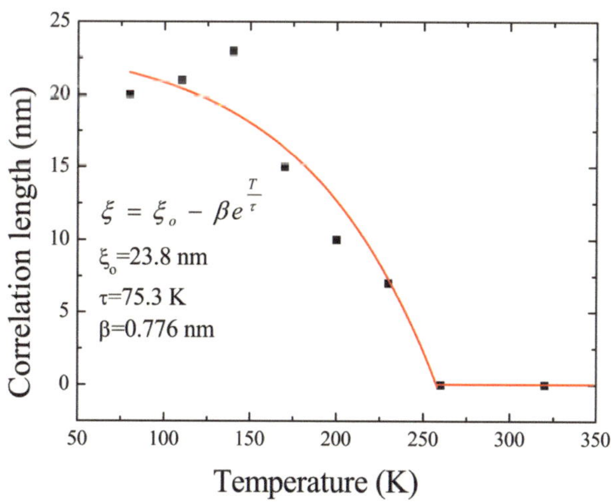

Fig. 14. Temperature dependence of the obtained correlation lengths, revealing a growth rate of 0.0103 Å/K and self-terminated length if ξ_0=23.8 nm.

5. Conclusion

In conclusion, we have fabricated GaN nanowires employing the LPTCVD method, which we take the advantage of the reaction of gallium with NH_3. The mean diameter of the GaN nanowires fabricated was 40(3) nm, and their crystallized into the known wurtzite GaN structure at ambient temperatures. Profile refining of the diffraction patterns shows that the low temperature patterns cannot be described using the hexagonal α-GaN solely. The ZB-GaN phase was found to develop below 260 K. A new short range ordered ZB-GaN phase was observed. The width of the diffraction profile associated to ZB-GaN is noticeably larger than that of the WZ-GaN phase. Short range ordering effect and the phase distribution of random ZB-GaNWs must be taken into account. A short range modeling was employed to identify the correlation lengths of the temperature dependence to the ordered domains [31]. The short-range ordered domains observed are not only of great interest for understanding the thermal effect of the phase separation in the GaNWs system (e.g., for CuO [32, 33], WO_2 [34], MoO_2 [35] and Ta_2O_5 nanowires [36-41]) but also for investigating fundamental physics and mechanisms in the future.

6. Acknowledgement

We would like to thank the National Science Council of the Republic of China for the financial support through project numbers NSC 97-2112-M-259-004-MY3.

7. References

[1] Martin, C. R. (1994). Nanomaterials: A Membrane-Based Synthetic Approach. *Science*, Vol. 266, (December 1994) pp. 1961-1966, ISSN 1095-9203

[2] Duan, X.; Huang, Y.; Cui, Y.; Wang, J. & Lieber, C. M. (2001). Indium phosphide nanowires as building blocks for nanoscale electronic and optoelectronic devices. *Nature*,Vol. 409, (January 2001) pp. 66-69, ISSN 0028-0836

[3] Xia, Y.; Yang, P.; Sun, Y.; Wu, Y.; Mayers, B.; Gates, B.; Yin, Y.; Kim, F. & Yan, H. (March 2003). One-Dimensional Nanostructures: Synthesis, Characterization, and Applications. *Advanced Materials*,Vol. 15, (March 2003) pp. 353-389, ISSN 1687-4110

[4] Shi, B. M.; Xie, M. H.; Wu, H. S.; Wang, N. & Tong, S. Y. (2006). Transition between wurtzite and zinc-blende GaN: An effect of deposition condition of molecular-beam epitaxy. *Applied Physics Letters*,Vol. 89, (October 2006) pp. 151921-1-151921-3, ISSN 0003-6951

[5] Yang, J. W.; Kuznia, J. N.; Chen, Q. C.; Asif Khan, M.; George, T.; De Graef, M. & Mahajan, S. (1995). Temperature-mediated phase selection during growth of GaN on (111)A and $(\bar{1}\,\bar{1}\,\bar{1})$ B GaAs substrates. *Applied Physics Letters*,Vol. 67, (December 1995) pp. 3759-3761, ISSN 0003-6951

[6] Kuykendall, T.; Pauzauskie, P. J.; Zhang, Y.; Goldberger, J.; Sirbuly, D.; Denlinger, J.; & Yang, P. (2004). Crystallographic alignment of high-density gallium nitride nanowire arrays. *Nature Materials*,Vol. 3, (July 2004) pp. 524-528, ISSN 0003-6951

[7] Cavallini, A.; Polenta, L.; Rossi, M.; Richter, T.; Marso, M.; Meijers, R.; Calarco, R. & Lüth, H. (2006). Defect Distribution along Single GaN Nanowhiskers. *Nano Letters*,Vol. 6, (June 2006) pp. 1548–1551, ISSN 1530-6984

[8] Dahara, S.; Datta, A.; Wu, C. T.; Lan, Z. H.; Chen, K. H.; Wang, Y. L.; Hsu, C. W.; Shen, C. H.; Chen, L. C. & Chen, C. C. (2004). Hexagonal-to-cubic phase transformation in GaN nanowires by Ga+ implantation. *Applied Physics Letters*, Vol. 84, (June 2004) pp. 5473-5475, ISSN 0003-6951

[9] McMahon, M. I. & Nelmes, R. J. (2005). Observation of a Wurtzite Form of Gallium Arsenide. *Physical Review Letters*, Vol. 95, (November 2005) pp. 215505-215508, ISSN 1063-7850

[10] Rietveld, H. M. (1969). A profile refinement method for nuclear and magnetic structures. *Journal of Applied Crystallography*,Vol. 2, (June 1969) pp. 65-71, ISSN 1095-9203

[11] Liang, C. H.; Chen, L. C.; Hwang, J. S.; Chen, K. H., Hung, Y. T. & Chen, Y. F. (2002). Selective-area growth of indium nitride nanowires on gold-patterned Si(100) substrates. *Applied Physics Letters.*, Vol. 81, (July 2002) pp. 22-24, ISSN 0003-6951

[12] Ponce, F.A.& Bour, D. P. (1997). Nitride-based semiconductors for blue and green light-emitting devices. *Nature*, Vol. 95, (March 1997) pp. 351-359, ISSN 0028-0836

[13] Nakamura, S. (1998). The Roles of Structural Imperfections in InGaN-Based Blue Light-Emitting Diodes and Laser Diodes. *Science*, Vol. 281, (August 1998) pp. 956-961, ISSN 1095-9203

[14] Maruska, H. P. & Tietjent, J. J. (1969). THE PREPARATION AND PROPERTIES OF VAPOR-DEPOSITED SINGLE-CRYSTAL-LINE GaN. *Applied Physics Letters*, Vol. 281, (November 1969) pp. 327-329, ISSN 0003-6951

[15] Alivisatos, A. P. (1996). Semiconductor Clusters, Nanocrystals, and Quantum Dots. *Science*, Vol. 271, (February 1996) pp. 933-937, ISSN 1095-9203

[16] Zolpher, J. C.; Shul, R. J.; Baca, A. G.; Wilson, R. G.; Pearton, S. J.; Stall, R. A. (1996). Ion-implanted GaN junction field effect transistor. *Applied Physics Letters*, Vol. 68, (April 1996) pp. 2273-2275, ISSN 0003-6951

[17] Duan, X. & Leiber, C. M. (2000). Laser-Assisted Catalytic Growth of Single Crystal GaN Nanowires. *Journal of the American Chemical Society*, Vol. 68, (December 1999) pp. 188–189, ISSN 0002-7863

[18] Han, W.; Fan, S.; Li, Q. & Hu, Y. (1997). Synthesis of Gallium Nitride Nanorods Through a Carbon Nanotube-Confined Reaction. *Science*, Vol. 277, (August 1997) pp. 1287-1289, ISSN 1095-9203

[19] Shi, W. S.; Zheng, Y. F.; Wang, N.; Lee, C. S.; Lee, S. T. (2001). Microstructures of gallium nitride nanowires synthesized by oxide-assisted method. *Chemical Physics Letters*, Vol. 345, (September 2001) pp. 377-380, ISSN 0009-2614

[20] Li,J.Y.; Chen,X. L.; Qiao, Z. Y.; Cao,Y. G.; Lan, Y. C. (2000). Formation of GaN nanorods by a sublimation method. *Journal of Crystal Growth*, Vol. 213, (June 2000) pp. 408-410, ISSN 0022-0248

[21] Chen, C. C.; Yeh, C. C.; Chen, C. H.; Yu, M. Y.; Liu, H. L.; Wu, J. J.; Chen, K. H.; Chen, L. C.; Peng, J. Y.; Chen, Y. F. (2001). Catalytic Growth and Characterization of Gallium Nitride Nanowires. *Journal of the American Chemical Society*, Vol. 123, (February 2001) pp. 2791–2798, ISSN 0002-7863

[22] Chang, K. W.; Wu, J. J. (2002). Low-Temperature Catalytic Synthesis of Gallium Nitride Nanowires. *Journal of Physical Chemistry B*, Vol. 106, (July 2002) pp. 7796–7799, ISSN 1520-5207

[23] Lyu, S. C.; Cha, O. H.; Suh, E. K.; Ruh, H.; Lee, H. J.; Lee, C. J. (2002). Catalytic synthesis and photoluminescence of gallium nitride nanowires. *Chemical Physics Letters*, Vol. 367, (January 2003) pp. 136-140, ISSN 0009-2614

[24] Pearton, S. J. (1997). GaN and Related Materials, CRC Press, ISSN 978-9056995171, Amsterdam

[25] Larson, A. C. & Von Dreele, R. B. (1990). General Structure Analysis System, Report LA-UR-86-748, Los Alamos National Laboratory, Los Alamos, NM.

[26] Muñoz, A. & Kunc, K. (1991). High-pressure phase of gallium nitride. *Physical Review B*, Vol. 44, (July 1991) pp. 10372 - 10373, ISSN 1098-0121

[27] Yeh, C. Y.; Lu, Z. W.; Froyen, S. & Zunger, A. (1992). Zinc-blende–wurtzite polytypism in semiconductors. *Physical Review B*, Vol. 46, (July 1992) pp. 10086 - 10097, ISSN 1098-0121

[28] Li, W. -H.; Hsieh, W. T. & Lee, K. C. (1995). Dependence of powder neutron scattering on the dimensionality of magnetic order. *Journal of Physics : Condensed Matter*,Vol. 7, (August 1995) pp. 6513-6522, ISSN 0953-8984

[29] Wu, S. Y.; Li, W. -H.; Lee, K. C.; Lynn, J. W.; Meen, T. H. & Yang, H. D. (1996). Two- and three-dimensional magnetic correlations of Tb in Pb2Sr2TbCu3O8. *Physical Review B* ,Vol. 54, (June 1995) pp. 10019 - 10026, ISSN 1098-0121

[30] Lu, Y. S., Wu, H. I., Wu, S. Y., Ma, Y. -R. (2007) Tip-induced large-area oxide bumps and composition stoichometry test via atomic force microscopy. *Surface Science*,.Vol. 601, (October 2007) pp.3788-3791, ISSN 0039-6028

[31] Wu, S. Y., Ji, J. -Y., Chou, M. H., Li, W. -H. & Chi, G. C. Low-temperature phase separation in GaN nanowires : An in situ x-ray onvestigation. *Applied Physics Letters*, Vol. 92, (April 2008) p. 161901, ISSN 0003-6951

[32] Zheng, X. G, Kubozono, H., Yamada, H., Kato, K., Ishiwata, Y. & Xu, C. N. Giant negative thermal expansion in magnetic nanocrystals. *Nature Nanotechnology*, Vol. 3, (October 2008) pp.724-726, ISSN 1748-3387

[33] Cheng, C. -L., Ma, Y. -R., Chou, M. H., Huang, C. Y., Yeh, V. & Wu, S. Y. Direct observation of short-circuit diffusion during the formation of a single cupric oxide nanowire, *Nanotechnology*,Vol. 18, (May 2007) p. 265604, ISSN 0957-4484

[34] Lu, D. Y.; Chen, J.; Chen, H. J.; Gong, L.; Deng, S. Z.; Xu, N. S. & Liu, Y. L. (2007). Raman study of thermochromic phase transition in tungsten trioxide nanowires. *Applied Physics Letters*,Vol. 90, (January 2007) p. 041919, ISSN 0003-6951

[35] Kumari, L.; Ma, Y. -R.; Tsai, C. C.; Lin, Y. W.; Wu, S. Y.; Cheng, K. W. & Liou, Y. (2007). X-ray diffraction and Raman scattering studies on large-area array and nanobranched structure of 1D MoO_2 nanorods. *Nanotechnology*,Vol. 18, (March 2007) pp. 115717-115723, ISSN 0957-4484

[36] Deavn R. S., Ho, W. -D., Lin, J. -H., Wu, S. Y., Ma, Y. -R.; Lee & Liou, Y. (2008). X-ray diffraction study of a large-scale and high-density array of one-dimensional crystalline Tantaium Oentoxide nanorods. *Grystal Growth & Design*,Vol. 8, (October 2008) pp. 4465-4468, ISSN 1528-7483

[37] Wu, S.; Chan, H. M. & Harmer, M. P. (2005) Effect of Alumina Additions on Microstructural Aspects of the β to α Transformation in Tantalum (V) Oxide. *Journal of the American Ceramic Society*,Vol. 88, (June 2005) pp. 2369-2373. ISSN 0002-7820

[38] Hale, J. S.; DeVries, M.; Dworak, B. & Woollam, J. A. (1998) Visible and infrared optical constants of electrochromic materials for emissivity modulation applications. *Thin Solid Films*, Vol. 313-314, (February 1998) pp.205-209 ISSN 0040-6090

[39] Yoshimura, H.& Koshida, N. (2006) Thermally crosslinked hole-transporting layers for cascade hole-injection and effective electron-blocking/exciton-confinement in phosphorescent polymer light-emitting diodes. *Applied Physics Letters*, Vol. 88, (Feburary 2005) p. 093509, ISSN 0003-6951

[40] González, J.; Ruiz, M. D. C.; Rivarola, J. B.; Pasquevich, D. (1998) Effects of heating in air and chlorine atmosphere on the crystalline structure of pure Ta2O5 or mixed with carbon. *Journal of Material Science*,. Vol. 33, (August 1998) pp.4173. -4180, ISSN 0022-2461

[41] Kukli, K.; Ritala, M.; Matero R. & Leskela, M. (2000) Influence of atomic layer deposition parameters on the phase content of Ta_2O_5 films. *Journal of Crysta. Growth*, Vol 212, (May 2000) pp.459-568, ISSN 0022-0248

12

Raman Spectroscopy on Semiconductor Nanowires

Ilaria Zardo[1], Gerhard Abstreiter[1] and Anna Fontcuberta i Morral[1,2]
[1]*Walter Schottky Institut and Physik Department, Technische Universität München*
[2]*Laboratoire des Matériaux Semiconducteurs, Ecole Polytechnique Fédérale de Lausanne*
[1]*Germany*
[2]*Switzerland*

1. Introduction

Raman scattering is an inelastic light scattering non-destructive technique which allows the access mainly to the phonon modes at the Γ point of materials and in some cases to the dispersion (Goni A.R. et al., 2001; Zunke et al., 1995; Weinstein et al., 1975). Since its discovery, Raman has been used both for the characterization of materials and for the understanding of basic interactions such as plasmonic excitations (Raman et al., 1928; Szymanski H.A. et al., 1967; Otto et al., 1992; Schuller et al., 1996; Steinbach et al., 1996; Ulrichs et al., 1997, Sood et al 1985, Abstreiter et al. 1979, Roca et al. 1994, Pinczuk et al 1977, Pinczuk et al., 1979). Raman spectroscopy can be experimentally performed at the nanoscale by using a confocal microscope or even a tip enhanced scanning microscope. It is possible to obtain lateral submicron resolutions of the properties of a material (Hartschuh et al., 2003). Nowadays Raman spectroscopy is a versatile and relative standard tool for the characterization of materials giving detailed information on crystal structure, phonon dispersion, electronic states, composition, strain and so-on bulk materials, thin film and nanostructures (Cardona, 1982; Anastassakis, 1997; Reithmaier et al., 1990; Spitzer et al., 1994; Pinczuk et al., 1977; Pinczuk et al., 1979; Baumgartner et al., 1984; Schuller et al., 1996; Pauzauskie et al., 2005; Long, 1979).

In the last decade Raman spectroscopy has been increasingly used to study nanowires and quantum dots (Abstreiter et al., 1996; Roca et al., 1994). Several new phenomena have been reported to date with respect to one-dimensional structures. For example, the high surface-to-volume ratio has enabled the measurement of surface phonon modes (Gupta et al., 2003a; Krahne et al 2006; Adu et al., 2006; Spirkoska et al., 2008). Some authors report a increase in the scattered intensity for nanoscale structures with respect to their bulk counterpart, effect denominated as 'Raman antenna effect' (Xiong et al., 2004; Xiong et al., 2006; Cao et al., 2007). Additionally, polarization dependent experiments on single carbon nanotubes and/or nanowires have shown that the physics behind Raman scattering of such one-dimensional nanostructures can differ significantly from the bulk (Frechette et al., 2006; Livneh et al., 2006; Cao et al., 2006). Indeed, the highly anisotropic shape of the nanowires can lead to angular dependencies of the modes which otherwise would not be expected from selection rules (Frechette et al., 2006; Livneh et al., 2006; Cao et al 2006).

Overall, Raman spectroscopy of nanostructures represents an extremely active and exciting field for the benefit of science and technology at the nanoscale. The arising new phenomena and technical possibilities open new avenues for the characterization of materials but also for the understanding of fundamental process in nanoscale matter. In this chapter, we provide a review of Raman spectroscopy on nanowires, in which an overview of the selection rules, appearance of new modes and size effects will be given.

2. Selection rules in Raman scattering of nanowires

2.1 Raman selection rules, application to the geometry of nanowires

Raman scattering is a manifestation of the interaction between the electromagnetic radiation and vibrational and/or rotational motions in a material. It provides information about the symmetry and composition of the system, the lattice dynamics, structural transitions, strain and electronic states. The scattering process involves two energy quanta simultaneously. It is usually schematized in two steps:

i. A photon with energy $h\nu_o$ and wave vector $\vec{q}$ is absorbed, exciting the system from a initial state *1* to a state *n*;
ii. The system emits a photon with energy $h\nu'$ and wave vector $\vec{q}'$ and relaxes from the state *n* to a final state 2.

In the case where the final state is identical to the initial one the incident and scattered light have the same frequency ($\nu' = \nu_o$). This process is called elastic or Rayleigh scattering. When the final state is different from the initial one, the scattering process is inelastic. In this case, the creation or the annihilation of an excited state of the system occurs, and the emitted photon has lost or gained energy. These processes are called respectively Stokes scattering and anti-Stokes Raman scattering. For the conservation of energy, the frequency of the scattered radiation is given by:

$$\nu' = \nu_o \pm \frac{E_1 - E_2}{h} = \nu_o \pm \nu_v \tag{1}$$

with ν_o the frequency of the incoming photon, ν' the frequency of the emitted photon,E_1 and E_2 the energies of the initial and final state of the system. These scattering processes are schematically illustrated in Fig. 1.

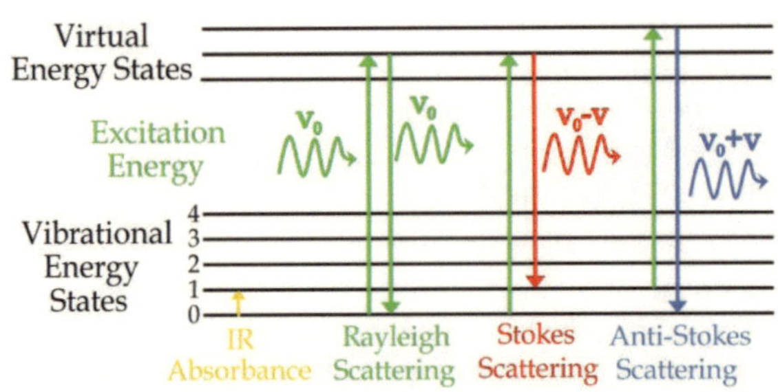

Fig. 1. Schematic drawing of transitions between generic vibrational energy states due to, from left to right, infrared absorbance, Rayleigh Scattering Stokes Raman Scattering and anti-Stokes Raman Scattering.

The difference between the Raman scattering frequency ν' and the excitation frequency ν_o is independent from the last one and it's equal to $\pm\nu''$. The probability of the Stokes scattering and the anti-Stokes scattering is different, because of the different population of the two energy levels at a temperature T. The intensity of the Raman lines is proportional to the transition probability. For this reason the Stokes lines are more intense than the anti-Stokes. In the following, a short description of the Raman effect is given within the classical picture (Turrel & Corset, 1996). In this frame, spatial and temporal fluctuations of the electronic contributions of the polarizability are at the origin of the Raman scattering. The electric field originated by a plane monochromatic wave with wave vector $\vec{K}_o$ in a point $\vec{r}$ in space in a transparent crystal is given by:

$$\vec{E} = E_o\{\exp - 2\pi i(\vec{K}_o \cdot \vec{r} - \nu_o t)\} \tag{2}$$

with ν_o the frequency of the light and t the time. The electric field causes an induced dipole moment given by:

$$\vec{\mu} = \alpha\vec{E} \tag{3}$$

where α is the polarizability. Both $\vec{\mu}$ and $\vec{E}$ are vectors, while α is a 3x3 tensor with real elements, unless magnetic phenomena are involved. The coordinate system and the symmetry of the crystal determine the tensor form. Due to the time dependency of $\vec{\mu}$ and $\vec{E}$, the induced dipole moment will oscillates in time, with consequent radiation emission. The polarizability can be expanded in as a Taylor series in the normal coordinates $Q_K = Q_K^0 \exp^{-2\pi i\left[\vec{K}_K \cdot \vec{r} - \nu_K t\right]}$, with $\vec{K}_K$ the wave vector of lattice wave K. Equation (3) becomes:

$$\vec{\mu} = \alpha_0 E_o \exp^{-2\pi i(\vec{K}_o \cdot \vec{r} - \nu_o t)} + E_o \sum_K \alpha_K' Q_K \exp^{-2\pi i\left[\left(\vec{K}_0 \pm \vec{K}_K\right)\cdot\vec{r} - (\nu_0 \pm \nu_K)t\right]} \tag{4}$$

with $\alpha_K' = \left(\dfrac{\partial\alpha}{\partial Q_K}\right)_0$.

The first term of equation (4) describes the Rayleigh scattering, the oscillation of the induced dipole at the same frequency of the incident light. The second term represents the dipole oscillating at a frequency shifted by the frequency of the normal modes. Therefore, $\nu_o \pm \nu_K$ is the frequency of the scattered light, which propagates in the direction $\vec{K}_0 \pm \vec{K}_K$.

The Raman scattering is governed by the conservation of energy and by conservation of momentum, which implies that $\vec{K}_0 = \vec{K}_S \pm \vec{K}_K$, being $\vec{K}_S$ the wave vector of the scattered light. Namely, the orientation of the crystallographic axes with respect to the direction and polarization of the scattered light affects the Raman spectrum. In this respect, it is evident that Raman spectroscopy on single crystals gives information about the crystal symmetry. The intensity of the scattered light I_s, which is the scattered energy per unit time, into a solid angle $d\Omega$ is given by:

$$I_S = I_i \cdot k \cdot \left|\hat{e}_i \cdot \vec{R} \cdot \hat{e}_s\right|^2 d\Omega \tag{5}$$

with $\boldsymbol{I_i}$ the irradiance - energy per unit area per unit time - of the excitation incident on the sample, $k = 4\pi^2 a^2 \nu_s^{-4}$, $a \approx 1/137$, ν_s the wavenumber of the scattered light, $\hat{e}_i$ ($\hat{e}_s$) the polarization unitary vectors of the incident (scattered) light and $\ddot{R}$ the scattering tensors. It is worth to note that the symmetry properties of the polarizability and the scattering tensors are the same. In the case where Raman scattering is realized at the submicron scale, a microscope objective has to be used. For simplicity, a backscattering configuration is preferred. In this configuration the scattered light is collected along the same direction of the excitation, as shown schematically in Fig. 2.

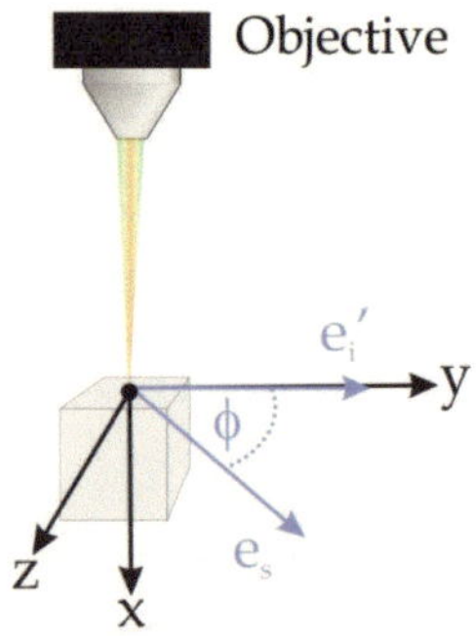

Fig. 2. Schematic drawing of the backscattering geometry. The incoming light is directed along the x direction, with the polarization directed along the y direction. The scattered light is collected along the x direction and its polarization has components along the y and z directions.

As an example, we consider the backscattering geometry with the set of axes as depicted in Fig. 2 and $\hat{e}_i \,||\, y$, the Raman intensity is calculated as follows:

$$I_s \propto \left| (010) \begin{pmatrix} R_{xx} & R_{xy} & R_{xz} \\ R_{yx} & R_{yy} & R_{yz} \\ R_{zx} & R_{zy} & R_{zz} \end{pmatrix} \begin{pmatrix} 0 \\ 1 \\ 1 \end{pmatrix} \right|^2 = \left| R_{yy} + R_{yz} \right|^2 \tag{6}$$

The use of a determined incident and analyzed polarization in the Raman scattering experiments results in the selection of certain elements of $\ddot{R}$. In this way, polarized Raman spectroscopy enables the determination of the Raman selection rules and the tensor symmetry. In the measurements on a single bulk crystal, it is much more convenient to use the crystallographic axis as a basis and express the polarizability tensor in this basis.

As an example, we consider zinc-blende GaAs. The phonon dispersion is composed of 6 different branches: two transverse and one longitudinal acoustical modes (TA and LA) as well as two transverse and one longitudinal optical modes (TO and LO). The optical photon modes are usually indicated E_1(TO) and A_1(LO). The notation E_1 and A_1 denote respectively to modes vibrating perpendicular and along the z axis. The Raman tensors for zinc blende GaAs are usually given in the base $x = (100)$, $y = (010)$ and $z = (001)$, resulting in:

$$R(x) = \begin{pmatrix} 0 & 0 & 0 \\ 0 & 0 & 1 \\ 0 & 1 & 0 \end{pmatrix},\ R(y) = \begin{pmatrix} 0 & 0 & 1 \\ 0 & 0 & 0 \\ 1 & 0 & 0 \end{pmatrix} \text{ and } R(z) = \begin{pmatrix} 0 & 1 & 0 \\ 1 & 0 & 0 \\ 0 & 0 & 0 \end{pmatrix} \tag{7}$$

In this set of axes and in backscattering geometry with the direction of excitation and collection perpendicular to a {001} plane, the TO is forbidden while the LO mode is allowed. The intensity of the scattered light polarized along the x or y direction as a function of the angle between the polarization of the excitation and the $x=(100)$ can be calculated using equation (6). The theoretical azymutal dependence of the intensities is plotted in Fig. 3.

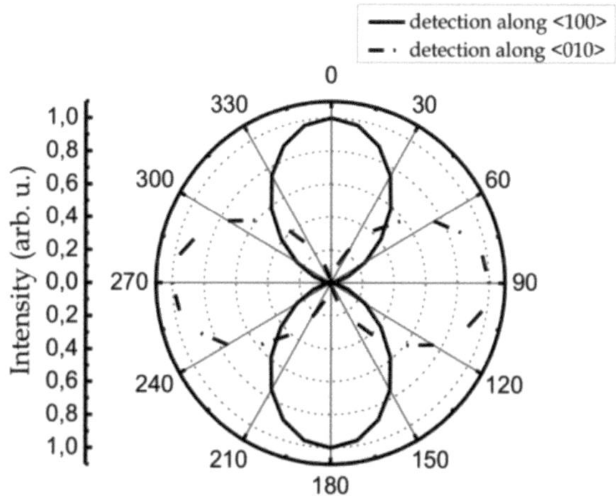

Fig. 3. Theoretical azimuthal dependence of the LO mode of a bulk GaAs (001). Continuous and dashed lines represent the components along the [100] and [010] of the Raman signal, respectively.

The selection rules and the dependency of the intensity on the polarization direction can be calculated even from the other surfaces in the same way. In backscattering geometry both the A_1(LO) or E_1(TO) are observed from the (111) surfaces, while only E_1(TO) is observed from the (110) planes.

If we now consider the geometry of the nanowires, the important crystallographic axis correspond to the directions $x=(0-11)$, $y=(211)$ and $z=(-111)$, which should be used as a basis. A schematic drawing of the relevant axes on a reference bulk sample is shown in Fig. 4a: the x axis corresponds the direction of the incident and scattered light in the [0-11] direction, while y and z are the in plane axes respectively parallel to [211] and [-111]. The selection rules are obtained by transforming the Raman tensor and by expressing the polarization vectors into the new basis and, using eq. (6). The values of the Raman tensor for the transversal modes in that configuration for incident light along the x axis are:

$$R'(y)=\begin{pmatrix} 0 & 1/\sqrt{3} & -1/\sqrt{6} \\ 1/\sqrt{3} & 2/3 & -1/3\sqrt{2} \\ -1/\sqrt{6} & 1/3\sqrt{2} & -2/3 \end{pmatrix},\quad R'(z)=\begin{pmatrix} 0 & -1/\sqrt{3} & 1/\sqrt{6} \\ -1/\sqrt{3} & 2/3 & 1/3\sqrt{2} \\ 1/\sqrt{6} & 1/3\sqrt{2} & -2/3 \end{pmatrix} \qquad (3)$$

The intensity of the scattered light polarized parallel or perpendicular to the [-111] direction, $I_s(\|)$ and $I_s(\perp)$, as a function of the angle α between the polarization of the excitation with the [-111] axis is:

$$I_s(\perp) = \left| (0 \quad \sin\alpha \quad \cos\alpha) \left[R'(y) \begin{pmatrix} 0 \\ 1 \\ 0 \end{pmatrix} + R'(z) \begin{pmatrix} 0 \\ 1 \\ 0 \end{pmatrix} \right] \right|^2 = \left| \frac{4}{3}\sin\alpha + \frac{2\cos\alpha}{3\sqrt{2}} \right|^2 \quad (4)$$

$$I_s(\|) = \left| (0 \quad \sin\alpha \quad \cos\alpha) \left[R'(y) \begin{pmatrix} 0 \\ 0 \\ 1 \end{pmatrix} + R'(z) \begin{pmatrix} 0 \\ 0 \\ 1 \end{pmatrix} \right] \right|^2 = \left| \frac{2}{3\sqrt{2}}\sin\alpha + \frac{4}{3}\cos\alpha \right|^2 \quad (5)$$

Fig. 4b and c show the theoretical and experimental dependency of the two components of the scattered light, respectively.

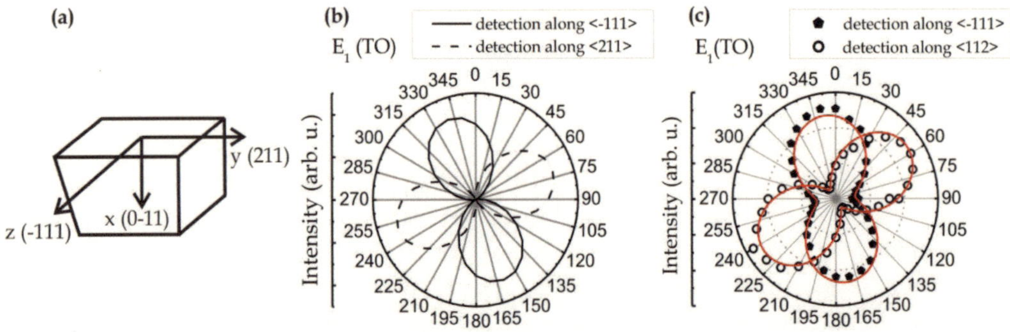

Fig. 4. (a) Crystal facets of the reference used for the measurement of the selection rules in GaAs. The axis correspond to the crystallographic directions: x=(0-11), y=(211) and z=(-111). (b) Theoretical azimuthal dependence of the TO mode of a bulk GaAs (0-11), as in (a). Continuous and dashed lines represent the components along the [-111] and [211] of the Raman signal, respectively. (c) Measured azimuthal dependence of the TO mode of a bulk GaAs (0-11). Diamonds and open circles represent the components along the [-111] and [211] of the Raman signal, respectively. The continuous line is a squared sine fit to the data, which describes polar behaviour.

As mentioned above, this set of axis is the one that should be used for the investigation of single GaAs nanowires with [111] growth axis. For further clarity, a schematic drawing of the nanowire with the corresponding set of axis, as used in Raman backscattering experiment, is presented in Fig. 5a. We studied nanowires presenting a mixture of zinc blende and wurtzite structure. In this case, a further optical mode can be observed at k=0, namely the E_2^H (see section 4). Fig. 5b shows representative Raman spectra realized under the main four polarization configurations. The azymuthal dependence of E_1(TO) and E_2^H is presented in Fig. 5c and d. The scattered light has been analyzed selecting the components with polarization parallel - $I_s(\|)$ - and perpendicular - $I_s(\perp)$ - to the z axis. The E_1(TO) mode is polarized along the axis of the nanowire. Interestingly, also $I_s(\perp)$ seems to have a slighty higher intensity when the incident light is polarized along the nanowire axis. The scattered light with polarization perpendicular to the z axis, exhibits a drop in the intensity, compared to the measurements on GaAs bulk (Fig. 4c). Indeed, the ratio of intensity between $I_s(\|)$ and $I_s(\perp)$ is about 5. Interestingly, the azymuthal dependence of the E_2^H mode associated with the wurtzite phase exhibits a quite different behavior. In this case the maximum intensity of the scattered light is observed when the incident light is perpendicular to the nanowire axis, both for Is(||) and Is(⊥) -though for the latter the dependence is less clear due to the low intensity-.

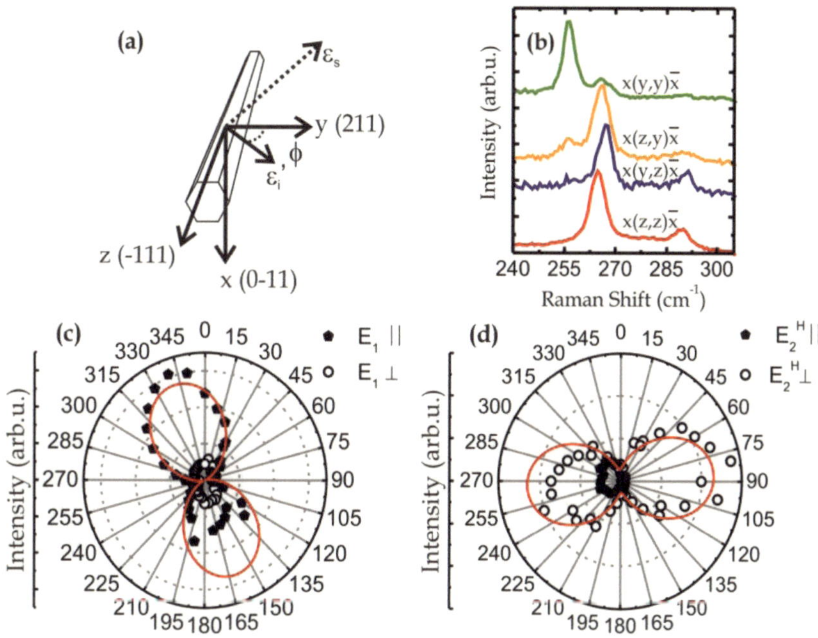

Fig. 5. (a) Sketch of the configurations used for the measurement of GaAs nanowires in backscattering geometry. The crystal facets of the nanowire and the corresponding set of axis used as indicated: x=(0-11), y=(211) and z=(-111). (b) Representative Raman spectra realized under the main four configurations. For better illustration, the spectra have been normalized and shifted vertically. All spectra have been collected in the same position of the nanowire. Azimuthal dependence of the E_1(TO) mode (c) and of the E_2^H mode (d), related to the wurtzite structure. Diamonds and open circles represent the parallel and perpendicular components of the Raman signal collected, respectively. The continuous lines are squared sine fit.

→ Effect of the dielectric mismatch

In the case of nanowires, it is worth noticing that there is an enhanced response of the Raman scattering for polarizations along the nanowire axis. As it will be shown in the following, this is partly due to the one-dimensionality and to the small diameter of the nanowires, as it has been reported in literature (Cao et al., 2007; Livneh et al., 2007; Papadimitriou & Nassiopoulou, 1998; Pauzauskie et al., 2005; Duesberg et al., 2006; Fréchette & Carraro, 2006; Cao et al., 2006; Xiong et al., 2006). Xiong et al. found that nanowires with a diameter $d \ll \lambda/4$, with λ the wavelength of the excitation, show a dipolar behavior. Namely, the Raman scattering intensity is $\sim I_0 \cos^2\alpha$, with I_0 the incident laser intensity and α the angle between the electric field of the laser and the nanowire axis. For larger diameters, $d \gg \lambda/4$, the nanowires present a multipolar character. The authors address the origin of this effect to the scattering of the electromagnetic field from a dielectric cylinder of nanoscale dimensions. The calculations show that the electric field inside the nanowires with bigger diameter is increased when the electric field of the excitation is either parallel or perpendicular to the nanowire axis. Instead, for smaller diameters, the electric field inside the nanowire is strongly suppressed when the electric field of the excitation is perpendicular to the nanowire axis. Experiments on silicon nanocones showed that the enhancement in the Raman scattering, due to the enhanced internal field, decreases with increasing the nanowires diameter and increases with the wavelength of the excitation, features which suggest a resonant nature (Cao et al., 2006). This enhancement in the Raman scattering is in analogy with absorption, photoluminescence and photocurrent measurements (Cao L., 2009; Wang J. 2009; Thunich S., 2009).

2.2 Appearance of new modes: surface and breathing modes.

Studies comparing Raman scattering experiments of bulk and nanostructured materials have been reported in literature for several different kind of systems. It is usually observed that the transversal optical (TO) and the longitudinal optical (LO) modes have a position in energy close to that observed in bulk. When scaling down the size and the dimensionality of the structures, the position can change (see section 3.1). Additionally, new Raman modes can be found. Effects related to the shape of the system can become significant. The existence of boundary conditions at the nanoscale gives rise to electric and polarization forces. The surfaces represent a new mechanical boundary, since the surface atoms are "less bound" and "feel" a different local field from the bulk. This has consequences even in the propagation of an optical phonon, where the oscillating dipoles - created by the out of phase oscillation of ions and cations - interact by a dipole-dipole interaction. Mahan et al. developed and presented a model which describes the variation of the long range dipolar interactions due to the nanowires geometry, leading to the split of the TO and LO modes in polar semiconductor nanowires (Mahan et al., 2003). Indeed, the highly anisotropic shape of the nanowires determines different contribution in the dipolar sums for the components in the cross sectional plane - x and y - which are truncated by the finite size, from the one along the nanowire growth axis - z -. Accordingly, the local electric field is modified too. The dispersion relation of the optical phonon can be related to the local spring constant (ω_o^2) and to the local electric field (E_μ):

$$\omega^2 q_\mu = \omega_o^2 q_\mu - \frac{e^*}{M} E_\mu \tag{6}$$

where e^* is the Szigeti charge and M is the reduced mass of the ion pair. The local field can be expressed as:

$$E_\mu = -T_{\mu\nu}\left[e^* q_\nu + \alpha E_\nu\right] \tag{7}$$

with $T_{\mu\nu}$ the components of the dipole-dipole interaction, ***a*** the polarizability of the unit cell and αE_ν the induced dipole in the same cell from core polarization. The anisotropy in the dipole sums ($T_{xx} = T_{yy} = \frac{4\pi}{6v_o}$ and $T_{zz} = \frac{-4\pi}{3v_o}$) for a thin wire with $L >> R$, with L length and $2R$ diameter of the wire, results into the anisotropy of the dielectric function, whose tensor has now two different components, $\varepsilon_{xx} = \varepsilon_{yy}$ and ε_{zz}, expressed by:

$$\varepsilon_{jj}(\omega) = \varepsilon_{jj}(\infty)\frac{\omega^2 - \omega_{Lj}^2}{\omega^2 - \omega_{Tj}^2}, \quad j = x, z \tag{8}$$

If we consider the optical phonons in a nanowire, equations (6) and (7) can be solved considering the uniaxial geometry, thereby obtaining:

$$\omega_{Lz}^2 = \omega_{LO}^2 = \omega_O^2 + \omega_p^2 \frac{2(\varepsilon_z(\infty)+2)}{9\varepsilon_z(\infty)} \tag{9}$$

$$\omega_{Tz}^2 = \omega_{TO}^2 = \omega_O^2 - \omega_p^2 \frac{(\varepsilon_z(\infty)+2)}{9} \tag{10}$$

with $\varepsilon = 1 + \frac{4\pi\alpha/V_o}{1-4\pi\alpha/3V_o}$ high frequency dielectric constant and $\omega_p^2 = \frac{4\pi e^{*2}}{MV_o}$ ion plasma frequency, for the z direction, and

$$\omega_{Lx}^2 = \omega_O^2 + \omega_p^2 \frac{(\varepsilon_x(\infty)+2)}{9(\varepsilon_x(\infty)+1)} \tag{11}$$

$$\omega_{Tx}^2 = \omega_O^2 + \omega_p^2 \frac{7(\varepsilon_x(\infty)+2)}{9(\varepsilon_x(\infty)-1)} \tag{12}$$

with $\varepsilon_x(\infty) = \frac{3\varepsilon_z(\infty)-1}{\varepsilon_z(\infty)+1}$ for the other directions. The predicted positions of the triplet arising from the split of the optical phonon due to the nanowire geometry are very close and therefore not always easily distinguishable. For example, in the case of GaAs or GaP nanowires, the ω_{Lx} and ω_{Tx} modes are about 2 cm^{-1} shifted from the ω_{Lz} mode (Cao et al., 2007). Nevertheless, an indication of the split can be given by the different position of the LO band in the nanowire spectra respect to the bulk. It has been shown that this shape dependence can explain even the occurrence of an angular dependencies of the phonon modes which otherwise would not be expected from the selection rules (Livneh et al., 2006, Fréchette & Carraro, 2006; Cao et al., 2006).

The reduction in the dimensionality and the presence of edge/boundaries in the crystal can also lead to the appearance/activation in the Raman spectra of inactive Raman modes (silent modes) at the Γ point of the Brillouin zone. This is due to the fact that the symmetry is changed by the existence of the edges, which leads to a rearrangement of the lattice structure. This has been especially observed in nanocrystals (Li et al., 2002; Kawashima & Katagiri, 1999).

Furthermore, there are other size-related phonons appearing when dealing generally with nanostructures, such as the surface optical phonons (SO) and breathing modes. Several works have reported the presence of a further peak in the Raman spectra of semiconductor nanowires or nanoparticles which have been assigned to SO phonons (Gupta et al., 2003a; Shan et al., 2006; Lin et al., 2003; Zeng et al., 2006; Spirkoska et al., 2008). The surface optical phonons are generated at the interface between different materials with different dielectric functions and propagate along the interface. The atoms involved in their propagation are

those close to the surface, so that the amplitude of the oscillations decays exponentially with the distance from the surface. This mode is activated by a breaking of the translational symmetry of the surface potential, which in the case of the nanowire can be addressed to the presence of roughness, sawtooth faceting on the nanowire sidewall or to a diameter oscillation along the nanowire length.
There are two characteristics which are distinctive of the SO modes and can therefore allow a reliable assignment of the mode: the dependence of the position (1) on the dielectric constant of the medium surrounding the wires and (2) on the diameter (or on the period of the diameter oscillation) of the wires. Indeed, it has already been shown that the SO mode position down shift increasing the dielectric constant of the surrounding optical medium and decreasing the nanowire diameter (Shan et al., 2006; Adu et al., 2006a; Spirkoska et al., 2008). Furthermore, the frequency of the SO modes at the center of the Brillouin zone is located between those of the TO and the LO.
The SO modes dispersion at the interface between a semiconductor and a dielectric material can be calculated imposing the condition:

$$\varepsilon(\omega) + \varepsilon_m = 0 \tag{13}$$

with $\varepsilon(\omega)$ the dielectric function of the semiconductor and ε_m the dielectric constant of the medium. In the case of an infinitely long cylinder equation (13) becomes:

$$\varepsilon(\omega) + \varepsilon_m f(qr) = 0 \tag{14}$$

where $f(qr)$ is given by

$$f(qr) = \frac{I_o(qr)K_1(qr)}{I_1(qr)K_o(qr)} \tag{15}$$

With q the phonon wavevector, r the nanowire radius and $I_i(qr)$ and $K_j(qr)$ the modified Bessel functions. Indeed, the dispersion relation for a SO mode for an infinitely long cylinder can be expressed by:

$$\omega_{SO}^2(q) = \omega_{TO}^2 + \frac{\tilde{\omega}_p^2}{\varepsilon_\infty + \varepsilon_m f(qr)} \tag{16}$$

with ω_{TO} the TO mode frequency, $\tilde{\omega}_p$ the screened ion plasma frequency given by $\tilde{\omega}_p^2 = \varepsilon_\infty(\omega_{LO}^2 - \omega_{TO}^2)$, ε_∞ the high frequency dielectric constant of the bulk material, ε_m the dielectric constant of the surrounding medium. Equation (16) establishes, therefore, the dependency of the surface phonon energy on the external medium and on the size of the wire, since the position of the surface optical phonon can be related to the dielectric constant of the surrounding medium as well as to the nanowire radius. Furthermore, values of q for the activation of the SO mode can be determined experimentally (Gupta et al., 2003a). Instead, the line width of the surface mode has not been yet well understood.
The effect of the position of the SO modes can be clearly observed by comparing semiconductor nanowires with various diameters. As an example, we show the Raman

spectra of GaAs nanowires of diameters 160 and 69 nm in Fig. 6a. For the nanowire with the largest diameter, the SO mode is barely observed as the position is very close to the LO phonon. As expected, the SO mode shifts to lower wavenumbers for smaller diameters, as it can be seen in the spectra obtained for nanowires with an average diameter of 69 nm. The entire trend of the position as a function of the diameter is shown in Fig. 6b. There, the line indicates what would be expected for nanowires with a circular section–GaAs nanowires exhibit a hexagonal section, which explains the discrepancy with the experimental data-. More details on the experiments can be found elsewhere (Spirkoska et al., 2008).

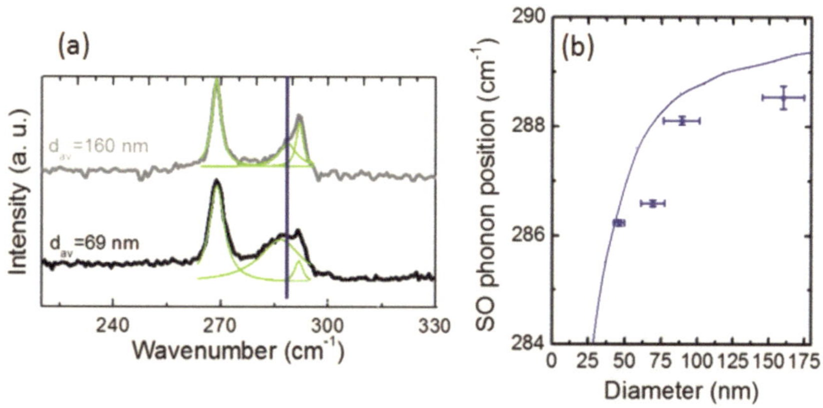

Fig. 6. a) Raman spectra of GaAs nanowire bundles with respectively an average diameter of 160 and 60 nm. The SO mode can be observed on the left of the LO mode b) Evolution of the SO phonon position as a function of the diameter of the nanowires. The line corresponds to the theoretical values expected for cylindrical GaAs nanowires.

Indeed, it has been proved that the cross section of the nanowires influences the surface mode dispersion (Adu et al., 2006a; Xiong et al., 2006). In these works, a model for nanowire with rectangular cross section has been developed. By setting z as growth direction and defining $L_i \quad (i = x, y)$ the edges of the rectangular cross section, the SO dispersion can be found solving the equations:

$$\varepsilon_{nw}(\omega)\tanh\left(\frac{q_i L_i}{2}\right) + \varepsilon_m = 0 \tag{17a}$$

$$\varepsilon_{nw}(\omega)\coth\left(\frac{q_i L_i}{2}\right) + \varepsilon_m = 0 \tag{17b}$$

where $q_i \quad (i = x, y)$ is the phonon wavevector of the modes propagating along x or y, which are the directions affected by the size effects, assuming the wire infinitively long along the z direction. Equation (17a) gives the symmetric mode, while equation (17b) the asymmetric one. Two more conditions have to be fulfilled:

$$q_x^2 + q_y^2 = q^2 \tag{18a}$$

$$q_x L_x = q_x L_y \tag{18b}$$

the latter one imposing the same parity to the optical phonon potential in the x and y directions. The symmetric and asymmetric SO phonon dispersion can then be expressed by:

$$\omega_{SO\ S}^{2}(q) = \omega_{TO}^{2} \frac{\varepsilon_o \tanh\left(\frac{q_i L_i}{2}\right) + \varepsilon_m}{\varepsilon_\infty \tanh\left(\frac{q_i L_i}{2}\right) + \varepsilon_m} \quad (19a)$$

$$\omega_{SO\ A}^{2}(q) = \omega_{TO}^{2} \frac{\varepsilon_o \coth h\left(\frac{q_i L_i}{2}\right) + \varepsilon_m}{\varepsilon_\infty \coth\left(\frac{q_i L_i}{2}\right) + \varepsilon_m} \quad (19b)$$

The comparison between equations 16 and 19 for the circular and rectangular sections leads to the conclusion that both the shape and size of the cross section have an important influence on the positon of the SO phonon.

Beside the surface modes, it is worth shortly mentioning another mode found in nanowires which cannot be observed in bulk materials: the Radial Breathing Mode (RBM). This mode was first observed in carbon nanotubes, corresponding to the atomic vibration of the carbon atoms in the radial direction. Its frequency was found to be highly dependent on the nanotubes diameter (Alvarez et al., 2000; Jorio et al., 2003; Maultzsch et al., 2005). The same mode has been observed even in semiconductor nanowires (Thonhauser & Mahan, 2005; Lange et al., 2008), and in both cases the inverse dependence on the nanowire diameter has been found. Assuming the nanowire as an infinitely long isotropic cylinder, the linear elasticity theory furnishes an expression of the RBM:

$$\omega = \frac{2\tau_n}{d} \sqrt{\frac{E(1-\nu)}{\rho(1+\nu)(1-2\nu)}} \quad (17)$$

with v the Poisson's ratio, ρ the nanowire specimen density, E the Young's modulus of the nanowire material and τ_n given by $\sqrt[n]{\tau J_o(\tau)} = \sqrt[n]{\frac{(1-2\nu)}{(1-\nu)}} J_1(\tau)$ where J_i are the Bessel functions. Equation (14) establishes the $1/d$ dependence of the radial mode frequency, being all the other terms dependent only on the material properties.

3. Confinement, heating effects and Fano resonance scattering

3.1 Phonon quantum confinement in nanowires

Some of the novel fundamental properties found in nanostructures are related to carrier and phonon confinement (Fischer et al., 2006; De Franceschi et al., 2003; Wanwees et al., 1988; Samuelson et al., 2004; Hu et al., 2007; Shorubalko et al., 2008; Rao et al., 1997; Bawendi et al. 1990; Lehmann et al., 1991). Confinement is usually correlated with tailoring novel physical properties, often giving rise to novel applications (Faist et al., 1994; Somers et al., 2008; Steckel et al., 2003). Raman spectroscopy is an ideal and relatively straightforward technique to test quantum confinement. Moreover, it can be realized under many extreme and non-extreme conditions, leading to an ideal technique for the investigation of processes in matter at low/high temperatures and/or high pressures (Iwasa et al., 2004; Kim et al., 1996; Wright et al., 1997; Weinstein et al., 1975; Congeduti et al., 2001). Indeed, phonon scattering in crystals of small dimension leads to a redshift and broadening of the first order Raman line. This is due to the relaxation of the q=0 selection rule when the volume objects becomes of the order of few phonon wavelengths. For nanoscale object such as nanocrystals or nanowires, the exact shape of the Raman peak becomes a convolution of the dispersion relation of phonons in the material (Richter et al., 1981; Campbell et al., 1986). Such effect was initially observed in nanocrystals and more recently in nanowires (Fauchet et al., 1988; Adu et al., 2006a; Jalilian et al., 2006; Fukata et al., 2006). In the particular case of nanowires, the confinement occurs in the diameter direction. It has been predicted and experimentally confirmed that the Raman scattering intensity $I_S(\omega,d)$ for a diameter d at a photon frequency ω relative to the laser frequency is given by (Campbell et al., 1986):

$$I_S(\omega,d) = I_O \cdot \int_0^1 2\pi q_\perp \cdot dq_\perp \cdot \frac{|C(q_\perp)|^2}{(\omega - \omega_O(q_\perp,T))^2 + \Gamma(T)^2/4} \tag{18}$$

Where $|C(q_\perp)|^2 \prec \exp(-\alpha a_O/\sqrt{2} \cdot q_\perp d)$ is the confinement function, a_0 the lattice constant of the material, α a material dependent constant, $q_\perp$ is the phonon wave vector in perpendicular to the nanowire axis, $1/\Gamma$ the phonon lifetime. In the case where Raman spectroscopy is realized on nanowire pads ensembles, it is essential to take into account of the nanowire diameter distribution for the exact modeling of the experimental curves (Adu et al., 2005). These observations extend to most of materials systems from silicon, germanium, zinc oxide, gallium phosphide, zinc sulfide... Equation (18) suggests that reducing the diameter of a cylindrical nanowire results in a redshift (towards lower frequencies) and a broadening of the Raman line. The exact shape of the spectrum is given by the equation and it would vary for other types of geometry, such as spherical or cubical nanoparticles or cylindrical or prismatic nanowires. In the case of silicon, a maximum shift of 8 cm^{-1} is observed for 4 nm nanowires (Adu et al., 2005).

In the following, we present an example of spatially resolved Raman spectroscopy measurements, indicating regions of the nanowire where the functional material achieves nanometer dimensions. Thereby, it helps to predict if it will be possible to obtain functional electronic devices with the nanowires. The samples consisted of germanium nanowires grown by chemical vapor deposition by using indium as a catalyst, the details reported elsewhere (Xiang et al., 2009). Structural analysis of the nanowires evidenced that they consisted in a crystalline core, surrounded by an amorphous shell, as shown in Fig.7a. Interestingly, it was shown that the crystalline core was not continuous along the nanowire and that it could shrink down to ~10 nm in diameter –see Fig. 7b-. The shrinking of the core poses many problems if these nanowires are to be used for electronic devices, as they will inevitably be short-circuited. A non-destructive diagnosis such as Raman can provide the information on what regions of the nanowire can be used for the devices. For that, it is necessary to realize scanning Raman spectroscopy measurements along the whole length of the nanowire. An example is shown in Fig. 7c. There, 100 nm spaced Raman spectra along an 86 nm wide germanium nanowire are shown -the diameter was obtained by measuring the height in an Atomic Force Microscope scan-. In the measurements, only the peak corresponding to the TO/LO phonon mode of crystalline germanium is observed. In the small diameter nanowires we do not observe the contribution of the amorphous band. This could be due to various effects: 1) due to the small diameter, the fraction of amorphous germanium is significantly smaller than for a nanowire with larger diameter 2) the density of the amorphous shell is smaller than that of the crystalline core. For reference, we have plotted the position of the unstrained germanium. Along the 2 µm of the scan, a recurrent shift towards lower frequencies is observed. The shift can be attributed to the phonon confinement in the core of the nanowire. The data fit well with the model in which the nanowire is assumed to have a spherical nancrystal shaper shape. According to this, the observed downshift of 6 cm^{-1} corresponds to a diameter smaller than 30 nm. The spatially resolved Raman scattering measurements indicate a variation of the core diameter along the nanowire, which are in agreement with the transmission electron micrographs realized.

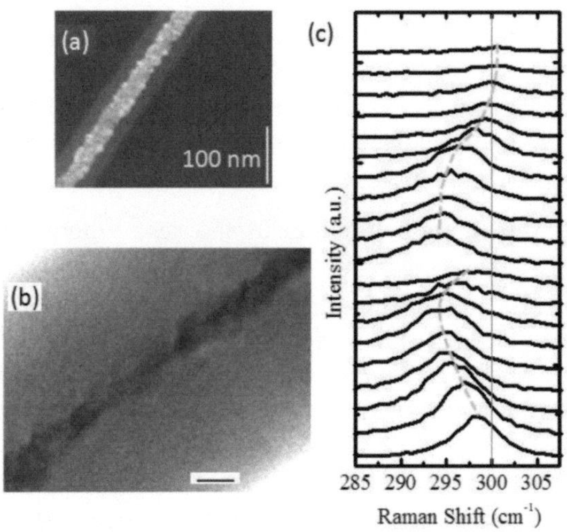

Fig. 7. a) Scanning TEM annular dark field micrograph obtained in one part of a 10 micron long germanium nanowire, showing a 40 nm multi-crystalline core, capped with a 21 nm thick amorphous layer b) Bright field TEM micrograph of a part of a germanium nanowire where the crystalline core is 5 nm. The amorphous shell is 40 nm thick,
c) Waterfall plot of Raman spectra taken every 100 nm of a thin nanowire. As a guide to the eye, the light grey line indicates the position of the TO/LO unstrainedGe mode, and the thick dashed line indicates the position of the Raman mode in the nanowire.

Relatively recent studies have shown that one should be very precautious in the analysis of quantum confinement measurements (Campbell et al., 1986; Fauchet et al., 1988). Indeed, note that in the eq. (18) the phonon frequency $\omega_0(q_\perp)$ and the phonon lifetime $1/\Gamma$ are a function of temperature. One should also note that the radiation power density incident on the nanowire increases dramatically for small diameter nanowires. Indeed, for equal diameter spot the volume of sample illuminated is proportional to d^2. As a consequence, the power density received by the sample is proportional to $1/d^2$. This immediately points out the possibility of inevitable heating in the case of extremely small diameter nanowires (the ones expected to exhibit quantum confinement). Additionally, high excitation power densities create a high density of free carriers. This is especially true for indirect bandgap semiconductors such as silicon which exhibit long recombination times. The carriers can also interfere with the phonons giving rise to Fano phenomenon and create an asymmetric line shape (Compaan et al., 1985). In the next sections we discuss these effects on the shape of the Raman spectra.

3.2 Heating effects during Raman spectroscopy measurements

A typical effect of laser irradiation on nanoscale samples is heating. This effect is amplified due to the relative increase in the power density, consequence of the sample geometry. It is also a consequence of the lower thermal conductivity of nanowires and of thermal insulation between the nanoscale object -e.g. nanowire- and the substrate (Li et al., 2003). The usual way to increase the temperature of samples during Raman spectroscopy measurements is to increase the incident irradiation power. An example of the effect on heating on the Raman spectrum is shown in Fig. 8. There, Raman spectroscopy measurements of a single GaAs nanowire as a function of the incident power density are shown. Clearly, both the TO and the LO modes become increasingly asymmetric as the incident power density is increased. A shift of the peak position towards lower wavenumbers is also clear. Between the two effects, the asymmetric broadening is the first one that arises. This tendency can be clearly seen in the graph of Fig. 8b, where the evolution of the peak positon and FWHM is shown for each excitation power.

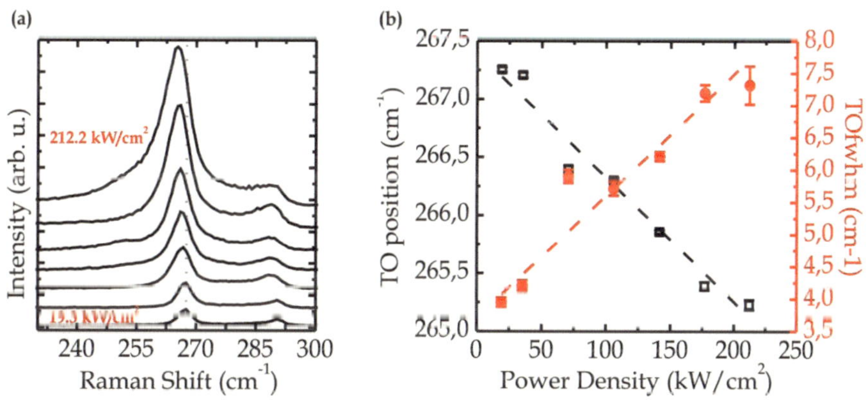

Fig. 8. Raman spectra of zinc-blende GaAs nanowire bundles collected increasing the power density from 19.3 till 212.2 kW/cm². b) Position and FWHM of the TO mode of the spectra shown in a), as a function of the power density. The dashed lines are linear fit to the data.

The temperature of the nanowire $\boldsymbol{T}$ upon laser heating is usually estimated by calculating the ratio of the integrated intensity between the Stokes and Anti-Stokes peaks -$\boldsymbol{I_S}$ and $\boldsymbol{I_{AS}}$- at the phonon frequency $\boldsymbol{\omega_o}$, which is (Balkanski et al., 1983):

$$\frac{I_S}{I_{AS}} = \exp\left(\frac{\hbar\omega_o}{KT}\right) \tag{19}$$

In the case of homogeneous heating of a material, the effect of temperature on the Raman line shape is due to: 1) the decrease in the phonon frequencies $\boldsymbol{\omega_o}$ because of thermal expansion and 2) to the increase in the inverse optical phonon lifetime $1/\Gamma$ for q=0 (Balkanski et al., 1983). For a uniform heating, the effect of temperature increase should homogeneously broaden and shift the Raman line. However, experimentally an asymmetric broadening is always observed (Jalilian et al., 2006; Piscanec et al., 2003). This observation can only be explained by the existence of temperature gradients along the nanowire. Indeed, one should consider the laser intensity distribution $I(z) = I_o e^{\left(-z/a\right)}$ - Gaussian- and the induced temperature response ***T(z)*** due to the thermal conductivity and capacity of the nanowire. Then, equation 18 is transformed in the following expression for the description of the line shape function (Adu et al. 2006b):

$$I(\omega) = \int_{-c}^{c} dz I_o e^{\left(-z/a\right)^2} \int_0^1 2\pi q_\perp dq_\perp \cdot \frac{|C(q_\perp)|^2}{(\omega - \omega_O(q_\perp, T))^2 + \Gamma(T)^2/4} \tag{20}$$

It has been demonstrated that the use of this equation for the fitting of the Raman spectra is essential to decouple the effect of thermal heating and confinement (Adu et al., 2006a, Adu et al., 2006b).

An additional consequence of the heating may be structural phase transformation. Indeed, other crystalline structures may be rendered stables at higher temperatures. This phenomenon may be accentuated by the geometry of nanoscale objects, which exhibit a much larger surface-to-volume ratio (Wickham et al., 2000). As an example, we have investigated the evolution of Raman spectra of silicon nanowires with hexagonal structure after heating them with the excitation laser –the measurements were realized at room temperature-. The hexagonal or lonsdaleite structure is a metastable phase of silicon under normal conditions, also denominated as Si-IV. It has been reported by several authors in the form of nanowires (Fontcuberta i Morral et al., 2007; Lopez et al., 2009; Arbiol et al.; 2008). Being Si-IV a metastable phase, it is expected that it may transform into diamond structure (Si-I) upon heating. In Fig. 9, the spectra of an ensemble of silicon nanowires exhibiting the Si-IV phase is shown. The Raman spectra after three annealing treatments of 200s at 60, 100 and 140 kW/cm^2 are also shown. The Raman spectra at the end of the irradiation are fitted to obtain the temperature, which corresponds to 200, 440 and 600°C. After the first

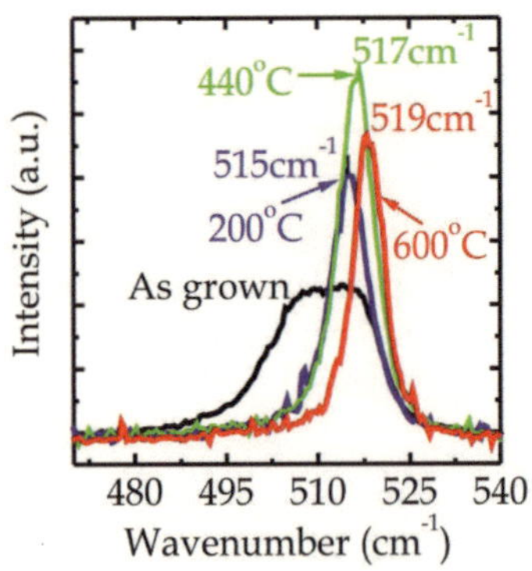

Fig. 9. Raman spectra of Si-IV nanowire bundles as grown and after heating them by illumination with the Raman objective at temperatures of 1) 200°C 2) 440°C and 3) 600°C.

treatment, the Raman peak related to the Si-IV phase disappears completely, and a new unique peak at 515cm^{-1} appears. By successive annealing, the peak shifts to 517 and 519 cm^{-1}. This indicates that the sample structure continues to consist in diamond silicon. Additionally, the correlation length of the phonons increases, in agreement with the TEM results showing an improvement in the crystalline structure -increase in grain size- (Prades et al., 2007).

3.3 Fano resonant scattering

Finally, we address another effect resulting from the use of high excitation power densities in Raman scattering experiment. High excitation power densities create a high density of free carriers, which can interfere with the phonon scattering. Fano interference in Raman scattering has been extensively studied in highly doped bulk silicon samples (Belitsky et al., 1997; Arya et al., 1979). It results in an asymmetric line shape of the first order phonon Raman peak, following the equation (Madidson et al., 2002):

$$I(\omega) = C + \sigma_o \frac{(q+\varepsilon)^2}{1+\varepsilon^2} \qquad \varepsilon = \frac{\omega - \omega_o}{\Gamma} \tag{21, 22}$$

Where $\boldsymbol{\omega}$ is the scattered photon energy, $\boldsymbol{\omega_o}$ and $\boldsymbol{\Gamma}$ are respectively the resonance frequency and width, and $\boldsymbol{\sigma_o}$ and $\boldsymbol{C}$ are constants. The influence of Fano scattering on the Raman spectra is determined by $\boldsymbol{q}$, the asymmetry parameter. In fact, it has been generally found in bulk silicon that $1/q$ is proportional to the free carrier concentration. The curve becomes Lorenzian for $q \to \infty$ and the asymmetry increases as the value gets smaller. In Fig. 10, we have plotted the shape of the Raman peak of germanium for different values of $\boldsymbol{q}$. There, it is clear that values of q of 10 start to be enough to create an asymmetry in the Raman spectrum. Experimentally, values of q between 35 and 4 have been measured for highly doped p-type bulk silicon samples (Madidson et al., 2002). In the case of undoped silicon nanowires, values of 8 and 17 have been reported (Gupta et al., 2003b). These studies have demonstrated that taking into account the effect of Fano interference, when fitting the measured Raman spectra. Indeed, Raman scattering of small diameter nanowires is not a straightforward measurement. Effects like quantum confinement, diameter distribution, inhomogeneous heating and Fano interference have to be taken into account correctly for the accurate interpretation.

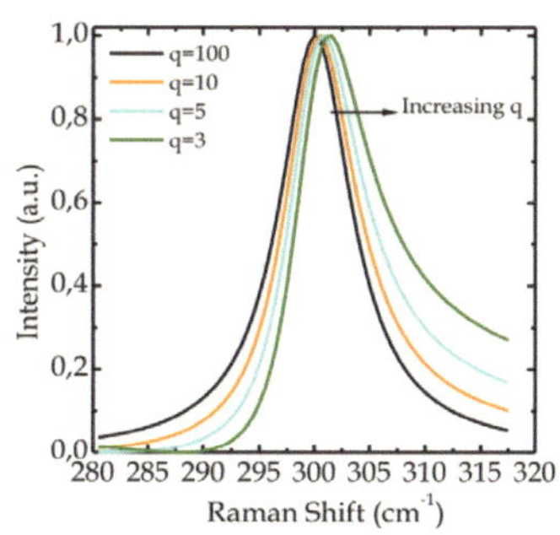

Fig. 10. Illustration of the effect of decreasing q in the asymmetry of one phonon Raman spectra of germanium -calculations following eq. 21).

4. Existence of different crystallographic phases in a nanowire: Study of GaAs nanowires with wurtzite/zinc-blende structures

Most of the binary octet semiconductors such as GaN and SiC present either zinc-blende or wurtzite structure, which correspond to the cubic and hexagonal structure with two atoms per basis. From the crystallographic point, the two structures differ only in the stacking periodicity of the atomic layers along the c-axis of the hexagonal structure. The stacking sequence is 'abcabc' for the cubic structure and 'abab' for the hexagonal one, as shown in Fig. 11 (Park et al., 1994).

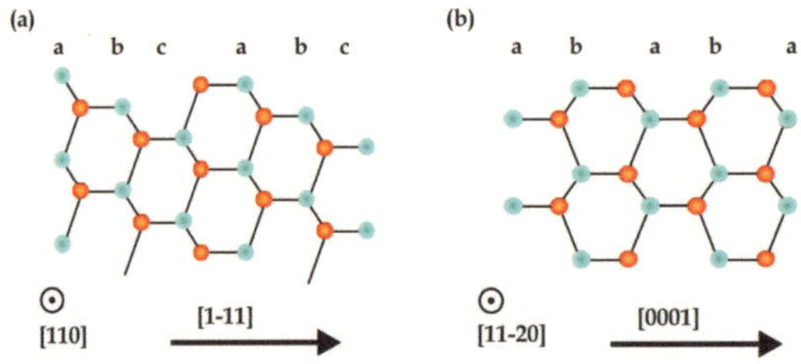

Fig. 11. Schematic drawing of the atomic arrangement in zinc-blende **(a)** and wurtzite **(b)** structures. The arrows indicate the [1-11] and the [0001] nanowire growth axes, respectively.

The spectroscopic, electronic and thermal conductivity properties of these two structures and of their polytypisms can be very different (Yeh et al., 1992). Especially interesting are structures formed by the two crystallographic phases, a sort of homo-heterostructure, which exhibit novel optical and electronic properties (Spirkoska et al., 2009). Controlled reproduction of polytypisms in materials give new degrees of freedom in the realization of electronic devices and in the structural bandgap engineering (Raffy et al., 2002; Algra et al., 2008; Mishra et al., 2007; Arbiol et al. 2009).

The different stacking order of the planes implies different symmetry groups. This, together with the slightly different lattice parameter should lead to different vibrational properties. Nevertheless - as it has been shown in the case of GaN, SiC and Si - the phonon dispersion of hexagonal structure can be deduced with good accuracy from the phonon dispersion of the cubic one by just considering the different stacking of the 'abc' and 'ab' layers (Harima, 2002; Loudon, 2001; Kobliska & Solin, 1973). The phonon dispersion of the cubic structure along the [111] direction corresponds to the **Γ→L** direction in the Brillouin zone. For clarity, we remind that the c axis of the hexagonal structure can be indexed in the 4 index Miller notation as [0001], and it is equivalent to the [111] axis of the cubic structure. The unit cell length along the [0001] axis of the hexagonal structure is double than that of the cubic structure along the [111] direction, since they correspond to the width of two and one bilayer, respectively. Consequently, the phonon dispersion of the hexagonal structure along the [0001] axis can be approximated by folding the one of the cubic structure along the [111] axis, as shown in Fig. 12 for the cases of GaN, GaAs and Si (Harima, 2002; Zardo, 2009b; Giannozzi, 1991).

As a consequence of the folding, the phonon modes at the L point are taken back at the Γ point of the Brillouin zone, giving rise to four new modes. As an example, in the case of GaAs we have the appearance of the E_2 and B_1 modes in the optical branches. In backscattering geometry Raman spectroscopy only the E_2 mode can be observed and it should be located at lower frequencies than the E_1(TO) mode. For silicon, a new optical branch appears down from the degenerate TO/LO one. One expects to observe a novel vibrational mode around ~500 cm^{-1}, 20cm^{-1} below the $q = 0$ TO/LO mode.

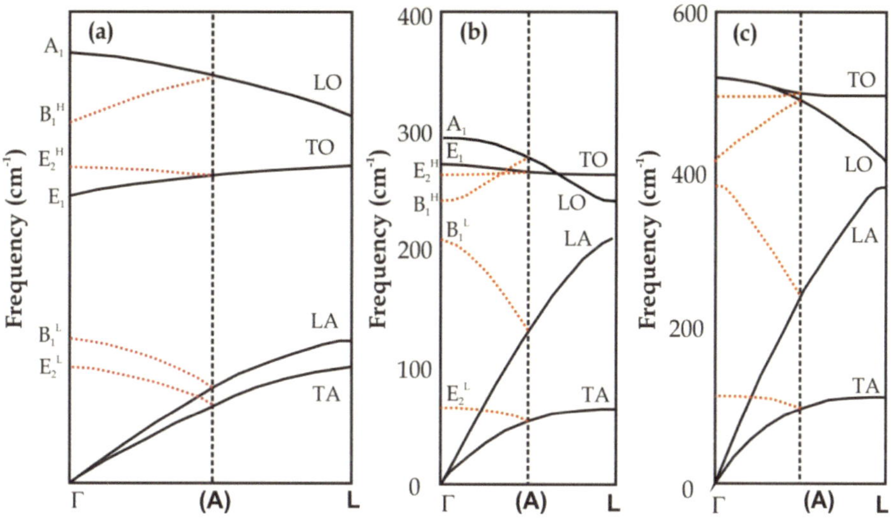

Fig. 12. Schematic representation of the phonon dispersion in GaN (a), GaAs (b) and Si (c). Phonon branches along [111] in the zinc-blende structure are folded to approximate those of wurtzite structure along [0001].

In this context, even the incidence of stacking faults and twins in nanowires gains attention and it is currently under deep investigation (Bandet et al., 2002; Lopez et al., 2009; Algra et al., 2008; Caroff et a., 2009; Zardo et al., 2009a; Conesa-Boj et al., 2009; Arbiol et al., 2009; Spirkoska et al., 2009). Indeed, the atomic stacking can be altered locally from a rotationally twin plane, so that when it occurs in a cubic nanowire gives rise to the occurrence of a monolayer of the hexagonal phase (Arbiol et al., 2009). Furthermore, twins can also cross or exist in high density, resulting into the formation of different structures, localized superstructures or heterostructure phase domains. For example, twinning superlattices are formed whenever twins occur with a certain periodicity. Additionally, the intersection of transversal and lateral twins (twins respectively along or with an angle with the growth axis) can lead to the formation of nanoscale domains with diamond hexagonal phase in the typical silicon cubic structure (Conesa-Boj et al., 2009). As already mentioned above, one should keep in mind that even their polytypisms can have very different physical properties from the pure crystalline phases (Lopez et al., 2009). As it will be shown in the following, Raman spectroscopy is a versatile technique that helps identify materials and areas in the materials with different crystal structures and/or polytypisms. The correlation with Transmission Electron Microscopy measurements can sustain and complement the information.

As an example, we show the case of GaAs nanowires with crystalline structures not stable in the bulk. The stable crystal structure for bulk GaAs is the zinc-blende. However, it has been shown GaAs nanowires can crystallize in the wurtzite structure, as shown in Fig. 13.

Fig. 14 contains an intensity map of the polarization dependent Raman spectra measured with a spacing of 100 nm along the nanowire. The incident and analyzed polarization are parallel respect to each other, and both perpendicular (Fig.14a) or parallel (Fig. 14b) to the nanowire growth axis z.

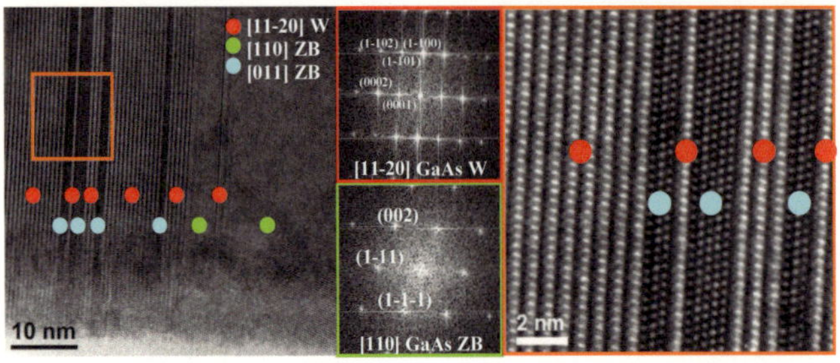

Fig. 13. HRTEM micrographs and power spectra analyses corresponding to GaAs NWs from a sample showing high content of Wurtzite and ZB regions with few monolayers.

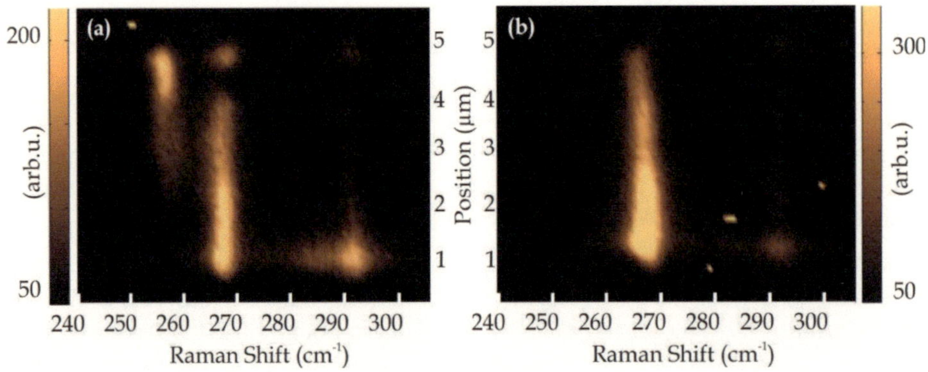

Fig. 14. Color plots showing polarized Raman scans from a nanowire consisting of 30% of wurtzite structure, obtained using different polarization direction of the incident light: a) Perpendicularly polarized Raman scan from perpendicularly polarized incident light: $x(y,y)\bar{x}$ b) parallel polarized Raman scan from parallel polarized incident light: $x(z,z)\bar{x}$.

The E_1^H (TO) mode is observable for both polarization configurations at 266.7 cm^{-1}, as expected for GaAs nanowire. When the polarization of the incident light is perpendicular to the nanowire axis, a further peak appears. This peak is positioned at about 256cm^{-1}, which corresponds to the E_2^H (TO) mode from the wurtzite GaAs phase, as a result of the folding of the E_1 (TO) branch of the phonon dispersion in the zinc blende structure, as illustrated above. The E_2^H (TO) mode intensity is higher at one end of the nanowire and decreases towards the middle, in good agreement with the percentage of wurtzite phase in the nanowire. Furthermore, in confirmation of the assignment of this peak to the E_2^H (TO) mode of the wurtzite structure, its dependence on the polarization of the excitation follows the Raman selection rules (see Fig. 5).

Another interesting feature of the measurements presented in Fig. 14 is the presence of the A_1(LO) mode. Even though the A_1 (LO) mode is not allowed for the backscattering configuration on {110} family surfaces, to which the nanowires side facets belong, it is weakly present at 290.9 cm^{-1} at one end of the nanowire. Its presence is related to the occurrence of highly dense twins in the zinc blende crystal structure, which cause that the facets of the nanowire are not of the family {110} anymore, but {111}. The A_1 (LO) mode is allowed for backscattering from {111}. The small {111} faceting at the end of the nanowire can explain the increased intensity of the A_1 (LO) phonon mode.

As a further example, we present Raman spectroscopy measurements on nanowires with a relatively high density of twins. Indeed, Raman spectroscopy is extremely sensitive even to structural defects such as the presence of dense stacking faults or twins (Lopez, 2009). Fig. 15 shows Raman spectra from silicon nanowires grown using Indium as catalyst. The nanowires present the [112] growth direction with very high density of twin defects along the {111} planes and consequent formation of hexagonal domain. In addition to the peak related to the TO/LO phonon, the presence of an additional peak at about 495 cm^{-1} can be observed (Prades et al. 2007). The existence of this peak has been explained by the presence of the highly twinned domains and hexagonal phase in nanowires (Kikkawa et al., 2005; Fontcuberta i Morral et al., 2007; Prades et al., 2007; Algra et al., 2008). The position of the band at 495 cm^{-1} coincides with the energy with the zone boundary of the phonon dispersion of silicon in the L point.

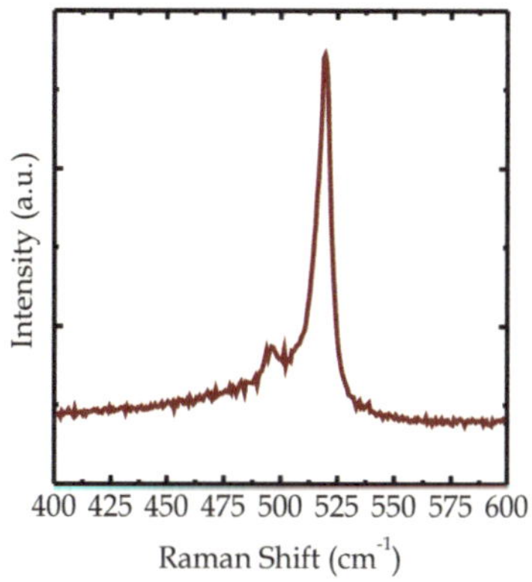

Fig. 15. Raman spectra of indium catalyzed silicon nanowires presenting high density of twins defects. Beside the TO/LO degenerate peak at 520 cm^{-1}, another peak at about 495 cm^{-1} appears.

5. Conclusions

We have presented the fundamentals for understanding Raman scattering on semiconductor nanowires. The basic physical principles of the specific phenomena related to the nanowire nature were presented. We have developed the theory and presented some experimental data on novel phenomena such as inhomogeneous heating, quantum confinement, Fano effect, the existence of surface and breathing modes and the existence of novel crystalline phases.

6. Acknowledgements

We thank the courtesy of Dance Spirkoska and J. Daniel Prades for lending us their raw data for the manuscript. Jordi Arbiol, Sonia Conesa-Boj, Francesca Peiro, Sonia Estradé and Joan Ramon Morante are greatly acknowledged for the TEM measurements and analysis. Max Bichler, Emanuele Uccelli, Sara Yazji, Norman Hauke, Ying Xiang and Mark Brongersma are kindly acknowledged for their precious technical help and fabrication of samples. We also greatly thank funding from the Marie Curie Excellence Grant 'SENFED', the DFG excellence initiative Nanosystems Initiative Munich and the California-Baviera cooperation program Bacatec.

7. References

Abstreiter G. & Ploog K. (1979). Inelastic light-scattering from a quasi-2-dimenstional electron system in GaAs-AlxGa1-xAs heterojunctions. *Phys. Rev. Lett.* 42, 1308-11, ISSN: 0031-9007

Abstreiter G., Schittenhelm P., Engel C., Silveira E., Zrenner A., Meertens D. & Jager W., (1996). Growth and characterization of self-assembled Ge-rich islands on SiSemicond. *Sci. Technol.* 11, 1521-28, ISSN: 0268-1242

Adu K.W., Gutierrez H.R., Kim U.J., Sumanasekera G.U. & Eklund P.C. (2005). Confined phonons in Si nanowires. *Nano Lett.* 5, 409 , ISSN: 1530-6984

Adu K.W., Xiong Q., Gutierrez H.R., Chen G. & Eklund P.C. (2006a). Raman scattering as a probe of phonon confinement and surface optical modes in semiconducting nanowires. *Appl. Phys. A* 85, 287, ISSN: 0947-8396

Adu K.W., Gutierrez H. R., Kim U.J. & Eklund P.C., (2006b). Inhomogeneous laser heating and phonon confinement in silicon nanowires: A micro-Raman scattering study *Phys. Rev. B* 73, 155333, ISSN: 1098-0121

Algra R.E., Verheijen M.A., Borgström M.T., Feiner L.F., Immink G., van Enckevort W.J.P., Vlieg E. & Bakkers E.P.A.M. (2008). Twinning superlattice in indium phosphide nanowires. Nature, 456, (Nov 2008) 369-372, ISSN: 0028-0836

Alvarez L., Righi A., Guillard T., Rols S., Anglaret E., Laplaze D. & Sauvajol J. (2000). Resonant Raman study of the structure and electronic properties of single-wall carbon nanotubes. *Chem. Phys. Lett.*, 316, (Jan 2000) 186-190, ISSN: 0009-2614

Anastassakis E. (1997). Angular dispersion of optical phonon frequencies in strained cubic crystals *J. Appl. Phys.*, 82, (Apr 1997) 1582, ISSN: 0021-8979

Arbiol J., Fontcuberta i Morral A., Estrade S., Peiro F., Kalache B., Roca i Cabarrocas P. & Morante J.R. (2008). Influence of the (111) twinning on the formation of diamond cubic/diamond hexagonal heterostructures in Cu-catalyzed Si nanowire. *J. Appl. Phys.* 104, (Sept 2008) 064312, ISSN: 0021-8979

Arbiol J., Estradé S., Prades J.D., Cirera A., Furtmayr F., Stark C., Laufer A., Stutzmann M., Eickhoff M., Gass M.H., Bleloch A.L., Peiró F. & Morante J.R. (2009). Triple-twin domains in Mg doped GaN wurtzite nanowires: structural and electronic properties of this zinc-blende-like stacking. *Nanotechnology*, 20, 14, (Apr 2009) 145704, ISSN: 0975-4484

Arya K., Kanehisa M., Jouanne M., Jain K. & Balkanski M., (1979). Pseudopotential calculation of the discrete-continuum interference in p-Si Raman spectra. *J. Phys. C* 12, 18, 3843-8, ISSN: 0022-3719

Balkanski M., Wallis R.F. & Haro E. (1983). Anharmonic effects in light-scattering due to optical phonons in silicon *Phys. Rev. B* 28, 4, 1928-34, ISSN: 1098-0121

Bandet J., Despax B. & Caumont M. (2002). Vibrational and electronic properties of stabilized wurtzite-like silicon. J. Phys. D: Appl. Phys., 35, (Jan 2002) 234-9, ISSN: 0022-3727

Baumgartner M. & Abstreiter G. (1984). Interaction between electronic and phonon Raman-scattering in hole space-charge layers on silicon. *Surf. Sci.*, 142, 357-60, ISSN: 0039-6028, ISSN: 0031-9007

Bawendi M.G., Wilson W.L., Rothberg L., Carroll P.J., Jedju T.M., Steigerwald M.L. & Brus . (1990). Electronic-structure and photoexcited-carrier dynamics in nanometer-size CdSe clusters. *Phys. Rev. Lett.*, 65, (Sep 1990) 1623-26, ISSN: 0031-9007

Belitsky V.I., Cantarero A., Cardona M., Trallero-Giner G. & Pavlov S. (1997). Feynman diagrams and Fano interference in light scattering from doped semiconductors. *J. Phys. Condens. Matter*. 9, (Jul 1997) 5965, ISSN: 0953-8984

Cao L.Y., Laim L., Valenzuela P.D., Nabet B. & Spanier J.E. (2007). On the Raman scattering from semiconducting nanowires. *J. Raman. Spectrosc.* 38, (Jun 2007) 697-703, ISSN: 0377-0486

Cao L.Y., Nabet B. & Spanier J.E. (2006). Enhanced raman scattering from individual semiconductor nanocones and nanowires. *Phys. Rev. Lett.*, 96, (Apr 2006) 157402-5, ISSN: 0031-9007

Cao L., White J. S., Park J. S., Schuller J. A., Clemens B. M. & Brongersma M. L. (2009). Engineering light absorption in semiconductor nanowire devices. *Nat. Mat.*, 8, (Aug 2009) 643-647, ISSN: 1476-1122

Campbell I.H. & Fauchet P.M. (1986). The effects of microcrystal size and shape on the one phonon Raman-spectra of crystalline semiconductors. *Solid State Commun.*, 58, (June 1986) 739-41, ISSN: 0038-1098

Caroff P., Dick K.A., Johansson J., Messing M.E., Deppert K. & Samuelson L. (2009). Controlled polytypic and twin-plane superlattices in III-V nanowires. *Nat. Nanotechnology,* 4, (Jan 2009) 50-55, ISSN: 0957-4484

Compaan A., Lee M.C. & Trott G.J., (1985). Phonon populations by nanosecond pulsed Raman-scattering in Si. *Phys. Rev. B* 32, 10, 6731-41, ISSN: 0163-1829

Conesa-Boj S., Zardo I., Estradé S., Wei L., Alet P.J., Roca i Cabarrocas P., Morante J.R., Peiró F., Fontcuberta i Morral A. & Arbiol J. (2009).Gallium catalyzed Silicon Nanowires: formation of lateral and transversal twin domains. Submitted to Crystal Growth & Design (2009)

Congeduti A., Postorino P., Caramagno E., Nardone M., Kumar A. & Sarma D.D. (2001). Anomalous high pressure dependence of the Jahn-Teller phonon in La0.75Ca0.25MnO3. *Phys. Rev. Lett.*, 86, (Feb 2001) 1251-4, ISSN: 0031-9007

De Franceschi S., van Dam J.A., Bakkers E.P.A.M., Feiner L.F., Gurevich L. & Kouwenhoven L.P. (2003). Single-electron tunneling in InP nanowires. *Appl. Phys. Lett.*, 83, (Jul 2004) 344-6, ISSN: 0003-6951

Duesberg G.S., Loa I., Burghard M., Syassen K. & S Roth S. (2000). Polarized Raman Spectroscopy on Isolated Single-Wall Carbon Nanotubes. *Phys. Rev. Lett.* 85, 25, (Dec 2000) 5437-9, ISSN: 0031-9007

Faist J., Capasso F., Sivco D.L, Sirtori C., Hutchinson A.L. & Cho A.Y. (1994). Quantum cascade laser. *Science,* 264 , (Apr 1994) 553-6, ISSN: 0036-8075

Fauchet P.M. & Campbell I.H. (1988). Raman-spectroscopy of low-dimensional semiconductors, *Crc. Critical reviews in solid state and materials sciences* 14, S79-101, ISSN: 0161-1593

Fischer S.F., Apetrii G., Kunze U., Schuh D. & Abstreiter G. (2006). Energy spectroscopy of controlled coupled quantum-wire states. *Nat. Phys.*, 2, (Feb 2006) 91-6, ISSN: 1745-2473

Fontcuberta i Morral A., Arbiol J., Prades J.D., Cirera A. & Morante J.R., (2007). Synthesis of silicon Nanowires with Wurtzite Crystalline Structure by Using Standard Chemical Vapor Deposition. *Adv. Mater.*, 19, (May 2007) 1347-1351, ISSN: 0935-9648

Fukata N., Oshima T., Okada N., Murakami K., Kizuka T., Tsurui T. & Ito S. (2006). Phonon confinement and self-limiting oxidation effect of silicon nanowires synthesized by laser ablation. *J. Appl. Phys.*, 100, (Jul 2006) 014311, ISSN: 0021-8979

Fréchette J. & Carraro C. (2006). Diameter-dependent modulation and polarization anisotropy in Raman scattering from individual nanowires. *Phys. Rev. B*, 74, (Oct 2006) 2161404, ISSN: 1098-0121

Giannozzi P., de Gironcoli S., Pavone P. & Baroni S., (1991). Ab initio calculation of phonon dispersions in semiconductors. *Phys. Rev. B*, 43, (March 1992) 7231-7242, ISSN: 0163-1829

Goni A.R. A. Pinczuk, Weiner J.S., Calleja J.M., Dennis B.S., Pfeiffer L.N. & West K.W., (1991). One-dimensional plasmon dispersion and dispersionless intersubband excitations in GaAs quantum wires. *Phys. Rev. Lett.*, 67, 3298-last pp, ISSN: 0031-9007

Gupta R., Xiong Q., Mahan G.D. & Eklund P.C., (2003a). Surface optical phonons in gallium phosphide nanowires. *Nano Lett.*, 3, (Dec 2003) 1745-50, ISSN: 1530-6984

Gupta R., Xiong Q., Adu C.K.,. Kim U.J & Eklund P.C. (2003b). Laser-induced Fano resonance scattering in silicon nanowires. *Nano Lett.*, 3, (May 2003) 627-31, ISSN: 1530-6984

Harima H. (2002). Properties of GaN and related compounds studied by means of Raman scattering. *J. Phys. Condens. Matter*,14, (Sep 2002) R967-R993, ISSN: 0953-8984

Hartschuh A., Sanchez E.J., Xie X.S. & Novotny L. (2003). Near-field second-harmonic generation induced by local field enhancement. *Phys. Rev. Lett.*, 90, (Jan 2003) 095503, ISSN: 0031-9007

Hartschuh A., Pedrosa H.N., Novotny L. & Krauss T.D. (2003). Simultaneous fluorescence and Raman scattering from single carbon nanotubes. *Science*, 301, 1354-6, ISSN: 0036-8075

Hu Y., Churchill H.O.H., Reilly D.J., Xiang J., Lieber C.M. & Marcus C.M. (2007). A Ge/Si heterostructure nanowire-based double quantum dot with integrated charge sensor *Nature Nanotech.*, 2, (Oct 2007) 622-5, ISSN: 1748-3387

Iwasa Y., Arima T., Fleming R.M., Siegrist T., Zhou O., Haddon R.C., Rothberg L.J., Lyons K.B., Carter H.L., Hebard A.F., Tycko R., Dabbagh G., Krajewski J.J., Thomas G.A. & Yagi T. (1994). New phases of C-60 synthesized at high pressure, *Science*, 264, (Jun 1994) 1570-2, ISSN: 0036-8075

Jalilian R., Sumanasekera G.U., Chandrasekharan H. & Sunkara M.K. (2006). Phonon confinement and laser heating effects in Germanium nanowires. *Phys. Rev. B* 74, (Oct 2006) 155421 ISSN: 1098-0121

Jorio A., Pimenta M., Souza A., Saito R., Dresselhaus G. & Dresselhaus M. (2003). Characterizing carbon nanotubes samples with resonance Raman scattering. *New J. Phys.*, 5, 139, (Oct 2003) 1-17, ISSN: 1367-2630

Kawashima Y. & Katagiri G. (1999). Observation of the out-of-plane mode in the Raman scattering from graphite edge plane. *Phys. Rev. B*, 59, 1, (Jan 1999) 62-4, ISSN: 0163-1829

Kikkawa J., Ohno Y. & Takeda S. (2005). Growth rate of Silicon nanowires. *Appl. Phys.Lett.*, 86, 123109, ISSN: 0003-6951

Kim K., Lambrecht W.R.L. & Segall B. (1996). Elastic constants and related properties of tetrahedrally bonded BN, AlN, GaN, and InN. *Phys. Rev. B*, 53, (Jun 1996) 16310-26 ISSN: 0163-1829

Kobliska R.J. & Solin S.A. (1973). Raman Spectrum of Wurtzite Silicon. *Phys. Rev. B*, 8, (Oct 1973) 3779, ISSN: 0163-1829

Krahne R., Chilla G., Schuller C., Carbone L., Kudera S., Mannarini G., Manna L., Heitmann D. & Cingolani R. (2006). Confinement effects on optical phonons in polar tetrapod nanocrystals detected by resonant inelastic light scattering. *Nano Lett.*, 6, (Mar 2006) 478-82, ISSN: 1530-6984

Lange H., Mohr M., Artemyev M., Woggon U. & Thomsen C. (2008). Direct Observation of the Radial Breathing Mode in CdSe Nanorods. *Nano Lett.*, 8, 12, (Nov 2008) 4614-7, ISSN: 1530-6984

Lehmann V. & Gosele U. (1991). Porous silicon formation - a quantum wire effect. *Appl. Phys. Lett.*, 58, (Feb 1991) 856-8, ISSN: 0003-6951

Li D.Y., Wu Y., Kim P.. Yang P. & Majumdar A. (2003). Thermal conductivity of Si/SiGe superlattice nanowires. *Appl. Phys. Lett.*, 83, (Oct 2003) 2394-8, ISSN: 0003-6951

Li H.D., Zahng S.L, Yang H.B, Zou G.T., Yang Y.Y., Yue K.T., Wu X.H. & Yan Y. (2002). Raman spectroscopy of nanocrystalline GaN synthesized by arc plasma. *J. Appl. Phys.*, 91, 7, (Apr 2002) 4562-7, ISSN: 0021-8979

Lin H.M., Chen Y.L., Yiang J., Liu Y.C., Yin K.M., Kai J.J., Chen L.C., Chen Y.F. & Chen C.C. (2003). Synthesis and Characterization of Core-Shell GaP@GaN and GaN@GaP Nanowires. *Nano Lett.*, 3, 4, (Mar 2003) 537-541, ISSN: 1530-6984

Livneh T., Zhang J.P., Cheng G.S. & Moskovits (2006). Polarized raman scattering from single GaN nanowires. *Phys. Rev. B*, 74, (Jul 2006) 035320, ISSN: 1098-0121

Long D.A., Raman Spectroscopy. McGraw-Hill, 1977

Lopez F.J., Hemesath E.R. & Lauhon L.J., (2009) Ordered Stacking Fault Arrays in Silicon Nanowires. *Nano Lett.*, 9, (Jul 2009) 2774-9 ISSN: 1530-6984

Loudon R. (2001). The Raman effect in crystals. *Adv. Phys.*,50, 7, 813-864, ISSN: 0001-8732

Magidson V. & Beserman R., (2002). Fano-type interference in the Raman spectrum of photoexcited Si. *Phys. Rev. B*, 66, (Nov 2002) 195206, ISSN: 1098-0121

Mahan G.D., Gupta R., Xiong Q., Adu C.K. & Eklund P.C., (2003). Optical phonons in polar semiconductor nanowires. *Phys. Rev. B*, 68, (Aug 2003) 073402, ISSN: 0163-1829

Maultzsch J., Telg H., Reich S. & Thomsen C. (2005). Radial breathing mode of single-walled carbon nanotubes: Optical transition energies and chiral-index assignment. *Phys. Rev. B*, 72, (Nov 2005) 205438, ISSN: 1098-0121

Mishra A., Titova L.V., Hoang T.B., Jackson H.E., Smith L.M., Yarrison-Rice J.M., Kim Y, Joyce H.J., Gao Q., Tan H.H. & Jagadish C. (2007). Polarization and temperature dependence of photoluminescence from zincblende and wurtzite InP nanowires. *Appl. Phys. Lett.*, (Dec 2007) 91, 263104, ISSN: 0003-6951

Otto A., Mrozek I., Grabhorn H. & Akemann W., J. (1992). Surface-enhanced Raman scattering. *J. Phys.: Condens. Matter*, 4, (Feb 1992) 1143-1212, ISSN: 0953-8984

Papadimitriou D. & Nassiopoulou A.G. (1998). Polarized Raman and photoluminescence study on silicon quantum wires. *J. Appl. Phys.* 84, 2, (Jul 1998) 1059-1063, ISSN: 0021-8979

Park C.H., Cheong B.H., Lee K.H. & Chang K.J. (1994). Structural and electronic properties of cubic, 2H, 4H, and 6H SiC. *Phys. Rev. B*, 49, 7, (Feb 1994) 4485-93, ISSN: 0163-1829

Pauzauskie P.J., Talaga D., Seo K., Yang P.D. & Lagugne-Labarthet F. (2005). Polarized Raman confocal microscopy of single gallium nitride nanowires. *J.A.C.S.*, 127, 49, (Dec 2005) 17146-7, ISSN: 0002-7863

Pinczuk A., Abstreiter G., Trommer R. & Cardona M. (1979). Resonance enhancement of Raman-scattering by electron-gas excitations of n-GaAs. *Sol. State Comm.*, 30, 429-32 ISSN: 0038-1098

Pinczuk A., Abstreiter G., Trommer R. & Cardona M. (1977). Raman-Scattering by wave-vector dependent coupled plasmon LO phonons of N-GaAs. *Sol. State Comm.*, 21, 959-62, ISSN: 0038-1098

Piscanec S., Cantoro M., Ferrari A.C., Zapien J. A., Lifshitz Y., Lee S.T., Hofmann S. & Robertson J. (2003). Raman spectroscopy of silicon nanowires. *Phys. Rev. B*, 68, (Dec 2003) 241312, ISSN: 1098-0121

Prades J.D., Cirera A., Arbiol J., Morante J.R. & Fontcuberta i Morral A. (2007). Concerning the 506 cm^{-1} band in the Raman spectrum of silicon nanowires. *Appl. Phys. Lett.*, 91, (Sep 2007) 123107 , ISSN: 0003-6951

Raffy C., Furthmüller J. & Bechstedt F. (2002). Properties of interfaces between cubic and hexagonal polytypes of silicon carbide. *J. Phys.: Condens. Matter*, 14, (Nov 2002) 12725-31, ISSN: 0953-8984

Raman C.V. & Krishnan K.S. (1928). A new type of secondary radiation. *Nature*, 121, (Jan 1928) 501-2, ISSN: 0028-0836

Rao A.M., Richter E., Bandow S., Chase B., Eklund P.C., Williams K.A., Fang S., Subbasawamy, M. Menon, A. Thess, R.E. Smalley, G. Dresselhaus & M.S. Dresselhaus K.R (1997). Diameter-selective Raman scattering from vibrational modes in carbon nanotubes. *Science*, 275, (Jan 1997) 187-191 ISSN: 0036-8075

Reithmaier J.P., Hoger R., Riechert H., Heberle A., Abstreiter G. & Weimann G. (1990). Band offset in elastically strained InGaAs GaAs multiple quantum wells determined by optical-absorption and electronic Raman-scattering. *Appl. Phys. Lett.*, 56, (Feb 1990) 536-8, ISSN: 0003-6951

Richter H., Wang Z.P. & Ley L. (1981). The one phonon Raman-spectrum in microcrystalline silicon. *Solid State. Commun.*, 39, 625-9, ISSN: 0038-1098

Roca E., Tralleroginer C. & Cardona M. (1994). Polar optical vibrational modes in quantum dots. *Phys. Rev. B*, 49, (May 1994) 13704 ISSN: 0163-1829

Shorubalko I., Leturcq R., Pfund A., Tyndall D., Krischek R., Schön S. & Ensslin K. (2008). Self-aligned charge read-out for InAs nanowire quantum dots. *Nano letters*, 8, (Feb 2008) 382-5, ISSN: 1530-6984

Schuller C., Biese G., Keller K., Steinebach C., Heitmann D., Grambow P. & Eberl K. (1996). Single-particle excitations and many-particle interactions in quantum wires and dots. *Phys. Rev. B*, 54, (Dec 1996) 17304-7, ISSN: 1098-0121

Shan C.X., Liu Z., Zhang X.T., Wong C.C. & Hark S.K.. (2006). Wurtzite ZnSe nanowires: growth, photoluminescence, and single-wire Raman properties. *Nanotechnology*, 17, (Oct 2006) 5561-4, ISSN: 0957-4484

Steinbach C., Krahne R., Biese G., Schuller C., Heitmann D. & Eberl K. (1996). Internal electron-electron interactions in one-dimensional systems detected by Raman spectroscopy. *Phys. Rev. B*, 54, (Nov 1996) 14281-4, ISSN: 0163-1829

Sood A.K., Menendez J., Cardona M. & Ploog K. (1985). Resonance Raman-scattering by confined LO and TO phonons in GAAs-AlAs superlattices. *Phys. Rev. Lett.* ,54, 2111-4, ISSN: 0163-1829

Somers R.C., Bawendi M.G. & Nocera D.G. (2007). CdSe nanocrystal based chem-/bio-sensors. *Chem. Soc. Rev.*, 36, 579-591 ISSN: 0306-0012

Spitzer J., Ruf T., Cardona M., Dondl W., Schorer R., Abstreiter G. & Haller E.E. (1994). Raman-scattering by optical phonons in isotopic 70Ge(n)74Ge(n) superlattices. *Phys. Rev. Lett.*, 72, (Mar 1994) 1565-8, ISSN: 0306-0012

Spirkoska D., Abstreiter G. & Fontcuberta i Morral A. (2008). Size and environment dependence of surface phonon modes of gallium arsenide nanowires as measured by Raman spectroscopy. *Nanotechnology*, 19, (Oct 2008) 435704, ISSN: 0957-4484

Spirkoska D., Arbiol j., Gustafsson A., Conesa-Boj S., Zardo I., Heigoldt M., Gass M.H., Bleloch A.L., Estrade S., Kaniber M., Rossler J., Peiro F., Morante J.R., Abstreiter G., Samuelson L. & Fontcuberta i Morral A. (2009). Structural and optical properties of high quality zinc-blende/wurtzite GaAs heteronanowires. *Submitted to Phys. Rev. B, (2009)* arXiv/cond-mat 0907.1444

Steckel J. S., Coe-Sullivan S., Bulovic V. & Bawendi M.G. (2003). 1.3 mu m to 1.55 mu m tunable electroluminescence from PbSe quantum dots embedded within an organic device. *Adv. Mater.*, 15, (Nov 2003) 18626, ISSN: 0935-9648

Thonhauser T. & Mahan G.D. (2005). Predicted Raman spectra of Si [111] nanowires. *Phys. Rev. B*, 71, (Feb 2005) 081307, ISSN: 1098-0121

Thunich S., Prechtel L., Spirkoska D., Abstreiter G., Fontcuberta i Morral A. & Holleitner A. W. (2009). Photocurrent and photoconductance properties of a GaAs nanowire. *Appl. Phys. Lett.*, 95, (Aug 2009) 083111, ISSN: 0003-6951

Turrel G. & Corset J. (1996). *Raman Microscopy Developments and Applications*, Academic Press, Harcourt Brace and Company, Malta, ISBN: 0121896900

Ulrichs E., Biese G., Steinebach C., Schuller C., Heitmann D. & Eberl K. (1997). One-dimensional plasmons in magnetic fields. *Phys. Rev. B*, 56, (Nov 1997) 12760-3, ISSN: 1098-0121

Vanwees B.J., Vanhouten H., Beenakker C.W.J., Williamson J.G., Kouwenhoven L.P., Vandermarel D. & Foxon C.T. (1988). Quantized conductance of point contacts in a two dimensional electron gas. *Phys. Rev. Lett.*, 60, (Aug 1988) 848-50, ISSN: 00031-9007

Wang J., Gudiksen M. S., Duan X., Cui Y. & Lieber C. M. (2001). Highly Polarized Photoluminescence and Photodetection from Single Indium Phosphide Nanowires. Science, 293, (Aug 2001) 1455-7, ISSN: 0036-8075

Weinstein B.A. & Piermarini G.J. (1975). Raman-scattering and phonon dispersion in si and GaP at very high-pressure. *Phys. Rev. B*, 12, 1172-86 ISSN: 0163-1829

Wickham J.N., Herhold A.B. & Alivisatos A.P. (2000). Shape change as an indicator of mechanism in the high-pressure structural transformations of CdSe nanocrystals. *Phys. Rev. Lett.*, 84, (Jan 2000) 923-6, ISSN: 0031-9007

Wright A.F. (1997) Elastic properties of zinc-blende and wurtzite AlN, GaN, and InN *J. Appl. Phys.*, 82, (Sep 2007) 2833-9, ISSN: 0021-8979

Xiang Y., Cao L., Conesa-Boj S., Estrade S., Arbiol J., Peiro F., Heiβ M., Zardo I., Morante J.R., Brongersma M.L., & Fontcuberta i Morral A. (2009). Single crystalline and core-shell indium-catalyzed germanium nanowires-a systematic thermal CVD growth study. *Nanotechnology*, 20, (Jun 2009) 245608-11, ISSN: 0957-4484

Xiong Q., Wang J., Reese O., Yoon L. C. & Eklund P.C.. (2004). Raman scattering from surface phonons in rectangular cross-sectional w-ZnS nanowires. *Nano Lett.* , 4, 10, (Oct 2004) 1991-6, ISSN: 1530-6984

Xiong Q., Chen G., Gutierrex H.R. & Eklund P.C., (2006). Raman scattering studies of individual polar semiconducting nanowires: phonon splitting and antenna effects *Appl. Phys. A* Vol. 85, (Nov 2006), 299-305, ISSN: 0947-8396

Yeh C.Y., Lu Z.W., Froyen S. & Zunger A. (1992). Zinc-blende – wurtzite polytypism in semiconductors. *Phys. Rev. B*, 46, 16, (Oct 1992)1086-97, ISSN: 0163-1829

Zardo I., Yu L., Conesa-Boj S., Estradé S., Alet P.J., Rössler J., Frimmer M., Roca i Cabarrocas P., Peiró F., Arbiol J., Morante J.R. & Fontcuberta i Morral A. (2009a). Gallium assisted plasma enhanced chemical vapor deposition of silicon nanowires. Nanotechnology, 20, (Mar 2009) 155602, ISSN: 0957-4484.

Zardo I., Conesa-Boj S., Peiró F., Morante J.R., Arbiol J., Abstreiter G. & Fontcuberta i Morral A. (2009b). Raman spectroscopy of wurtzite and zinc-blende GaAs nanowires: polarization dependence, selection rules and strain effects. Submitted to Phys. Rev. B, (2009) arXiv/cond-mat 0910.5266

Zeng H., Cai W., Cao B., Hu J., Li Y. & Liu P. (2006). Surface optical phonon Raman scattering in Zn/ZnO core-shell structured nanoparticles. *Appl. Phys. Lett.*, 88, (May 2006), 181905, ISSN: 0003-6951.

Zunke M., Schorer R., Abstreiter G., Klein W., Weimann G. & Chamberlain M.P. (1995). Angular-dispersion of confined optical phonons in GaAs/AlAs superlattices studied by micro-Raman spectroscopy. *Sol. State. Comm.*, 93, (Mar 1995) 847-51, ISSN: 0038-1098

13

Phonon Confinement Effect in III-V Nanowires

Begum N[1], Bhatti A S[1], Jabeen F[2], Rubini S[2], Martelli F[2]
[1]*Department of Physics, COMSATS Institution of Information Technology, Islamabad,*
[2]*Laboratorio Nazionale TASC INFM-CNR, S.S. 14, km 163,5, 34012 Trieste,*
[1]*Pakistan*
[2]*Italy*

1. Introduction

Novel properties offered by self assembled nanowires (NWs) based on III-V materials, make them a potential candidate for electronic and optoelectronic industry. An emerging field, "spintronic" is of marvellous importance in scientific and technological applications. Dilute magnetic semiconductors (DMS) specially the discovery of Mn doped III-As [1,2] leads the way to fabrication of semiconductor spin devices[3].
A wide number of III-V NWs have been grown by several techniques, including, metal-organic vapor phase epitaxy [8, 9] and molecular beam epitaxy (MBE) [10]. One major problem of self assembled nanowires is the presence of defects in these NWs.
Work has been reported on the analysis of defects and disorders in heterostructures, thin films and in crystals by using the phonon confinement model [11-13]. All these authors gave the qualitative analysis of defects and disorders in thin films or single crystals but to the best of our knowledge no one has estimated the defect density in a single NW.
Here, we study the structural defects present along the body of NWs carried out by means of μ-Raman scattering. A detailed analysis of the defect density in GaAs and InAs NWs and surface phonons will be presented. Phonon confinement model (PCM) will be used to fit the LO phonon peaks, which also takes into account contribution for asymmetry in the line shape due to presence of surface optical (SO) phonons and structural defects. This also allows us to determine the correlation lengths in these wires, the average distance between defects and defect density in these nanowires. Influence of these defects on SO phonon will also be investigated. Behavior of SO phonon modes by using the model presented by Ruppin and Englman will be discussed in detail.

2. Raman spectroscopy

A typical micro-Raman system consists of a microscope, excitation laser, filters, slits, diffraction grating, necessary optics, detector and post-processing software. Selection of appropriate light source for micro-Raman spectroscopy is of great importance because the Raman scattered photon have frequencies shift relative to the excitation frequencies. Raman signals are usually much weaker than the excitation intensity. For strong scattering material only one Raman scattered photon can be explored for every 10^7 incident photons. For the purpose coherent light source (laser) is preferred due to its high power, monochromatic and collimated beam nature. Laser light after entering the system is governed by neutral density (ND) filters. After passing through the spatial filter, which removes the higher order spatial modes so that the beam achieves a better focus, the light is directed to the microscope to focus on the sample. Scattered light from the sample is then collected by the objective and directed to the spectrometer. On its return path elastically scattered light is filtered by the holographic filter. Raman light is then separated by the diffraction grating into discrete wavelengths, where each frequency is measured simultaneously. Final Raman signal is then directed to the CCD detector. Assuming a static scan is made, only a finite range of wavelengths reach the CCD via the diffraction grating. Photoelectrons are created in the CCD upon exposure to the scattered beam. The dispersed beam is spread vertically across horizontal lines of pixels, which are binned, or summed, to integrate each signal. This electric signal is then processed by the integrated system software.

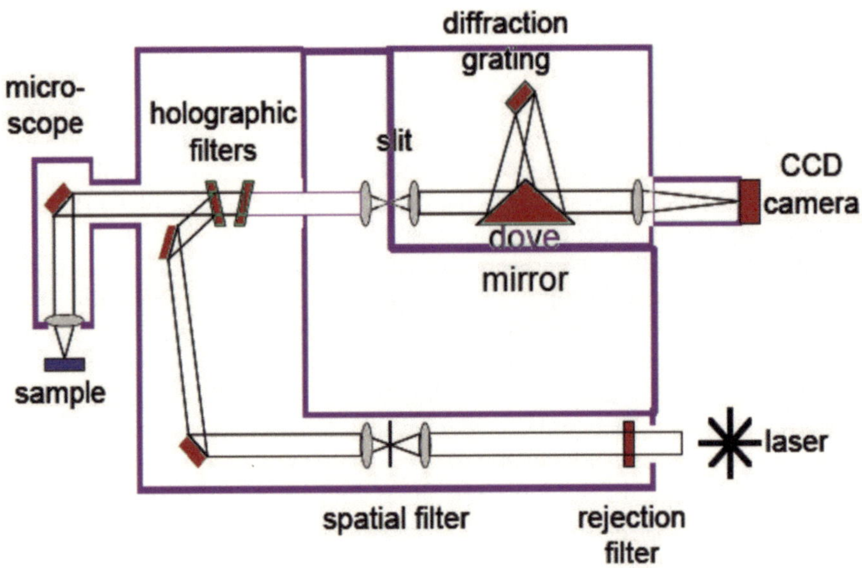

Fig. 1. Schematic of micro-Raman spectrometer. Where a collimated monochromatic light beam is focused on the sample and scattered light is directed through the optics of spectrometer and measured by CCD detector.

In the present case Raman spectra from the GaAs and InAs NWs grown on different substrates (SiO_2, GaAs (001) and GaAs (111)B substrates) with different catalysts (Au, Mn) by using MBE growth technique, were recorded at room temperature by using the Renishaw 1000 micro-Raman system equipped with CCD detector in pure backscattering geometry, with the wires lying in the plane of incidence of the incoming light. The excitation source was 514.5 nm line of an Ar^+ ion laser with a spot diameter of about 400 nm and excitation density of 1.67 mW/μm^2 for GaAs NWs and 0.56 mW/μm^2 for InAs NWs. Other details of NWs growth and Raman experimentation can be found in reference [7, 18- 21]. In order to measure the light scattered by single nanowires, after the growth, the wires were mechanically transferred onto a Si substrate. All the data presented here was obtained from single nanowire scattering. To achieve a statistically significance into the NW quality Raman spectra were taken on hundreds of wires.

3. Growth of the nanowires

GaAs and InAs NWs were synthesized by means of solid source molecular beam epitaxy at different growth temperatures ((540-620) °C for Mn-assisted GaAs NWs; (580-620) °C for Au-assisted GaAs NWs, and (390–430) °C for InAs NWs independently of the catalyst used) on SiO_2, ox-GaAs (100), and GaAs(111)B substrates. For the growth of NWs 1 nm of Mn or Au-catalyst was used. High density of NWs has been obtained on SiO_2 for either Au or Mn-catalyzed GaAs NWs (Figure 2 (a) & (b)). NW length was found as high as $20\,\mu m$ with the diameter in the range from few tens to about 200 nm. GaAs (100) and GaAs (111)B substrate gives the oriented NWs, which are much shorter than the NWs grown on SiO_2 (Figure 2 (c) & (d)). Figure 2(e) and 2(f) show the InAs NWs obtained on SiO_2 with the use of Au or Mn as catalyst, respectively.

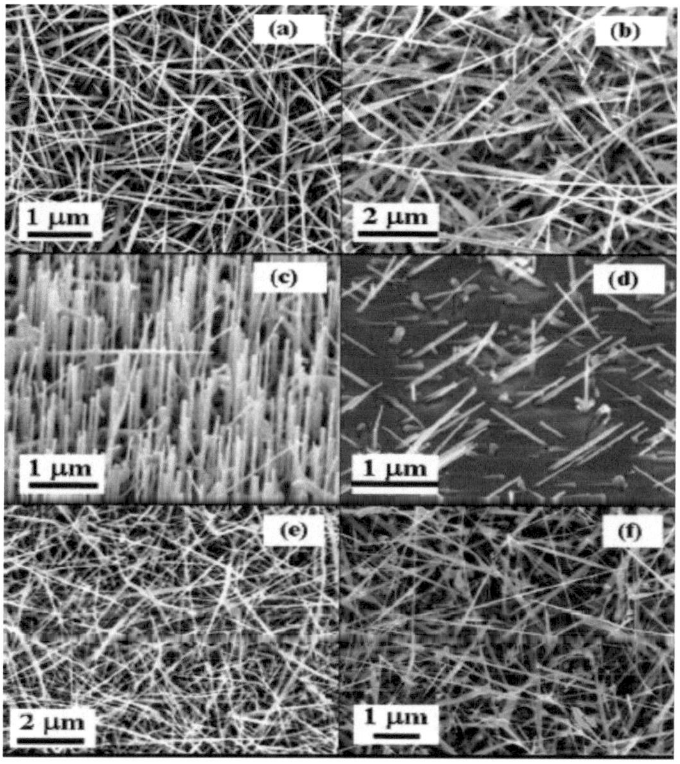

Fig. 2. SEM images of Au- and Mn-catalyzed NWs grown on different substrates: (a) Au-catalyzed GaAs NWs grown on SiO_2 substrate, (b) Mn-catalyzed GaAs NWs grown on SiO_2 substrate, (c) Au-catalyzed GaAs NWs grown on GaAs (111) *B*, (d) Au-catalyzed GaAs NWs grown on GaAs (100) substrate, (e) Au-catalyzed InAs NWs grown on SiO_2 substrate, and (f) Mn-catalyzed InAs on SiO_2. All images are plan views except (c) and (d), which were taken with the sample tilted by 45°.

4. Raman line shape analysis of GaAs and InAs semiconductor NWs

Raman spectra provide not only the basic structural information about the structure under investigation but also subtle spectra alterations can be used to access the nano-scale structural changes. In III-V and other crystals with zincblende structure, the long range Coulomb forces split the LO and TO phonons at small *k* [22]. according to Raman selection rules only LO phonon can be observed in the backscattering geometry from <100> surface. But defects and electric field affects the scattering efficiency and the line shape of LO phonons. Therefore the line shape and FWHM of Raman mode provide useful information about the quality of semiconductor nanostructures. TO phonon scattering is forbidden in the backscattering geometry. However, structural disorder, alloy disorder and impurities can break the selection rule and activate the TO phonon. A Raman spectrum is therefore, a unique tool for probing the structural defects in nanostructures. In nanostructures the Raman spectrum remains sufficiently similar to that of the corresponding bulk material to facilitate the direct identification of the material. Once the Raman spectra are known, structural defects can be characterized through the mode variation [11, 23]. Besides, the observation of any theoretically forbidden mode is a very sensitive probe of lattice distortions.

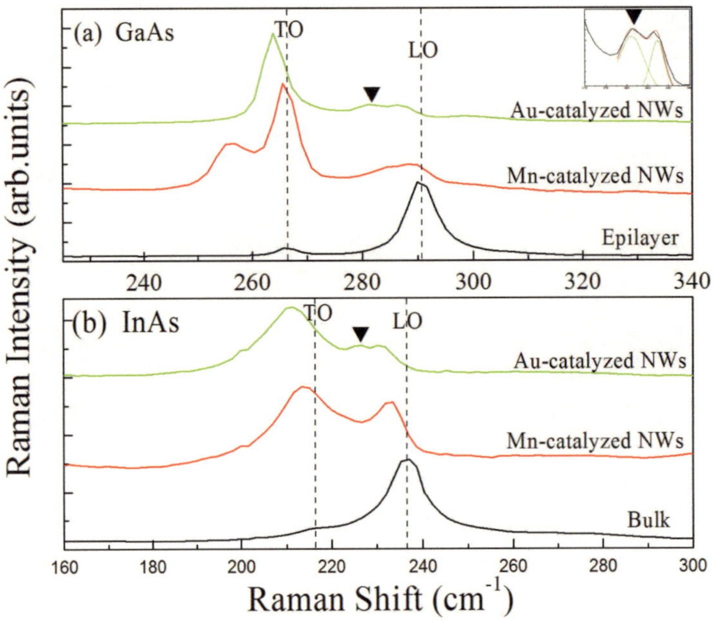

Fig. 3. Raman spectra of (a) GaAs and (b) InAs NWs in comparison of their respective bulk spectra. Surface phonon mode activated in NWs is marked by solid triangle. Inset shows the clear splitting of LO phonon mode indicating the SO phonon by solid triangle. Green curves in inset indicate the lorentzian fitting [20].

Micro-Raman spectroscopy is particularly well suited to study the III-V semiconductor NWs, because it gives good signal of the 1st order Raman spectra for both LO and TO phonons of III-V semiconductor material. Figure 3 illustrate this fact, where the representative Raman spectra of (a) GaAs NWs and (b) InAs NWs grown with two different catalysts, Au-catalyzed NWs (top spectrum), Mn-catalyzed NWs (middle spectrum) and GaAs epilayers or InAs bulk (bottom-most spectrum) are shown [20]. Raman spectra shown in Figure 3 are for the NWs grown on SiO_2 substrate. Raman spectra of GaAs epilayers gives TO peak at ~266.15 cm^{-1} and LO peak at ~ 290.59 cm^{-1}, while InAs bulk gives TO peak at

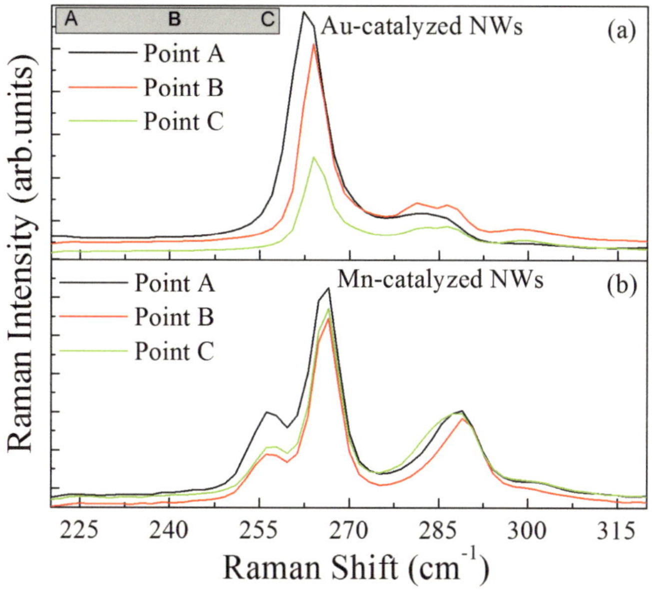

Fig. 4. Raman spectra taken at different points of single NW of (a) Au- and (b) Mn-catalyzed GaAs NWs. The inset schematically shows the single NW indicating the points where Raman spectra are taken.

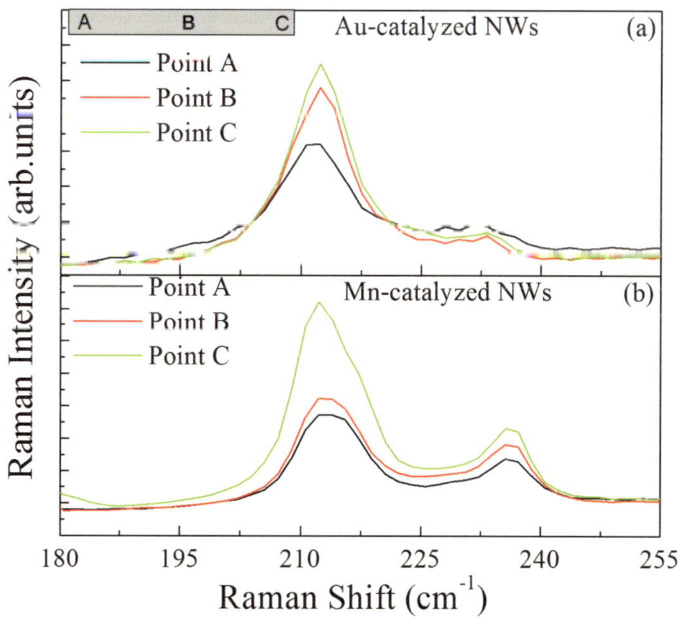

Fig. 5. Raman spectra taken at different points of single NW of (a) Au- and (b) Mn-catalyzed InAs NWs. The inset schematically shows the single NW indicating the points where Raman spectra are taken.

~217.14 cm^{-1} and LO peak at ~ 237.50 cm^{-1}. Raman spectrum of NWs shows that both TO and LO peaks are shifted toward lower energy with asymmetrical broadening. There is another peak observed between LO and TO phonon peak marked by inverted solid triangle in the Figure 3 (a) & (b), which can be related to the surface optical phonon (SO). Inset shows the clear splitting of LO phonon mode indicating the SO phonon by solid triangle. Solid black lines in the inset are the experimentally recorded data while green lines are results from a multiple Lorentzian fit. This peak is not resolvable in all the cases. Where this peak is not resolvable, LO peak shows more broadening and asymmetry due to weak contribution from the SO phonon. In those cases where this peak was not resolvable multiple Lorentzian fitting was done to resolve the LO peak in two peaks for having the contribution of SO phonons in LO peak. LO phonon lines were fitted with PCM to determine the correlation length, which also include SO phonon contribution. A detailed analysis of SO phonons and the influence of defects (in terms of correlation length) on SO phonons will be discussed later in this paper. In case of GaAs NWs there is another peak observed at ~ 255 cm^{-1} on low energy side of TO phonon.

The downshift in Raman lines observed in the wires is always larger for the LO phonon than for the TO phonon at low excitation intensities. The downshift with respect to epilayer shows variation from wire to wire and even within single nanowire. Figure 4 shows the single NW Micro-Raman spectra taken along the axis of the NW for (a) Au-catalyzed and (b) Mn-catalyzed GaAs NWs. Inset shows the schematic of the NW indicating the points where spectra has been taken. The spectra confirm that downshift varies even within a single NW. It also shows that the peak at 255 cm^{-1} is not found in all GaAs NWs. Figure 5 shows the single NW Micro-Raman spectra for (a) Au and (b) Mn-catalyzed InAs NWs. As for the shift measured for the LO phonon, however, in all those cases where no SO line is clearly distinguishable, the real LO downshift is smaller by a few cm^{-1} because of the presence of SO contribution that could not be resolved. This is in particular important for all the largest values given. Within a single wire the variation in the LO energy downshift is in general smaller, varying less than a factor of 2, except for very few wires, where the measured variation is larger.

In all of the spectra we have observed that the TO phonon appears much more intense than the LO phonon, independently of the wire orientation with respect to the plane of measurement. The intensity ratio I_{TO}/I_{LO} for NWs was found to be much larger (>1.5) than the I_{TO}/I_{LO} ratio of the GaAs epilayers (< 0.1). Similar ratios have also been reported by other groups for III-V NWs [8, 24]. The I_{TO}/I_{LO} intensity ratio strongly depends on the crystal orientation, geometry of the measurement but also on the surface electric field [see, *e.g.*, 25]. Our wires have a wurtzite crystal lattice and grown in the (0001) direction [7, 26] with large surface-to-volume ratio. Therefore, surface electric field may play an important role in determining the TO and LO line intensity.

4.1 Factors affecting the downshift and asymmetrical broadening of LO and TO phonons

Peak shift (red shift or blue shift) and asymmetrical broadening of Raman line shape may lead to the important information related to the system under investigation. Many factors can be involved in this shift and broadening of the Raman line shape. Tensile and compressive stress causes the Raman band to red shift and blue shift, respectively. Disordered and low dimensional systems (nanowires and quantum dots) lead to the asymmetrical broadening and downshift of the one-phonon Raman band. But another

factor that results in downshift and asymmetrical broadening of 1st order Raman band is the thermal heating process, which results in softening of the lattice parameter with increasing temperature and leading to the fano line shape. Detailed studies of the temperature dependence of the first order Raman band have been reported by many authors [27].
Piscanec and co-workers pointed out for Si NWs [30], heating of the wires by the impinging light lead to a downshift of the phonon energy. Similar effect was observed in present case of III-V NWs. This is shown in Figure 6(a) where we depict two spectra taken on Au-catalysed GaAs wire at two different excitation intensities. Figure 6(b) and 6(c) shows the energy downshift as a function of the excitation intensity representative for GaAs and InAS NWs, respectively. A clear saturation is observed at low intensities and below $I_0/10$ (where I_0 is the incident laser intensity) no further reduction of the downshift is observed. This

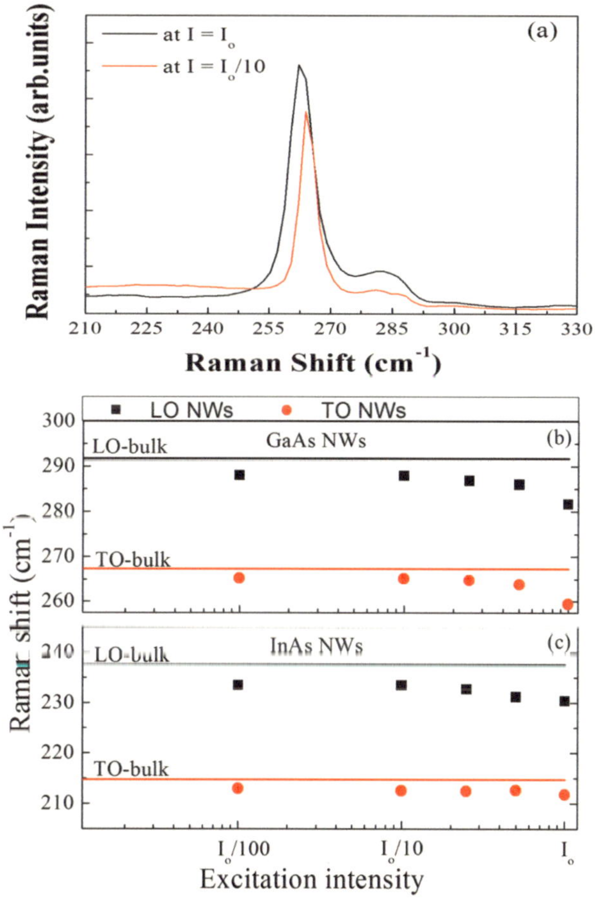

Fig. 6. (a) Raman spectra of an Au-catalyzed GaAs NW at different excitation intensities showing the TO peak shift due to heating effect: the red solid curve is taken with an excitation intensity ten times higher than the blue dotted curve. The red curve has been shifted upward for sake of clarity. (b) LO (black squares) and TO (red rhombuses) phonon line energies vs the excitation intensity. The data, given for a single wire, are representative of the general behavior. The black and red lines are the energy position of LO and TO phonons in the epitaxial GaAs. (c) The same as in (b) but for InAs.

behaviour has been observed to be quantitatively similar in all of the several wires investigated by μ-Raman belonging to both GaAs and InAs. Therefore, in order to avoid heating-induced downshift, all of the other data shown, analysed and discussed here have been recorded at excitation intensities that do not induce a measurable lattice heating in the wires. We notice that weaker light intensities are necessary to avoid heating for InAs NWs compared to GaAs NWs. This is easily explained by the fact that the band-gap of InAs is smaller than that of GaAs. The e-h pairs excited by the green light hence have a higher extra energy to be dissipated through the crystal lattice.

Small physical dimensions of the material may lead to a downshift and a broadening of the LO-Raman line because of the relaxation of the $q = 0$ selection rule. The size of our wires with diameter 20 nm to 200 nm, however, is too large to explain the observed downshift by the reduced size of the NWs. Calculations by Campbell and Fauchet [34] in the case of an infinitely long cylindrical microstructure showed that a diameter below 3 nm is required to downshift the LO feature by 4 cm^{-1}. The diameter of our NWs is 10 to 70 times larger, ruling out this type of explanation. As discussed below, we suggest that the relaxation of the $q = 0$ selection rule that can explain our data is due to the presence of lattice defects in the nanowires.

Downshift and broadening of the LO phonon line can be understood in the framework of the same "Phonon confinement model" proposed by Richter *et al.* [11] and by Tiong and coworkers [35] and generalized by Campbell and Fauchet [34] that has also been used to explain the size-induced energy downshift observed for very small NWs. This model was proposed to describe the crystalline quality by introducing a parameter known as correlation length, which is defined as the average size of the material homogeneity region. Correlation length may correspond to the actual grain size, average distance between defects, distribution of atoms in nominally disordered semiconducting alloys, and short range clustering in semiconducting alloys [11, 23, 36-37]. However, in our case the downshift of the Raman line gives account for the average size of defect-less regions, the smaller giving rise to larger downshifts.

The peak on low energy side of the TO phonons (at ~ 255cm^{-1}) can be assigned to the Raman scattering from the oxidized GaAs [38-39]. The assignment is also supported by the remark that this peak is observed with higher probability in Mn-catalyzed wires that show larger oxidized sidewalls than those catalyzed by Au [7].

5. Phonon confinement model (PCM)

One way to investigate the lattice disorders is the Raman line shape analysis using "the phonon confinement model (PCM)". This model describe the crystalline quality by the introduction of an important parameter known as "correlation length", defined as the average size of the material homogeneity region *i.e.,* the average distance between two defects [11].

Malayah *et al.* [23] study the misfit dislocation density near the interface in a GaAs/Si hetrostructure by using the Raman line-shape analysis. They calculate the broadening and frequency shift of the LO and TO phonons corresponding to the correlation length (the average distance between two defects) by using PCM as shown in Figure 7. By locating the experimental values of frequency shift corresponding value of the correlation length can easily be calculated. They also compare the values of the defect density calculated by the PCM with that of the found by TEM and prove that Raman spectroscopy is as good as that of TEM for finding the defect density in any structure.

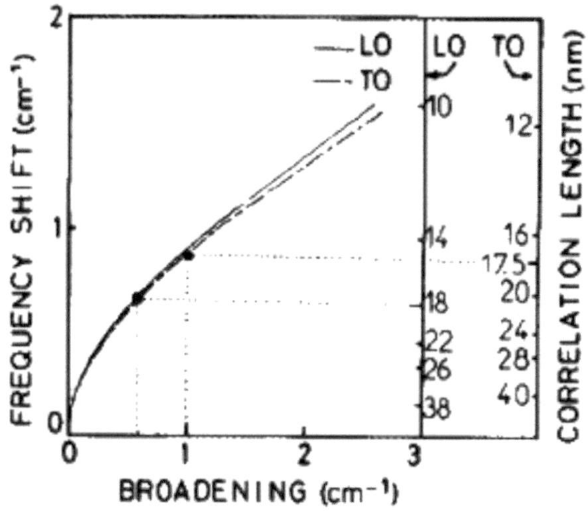

Fig. 7. Calculation performed using the PCM. The peak position shift and broadening scales are for both LO and TO modes. The dot refer to values of the broadening at the interface deduced from the fitting procedure [23].

The phonon confinement introduces additional contributions on the low frequency side of the single crystal phonon mode and resulting peak becomes asymmetric. Normally momentum conservation in single crystals allows only phonon modes with a wave vector $q \sim 0$ (close to the centre of the Brillouin zone). In phonon confinement model broadening is ascribed to the relaxation of the $q \sim 0$ selection rule in the small confinement region when vibrational modes away from the BZ centre contribute to the Raman scattering. The shift in LO peak energy is proportional to the phonon confinement region. Phonon can be confined by any "spatially limiting" feature in the confinement region (*e.g.*, twins, stacking faults, vacancies, boundaries, pores *etc*). According to the phonon confinement model (PCM), the Raman line intensity, I (ω) at the frequency ω can be written as [11]:

$$I(\omega) \cong \int_{BZ} \frac{|C(0,q)|^2 \, d^3q}{(\omega - \omega(q))^2 + \left(\frac{\Gamma_0}{2}\right)^2}, \tag{1}$$

where $|C(0,q)|^2 = e^{-\frac{q^2L^2}{16\pi^2}}$ is the Fourier coefficient, q is wave vector, L is the average distance between defects, Γ_0 is the natural line width (FWHM) and $\omega(q)$ is LO phonon dispersion. For the LO phonon frequency dispersion, the linear-chain model has been used [40], which gives

$$\omega(q) = C' \sqrt{\frac{M_1 + M_2 + \sqrt{M_1^2 + M_2^2 + 2M_1M_2Cos(qa)}}{M_1M_2}}. \tag{2}$$

For GaAs, a is the lattice constant, M_1 and M_2 are the atomic masses of Ga and As, respectively and C' is the fitting parameter. For semiconductor quantum dots (QDs), nanowires (NWs) or slabs, PCM is easily adapted using the appropriate expressions for the d^3q integration volume in equation (1).

Figure 8 shows calculated line shape of LO-phonon lines of GaAs by using the equation (1) and equation (2). The parameters used in equation (2) were M_1=69.723amu for Ga, M_2=74.922 amu for As and a=5.6533Å at 300 K for GaAs. The curves shown in Figure 8 were calculated for correlation lengths (*L*) of 30 nm, 20 nm, 10 nm, 5 nm and 3 nm. The value of Γ_0 was taken ~4.5cm^{-1} in the present case. Figure 8 shows that as the correlation length is decreasing (*i.e.*, confinement is increasing), LO line shape is more and more asymmetric along with shift towards low energy side. This clearly indicates that the downshift of the peak energy and asymmetrical broadening of the peak is mainly due to the contribution of the low energy phonons.

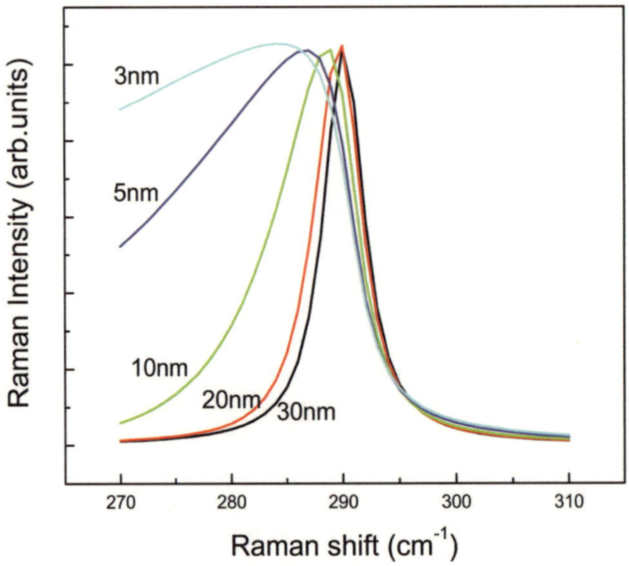

Fig. 8. Calculated Raman line shape of GaAs for different correlation lengths as calculated by using phonon confinement model [11] for GaAs. At 20nm and above 20nm the line shape is almost Lorentzian as the correlation length decreases below 20nm the line shape becomes asymmetric and downshifts

5.1 Defect density in GaAs and InAs NWs

Figures 9 (a) and (b) show the representative of the fitting of experimental LO phonon mode using equation 1 for the case of GaAs NWs (Figure 9 (a)) and InAs NWs (Figure 9 (b)). The fit shown in Figure 9 (a) gives the correlation length " $L=(10.00\pm0.41)nm$ " in Mn-catalyzed GaAs NWs. The fit shown in Figure 9 (b) gives the correlation length " $L=(16.00\pm0.35)nm$ " in Mn-catalyzed InAs NWs showing the good agreement of experimental data with theoretical curves.

The correlation lengths (*L*) obtained from LO fit for different samples of GaAs NWs were found in a range of 2 nm to 20 nm as summarized in Figure 10 (a) along with curves calculated by using equation 1. In our case, NWs lengths are as large as 20 µm while the diameters range from few tens to about 200 nm, therefore, by considering the nanowire length equal to 20 µm and the range of correlation length from 2 nm to 20 nm, defect (stacking faults) density would correspond to 0.5 to 0.05 defects/nm, respectively for GaAs NWs.

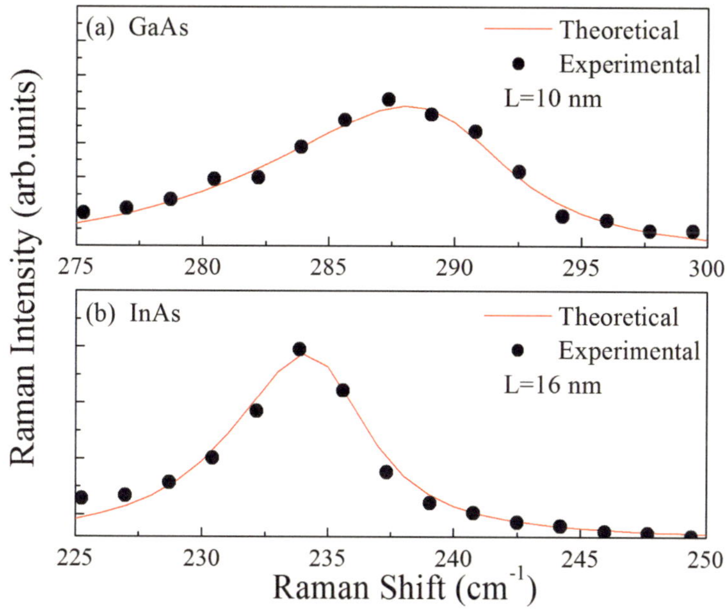

Fig. 9. Representative of the LO band in the Raman spectrum of Mn-catalyzed (a) GaAs NWs (b) InAs NWs along with the theoretical fitted curve by using the "phonon confinement model". The correlation length (L) =10 nm for GaAs and 16 nm for InAs NWs.

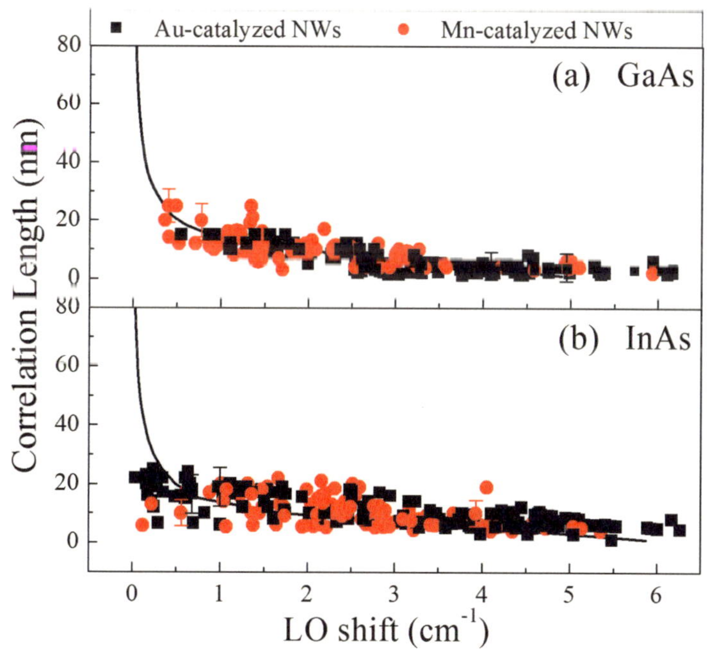

Fig. 10. Correlation length versus LO peak shift for Mn-catalyzed (dots) and Au-catalyzed (squares) (a) GaAs NWs & (b) InAs NWs in comparison of the theoretically calculated curve for GaAs and InAs bulk, respectively.

Figure 10 (b) shows the correlation length versus LO peak shift for Mn-catalyzed (dots) and Au-catalyzed (squares) InAs NWs in comparison of the theoretically calculated curve for InAs bulk. In case of InAs NWs correlation length was found to be in the range of $(1.0 \pm 0.8)nm$ to $(25.0 \pm 1.5)nm$ for Au-catalyzed and $(1.3 \pm 1.0)nm$ to $(22.0 \pm 0.8)nm$ for Mn-catalyzed NWs. Similar to GaAs NWs taking the InAs NW length equal to 20 μm and correlation length from 1 nm to 25 nm defect density (staking faults) corresponds to 1.00 to 0.04 defects/nm, respectively.
The fluctuations of the downshifts of the Raman line can then be understood to be due to variations in the defect density within the same growth batch or different batches, and in the same wire [18]. A similar explanation of the downshift of the Raman lines in Si NWs was given in the paper by Li and co-workers [41]. The smaller shift of the TO phonon is due to its smoother dispersion curve [42]. From the energy downshift the average dimension of the defect-free regions can be deduced.

6. Surface optical (SO) phonons

Surface optical (SO) phonons, activated due to vibrations confined to near-surface region, are observed in the compound semiconductors when the translational symmetry of the surface potential is broken. So far, SO phonon modes have been studied in thin films [43] and in micro-crystals [44], but not much literature is available on SO phonons in nanowire Gupta *et al.* [45] studied SO phonons in GaP nanowires. They clearly show that the SO phonon peak becomes more pronounced and downshifted with the increase of dielectric constant of surrounding medium. They also identified the symmetry breaking mechanism necessary for activating the SO mode with a diameter modulation along the wire, which arises from instability in the vapor-liquid-solid (VLS) growth mechanism. The dominant wavelength of this modulation was observed in TEM and found to be consistent with the theory for the band position.
Xiong et al. [46] studied the SO phonons in wurtzite ZnS nanowires. They studied the highly crystalline, rectangular cross-sectional W-ZnS nanowires grown by laser ablation. Similarly to the work of Gupta, they show that the downshift of the SO phonon frequency depends on the dielectric constant of surrounding medium. They also show that the dispersion of SO modes in wires with rectangular or circular cross section differs.
Spirkoska et al. [47] studied the effect of size and surrounding medium on the self catalytic GaAs NWs grown by MBE technique. They found that SO phonons are activated in the NWs with average diameter of 160 nm. By comparing the experimental data with that of theoretical curve they observe the larger shift in the SO phonon position especially for larger diameter. This larger shift was related to the hexagonal cross-section of the nanowires.
Above mentioned studies refer to the SO phonon mode activation due to the confinement along the diameter of the nanowires using the simple model based on electromagnetic theory proposed by Ruppin and Englman [49]. According to that model surface phonon frequencies in the cylinder can be calculated by the expression [49]:

$$\omega_{SO} = \omega_{TO}\sqrt{\frac{\varepsilon_o - \varepsilon_m \rho}{\varepsilon_\infty - \varepsilon_m \rho}}, \tag{3}$$

where ω_{SO} is surface phonon frequency; ω_{TO} is TO phonon frequency; ε_o, ε_∞ are static dielectric and dynamic dielectric constants, respectively; ε_m is dielectric constant of the surrounding medium and

$$\rho = \frac{[K_1(x)I_0(x)]}{[K_0(x)I_1(x)]} \tag{4}$$

where I_n, K_n, are the modified Bessel functions and $x = qr$ (r being the radius of the NW). This model was successfully applied to describe the SO phonon spectra of microcrystalline spheres [44], GaAs NWs [47], Si NWs [48], and GaAs cylinders [50].

6.1 Surface optical phonons in GaAs and InAs NWs

There is another peak observed between LO and TO phonon peak marked by inverted solid triangle in the Figure 3 (a) & (b). We ascribe it as the surface optical phonon (SO). Inset shows the clear splitting of LO phonon mode and the SO phonon. Solid black lines in the inset are the experimentally recorded data while green lines are multiple peaks Lorentzian fit. This peak is not resolvable in all the cases. In case, this peak is not resolvable, LO peak shows more broadening and asymmetry probably due to weak contribution from the SO phonon. LO phonon lines were fitted with PCM to determine the correlation length, which also include SO phonon contribution.

SO modes are found to be activated in the cylinders with large surface to volume ratio and are simply related to the characteristic shape of the cylinder [47, 48, 50]. SO phonons are sensitive to the crystal defects and surface roughness. The position of the SO phonon modes depends on the dielectric constant of the medium that is surrounding the wires as well as on the diameter of the wires. In large crystals, phonons propagate to infinity and the 1st order Raman spectrum only consists of $q \approx 0$ phonon modes. When crystalline perfection is destroyed due to lattice disorder and defects, symmetry forbidden modes (like SO phonon modes) are activated and become stronger with increasing defect density.

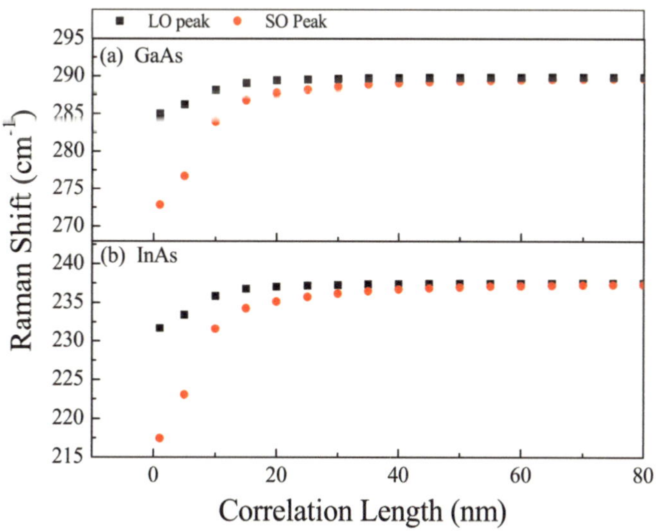

Fig. 11. Variation of LO and SO peaks of (a) GaAs and (b) InAs versus the correlation length as calculated by using the phonon confinement model (PCM).

SO phonons are activated when surface potential along the axis of the NW is perturbed by a strong component of wavevector q. Considering $q = 2k = 4\pi/\lambda_{ex}$ be the same as scattering vector in the back scattering geometry. Then for λ_{ex} = 514.5 *nm* corresponding value of q will be 0.0244 nm^{-1}. By using the peak position of SO phonons the corresponding diameter of NWs can easily be found from equations (3) and (4). If we take for GaAs NWs average value of $\omega_{SO(exp)}$ = 283.61 cm^{-1} then corresponding diameter of NW ($d = 2r = 2(x/q)$ calculated from equations (3) and (4) is ~45 *nm*, which lies in spread in the range of NWs diameter measured by SEM [7, 18, 26]. Similarly for InAs NWs with average $\omega_{SO(exp)}$ = 228.84 cm^{-1} it is found the corresponding diameter of NW ~88 *nm*.

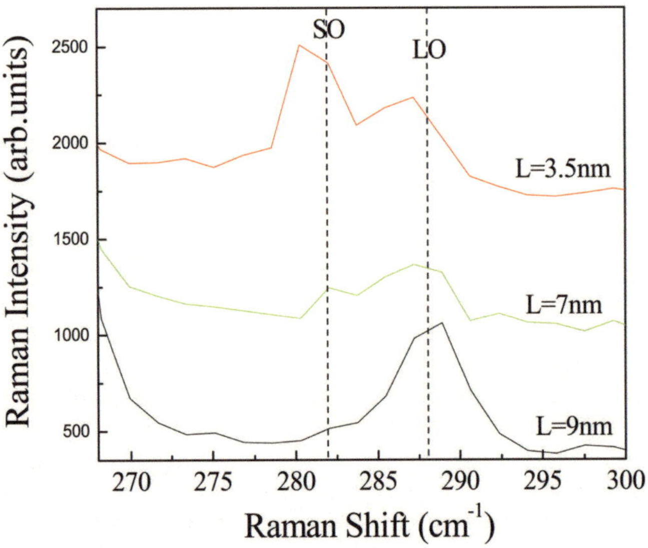

Fig. 12. Raman spectra of GaAs NWs for different correlation lengths showing that as the correlation length decreases, the SO phonon becomes stronger.

Many factors perturb the surface potential necessary for the SO phonon activation. This includes surface roughness of the NWs and the modulation of diameter along the growth direction during the VLS growth of the NWs [48]. During VLS growth the NWs grow continuously with the periodically modulated growth rate due to periodic change/modulation of available chemical species in the catalyst droplet. As a result of this modulation droplet will reshape periodically between nearly spherical and nearly ellipsoidal resulting in the formation of polytypic structure in the NWs and thus the stacking faults and twinning in the body of the NW [51]. Adu *et al.* [48] believe that the observation of SO modes should be a general indication of diameter modulation in the NWs.

In order to see the effect of defects (probably stacking faults in the present case) multiple curve fitting of the LO peak was carried out to determine the SO peak position. Correlation length corresponding to the curve containing both LO and SO elements was calculated by using the PCM. Figure 11 shows the behaviour of LO and SO phonons with correlation length for GaAs and InAs. The frequencies of SO phonon modes are found to be more sensitive to correlation length than LO phonon and for $L \leq 30$ *nm*, SO phonon energy can easily be separated from that of the LO phonon energy. When L varies from 10 nm to 1 nm

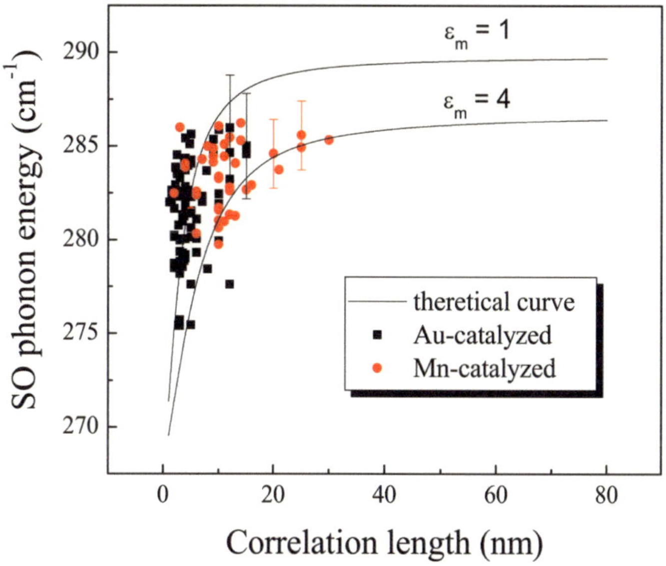

Fig. 13. Comparison of experimental and theoretical results for SO phonon vs correlation length. Solid line indicates the curve calculated by using the model presented by Ruppin et al [49] for two different values of the dielectric constant of the surrounding medium. Experimental data points for Mn-catalyzed GaAs NWs (dots) and Au-catalyzed GaAs NWs (squares) located on the theoretically calculated curve for the SO phonons in terms of correlation length

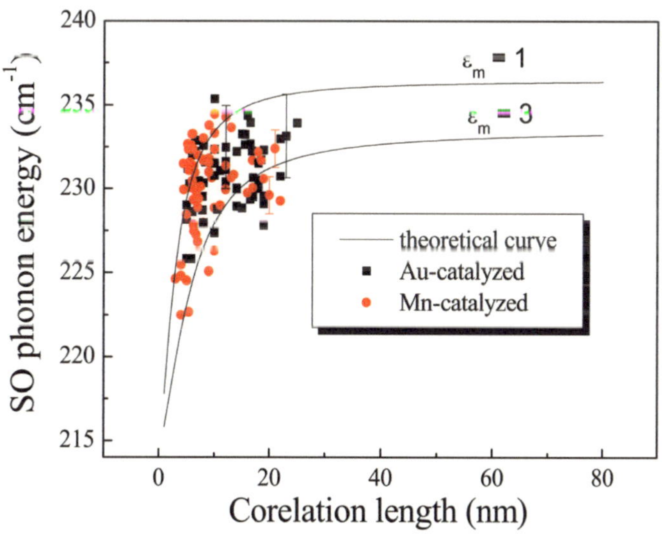

Fig. 14. Comparison of experimental and theoretical results for SO phonon vs correlation length. Solid line indicates the curve calculated by using the model presented by Ruppin *et al.* [49] for two different values of the dielectric constant of the surrounding medium. Experimental data points for Mn-catalyzed InAs NWs (dots) and Au-catalyzed InAs NWs (squares) located on the theoretically calculated curve for the SO phonons in terms of correlation length

SO phonon shows downshift of 11.1 cm^{-1}, while LO phonon shows downshift of only 3.2 cm^{-1} for GaAs. In case of InAs for variation of L from 10 nm to 1 nm, SO phonon shows downshift of 14.2 cm^{-1} and LO phonon shows downshift of 4.2 cm^{-1}. The energy separation between the LO and SO phonons of GaAs is found in the range of 0.10 cm^{-1} to 12.15 cm^{-1} (Figure 11) and 0.09 cm^{-1} to 14.22cm^{-1} for InAs, corresponding to correlation lengths of 100 nm to 1 nm. In case of NWs, where the energy separation was measurable, its value varies from (2.0 ± 0.4) cm^{-1} to (8.4 ± 0.3) cm^{-1} for GaAs NWs and (0.95 ± 0.51) cm^{-1} to (8.79± 0.53) cm^{-1} for InAs NWs, which lies in the calculated range. It also supports our argument that the peak located between the TO and the LO peaks can indeed be ascribed to Raman scattering from surface phonons.

Figure 12 shows the representative Raman spectra of the Au-catalyzed GaAs NWs indicating distinct appearance of SO phonons. The line shape analysis of these spectra gives correlation lengths of L = 9 *nm*, 7 *nm* and 3.5 *nm*. In Figure 12, it can be seen that for smaller correlation lengths the SO phonon peak is stronger than LO phonon peak and shifted downward in energy as was predicted by Ruppin *et al.* [49]. Figure 13 and 14 shows the experimental data points of GaAs NWs (Figure 13) and InAs NWs (Figure 14) for Mn-catalyzed NWs (dots) and Au-catalyzed NWs (squares) located on the theoretically calculated curves by using the equation (3) for the SO phonons as function of correlation length. In case of GaAs NWs maximum data of Au-catalyzed NWs shows good agreement for ε_m = 1, while for the case of Mn-catalyzed NWs, ε_m lies between 1 and 4. In case of InAs NWs, ε_m lies between 1 and 3 irrespective of the catalyst used. This variation of the dielectric constant can be justified by considering the oxide layer that surrounds the NWs.

As reported in Sec. 3, for the case of GaAs NWs, Au-catalyzed NWs have very thin oxide layer of only 1 nm thickness and Mn-catalyzed NWs have thick oxide layer of 4 nm [7, 18]. Similar to GaAs NWs thick oxide layer (2 nm to 3 nm) is surrounding the InAs NWs in both Au and Mn-catalyst [19]. This oxide layer consists of oxides of Ga, In and As, which have larger dielectric constant. This variation in the thickness of oxide layer may cause the variation in dielectric constant.

7. Conclusions

In this chapter, Raman characterization of GaAs and InAs NWs was carried out. GaAs and InAs NWs grown by MBE technique on different substrates at different temperatures by using Au and Mn as catalyst were characterized by Raman spectroscopy for their optical and structural quality. Raman spectra of GaAs and InAs NWs shows the downshift of LO and TO phonons along with asymmetrical broadening. This downshift and asymmetrical broadening of LO and TO phonons was related to the defects present in the body of the NWs. Variation in the downshift of LO and TO phonons was explained in terms of the defect density. Oxide related peaks on low energy side of the TO phonon were also observed in both Au and Mn-catalyzed GaAs NWs.

Phonon confinement model was used to calculate the average distance between the defects (*i.e.,* correlation length) and the defect density. Correlation lengths determined were (2.0 ± 0.1) nm to (15.0 ± 0.5) nm for Au-catalyzed and (2.0 ± 0.8) nm to (20.0 ± 1.5) nm for Mn-catalyzed GaAs NWs. In case of InAs NWs, correlation length was found to be in the range of (1.0 ± 0.8) *nm* to (25.0 ± 1.5) *nm* for Au-catalyzed and (1.3 ± 1.0) *nm* to (22.0 ± 0.8) *nm* for Mn-catalyzed NWs. The ccorrelation lengths in InAs and GaAs NWs were found comparable in both Au and Mn-catalyzed NWs.

SO phonons were also found to be activated in both types of NWs. In samples with shorter correlation length SO phonons are more pronounced and stronger as was predicted by electromagnetic theory. An important conclusion drawn from the present study is that the structural quality of Mn-catalyzed NWs is comparable to that of Au-catalyzed wires, a feature that confirms Mn as an interesting alternative catalyst, especially in view of the possible fabrication of dilute ferromagnetic III-V based nanowires. Our results prove that Raman spectroscopy is a useful tool for rapid screening of the structural quality of semiconductor nanowires. This type of screening is important to give a picture of the real quality of the NWs growth.

8. Acknowledgements

N.B is obliged to the Abdus Salam *International Centre for Theoretical Physics* (ICTP), Trieste, Italy, for providing her funding and opportunity to work at TASC LABS, under TRIL program. N.B. would also thank to National Engineering and Scientific Commission (NESCOM), Pakistan, for providing her financial assistance for her PhD program.

9. References

[1] H. Ohno, H. Munekata, T. Penney, S. von Molnar and L.L Chang, Phys. Rev. Lett. 68 (1992) 2664.
[2] H. Ohno, A. Shen, F. Matsukura, A. Oiwa, A. Endo, S. Katsumoto and Y. Iye, Appl. Phys. Lett. 69 (1996) 363.
[3] G.A Prinz, Phys. Today 48 (1995) 58.
[4] Doo Suk Han and Jeunghee Park, Kung Won Rhie, Soonkyu Kim and Joonyeon Chang, Appl. Phys. Lett. 86 (2005) 032506.
[5] Y. Q. Chang, D. B. Wang, X. H. Luo, X. Y. Xu, X. H. Chen, L. Li, C. P. Chen, R. M. Wang, J. Xu, and D. P. Yu, Appl. Phys. Lett. 83 (2003) 4020.
[6] S. J May, J. G. Zheng, B. W Wessels, L. J Lauhon, Adv. Mater 17 (2005) 598.
[7] F. Martelli, S. Rubini, M. Piccin, G. Bais, F. Jabeen, S. De Franceschi, V. Grillo, E. Carlino, F. D'Acapito, F. Boscherini, S. Cabrini, M. Lazzarino, L. Businaro, F. Romanato and A. Franciosi, Nano Letters 6 (2006) 2130.
[8] W. S. Shi, Y. F. Zheng, N. Wang, C. S. Lee, and S. T. Lee, Appl. Phys. Lett. 78 (2001) 3304.
[9] B. Mandl, J. Stangl, T. Mårtensson, A. Mikkelsen, J. Eriksson, L. S. Karlsson, G. Bauer, L. Samuelson, and W. Seifert, Nano Lett. 6 (2006) 1017.
[10] A. L. Roest, M. A. Verheijen, O. Wunnicke, S. Serafin, H. Wondergem, and E. P. A. M. Bakkers, Nanotechnology 17 (2006) S271.
[11] H. Richter, Z.P. Wang, and L. Ley, Solid State Commun. 39 (1981) 625
[12] J. Chen, W. Z. Shen , J. B. Wang, H. Ogawa and Q. X. Guo, Journal of Crystal Growth, 262 (2004) 435-441
[13] Masahiro Kitajima, Critical Reviews in Solid State and Materials Sciences, 22 (1997) 275 - 349
[14] C. V. Raman and K. S. Krishnan, Nature, 121 (1928) 501
[15] G. Abstreiter, E. Bauser, A. Fischer, K. Ploog, Appl. Phys. Lett. 16 (1978) 345
[16] A.J. Pal, J. Mandal, Journal of Alloys and Compounds 216 (1994) 265-267
[17] J. B. Hopkins, L. A. Farrow and G. J. Fisanick, Appl. Phys. Lett. 44 (1984) 535
[18] S. Rubini, M. Piccin, G. Bais, F. Jabeen, F. Martelli and A. Franciosi, Journal of Physics: Conference Series 61 (2007) 992–996

[19] F. d'Acapito, M. Rovezzi, F. Boscherini, F. Jabeen, M. Piccin, L. Felisari, V. Grillo, G. Bais, S. Rubini, and F. Martelli, unpublished
[20] N Begum, M Piccin, F Jabeen , G Bais , S Rubini, and F Martelli, and A S Bhatti, J. Appl. Phys., 104 (2008) 104311
[21] N Begum, A S Bhatti, M Piccin, G Bais, F Jabeen, S Rubini, F Martelli and A Franciosi, Advanced Materials Research, 31 (2008) 23
[22] G. Burns, C, R. Wie, F. H. Dacol, G. D. Pettit, and J. M. Woodall, Appl. Phys. Lett. 51 (1987) 1919
[23] A. Mlayah, R. Carles, G. Landa, E. Bedel, C. Fontaine, and A. Munoz-Yague, J. Appl. Phys. 68 (1990) 4777
[24] G. D. Mahan, R. Gupta, Q. Xiong, C. K. Adu and P. C. Eklund, Phys. Rev. B 68 (2003) 073402
[25] H. Brugger, F. Schäffler and G. Abstreiter, Phys. Rev. Lett. 52 (1984) 142
[26] M. Piccin, G. Bais, V. Grillo, F. Jabeen, S. De Franceschi, E. Carlino, M. Lazzarino, F. Romanato, L. Businaro, S. Rubini, F. Martelli and A. Franciosi, Physica E 37 (2007) 134.
[27] T. R. Hart, R.L. Aggarwal, and B. Lax, Phys. Rev. B, 1(1970) 638-642
[28] G. E. Jelison, D. H. Lowndes and R. F. Wood, Phys. Rev. B 28 (1983) 3272
[29] G. D. Pazionis, H. Tang, I. P. Herman, IEEE J. Quantum Electron, QE-25 (1989) 976
[30] S. Piscanec, M. Cantoro, A.C. Ferrari, J.A. Zapien, Y. Lifshitz, S.T. Lee, S. Hofmann, and J. Robertson, Phys. Rev. B 68 (2003) 241312
[31] A. Freundlich, G. Neu, A. Leycuras, R. Carles, and C. Verie, In Hetroepitaxy on Silicon: Fundamentals, Structure and Devices, Material Research society Symposia Proc. 116 (1988)
[32] S. A. Lyon, R. J. Nemanich, N. M. Johnson, and D. K. Biegelson, Appl. Phys. Lett. 40 (1982) 316
[33] G. Landa, R. Carles, C. Fontaine, E. Bedel, and A. Muñoz-Yagüe, J. Appl. Phys. 66 (1989) 196
[34] I. H. Campbell and P. M. Fauchet, Solid State Commun. 58 (1986) 739
[35] K.K. Tiong, P.M. Amirtharaj, F.H. Pollak, and D.E. Aspnes, Appl. Phys. Lett. 44 (1984) 122
[36] P. Parayanthal and Fred H. Pollak, Phys. Rev. Lett. 52 (1984) 1822.
[37] A. Fischer, L. Anthony, and A. D. Compaan, Appl. Phys. Lett. 72 (1998) 2559
[38] R. L. Farrow, R. K. Chang, S. Mroczkowski, and F. H. Pollak, Appl. Phys. Lett. 31 (1977) 768.
[39] J. A. Cape, W. E. Tennant, and L. G. Hale, J. Vac. Sci. Technol. 14 (1977) 921.
[40] C. Kittle, Introduction to Solid state physics (Wiley, New York, 2003), 7th edn. Chapt. 4.
[41] Bibo Li, Dapeng Yu, and Shu-Lin Zhang, Phys. Rev. B 59 (1999) 1645.
[42] R. Shuker and R.W. Gammon, Phys. Rev. Lett. 25 (1970) 222 S.
[43] S. Ushioda, A.Aziza, J. B. Valdez, and G. Mattei, Phys. Rev. B 19 (1979) 4012.
[44] S. Hayashi and H. Kanamori, Phys. Rev. B 26 (1982) 7079.
[45] Rajeev Gupta, Q. Xiong, G. D. Mahan, and P. C. Eklund, Nano Lett., 3 (2003) 1745
[46] Qihua Xiong, Jinguo Wang, O. Reese, L. C. Lew Yan Voon, and P. C. Eklund, Nano Lett., 4 (2004) 1991
[47] D Spirkoska, G Abstreiter and A Fontcuberta i Morral, Nanotechnology 19 (2008) 435704
[48] K.W. Adu, Q. Xiong, H.R. Gutierrez, G. Chen, P.C. Eklund, Appl. Phys. A 85 (2006) 287–297
[49] R. Ruppin and R. Englman, Rep. Prog. Phys. 33 (1970) 149
[50] M Watt, C M Sotomayor Torres, H E G Arnot and S P Beaumont, Semicond. Sci. Technol. 5 (1990) 285
[51] Yufeng Hao, Guowen Meng, Zhong Lin Wang, Changhui Ye, and Lide Zhang, Nano. Lett. 6 (2006) 1650.

14

Finite- and Quantum-size Effects of Bismuth Nanowires

Thomas W. Cornelius[1] and M. Eugenia Toimil-Molares[2]
[1]*European Synchrotron Radiation Facility (ESRF), B.P. 220, 38043 Grenoble Cedex*
[2]*GSI Helmholtz Institute for Heavy Ion Research GmbH, Planckstr. 1, 64291 Darmstadt*
[1]*France*
[2]*Germany*

1. Introduction

During the last decades, the miniaturization of electronic devices has strongly influenced the technological evolution. For instance, Moore's law states that roughly every 18 months the number of transistors per microchip doubles. This law has been valid remarkably well in the last 40 years and it is expected to hold for some more years to come. This process requires an ongoing reduction of feature sizes, and assumes that the components of a microchip can be scaled down arbitrarily without changes in their performance. However, conventional transport theories will fail on such small feature sizes. When the size is comparable to the mean free path l_e of the conduction electrons, the mesoscopic transport regime is entered leading to so-called finite-size (FSE) or classical-size effects. These effects include additional electron scattering processes both at the surface of the structure and at inner grain boundaries. When the structure size even reaches the Fermi-wavelength λ_F and, hence, the electronic wave function becomes confined, quantum-size effects (QSE) are expected to occur which involve a change of the density of states and, thus, affect the transport properties as well as the optical properties.

1.1 Bismuth

In order to study these confinement effects, bismuth is an ideal material because of its unique electronic properties. The electrons possess a large mean free path being in the range of 100 to 250 nm at room temperature that increases to the millimeter range at 4 K (Cronin et al., 2002). Further, they exhibit a long Fermi wavelength of 40 to 70 nm (Garcia et al., 1972; Duggal & Rup, 1969) which is more than one order of magnitude larger than in typical metals. Since both intrinsic length scales (l_e and λ_F) are large, finite-size as well as quantum-size effects can be investigated in comparatively large specimen. Furthermore, Bi is a semimetal with a very small indirect band overlap resulting in a low charge carrier density n compared with conventional metals (3×10^{18} cm^{-3} at 300 K, 3×10^{17} cm^{-3} at 4 K). The electron effective mass m^* in bismuth amounts to (0.001 – 0.26) m_e depending on the crystalline orientation with m_e being the free electron mass (Lin et al., 2000). This very small m^* facilitates the observation of QSE.
The current study of the transport properties of bismuth and bismuth compound nanowires is also partly motivated by theoretical studies, that predicted substantial enhancements of

the thermoelectrical figure of merit zT for low-dimensional materials (Lin et al., 2000). While in bulk materials the parabolic density of states $D(E)$ means that the electron density around the Fermi level is small, in one-dimensional structures such as nanowires the spikes in the $D(E)$ suggest that the thermal power factor $S^2\sigma$ can be intentionally increased in a controlled manner. In addition, thermal conductivity κ can be significantly reduced by phonon scattering at wire boundaries. Thus, a figure of merit $zT = 6$ at 77 K has been predicted for *n*-doped Bi nanowires with diameter 5 nm oriented along the trigonal axis (Lin et al., 2000). This efficiency approaches that of conventional vapor-compression unit and, thus, experimental evidence of such a zT enhancement would have a large impact on the thermoelectrics industry.

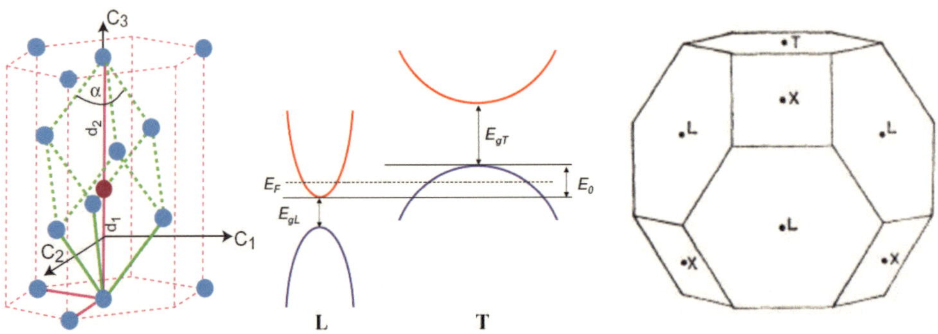

Fig. 1. a) Schematic of the rhombohedral lattice structure (dashed bold lines) of bismuth belonging to space group $R\bar{3}m$ together with the hexagonal unit cell (dashed thin lines) (Hofmann, 2005). The solid bold and thin lines are the vectors spanning the rhombohedral and hexagonal lattice, respectively. C_1, C_2, and C_3 are the bisectrix, binary, and trigonal axes. Schematics of b) the band structure and c) the Brillouin zone of bismuth.

Bismuth crystallizes in a rhombohedral lattice structure which belongs to the $R\bar{3}m$ space group with two atoms per unit cell. Alternatively, the structure can be described as hexagonal with six atoms per unit cell. The relation between these different unit cells is shown in Fig. 1(a) (Hofmann, 2005). The dashed bold and thin lines indicate the rhombohedral and hexagonal unit cell, respectively. Directions in the bismuth crystal are specified with respect to three mutually perpendicular directions which are marked by C_1, C_2, and C_3; C_2 – one of the three axes of twofold symmetry (binary direction), C_3 – the axes of threefold symmetry (trigonal direction), C_1 – an axis perpendicular to C_2 and C_3 forming a right-hand triad in the order 1-2-3 (bisectrix direction). The trigonal direction of the rhombohedral structure is the c-axis of the hexagonal lattice. The rhombohedral angle α = 57.23 ° is slightly distorted from its value of 60 ° in a perfect fcc lattice. This distortion leads to highly anisotropic Fermi surface and, thus, is largely responsible for bismuth's unique electronic properties. Due to the Fermi surface anisotropy all transport properties (electrical and thermal conductivities) and the effective masses of charge carriers depend on the crystalline orientation. Both thermal and electrical conductivities of bulk Bi are small compared to metals. The specific electrical resistivity amounts to about 135 and 110 μΩ cm at 300 K parallel and normal to the trigonal axis, respectively. At room temperature the thermal conductivity is about 10 and 6 W m^{-1} K^{-1} parallel and perpendicular to the trigonal direction, respectively (Gallo, 1962).

The band structure of bismuth is schematically depicted in Fig. 1(b). The electrons and holes are located at the L- and at the T-point of the Brillouin zone, respectively (Fig. 1(c)). A small indirect overlap E_0 is present between the valence band at the T-point and the conduction band at the L-point making Bi to a semimetal. Both at the L-point and at the T-point a direct band gap exists. The size of the gap at the L-point amounts to $E_{gL} \sim 15$ meV and ~ 36 meV at 0 and 300 K, respectively (Black et al., 2003). In contrast, the size of the gap at the T-point is rather unknown. In literature, values ranging from 50 to 400 meV can be found for room temperature (Bate & Einspruch, 1967; Isaacson & Williams, 1969; Omaggio et al., 1993). The Fermi energy E_F, whose value is defined with respect to the lower edge of the L-point conduction band, amounts to 26 and 56 meV at 0 and 300 K, respectively (Gallo et al., 1962). In the following sections the effect of low-dimensionality on the electronic properties will be described briefly.

1.2 Finite-size effects

Finite-size effects involve additional electron scattering from the wire surface as well as from grain boundaries within the wire leading to an increase of the specific electrical resistivity. Mayadas and Shatzkes developed a model for the influence of the electron scattering from grain boundaries on the electrical resistivity which depends on the mean grain size D, the mean free path l_e, and the reflexion coefficient R of the boundaries (Mayadas & Shatzkes, 1970).

$$\frac{\rho_0}{\rho_g} = 3\left[\frac{1}{3} - \frac{1}{2}\alpha + \alpha^2 - \alpha^3 \ln\left(1 + \frac{1}{\alpha}\right)\right] \text{ with } \alpha = \frac{l_e}{D}\frac{\mathcal{R}}{1-\mathcal{R}} \tag{1}$$

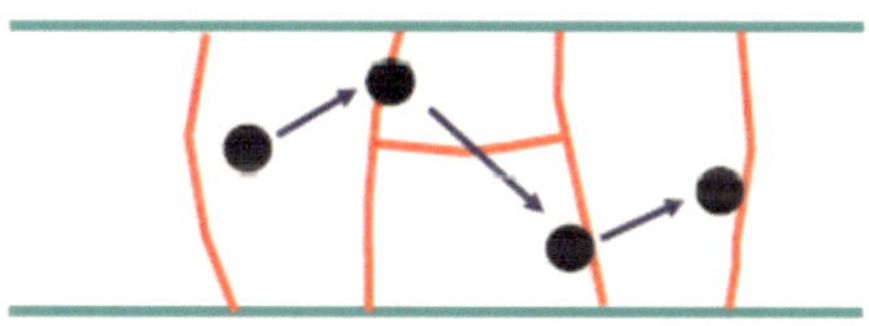

Fig. 2. Schematic of electron scattering from grain boundaries.

The influence of electron scattering at surfaces on the electrical transport properties was studied theoretically already in the first half of the last century. In the 1930's K. Fuchs developed a model to treat the electron scattering from the surface of thin films (Fuchs, 1938). This theory was extended to square shaped and cylindrical wires by MacDonald and Sarginson (MacDonald & Sarginson, 1950) and R.B. Dingle (Dingle, 1950), respectively. According to these theories the contribution of the electron scattering from the surface to the electrical resistivity depends on two parameters: (i) The specularity ε of the scattering processes with $0 < \varepsilon < 1$ and (ii) the ratio k of film thickness, respectively, wire diameter d and electron mean free path ($k = d/l_e$). The equations (2a) and (2b) are approximations for completely diffuse scattering ($\varepsilon = 0$) from the surface for cylindrical wires with diameters much smaller ($k \ll 1$) and much larger ($k \gg 1$) than the electron mean free path [4], respectively. Expressions (2c) and (2d) are the analogical approximations for partly diffuse scattering processes ($\varepsilon = 0.5$).

$$\frac{\sigma}{\sigma_0} = 1 - \frac{3}{4k} + \frac{3}{8k^3} \cdots \tag{2a}$$

$$\frac{\sigma}{\sigma_0} = k - \frac{3k^2}{8}\left(\ln\frac{1}{k} + 1.059\right) - \frac{2k^3}{15} \tag{2b}$$

$$\frac{\sigma}{\sigma_0} = 1 - \frac{0.375}{k} + \frac{0.109}{k^3} \cdots \tag{2c}$$

$$\frac{\sigma}{\sigma_0} = 3k - 11.22k^2\left(\log\frac{1}{k} - 0.245\right) - 9.74k^3 \tag{2d}$$

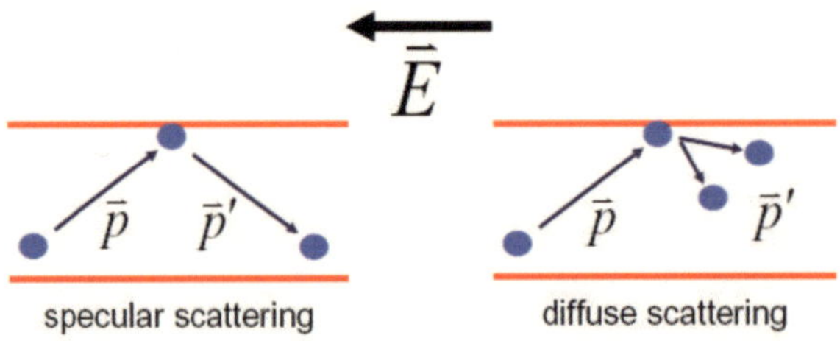

Fig. 3. Schematic of specular and diffuse scattering of electrons from the wire surface. E is the applied electrical field.

Note that the models are simplified in so far that they assume materials having a spherical Fermi surface and that they do not take into account the angle of incidence of the charge carriers on the surface. J.E. Parrott improved Fuchs' theory to materials having spheroidal and ellipsoidal Fermi surfaces and by introducing a critical wave vector K_0 which depends on the angle of incidence θ of the electrons on the surface (Parrott, 1965). By this approach the reflection coefficient becomes a function of the wave vector and, thus, of θ. When the wave vector is less than K_0 the reflection coefficient is $\varepsilon = 1$ and $\varepsilon = 0$ for $K > K_0$, i.e., the scattering from the surface becomes specular for $\theta < \theta_0$ while it becomes abruptly completely diffuse for $\theta > \theta_0$. Barati and Sadeghi extended this model to nanowires of materials with non-spherical Fermi surfaces, in particular, for bismuth nanowires (BS model) (Barati & Sadeghi, 2001). For bismuth nanowires oriented normal to the trigonal axis, the conductivity can be written as

$$\sigma = \frac{\sigma_{bulk} a_1}{\sqrt{\det a}}\left[1 - \frac{6}{k}\int_{\mu_0}^{1} \exp\left(-\frac{k}{2\mu}\right)\left[1 - \exp\left(-\frac{k}{2\mu}\right)\right]\mu^2\sqrt{1-\mu^2}\, d\mu\right] \tag{3}$$

where $\mu = \sin\theta$, $k = r/l_e$, σ_{bulk} is the electrical conductivity of bulk bismuth, and α is the reciprocal mass tensor.

1.3 Quantum-size effects

As the system size decreases and approaches nanometer length scales, it is possible to cause dramatic variations in the density of electronic states $D(E)$ which is schematically depicted for different dimensionalities in Fig. 3. This dramatic change is caused by the spatial confinement of the electronic wavefunction. While for 3-D materials $D(E) \sim \sqrt{E}$, in quasi-one-dimensional structures – such as nanowires – it exhibits a spike-like shape arising from discrete energy levels due to a splitting of the energy bands into subbands. This gives rise to manipulate the electronic properties of the material and to utilize the electronic transport properties of various quasi-1D materials systems for a wide range of practical devices. For instance, according to Sandormirskii (Sandormirskii, 1967) and Farhangfar (Farhangfar, 2006), semimetallic thin films and rectangular nanowires show oscillating transport properties caused by the subband splitting.

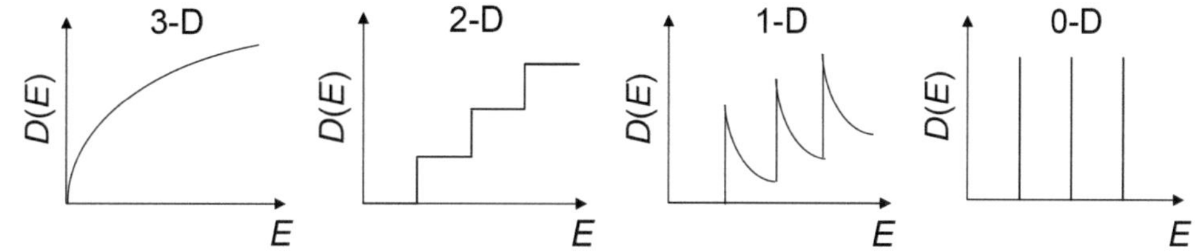

Fig. 4. Density of electronic states $D(E)$ for different dimensionalities

With decreasing structure size the energy separation between the subbands increases and, consequently, the edges of the energy bands shift away from each other. In the simplest case, a nanowire can be modelled as a two-dimensional infinitely deep potential well leading to the following equation for the energy shift that the band edges experience with N being the number of the subband and d the wire diameter.

$$\Delta E \approx \frac{N^2 \hbar^2 \pi^2}{m^* d^2} \tag{4}$$

Since the electron effective mass is comparatively small in bismuth, QSEs are strongly pronounced in this material. Note that the valence and the conduction band at the L-point are strongly coupled leading to an increasing electron effective mass with increasing band gap. The dispersion relation for the L-point bands is described by the Lax two-band model (Lax et al., 1960) (see equation (5)). For interband transitions, transitions between the n-th valence subband to the n-th conduction subband are allowed, where $|\langle v|p|c\rangle|$ is the coupling between the valence subband v and conduction subband c, and m_e is the free electron mass. At the band edge ($k = 0$), the momentum matrix element for the different subband states is the same, i.e., as the band gap enlarges, $|\langle v|p|c\rangle|$ remains constant, and the effective mass increases (Black et al., 2002).

$$\frac{1}{m^*} = \frac{1}{m_e} + \frac{2}{m_e^2}\frac{\left|\langle v|p|c\rangle\right|^2}{E_g} - \frac{12\hbar^2 k^2 \left|\langle v|p|c\rangle\right|^4}{E_g^3 m^2} \tag{5}$$

Owing to the spike-like $D(E)$ for 1D nanowires, the kinetic properties of the system will oscillate as a function of d, as long as the carrier gas is degenerate. These oscillations are connected with abrupt changes of the density of states on the Fermi surface as one subband after another passes E_F. According to Sandormirskiĭ (Sandormirskiĭ, 1967) and Farhangfar (Farhangfar, 2006), the periodicity for films and rectangular nanowires is given by the critical thickness and width, respectively. Assuming that the Fermi level does not shift with the wire diameter, and according to equation (4), the critical diameter d_c is approximately given by

$$d_c \approx \frac{N\hbar\pi}{\sqrt{\Delta E\, m^*}} = \frac{Nh}{\sqrt{4E_F m^*}} \tag{6}$$

Figure 5 displays the computed band-edge energies as a function of the nanowire diameter with respect to the energy of the band edge of the L-point conduction band in bulk Bi. For calculations, the values E_{gL} = 36 meV, E_0 = 98 meV, E_F = 56 meV, and E_{gT} = 200 meV are used. In Fig. 5(a), the electron effective mass is assumed to be constant, namely m^* = 0.002 m_e, whereas in 5(b) the Lax two-band model has been taken into account. The strong coupling of electrons and holes at the L-point leads to a smaller energy shift of the band edges at the L-point for a given wire diameter. In addition to the shift of the band edges, the Fermi energy increases with diminishing wire diameter. The rise of E_F occurs in order to obtain charge neutrality, i.e., for achieving the same number of electrons and holes. At a critical wire diameter d_c the lowest subband of the conduction and the highest subband of the valence band do not overlap anymore. Thus, bismuth nanowires undergo a transition from a semimetal to a semiconductor at d_c of about 35 nm and 70 nm for calculations with and without incorporating the dependence of m^* on the band gap, respectively. Note that d_c is a function of the crystalline orientation due to the highly anistropic Fermi surfaces, resulting in a dependence of the electron effective mass on the crystalline direction. Lin et al. predicted the semimetal-semiconductor-transition to take place at 77 K for nanowires of d_c = 45, 44, 33 and 20 nm oriented along $\left[10\bar{1}1\right]$, bisectrix, trigonal and binary axis, respectively (Lin et al., 2000).

Note that in order to observe QSEs two preconditions have to be satisfied: (i) The relaxation time τ must be sufficiently long, i.e., $\tau > \hbar/(E_{n+1} - E_n)$ with $(E_{n+1} - E_n)$ being the energy separation between two subbands at the Fermi level. (ii) The thermal broadening of the levels must be small compared to the subband splitting. In order to fulfil the first requirement, the specimens must consist of grains much larger than l_e, and should possess only a small number of defects. In order to meet the second precondition, the subband splitting at the Fermi level must exceed the thermal broadening (e.g. 25 meV which corresponds to kT at 300 K). A number of studies exist reporting QSE at room temperature for Bi thin films with thicknesses > 100 nm (Duggal & Rup, 1969; Rogacheva et al., 2003). Since nanowires confine the electronic wavefunction in two instead of only one dimension like in thin films, quantum size effects are expected to occur at diameters larger than the reported film thicknesses and, hence, QSE can be investigated in few hundred nanometer thick wires.

1.4 Wear-out failure

An important aspect for components based on nanowires or carbon nanotubes (CNTs) is the maximal current density such quasi one-dimensional structures are able to carry. Bulk metals fail because of Joule heating at current densities of 10^3 to 10^4 A/cm^2 (Ho & Kwok, 1989) while CNTs can carry 10^9 A/cm^2 (Dai et al., 1996). In the case of bulk bismuth, the specific electrical conductivity and the thermal conductivity are more than 50 times smaller than for metals and the melting temperature amounts only to T_m = 544 K. Consequently, Bi

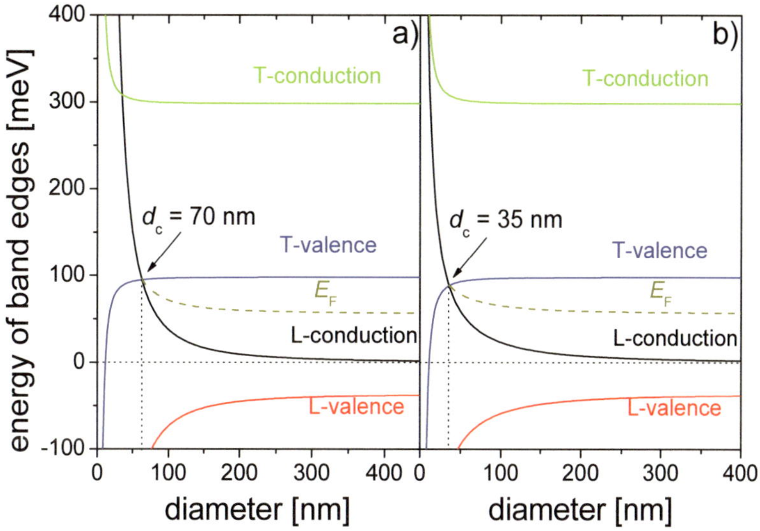

Fig. 5. Energy of the band edges as a function of the wire diameter: a) $m^* = 0.002\ m_e$ = const, b) employing the Lax two-band model with $m^* = 0.002\ m_e$.

is expected to fail at one to two orders of magnitude lower current densities. High critical current densities of the order of 10^6 A/cm^2 can be observed in microstructured metal lines, which serve as interconnects in microelectronic circuits. These metal lines are fabricated by photolithographic patterning on silicon wafers. Because of the intimate thermal contact of the film to the substrate with high thermal conductivity (150 W m^{-1} K^{-1} for silicon), the Joule heat can be dissipated immediately to the substrate and, therefore, higher current densities compared to suspended metal wires can be observed. In addition to Joule heat dissipation, another effect was identified in microelectronics as a potential wear-out failure for devices that employ metal film conductors, namely electromigration (Black et al., 1969). Electromigration is a mechanism for transport of matter by high electric current densities: when a metal ion is thermally activated and is at its saddle point, i.e., it is lifted out of its potential well and is essentially released from the metal lattice, two opposing forces act in an electrically conducting single-band metal: (i) the electric field applied induces a force on the activated positive ion in the direction opposite to the flow of the conduction electrons. (ii) The rate of momentum exchange between the electrons colliding with the ions exerts a force on the metal ions in the direction of the electron flow. Because of the shielding effect of the electrons, the force on the ions due to the electric field is quite small and, hence, the predominant force is that of the "electron wind". This causes dramatic failures in interconnects and it has been shown that is material, geometry and size-dependent.

2. Nanowire synthesis and structural characterization

Because of the low melting point of bismuth the synthesis of Bi nanowires is rather difficult. Most of the existing high-temperature approaches like laser ablation, plasma-arc or chemical vapour deposition are not appropriate for their fabrication. However, several different techniques exist for Bi nanowire preparation including solvothermal synthesis (Gao et al., 2003), the Taylor process (Glocker & Skove, 1977), electron beam writing techniques (Chiu &

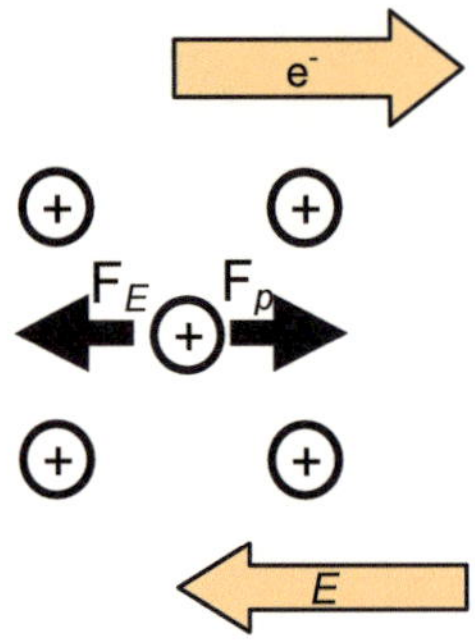

Fig. 6. Schematic of forces acting on a thermally activated ion at its saddle point in a conductor.

Shih, 2004), and the template method (Cornelius et al., 2005; Huber et al., 2003; Wang et al., 2001; Yin et al., 2001; Zhang et al., 1998). Solvothermal synthesis provides nanowires of good crystalline quality with diameters down to 40 nm and with lengths of several μm. By using the Taylor process, single-crystalline Bi nanowires were created by filling a glass tube with metal and, subsequently, drawn into a thin capillary. The drawback of this method is the fact that the temperature must exceed the softening point of glass and, as a consequence, the wires may be contaminated. Electron beam lithography enables the creation and electrical contacting of rectangular bismuth wires with lengths of several tens of micrometers, thickness 50 nm and width in the range of 70 - 200 nm. However, the wire crystallinity that strongly influences all transport properties cannot be controlled by this approach. In the case of polycrystalline materials electron scattering at the grain boundaries leads to higher resistivity values than for single-crystalline ones. Furthermore, grain boundaries are locations at which electromigration-driven failure predominantly occurs. Moreover, in the case of bismuth, due to its highly anisotropic Fermi surface, the properties of the material are highly dependent on the crystalline orientation. The most common technique is the template method which is a very general approach used to prepare arrays of nanomaterials. The most commonly used nanoporous templates are track-etched polycarbonate membranes (Martin, 1994; Toimil Molares et al., 2001; Cornelius et al., 2005; Karim et al., 2006) and porous anodic aluminum oxide (AAO) membranes (Masuda & Nishio, 2006; Heremans et al., 2000; Huber et al., 2003; Zhang et al., 1998). Common to all these templates is the fact that the pores are perpendicular to the surface of the membrane. Bismuth was synthesized in the pores by several techniques, most commonly by high-pressure injection (Huber et al., 2003; Zhang et al., 1998) and the electrochemical deposition of bismuth into the channels (Wang et al., 2001; Yin et al., 2001; Cornelius et al., 2005).

2.1 Template fabrication

- Anodic alumina oxide (AAO) templates (Masuda & Nishio, 2006)

Porous anodic aluminum oxide templates are fabricated by anodizing aluminum sheets in an oxalic acid solution. During this process, cylindrical pores with diameter 7 - 200 nm are self-assembled into a hexagonal array. The pore diameter and the distance between the pores are controlled systematically by varying the anodization voltage and the electrolyte used. The thickness of the alumina template determines the wire length, and can be controlled by the anodization time. By this method, templates containing a high density of pores of up to 10^{10} pores per cm^2 can be prepared. Thus, leading to densely packed nanowire arrays.

- Ion track-etched membranes

To fabricate ion track-etched membranes, polycarbonate foils tens of micrometers thick are irradiated with swift heavy ions. The charged particles modify the material along their trajectory creating cylindrical volumina - so-called ion tracks - which behave chemically different to the non-irradiated foil. Tracks in polycarbonate are selectively dissolved in aqueous sodium hydroxide. Prior to the chemical etching the foils are treated by UV light for sensitizing them to the leach. The pore density is given by the applied ion fluence and vary between 1 single pore per foil and 5 x 10^9 pores/cm^2. The pore diameter increases linearly with the etching time, and the etching rate depends on the etchant concentration and temperature.

Single nanochannel membranes are fabricated at the GSI Helmholtz Center for Heavy Ion Research GmbH, by single ion irradiation and chemical etching (Apel et al., 1999; Siwy et al., 2003; Cornelius et al., 2007). Single ion irradiation is performed by defocusing the ion beam so that ~ 10^4 ions/cm^2 reach the sample. A mask with a 100 micrometer aperture is placed in front of the sample, and a detector is placed behind the sample. As soon as an ion reaches the detector, a chopper deflects the ion beam. With this set up, single-ion irradiation or irradiation with a preset number of ions can be performed. For the etching of single-track membranes, the irradiated polymer foil is placed between two compartments of an electrolytical cell, using a 2M NaOH solution at 60 °C as etchant and a gold rod as electrodes is immersed on each side (Chtanko et al., 2004). During etching a potential is applied between the two electrodes and the current is recorded as a function of the etching time. After etching, the membrane is rinsed with distilled water. The effective pore diameter d_{eff} of a single nanopore is determined by conductivity measurement with 1 M KCl solution and two Ag/AgCl electrodes. By means of the following equation d_{eff} is calculated assuming an ideal cylinder

$$d_{eff} = \sqrt{\frac{4L}{\pi\kappa}\frac{I}{U}} \tag{7}$$

with L being the length of the pore, i.e., the membrane thickness, U is the applied voltage, κ is the conductivity of the 1 M KCl solution (κ= 10 S/m at 20 °C), and I is the recorded current.

Ion track-etched membranes can also be purchased e.g. from Whatman company. However, misalignment of the pores (tilted up to 34°) and cigar-like shapes of the commercial membrane pores have been reported (Schönenberger et al., 1997).

2.2 Nanowire growth and characterization

Bismuth nanowires have been grown in templates using different growth techniques, and few of them are shortly described in as following:

i. *Pressure injection*: Molten bismuth was pressed by high pressure into the pores of AAO membranes. Afterwards, the setup was cooled down leading to recrystallization of the molten material within the nanopores. Since the pressure needed for pushing the melt into the pores scales inverse with the pore diameter, this techique is limited to pores with a diameter of few ten nanometers (Huber et al., 2003; Zhang et al., 1998).

ii. *Vapor phase technique*: For this purpose, a heater containing bismuth and being covered by an AAO template is heated to temperatures of 400 to 500 °C. At this temperature bismuth evaporates and penetrates the nanopores. Then, a slow cool-down phase is intiated. Since the top plate is cooler than the crucible, bismuth starts to condense at the top-side of the porous plate downwards. When the temperature has fallen below the melting temperature of Bi (271 °C) the process is finished. This method allowed the creation of nanowires with diameters down to 7 nm (Heremans et al., 2000).

iii. *Electrochemical deposition (Fig. 7)*: Electrochemical deposition of bismuth in nanopores has been reported using two different electrolytes based on $Bi(NO_3)_3$ (Wang et al., 2001) and $BiCl_3$ (Cornelius et al., 2005). One side of the template was coated by a conductive layer that served as cathode during the subsequent electrochemical deposition of bismuth inside the nanopores. In most cases, depositions were performed potentiostatically. Reverse pulse deposition at both low and high frequency have also been reported. For potentiostatic and low frequency reverse pulse depositions the wires fill the complete nanopore cross-section. In contrast, for high-frequency depositions the needles possess a smaller diameter than the pores and, thus are free-standing within the template (Li et al., 2005).

Fig. 7. Schematic of the electrochemical deposition template method: a) nanoporous membrane, b) deposition of a conductive layer on one side of the template, c) growth of the naowires inside the pores.

2.3 Nanowire crystallinity

The crystallinity of nanowires prepared in templates by the different synthesis methods has been investigated by several groups by X-ray diffraction (XRD) as well as transmission electron microscopy (TEM) including selected-area electron diffraction (SAED). As expected, crystallinity and preferred orientation reported for the bismuth nanowires differs depending on the fabrication technique. Single-crystalline Bi nanowires oriented along the [102] direction were prepared by a low-temperature solvothermal processn (Gao et al., 2003). Nanowires with diameters 40 and 95 nm created by high-pressure injection in alumina templates exhibited a <202> and <012> texture, respectively, indicating that the preferred crystalline orientation is a function of the wire diameter (Lin et al., 2000). Nanowires grown by the vapor technique possess a <202> texture as 40-nm diameter wires grown by pressure injection (Heremans et al., 2000). Bismuth nanowires fabricated by potentiostatic electrodeposition exhibited in all cases a single-crystalline <110> texture independently of the template and of the bath that was used (Cornelius et al., 2005; Wang et al., 2001). When using a bismuth chloride electrolyte, a pronounced increase of the <110> texture with decreasing overpotential and rising temperature. Fig. 8(a) displays the θ-2θ scan for an array of nanowires deposited at -17 mV and 60 °C revealing a very pronounced <110> texture. The single-crystallinity was proven by rocking curves around the <110> direction (Fig. 8(b)). In (Cornelius et al., 2005), the authors reported an increase in the texture with diminishing wire diameter. The preferred orientation of the crystals can be varied by applying reverse pulses instead of a constant potential during the deposition process. As can be seen from the rocking curves around the <110> direction (inset of Fig. 8(b)) for short cathodic cycles t_c and high anodic potentials U_a, <100> textured wires were obtained. This preferential orientation becomes less pronounced for longer t_c and lower U_a while the <110> texture becomes stronger.

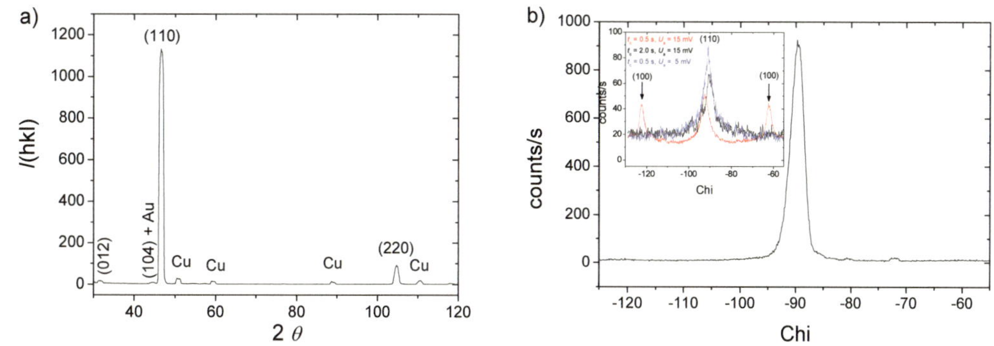

Fig. 8. a) θ-2θ scan and b) rocking curve for bismuth nanowire array deposited at -17 mV and 60 °C. The inset shows rocking curves for nanowires deposited applying reverse pulses.

For transmission electron microscopy studies, the template was removed and the wires were subsequently placed onto a TEM grid. Fig. 9(a) presents an overview of bismuth nanowires with diameter 80 nm grown potentiostatically at 60 °C and- 17 mV. In the image, the wires attach to the structures of the TEM grid exhibitng great flexiblility. Further, the micrograph demonstrates the smooth contour of the wires as well as the homogeneous diameter along the wire length. A high-resolution TEM (HRTEM) micrograph and a SAED pattern of a wire with diameter 30 nm are displayed in Fig. 9(b) and (c), respectively. The HRTEM image shows the atomic lattice of the wire demonstrating the single-crystalline character of the wire being supported by the SAED pattern which displays regular diffraction signals. Further, the diffraction pattern evidences that the wire is oriented along the <110> direction coinciding with the results obtained by XRD.

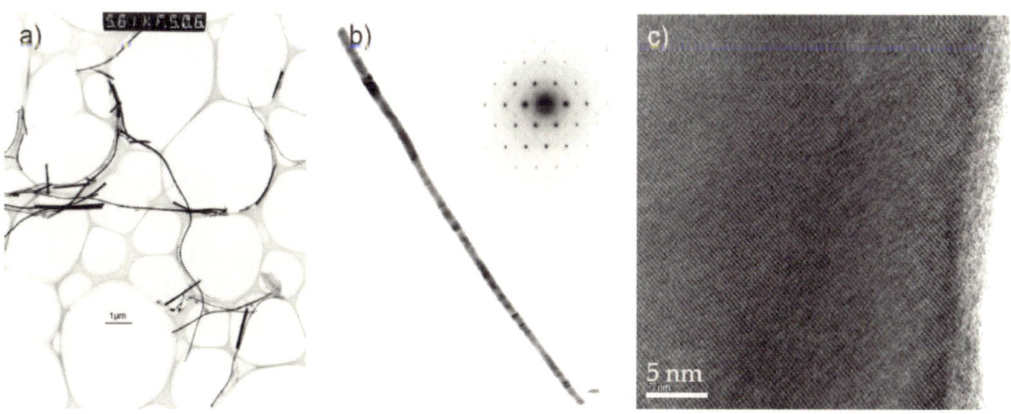

Fig. 9. a) TEM image of Bi nanowires with diameter 80 nm. b) TEM micrograph of Bi nanowire with diameter 50 nm and correspondent SAED pattern. c) High resolution TEM image of the same Bi nanowire.

Due to their high surface-to volume ratio of nanostructures, surfaces influence strongly the properties of nanowires. Alumina templates can be only removed with either strong acids or sodium hydroxide, and studies revealed that solvents used for dissolving the AAO template may attack and modify the wire surface. Thus, a comparatively thick oxide layer (~ 7nm) has been found on bismuth wire surfaces after dissolution of the template (Cronin et al., 2002). Polymer membranes are dissolved by organic solvents such a dimethylformamide (DMF, C_3H_7NO) and dichloromethane (CH_2Cl_2). Figures 10(a) and (b) display TEM images of two nanowires deposited in the same template, but from two different sample areas dissolved either in dichloromethane (Fig. 10(a)) or in dimethylformamide (Fig. 10(b)). In contrast to DMF, dichloromethane as solvent modifies the surface of the wires being marked by a bracket. The layer consists of a material that possesses a larger interlattice plane distance than bismuth. On top of this layer, an amorphous material is observed that contains most probably polymer residuals. Energy dispersive X-ray analysis (EDX) and the Fourier transform of the micrograph in Figure 10(c), reveal a modified surface layer consisting of bismuth oxychloride (BiOCl) piling up along the c-axis, i.e., normal to the wire axis, presumably developed during or after the dissolution of the template in dichloromethane (CH_2Cl_2).

For further processing such as lithographical contacting of single wires for electrical resistance measurements such surface layers must be avoided. Thus to avoid that large contact resistances hinder reliable measurements, a careful choice of the template solvent is required.

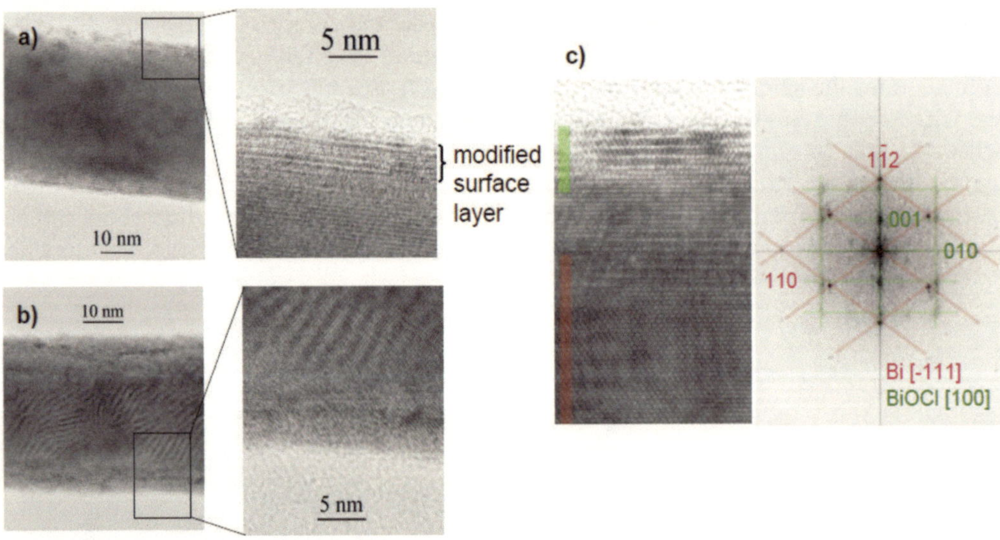

Fig. 10. High-resolution transmision electron micrographs. The templates were removed by a) dichloromethan (CH_2Cl_2) and b) dimethylformamide. c) HRTEM combined with SAED and EDX reveals that the surface layer consists of BiOCl.

3. Electrical transport properties

For investigating the electrical transport properties of low-dimensional systems, the electrical resistivity of bismuth nanowire arrays as well as of single wires were studied. In the case of arrays a pseudo-four-point contacting method was employed, i.e., the wires were left embedded in the template (AAO or ITPC membrane) and on each side of the membrane two leads were fixed by a conductive glue (Fig. 11(a)). One lead on each side serves for measuring the potential drop and the other one for the electrical current. The contact resistance cannot be determined by this method but at least the resistance of the leads does not affect the measurement. Since the exact number of contacted nanowires is unknown, absolute values of the electrical resistivity cannot be determined in the case of arrays. To measure absolute values of the nanowire resistance, measurements at the single nanowire level are needed. Different approaches have been reported for contacting single wires.

As mentioned above, nanowire arrays were created in alumina templates by high-pressure injection of molten bismuth into the pores, the matrix being subsequently removed using an oxidizing acid. Since the wire surface was oxidized by this strong acid, the wires could not be reliably contacted by lithography (Cronin et al., 2002). Single wires with rectangular cross-section were prepared on a Si substrate by an electron beam writing technique. Bonding pads were formed on the oxide employing filament evaporation and optical lithography. Such wires were subsequently treated by reactive ion etching in order to reduce their dimensions. With this method, electrical transport measurements on the same wire with incremental reductions in diameter. However, this approach may affect the surface roughness of the wire and, hence, influence the transport characteristics (Farhangfar, 2007). Both experiments used a contacting approach as schematically depicted in Fig. 11(b).

Single bismuth nanowires were grown in single-pore ion track-etched polycarbonate membranes (Toimil Molares et al., 2004; Cornelius et al., 2006; Cornelius et al., 2008). After growth of the nanowire and the formation of a cap on top of it, this wire was left embedded in the template. Thus, the wires were protected against oxidiation and, in particular, not endangered by chemical reactions with a matrix solvent. In order to contact the embedded needles electrically an additional gold layer was sputtered on the cap grown on top of the wire and the complete sample was placed between two macroscopic copper plates (Fig. 11(c)).

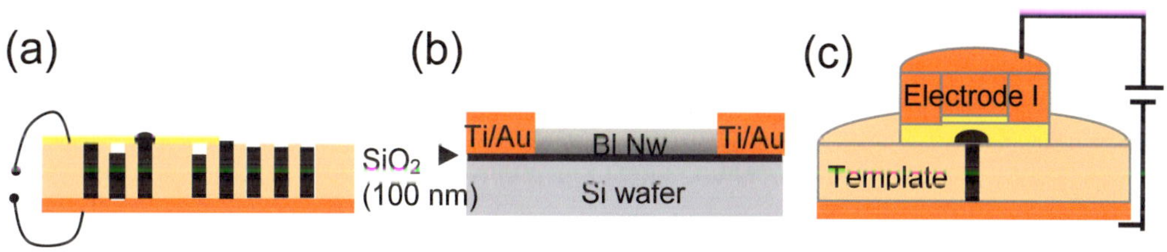

Fig. 11. Schematics of different methods for contacting single nanowires electrically: a) An undetermined number of wires in a multipore template. b) Single nanowire on a Si/SiO_2 substrate contacted by lithographic methods. c) Single nanowire electochemically grown in a single nanopore membrane and contacted by two macroscopic metallic layers.

3.1 Finite-size effects

Finite-size effects on the electrical transport properties of single bismuth nanowires were studied on wires grown electrochemically under different deposition conditions. By employing three different sets of deposition parameters (temperature and voltage), nanowires with diameters between 200 and 1000 nm with different average grain size *D* were successfully synthesized. The scanning electron micrographs (Fig. 12) of caps grown on top of the three types of wires show the variation of the mean grain size as a function of deposition parameters. Each single nanowire was electrically contacted using the setup shown in Fig. 11(c). The specific electrical resistivity of wires with diameter > 150 nm was calculated from the *I-U* characteristics and is plotted as a function of the diameter in Fig. 12. The data evidence that nanowires deposited at lower *U* and higher *T*, i.e. with larger grain size (red circles) exhibited the lower resistivity values, while nanowires deposited at higher *U* and lower *T*, i.e. with the smallest grain size (green squares), exhibited the larger resistivity values. The specific resistivity amount to 850, 440, and 315 μΩ cm, i.e. 7.5, 3.8, and 2.8 times larger than the bulk resistivity (solid black line), for the three types of nanowires, increasing with decreasing mean grain size. This is in good qualitatively agreement with the MS model presented by Mayadas and Schatzkes (Mayadas & Shatzkes, 1970). It is remarkable that even for nanowires built up by large grains the specific electrical resistivity is about threefold the bulk value.

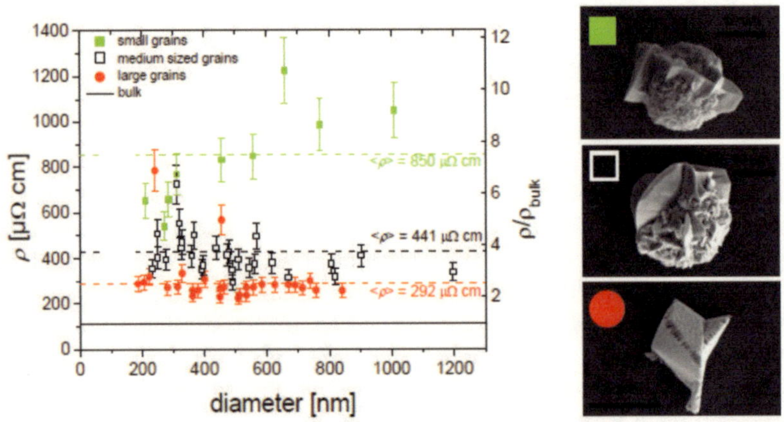

Fig. 12. Specific electrical resistivity of single Bi nanowires deposited under three different conditions leading to three different mean grain sizes. Scanning electron micrographs of nanowire caps representing the different grain size for the different types of wires.

Different groups reported measurements of the resistance of bismuth nanowires as a function of temperature, revealing different behaviours (Cornelius et al., 2006; Heremans et al., 2000; Liu et al., 1998; Zhang et al., 2000). For nanowire arrays prepared by high-pressure injection in alumina templates and for single wires electrochemically deposited in single-pore ion track-etched membranes with d > 200 nm a non-monotonic resistance versus temperature behaviour was reported. The resistance increased, passed a maximum, and declined when cooling down the wires from 300 to 10 K (Fig. 13(a)). The resistance maximum shifts to lower temperatures and the residual resistance becomes larger for increasing wire diameter. The same trend is observed for nanowire arrays (Kaiser et al., 2009) (Fig. 13(b)).

The non-monotonic behavior, when lowering the temperature, was attributed to the opposite contributions of decreasing charge carrier density n and increasing mobility μ to the electrical resistivity $\rho = ne\mu$. In bulk Bi, the gain in mobility surpasses the reduction of charge carrier density, leading to a monotonic decrease of resistivity with lowering T. In contrast, $l_e \sim \mu$ is limited in nanowires by both wire diameter and grain size, and thus the gain in mobility is limited due to finite-size effects (Fig. 14). These converse effects (decreasing number of phonons, increase and limitation of l_e, and decreasing n) lead to the

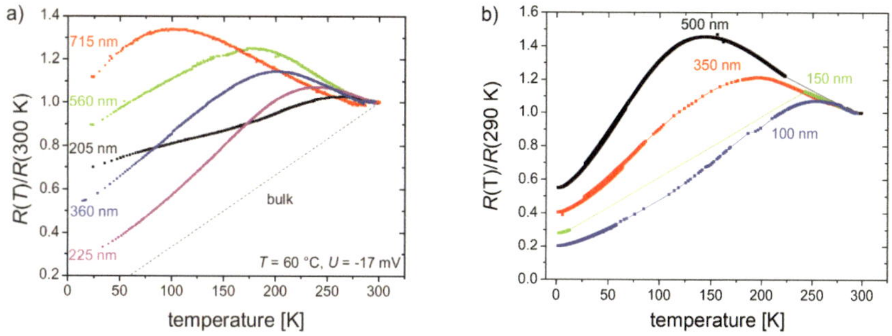

Fig. 13. Electrical resistance as a function of temperature for a) single Bi nanowires and b) nanowire arrays.

observed non-monotonic R vs T curve. In (Cornelius et al., 2006), the temperature dependence of the electron mobility for single Bi nanowires was calculated from the resistance data evidencing that μ saturates at few ten Kelvin. The saturation value was one to two orders of magnitude smaller than in bulk bismuth, and depended on both mean grain size and wire diameter. For wires of similar diameter, μ increases with growing mean grain size due to the decrease of electron scattering processes from grain boundaries when increasing D. For wires created under the same conditions, the smaller the wire diameter, the higher the mobility. This dependence can be explained either by the fact that thinner wires are formed by larger grains or that charge carrier density is affected by d.

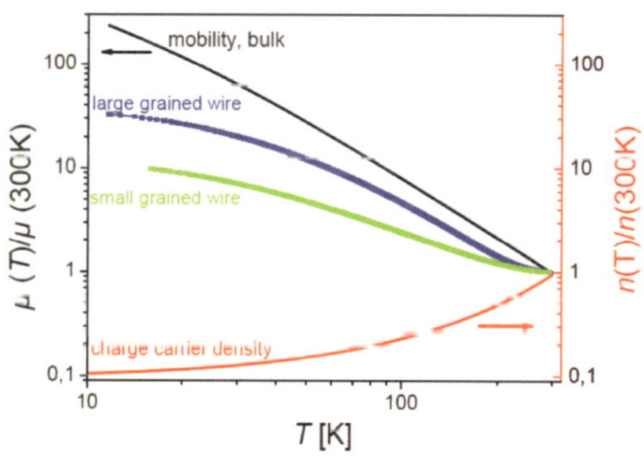

Fig. 14. Mobility of charge carriers in bulk bismuth and bismuth nanowires as well as the charge carrier density in Bi as a function of temperature.

In contrast to the non-monotonic $R(T)$ behaviour a monotonic increase of the resistance with decreasing temperature was reported for wires with diameters ranging from 200 nm to 2 μm prepared electrochemically in alumina templates (Liu et al., 1998). The authors attributed these observations to the fact that the wires possessed very small mean grain size, that in turn lead to a strong limitation of the carrier mobility. The same behaviour was found for wires with $d < 50$ nm created by the vapour-phase technique in alumina templates and for rectangular wires having a thickness of 50 nm and a width ranging from 75 to 200 nm created by the electron beam writing technique. This finding may be attributed either to the smallness of the grains like in the case before or to the fact that the wire width is close to the predicted critical diameter for the semimetal-semicondcutor transition. Thus, the observation may also be explained by a semiconductor like behaviour.

Numerous works concentrated on the magnetoresistance (MR) of bismuth nanowire arrays. In (Liu et al., 1998) the authors reported a very large positive magnetoresistance of 300% at low temperatures and 70% at room temperature with a quasilinear field dependence. Moreover, MR measurements on Bi nanowire arrays at low temperatures revealed a steplike increase that was attributed to a transition from one-dimensional to three-dimensional localization (Heremans et al. 1998; Kaiser et al., 2009). In addition, a dependence of the low temperature resistance of individual single-crystal bismuth nanowires on the Aharonov-Bohm phase of the magnetic flux threading the wire was reported (Nikolaeva et al., 2008).

Recently, also superconductivity was reported for nano-granular bismuth nanowires that were prepared electrochemically (Tian et al., 2006). The critical temperatures T_c amount to 7.2 and 8.3 K which coincide with T_c values of the high-pressure phases Bi-III and Bi-V formed at 2.7 and 7.7 GPa, respectively. The superconductive transition for the wires was attributed to structural reconstruction in the grain boundary areas. For bulk rhomobohedral bismuth superconductivity was not yet found down to 50 mK, while amorphous bismuth becomes superconductive at ~ 6 K.

3.2 Quantum-size effects

In low-dimensional materials the density of states is modified due to spatial confinement. This enables the manipulation of the transport properties of low-dimensional structures. For instance, the confinement of the electron gas in a low-dimensional system causes an oscillation of the electronic transport characteristics as a function of the structure size. Such kind of oscillations were found for bismuth thin films (Duggal & Rup, 1969), (Rogacheva et al., 2003) and very recently also for Bi nanowires by Farhangfar (Farhangfar, 2007) and Cornelius et al. (Cornelius et al., 2008). Also, oscillations with diffreent periodicity and magnitude were reported for rectangular and cylindrical nanowires. On the one hand, the conductivity of rectangular nanowires was measured on an individual wire whose width was reduced gradually by reactive ion etching. On the other hand, single cylindrical nanowires were prepared in a single-pore membrane. The different oscillatory characteristics can originate from the different wire cross-sections or from different surface properties. While the etching process applied to the rectangular nanowires may influence the surface roughness and, thus, the electron scattering at the surface, the cylindrical nanowires were left embedded in the polymer membrane in order to avoid any surface modification due to oxidation or to reactions with the solvent when removing the template. In the case of single cylindrical nanowires, the electrical conductivity oscillates as a function of the wire diameter (Fig. 15). These modulatione have been described by exponential functions $\sigma \sim \exp(-d/d_c)$. The periodicity is extracted from the intersection points of the extrapolated fitting curves (represented by dotted lines) with the x-axis. The first intersection point represents the critical diameter for the semimetal-semiconductor transition amounting to d_c = 40 nm. Fitting this experimentally found d_c value by equation (5) using E_F = 56 meV (Black et al., 2000), m^* = 0.0022 m_e is derived for bulk Bi, in reasonable agreement with the value correspondent to electrons at the L-point. The critical diameters amount to 40, 115, 182, 281, and 358 nm. This progression can be approximated by a $d_{c_n} \sim (n+1/2)d_0$ proportionality with d_0 = 80 nm and n being a natural number including zero similar to the energy levels of a harmonic oscillator.

The exponential behaviour of the oscillations may be explained by the thermal excitation of charge carriers. Only charge carriers close to the Fermi level contribute to the electrical transport. When a subband passes E_F, it is depleted and the charge carriers are transferred to lower states. At RT, they are excited thermally to higher subbands so that they participate in the electrical transport. With increasing band gap, i.e., decreasing wire diameter, the number of thermally excited carriers decreases and, thus, the electrical conductivity drops. As soon as a lower subband approaches E_F, the charge carrier density recovers and, hence, the conductivity increases to about 3000 S/cm being in agreement with the mean resistivity of 440 μΩ cm (σ = 2272 S/cm) reported in (Cornelius et al., 2006). The whole process occurs repeatedly until the lowest subband passed the Fermi level and the material is transferred into a semiconductor.

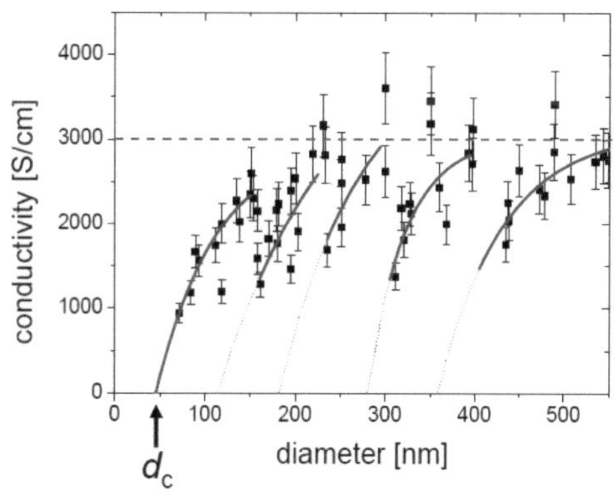

Fig. 15. Specific electrical conductivity of single Bi nanowires as a function of diameter.

4. Infrared optical properties

Infrared spectroscopy was performed on bismuth nanowire arrays embedded in alumina templates, on free-standing wire arrays, and on single nanowires that were placed on a Si wafer. The interpretation of spectroscopic features from arrays is rather complicated: the dielectric properties (in particular, vibration excitations) of an embedding material do not necessarily cancel out by measuring relative spectra and, furthermore, the spectra measured on nanowire arrays can be affected by the electromagnetic coupling of neighbouring needles. In order to separate the contributions of the wires and the alumina template an effective medium theory (EMT) was employed. In order to perform infrared spectroscopy on single bismuth nanowires the ion track-etched polymer membrane was dissolved in DMF and, subsequently, few drops of the solvent containing clean wires were put on a Si substrate. The single wires lying on the wafer were investigated at room temperature by spectroscopic microscopy at the IR beamline of the ANKA synchrotron light source at Forschungszentrum Karlsruhe. Single nanowires were located by means of optical microscopy and, subsequently, selecetd by an aperture with diameter 8 μm in the focal plane of the microsopce. IR transmission spectroscopy was performed at normal incidence of light from 800 to 5500 cm^{-1}. For this purpose, a Fourier transform IR spectrometer (Bruker IFS 66v/S) with a cooled mercury-cadmium-telluride detector was used. Nearby areas without any wires were selected for reference measurements. To obtain the relative IR transmission $T_{rel}(\omega)$ of a single wire, the substrate intensity $I_{ref}(\omega)$ is substracted from the wire signal $I_{wire}(\omega)$ by

$$T_{rel}(\omega) = \frac{I_{wire}}{I_{ref}} \tag{8}$$

and the absorption spectra $A(\omega)$ are calculated by means of equation (9).

$$A(\omega) = 1 - T_{rel}(\omega) \tag{9}$$

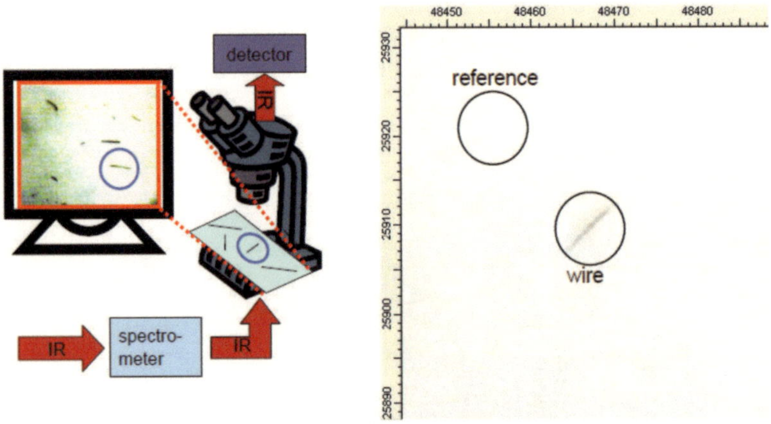

Fig. 16. Schematic of experimental setup. Optical microscopy image of single Bi nanowire of d = 200 nm on a Si wafer. The two circles correspond to the locations where IR spectroscopy has been performed: (1) single wire and (2) reference position.

Experiments on single <110> textured bismuth nanowires with various diameters lying on a Si wafer revealed a pronounced infared absorption (Fig. 17). The absorption onsets are defined by the intersection of the dashed lines (extrapolation of the strongest absorption increase) and the x-axis. The absorption edge was found to be blue-shifted with diminishing wire diameter. For instance, it is located at 1100 and 3500 cm^{-1} for wires with diameter of 400 and 30 nm, respectively. This absorption is ascribed to direct transitions in the vicinity of the L-point of the Brillouin zone and the blueshift is assigned to quantum size effects which involve

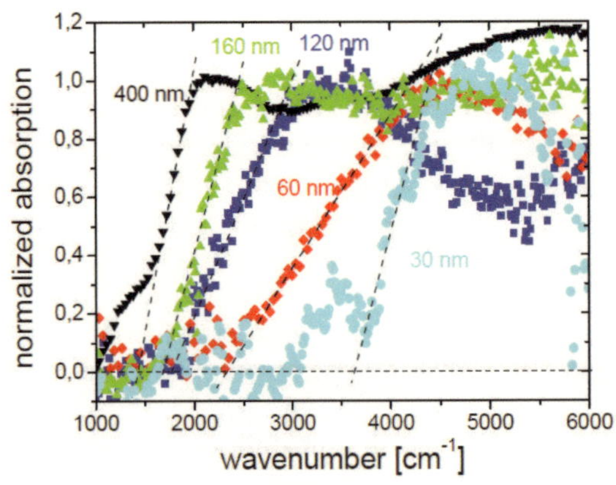

Fig. 17. Normalized infrared absorption spectra of single bismuth nanowires with different diameters.

a subband splitting of the energy bands. Hence, the band edges at the L-point shift away from each other leading to an increase of the forbidden band with decreasing specimen size.
For free-standing <012> textured Bi nanowires with diameter 45 nm a strong absorption at ~ 1050 cm^{-1} was found. In addition to this large peak, smaller absorption signals were observed at ~ 616, ~ 667, ~ 798, and ~ 986 cm^{-1}. These peaks were not seen for the nanowires embedded in alumina. The strong absorption signal at 1050 cm^{-1} was attributed to an indirect transition from the L-point valence band to the T-point valence band while no indication for direct transitions were found as for single wires located on a Si wafer. For bulk Bi this transition is very weakly pronounced due to the necessity of a phonon for changing the momentum of the excited electron while it surpasses all other signals in the case of these nanowires. Due to the spatial confinement in low-dimensional systems electrons may be effectively scattered at the surface and, thus, different points in the Brillouin zone are accessible by the excited charge carriers. Hence, no phnons are needed for those indirect transitions in low-dimensional systems. Note that the connections of different points in the Brillouin zone depend on the crystalline orientation of the system. This additional path for indirect transitions is also the explanation for the different absorption spectra of nanowires possessing different crystalline orientation. The other, smaller peaks in the spectrum are ascribed to intraband excitations which can only be found for low-dimensional structures. Because of quantum-size effects the energy bands split in sublevels.

5. Burn-out current density

Studies of the maximum current density of single bismuth nanowires were carried out. For this purpose, single Bi wires were electrochemically deposited in single-pore ion track-etched polycarbonate mebranes as described in CHAPTER 2. After electrical contacting (Fig. 11(c)), I-V characteristics were measured, ramping up the current in 50 µA steps until failure occurred. Figure 18(a) displays the *V-I* curve of a single bismuth nanowire with diameter 230 nm (deposition conditions: T = 50 °C, U = −25 mV). The wire exhibits an Ohmic behavior (linear *V-I* curve) for currents up to 800 µA. Shortly before breakage, the *V-I* curve deviates from the linear behavior. The wire fails at I = 960 µA, i.e., the wire is able to carry a current density of $j_{max} = 2.31 \times 10^6$ A/cm². This j_{max} is more than two orders of magnitude higher than in bulk metals such as Cu and Au and three to four orders of magnitude higher than expected for bulk bismuth.
In Fig. 18(b), the burnout current density is plotted as a function of the inverse wire diameter for several single bismuth nanowires, evidencing a direct proportionality between I_{max} and d^{-1}. Our experiments reveal that wires with diameter of 1 µm may carry $j_{max} \sim 5 \times 10^5$ A/cm² and that the maximum current density increases with decreasing wire diameter. For metal thin films grown on substrates, it was found that j_{max} amounts to 10^4 - 10^6 A/cm², i.e., these films can carry one to two orders of magnitude higher current densities than their bulk counterparts. The increase in j_{max} was attributed to the fact that the film dissipates the heat to the substrate. Similarly, nanowires embedded in a template may dissipate heat to the surrounding polymer. The heat dissipation will be more efficient for thinner nanowires that possess a higher surface-to-volume ratio. The results in Fig. 18(b) evidence that wires created at higher temperature and smaller overpotential (blue empty circles) can systematically carry higher current densities than wires created at higher voltages and lower temperatures (black filled squares). This is explained either by the fact that these wires contain less grain boundaries where mostly atom diffusion takes place, or that a smaller number of grain boundaries involves a smaller specific electrical resistivity and, hence, less Joule heat is generated.

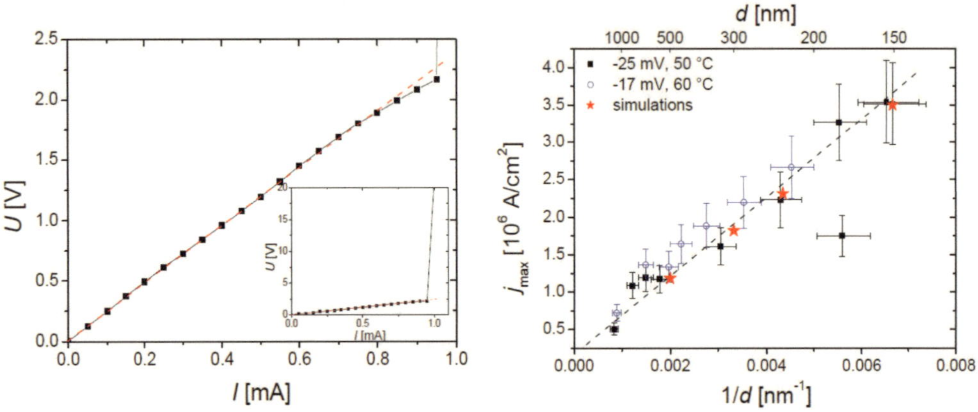

Fig. 18. (a) *U-I* characteristic of a single bismuth nanowire embedded in the PC membrane and contacted between two macroscopic contacts displaying the burn-out current density. The inset displays a zoom in at the maximal current density. (b) Maximal current density vs d^{-1} for single nanowires created under two different deposition conditions.

The failure of the nanowires can be attributed to either Joule heating or electromigration, or a combination of both. The temperature of a wire with diameter of 230 nm embedded in a template with a model diameter of 1 mm and thickness 30 μm was simulated by using the commercial software COMSOL MULTIPHYSICS. For computations, the thermal conductivities of the polymer template and the bismuth wires were 0.2 and 4 W m^{-1} K^{-1} (as known from Bi thin films (Völklein & Kessler, 1986)), respectively. We assumed a reduced thermal conductivity of the wires compared to bulk bismuth (κ_{bulk} = 7.87 W m^{-1} K^{-1}) due to finite-size effects. The thermal conductivity of the metal coatings on both sides of the template was 400 W m^{-1} K^{-1}. As mentioned above, the current density j_{max} of the nanowire with 230 nm diameter was 2.31 x 10^6 A/cm^2. Since the generated Joule heat is dissipated to the ambient by convection and radiation at the metallic surfaces, Neumann boundary conditions have been chosen at these surfaces with heat transfer coefficient α = 6.2 W/(m^2 K) for convection and emissivity ε = 0.05 for thermal radiation heat transfer according to the Stefan-Boltzmann law. The calculated temperatures along the nanowire axis and normal to the wire (through its center) along a radial line through the template are presented in Figs. 19(a) and (b), respectively. The temperature of the wire contacts with the metal coatings remains at room temperature while the central 10 μm of the wire are heated to 542 K. Moreover, the temperature decreases to room temperature radially around the wire within 30 μm in the template material. The computed maximal temperature of the wire reaches almost the melting temperature of bulk bismuth (T_m = 544 K) indicating that the wire fails due to Joule heating. The asterisks in Figure 18(b) represent the calculated maximum current densities assuming wire failure due to Joule heating. The computed j_{max} values are in excellent agreement with the exerperimentally determined current denisties.

6. Conclusion

In conclusion, bismuth is, due to its extraordinary electronic properties, namely a long mean free path of 100 nm, a large Fermi wavelength of 40 nm, and a very small electron effective mass (0.001 - 0.26 m_e), an excellent material to study the influence of finite-size and quantum-size effects on the transport properties of nanostructures. For the study of these

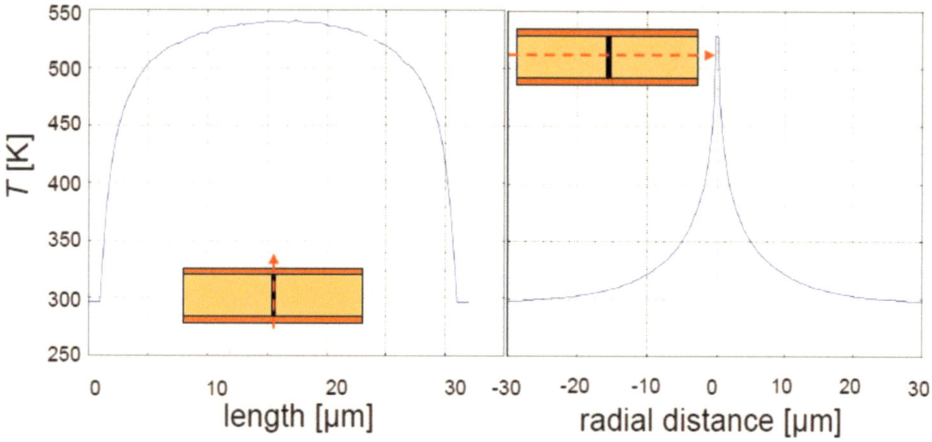

Fig. 19. Simulated temperature profiles for a Bi nanowire of diameter 230 nm embedded in the template along the wire axis and across the wire center. The insets indicate the scanning directions.

effects, nanowire arrays as well as individual nanowires have been created by numerous techniques including a solvothermal process, electron beam lithography, the Taylor process, and the template method. Depending on the fabrication method, polycrystalline and single-crystalline wires with different preferred orientations have been grown. Measurements of the electrical conductance of both nanowire arrays and individual nanowires have evidenced a very strong influence of the wire crystallinity on the electrical resistivity value. In particular, the resistivity increases with respect to bulk with decreasing grain size, as predicted by the Mayadas-Shatzkes model. Moreover, resistance measurements as a function of temperature revealed a strong reduction of the charge carrier mobility originating from additional electron scattering processes at grain boundaries. This reduction of the mobility leads to a non-monotonic resistance vs temperature behaviour for wires consisting of large grains and a monotonic increasing resistance with decreasing temperature for fine-grained wires. In addition, studies of the electrical resistivity on individual wires as a function of the wire diameter revealed an oscillatory behaviour that has been attributed to quantum-size effects causing a splitting of the energy bands in subbands and, thus, to a modulation of the charge carrier density at the Fermi level. These novel findings are further supported by recent infrared spectroscopy experiments. For <012> textured wire arrays infrared absorptions were reported stemming from intraband excitations, i.e., excitations of the electrons from one subband to another. In addition, for single <110> textured wires an infrared absorption was found whose edge shifts to higher energies with decreasing wire diameter due to the shift of the band edges away from each and, hence, increasing the direct band gap in the vicinity of the L-point. The fact that different absorption spectra were found for differently textured nanowires was attributed to the possibility that electrons can be effectively scattered at the surface of spatially confined materials and, thus, making different points of the Brillouin zone accessible. Last but not least, studies on single bismuth nanowires embedded in the polymer template evidenced burn-out current densities 3 to 4 orders of magnitude larger than for bulk bismuth. In addition, the thinner the wire diameter, the larger the maximal current supported by the single nanowire. This effect is attributed to a more efficient heat dissipation from the wire to the surrounding polymer, for nanowires displaying higher surface-to-volume ratio.

7. References

Apel, P.Yu.; Spohr, R.; Trautmann, C.; Vutsadakis, V. (1999). Track structure in polyethylene terephthalate irradiated by heavy ions: Let dependence of track diameter. *Radiat. Meas.* 31 (1999) 51 - 56.

Barati, M.; Sadeghi, E. (2001). Study of ordinary size effect in the electrical conductivity of Bi nanowires. *Nanotechnology* 12 (2001) 277 - 280.

Bate, R.T.; Einspruch, N.G. (1967). Galvanomagnetic studies of Sn-doped Bi. I. Positive Fermi energies. *Phys. Rev.* 153 (1967) 796 - 799.

Black, M.R.; Padi, M.; Cronin, S.B.; Lin, Y.-M.; Rabin, O.; McClure, T.; Dresselhaus, G.; Hagelstein, P.L.; Dresselhaus, M.S. (2000). Intersubband transitions in bismuth nanowires. *Appl. Phys. Lett.* 77 (2000) 4142 - 4144.

Black, M.R.; Lin, Y.-M.; Cronin, S.B.; Rabin, O.; Dresselhaus, M.S. (2002). Infrared absorption in bismuth nanowires resulting from quantum confinement. *Phys. Rev. B* 65 (2002) 195417.

Black, M.R.; Hagelstein, P.L.; Cronin, S.B.; Lin, Y.M.; Dresselhaus, M.S. (2003). Optical absorption from indirect transition in bismuth nanowires. *Phys. Rev. B* 68 (2003) 235417.

Black, J.R. (1969). Electromigration - A brief survey of recent results. *IEEE Trans. Electron. Devices ED*-16 (1969) 338 - 347.

Chiu, P.; Shih, I. (2004). A study of the size effect on the temperature-dependent resistivity of bismuth nanowires with rectangular cross-section. *Nanotechnology* 15 (2004) 1489 - 1492.

Chtanko, N.; Toimil Molares, M.E.; Cornelius, Th.; Dobrev, D.; Neumann, R. (2004). Etched single-ion-track templates for single nanowire synthesis. *J. Phys. Chem. B* 108 (2004) 9950 - 9954.

Cornelius, T.W.; Apel, P.Yu.; Schiedt, B.; Trautmann, C.; Toimil Molares, M.E.; Karim, S.; Neumann, R. (2007). Investigation of nanopore evolution in ion track-etched polycarbonate membranes. *Nucl. Instr. Meth. B* 265 (2007) 553 - 557.

Cornelius, T.W.; Brötz, J.; Chtanko, N.; Dobrev, D.; Miehe, G.; Neumann, R.; Toimil Molares, M.E. (2005). Controlled fabrication of poly- and single-crystalline bismuth nanowires. *Nanotechnology* 16 (2005) S246 - S249.

Cornelius, T.W.; Toimil-Molares, M.E.; Karim, S.; Neumann, R. (2008). Oscillations of electrical conductivity in single bismuth nanowires. *Phys. Rev. B* 77 (2008) 125425.

Cornelius, T.W.; Toimil-Molares, M.E.; Lovrincic, R.; Karim, S.; Neumann, R.; Pucci, A.; Fahsold, G. (2006). Quantum size effects manifest in infrared properties of single bismuth nanowires. *Appl. Phys. Lett.* 88 (2006) 103114.

Cornelius, T.W.; Toimil-Molares, M.E.; Neumann, R.; Karim, S. (2006). Finite-size effects in the electrical transport properties of single bismuth nanowires. *J. Appl. Phys.* 100 (2006) 114307.

Cronin, S.B.; Lin, Y.-M.; Rabin, O.; Black, M.R.; Ying, J.Y.; Dresselhaus, M.S.; Gai, P.L.; Minet, J.-P.; Issi, J.-P. (2002). Making electrical contacts to nanowires with a thick oxide coating. *Nanotechnology*, 13 (2002) 653 - 658.

Dai, H.; Wong, E.W.; Lieber, C.M. (1996). Probing electrical transport in nanomaterials: conductivity of individualcarbon nanotubes. *Science* 272 (1996) 523 - 526.

Dingle, R.B. (1950). The elctrical conductivity of thin wires. *Proc. Roy. Soc. A* 201 (1950) 545 - 560.

Duggal, V.P.; Rup, R. (1969). Thickness-dependent oscillatory behavior of resistivity and Hall coefficient in thin single-crystal bismuth films. *J. Appl. Phys.* 40 (1969) 492 - 495.

Farhangfar, S. (2006). Quantum size effects in a one-dimensional semimetal. *Phys. Rev. B* 74 (2006) 205318.

Farhangfar, S. (2007). Quantum size effects in solitary wires of bismuth. *Phys. Rev. B* 76 (2007) 205437.

Fuchs, K (1938). The conductivity of thin metallic films according to the electron theory of metals. *Proc. Cambridge Philos. Soc.* 34 (1938) 100.

Gao, Y.; Niu, H.; Zeng, C.; Chen, Q. (2003). Preparation and characterization of single-crystalline bismuth nanowires by a low-temperature solvothermal process. *Chem. Phys. Lett.* 367 (2003) 141 - 144.

Gallo, C.F.; Chandrasekhar, B.S.; Sutter, P.H. (1962). Transport properties of bismuth single crystals. *J. Appl. Phys.* 34 (1962) 144 - 152.

Glocker, D.A.; Skove, M.J. (1977). Field effect and magnetoresistance in small bismuth wires. *Phys. Rev. B* 15 (1977) 608 - 616.

Garcia, N.; Kao, Y.H.; Strongin, M. (1972). Galvanomagnetic studies of bismuth films in the quantum-size-effect region. *Phys. Rev. B* 5 (1972) 2029 - 2039.

Heremans, J.; Thrush, C.M.; Lin, Y.-M.; Cronin, S.; Zhang, Z.; Dresselhaus, M.S.; Mansfield, J.F. (2000). Bismuth nanowire arrays: Synthesis and galvanomagneitic properties. *Phys. Rev. B* 61 (2000) 2921 - 2930.

Heremans, J.; Thrush, C.M.; Zhang, Z.; Sun, X.; Dresselhaus, M.S. (1998). Magnetoresistance of bismuth nanowire arrays. A possible transition from one-dimensional to three-dimensional localization. *Phys. Rev. B* 58 (1998) R10091.

Ho, P.S.; Kwok, T. (1989). Electromigration in metals. *Rep. Prog. Phys.* 52 (1989) 301 - 348.

Hofmann, P. (2005). *The surfaces of bismuth: Structural and electronic properties.* University of Aarhus (2005)

Huber, T.E.; Celestine, K.; Graf, M.J. (2003). Magntoquantum oscillations and confinement effects in arrays of 270-nm-diameter bismuth nanowires. *Phys. Rev. B* 67 (2003) 245317.

Isaacson, R.T.; Williams, G.A. (1969). Alfvén-wave propagation in solid-state plasmas. III. Quantum oscillations of the Fermi surface of bismuth. *Phys. Rev.* 185 (1969) 682 - 688.

Kaiser, Ch.; Weiss, G.; Cornelius, T.W.; Toimil Molares, M.E.; Neumann, R. (2009). Low temperature magnetoresistance measurements on bismuth nanowire arrays. *J. Phys.: Condens. Matter* 21 (2009) 205301.

Karim, S.; Toimil-Molares, M.E.; Maurer, F.; Miehe, G.; Ensinger, W.; Liu, J.; Cornelius, T.W.; Neumann, R. (2006). Synthesis of gold nanowires with controlled crystallographic characteristics. *Appl. Phys. A* 84 (2006) 403 - 407.

Lax, B.; Mavroides, J.G.; Zeiger, H.J.; Keyes, R.J. (1960). Infrared magnetoreflection in bismuth. I. High fields. *Phys. Rev. Lett.* 5 (1960) 241 - 243.

Li, L.; Zhang, Y.; Li, G.; Wang, X.; Zhang, L. (2005). Synthetic control of large-area, ordered bismuth nanowire arrays. *Mater. Lett.* 59 (2005) 1223 - 1226.

Lin, Y.-M.; Cronin, S.B.; Ying, J.Y.; Dresselhaus, M.S.; Heremans, J.P. (2000). Transport properties of Bi nanowire arrays. *Appl. Phys. Lett.* 76 (2000) 3944 - 3946.

Lin, Y.-M.; Sun, X.; Dresselhaus, M.S. (2000). Theoretical investigation of thermoelectric transport properties of cylindrical Bi nanowires. *Phys. Rev. B* 62 (2000) 4610 - 4623.

Liu, K.; Chien, C.L.; Searson, P.C.; Yu-Shang, K. (1998). Structural and magneto-transport properties of electrodeposited bismuth nanowires. *Appl. Phys. Lett.* 73 (1998) 1436 - 1438.

MacDonald, D.K.C.; Sarginson, K. (1950). Size effect variation of the electrical conductivity of metals. *Proc. Roy Soc. A* 203 (1950) 223 - 240.

Martin, C.R. (1994). Nanomaterials: A membrane-based synthetic approach. *Science* 266 (1994) 1961 - 1966.

Masuda H.; Nishio, K. (2006). Synthesis and Applications of Highly Ordered Anodic Porous Alumina. In: *Self organized nanoscale materials*, 296 - 312, Springer New York, ISBN 978-0-387-27975-6

Mayadas, A.F.; Shatzkes, M (1970). Electrical-resistivity model for polycrystalline films: the case of arbitrary reflection at external surfaces. *Phys. Rev. B* 1 (1970) 1382 - 1389.

Nikolaeva, A.; Gitsu, D.; Konopko, L.; Graf, M.J.; Huber, T.E. (2009). Quantum interference of surface states in bismuth nanowires probed by the Aharonov-Bohm oscillation of the magnetoresistance. *Phys. Rev. B* 77 (2008) 075332.

Omaggio, J.P.; Meyer, J.R.; Hoffmann, C.A.; DiVenere, A.; Yi, X.J.; Hou, C.L.; Wang, H.C.; Ketterson, J.B.; Wong, G.K.; Heremans, J.P. (1993). Magneto-optical determination of the T-point energy gap in bismuth. *Phys. Rev. B* 48 (1993) 11439 - 11442.

Parrott, J.E. (1965). A new theory of the size effect in electrical conduction. *Proc. Phys. Soc.* 85 (1965) 1143 - 155.

Rogacheva, E.I.; Grigorov, S.N.; Nashchekina, O.N.; Lyubchenko, S.; Dresselhaus, M.S. (2003). Quantum-size effects in *n*-type bismuth thin films. *Appl. Phys. Lett.* 82 (2003) 2628 - 2630.

Sandormirskiĭ, V.B. (1967). Quantum size effects in a semimetal film. *Sov. Phys. JETP* 25 (1967) 101

Schönenberger, C.; van der Zande, B.M.I.; Fokkink, L.G.J.; Henny, M.; Schmid, C.; Krüger, M.; Bachtold, A.; Huber, R.; Birk, H.; Staufer, U. (1997). Template synthesis of nanowires in porous polycarbonate membranes: Electrrochemistry and morphology. *J. Phys. Chem. B* 101 (1997) 5497 - 5505.

Siwy, Z.; Apel, P.; Baur, D.; Dobrev, D.; Korchev, Y.E.; Neumann, R.; Spohr, R.; Trautmann, C.; Voss, K.O. (2003). Preparation of synthetic nanopores with transport properties analogous to biological channels. *Surf. Sci.* 532-535 (2003) 1061 - 1066.

Tian, M.; Wang, J.; Kumar, N.; Han, T.; Kobayashi, Y.; Liu, Y.; Mallouk, T.E.; Chan, M.H.W. (2006). Observation of superconductivity in granular Bi nanowires fabrictaed by electrodeposition. *Nano Lett.* 6 (2006) 2773 - 2780.

Toimil Molares, M.E.; Buschmann, V.; Dobrev, D.; Neumann, R.; Scholz, R.; Schuchert, I.U.; Vetter, J. (2001). Single-crystalline copper nanowires produced by electrochemical deposition in polymeric ion track membranes. *Adv. Mater.* 13 (2001) 62 - 65.

Toimil Molares, M.E.; Chtanko, N.; Cornelius, T.W.; Dobrev, D.; Enculescu, I.; Blick, R.H.; Neumann, R. (2004). Fabrication and contacting of single Bi nanowires. *Nanotechnology* 15 (2004) S201 - S207.

Völklein, F.; Kessler, E. (1986). Analysis of the lattice thermal conductivity of thin films by means of a modified Mayadas-Shatzkes model. *Thin Solid Films* 142 (1986) 169 - 181.

Wang, X.F.; Zhang, L.D.; Zhang, J.; Shi, H.Z.; Peng, X.S.; Zheng, M.J.; Fang, J.; Chen, J.L.; Gao, B.J. (2001). Ordered single-crystalline Bi nanowire arrays embedded in nanochannels of anodic alumina membranes. *J. Phys. D: Appl. Phys.* 34 (2001) 418 - 421.

Yin, A.J.; Li, J.; Jian, W.; Bennett, A.J.; Xu, J.M. (2001). Fabrication of highly ordered metallic nanowire arrays by electrodeposition. *Appl. Phys. Lett.* 79 (2001) 1039 - 1041.

Zhang, Z.; Ying, J.Y.; Dresselhaus, M.S. (1998). Bismtuh quantum-wire arrays fabricated by a vacuum melting and pressure injection process. *J. Mater. Res.* 13 (1998) 1745 - 1748.

Zhang, Z.; Sun, X.; Dresselhaus, M. S.; Ying, J. Y.; Heremans, J. (2000). Electronic transport properties of single-crystal bismuth nanowire arrays. *Phys. Rev. B* 61 (2000) 4850 - 4861.

15

Electronic States and Piezoresistivity in Silicon Nanowires

Koichi Nakamura, Dzung Viet Dao, Yoshitada Isono,
Toshiyuki Toriyama, and Susumu Sugiyama
Ritsumeikan University
Japan

1. Introduction

Recent progress in integrated-circuit processes can reduce the typical dimensions of the piezoresistive sensor to the microscale level to realize a micro-electro-mechanical systems (MEMS) device. However, size reduction causes a drastic loss in sensitivity for MEMS piezoresistive inertial sensors. For example, the sensitivity reduction of a MEMS piezoresistive accelerometer approximately is proportional to $1/L^3$–$1/L^4$ scaling, and 1 mm square may be the size limit for commercial applications (Sasaki et al., 2007). Therefore, a novel nanoscale piezoresistor with higher piezoresistance than bulk counterparts is necessary to fabricate microscale piezoresistive sensors free from sensitivity loss due to size reduction.

Single-crystal silicon has been one of the most important materials for piezoresistive sensors in MEMS or nano-electro-mechanical systems (NEMS) technology, because silicon has high crystallinity and mechanical stability even at the microscale level. These characters in silicon give a great expectation that some nanoscale mechanical sensors will be easily developed by the miniaturization of single-crystal silicon as component materials. However, the miniaturization should cause a drastic change in the electronic behavior in silicon even if the mechanical stability is maintained. In particular, low-dimensional materials such as nanosheets or nanowires have a radically different electronic state, that controls electrical conductivity, and accordingly, the piezoresistance coefficients in a low-dimensional nanomaterial would also be markedly different from those in the same material at the microscale level, combined with the large effect of strain and stress over the nanoscale material.

Silicon nanowire (SiNW) has been studied experimentally and theoretically as a particularly attractive candidate material composing MEMS/NEMS devices. We have succeeded in fabricating a p-doped single-crystal ⟨110⟩-oriented SiNW by electron beam (EB) direct writing and reactive ion etching (RIE) with a 50 × 50 nm^2 cross-sectional area in separation by implanted oxygen (SIMOX) substrate and in measuring the piezoresistance coefficients (Toriyama & Sugiyama, 2003). The SiNW we fabricated has a longitudinal piezoresistance coefficient $(\pi_{l\langle 110\rangle})^p$ of 38.7×10^{-11} Pa^{-1} at a surface concentration N_s of 9×10^{19} cm^{-3}, which is 54.8% larger than that of the p^+ diffused microscale piezoresistor obtained by Tufte and Stelzer, $\pi_l = 25 \times 10^{-11}$ Pa^{-1} (Tufte & Stelzer, 1963).

Figure 1 shows the latest SEM image of one SiNW fabricated by our group (Tung et al., 2009). The width dependence of $(\pi_{l\langle 110\rangle})^p$ is shown in Fig. 2. As the width of SiNW gets thinner, the value of $(\pi_{l\langle 110\rangle})^p$ increases clearly. This result proves that the miniaturization of SiNW piezoresistor is extremely effective to improve the piezoresistive-based sensing function.

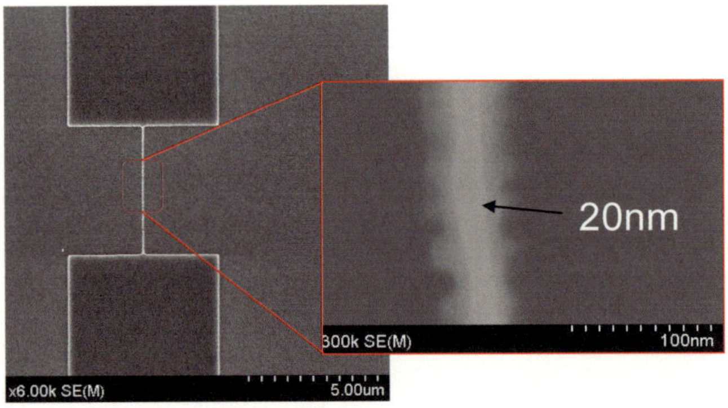

Fig. 1. SEM image of 2-terminal p-SiNW piezoresistor: Longitudinal direction of SiNW is ⟨110⟩

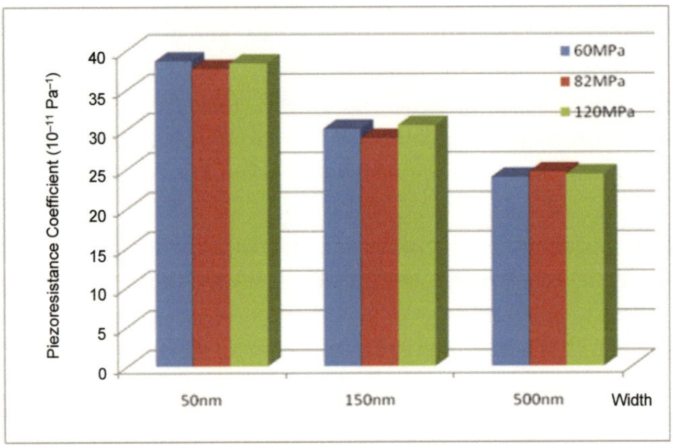

Fig. 2. Longitudinal piezoresistance coefficient $(\pi_{l\langle 110\rangle})^p$ with respect to wire width of p-SiNW

Recently, He and Yang reported that p-doped ⟨111⟩ SiNW has a giant piezoresistance coefficient, $(\pi_{l\langle 111\rangle})^p = -3{,}550 \times 10^{-11}$ Pa^{-1}, according to their experimental measurement (He & Yang, 2006), and their experimental result of high piezoresistivity was supported through first-principles calculations of ⟨111⟩ SiNW without termination atoms at the wire wall (Cao et al., 2007), where the wave functions that contribute to hole transport were obviously localized in dangling bonds on the wire wall. However, dangling bonds are generally too reactive to exist for a long time; in general, they are attacked by oxygen, and occasionally by hydrogen (H) or halogen; therefore, piezoresistive properties of Si without dangling bonds will be practical in the application of MEMS/NEMS devices.

H-terminated SiNW models are effective for representating SiNW without dangling bonds. In this chapter, we present the simulation of the strain response to electronic states and the piezoresistance coefficients in H-terminated single-crystal SiNWs by using first-principles calculations of model structures with ⟨001⟩, ⟨110⟩, and ⟨111⟩ wire crystallographic-orientations, in preparation for nanoscale piezoresistor applications (Nakamura et al., 2008; Nakamura et al., 2009a).

2. Method of calculation

2.1 First-principle calculation

We have carried out first-principles calculations of the periodic boundary models for bulk Si crystals and SiNW models using VASP (Kresse & Hafner, 1993; Kresse & Furthmüller, 1996) and FHI98md (Bockstedte et al., 1997) program packages, on the basis of the density functional theory (DFT) (Hohenberg & Kohn, 1964). For the DFT exchange-correlation interaction, the generalized-gradient approximation (GGA) method was used with the Perdew-Burke-Ernzerhof (PBE) functional (Perdew et al., 1996). We adopted the three-dimensional supercell approximation technique with norm-conserving pseudopotentials (Hamann, 1989). The cutoff energy for wave functions of electrons with plane-wave expansion was set at 40 Ry (544 eV) for bulk Si crystals and 30 Ry (408 eV) for SiNWs.

2.2 SiNW models

SiNW models have been devised by cutting out a fragment with a one-dimensional periodic boundary from the optimized bulk Si, and all dangling bonds of Si atoms were terminated with H atoms. The direction of the fragment with the one-dimensional periodic boundary can be defined as the longitudinal direction of SiNW, parallel to the z-axis in the three-dimensional supercell as shown in Fig. 3. We have devised ⟨001⟩, ⟨110⟩, and ⟨111⟩ SiNW models, according to the orientation of the fragment on the xy-plane, where the longitudinal direction is respectively set to [001], [110], and [111]. We have set up tetragonal supercells for ⟨001⟩ and ⟨110⟩ SiNW models, and a periodic boundary condition along the transverse directions, or perpendicular directions to the wire axis, was given by inserting sufficient space between H-terminated SiNWs with two large cell parameters of the supercell along x- and y-axes. For ⟨111⟩ SiNW models, hexagonal supercells have been introduced with two large cell parameters perpendicular to the z-axis.

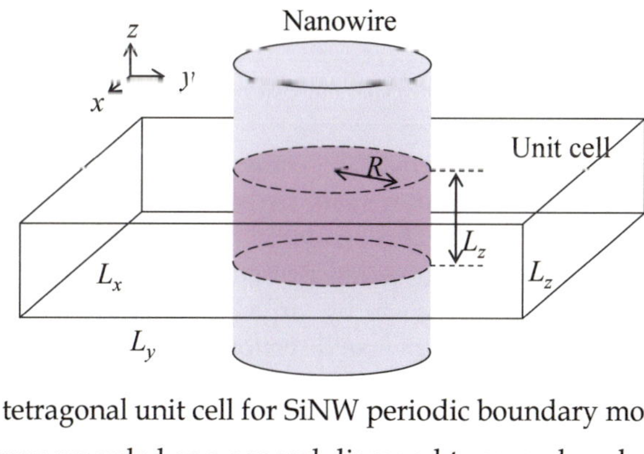

Fig. 3. Image of the tetragonal unit cell for SiNW periodic boundary model

The bulk Si crystal was regarded as a general diamond type, and we have optimized the cell parameters of a unit cell containing two Si atoms. We obtained the optimized cubic lattice constant of the diamond-type Si crystal as 5.463 Å. The effect of uniaxial tensile strain on structure was represented by partial optimization with a fixed lattice constant along the tensile direction. For 1% uniaxial tensile strain in the [001] direction, we set [001] lattice constant at 5.518 Å and optimized [100] and [010] lattice constants to 5.449 Å, reduced by 0.25%. For 1% uniaxial tensile strain in the [110] direction, [001] lattice constants were optimized to 5.447 Å and [1$\bar{1}$0] translational vectors were slightly reduced by less than 0.01%. These fully and partially optimized parameters have been applied to structures of strain-free and tensile-strained SiNW models, respectively, as shown in Fig. 4.

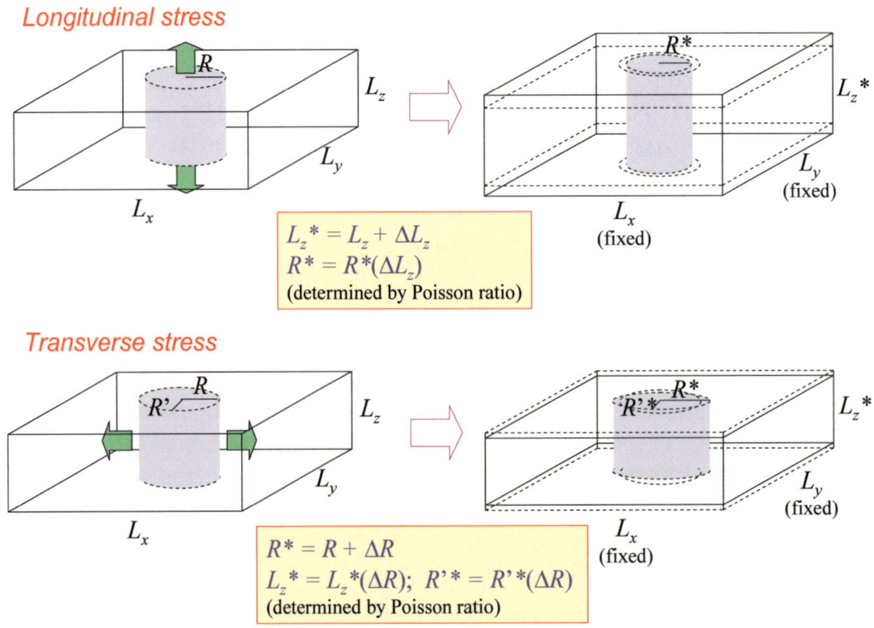

Fig. 4. Setting of tensile-strained SiNW models

Figure 5 displays the SiNW models we have devised in this chapter. Structural parameters of SiNW models and their unit cells are summarized in Table 1. The wire radius R is defined simply as the distance between the wire axis and the farthest nuclear including H atoms, and the volume of SiNW in the unit cell is determined by $V = \pi R^2 L_z$.

3. Band structures

The electronic band structure of each SiNW model was given in terms of the three-dimensional coordinates of the reciprocal space because we adopted the three-dimensional supercell approximation technique. An example of a band diagram for the three-dimensional band structure in the tetragonal Brillouin zone is shown in Fig. 6. Sufficient space between H-terminated SiNWs given by large cell parameters L_x and L_y causes band energies to be invariable along the transverse directions in the reciprocal space, for example, Z-R, X-Γ, Z-T, and M-Γ paths in the band diagram shown in Fig. 6. In other words, the invariance of band energies along these paths ensures "a piece of SiNW" without interaction between SiNWs. Thus, the band structure is reduced to be one dimensional, which is dependent on only one reciprocal coordinate, k_z, and therefore, the band diagram can only be represented along the Γ-Z path. For the hexagonal supercells, we have obtained the band structures that are dependent on only k_z with large transverse cell parameters as in the case of the tetragonal supercells. All of the band structures for strain-free SiNW models give a

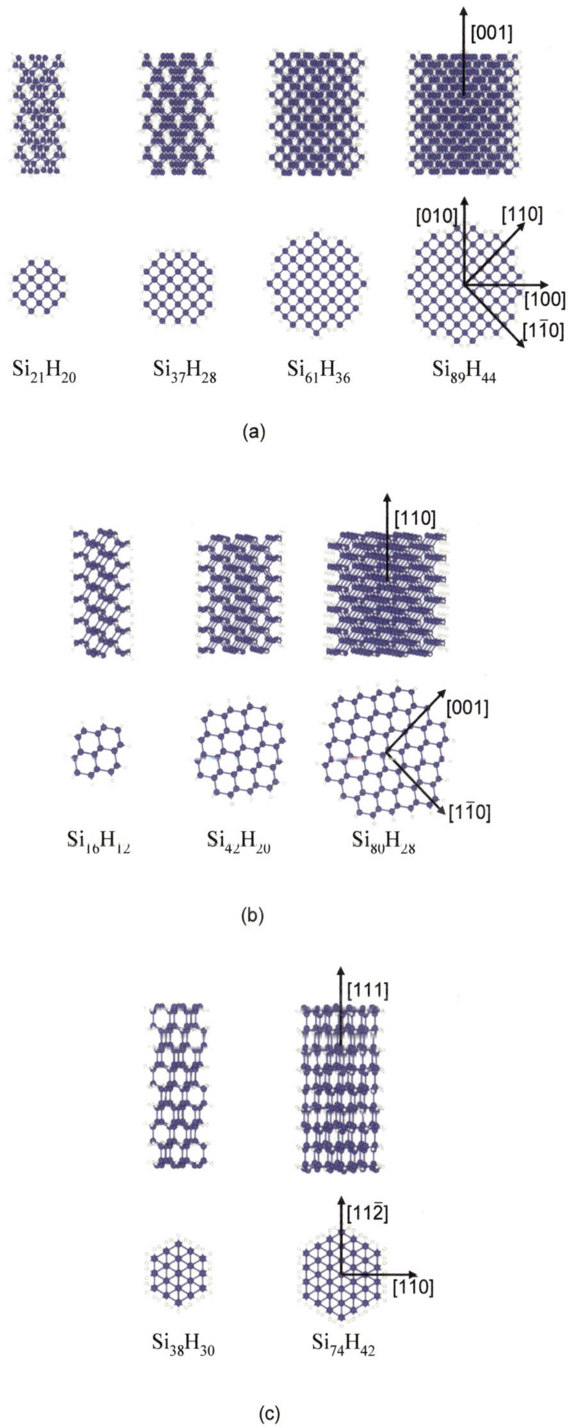

Fig. 5. Side and top views of H-terminated SiNW models: (a) ⟨001⟩ series with 4 unit cells, (b) ⟨110⟩ series with 6 unit cells, and (c) ⟨111⟩ series with 3 unit cells

	Radius	Cell parameters		
	R (Å)	L_x (Å)	L_y (Å)	L_z (Å)
⟨001⟩ series				
$Si_{21}H_{20}$	5.435	13.229	13.229	5.463
$Si_{37}H_{28}$	7.272	16.934	16.934	5.463
$Si_{61}H_{36}$	9.149	21.167	21.167	5.463
$Si_{89}H_{44}$	11.044	25.401	25.401	5.463
⟨110⟩ series				
$Si_{16}H_{12}$	5.641	13.229	13.229	3.863
$Si_{42}H_{20}$	9.075	21.167	21.167	3.863
$Si_{80}H_{28}$	12.899	27.517	27.517	3.863
⟨111⟩ series				
$Si_{38}H_{30}$	5.865	13.229	13.229	9.463
$Si_{74}H_{42}$	8.096	18.521	18.521	9.463

Table 1. Structural parameters (referred to Fig. 3) of SiNW models and their tetragonal (⟨001⟩ and ⟨110⟩ series) and hexagonal (⟨111⟩ series) unit cells

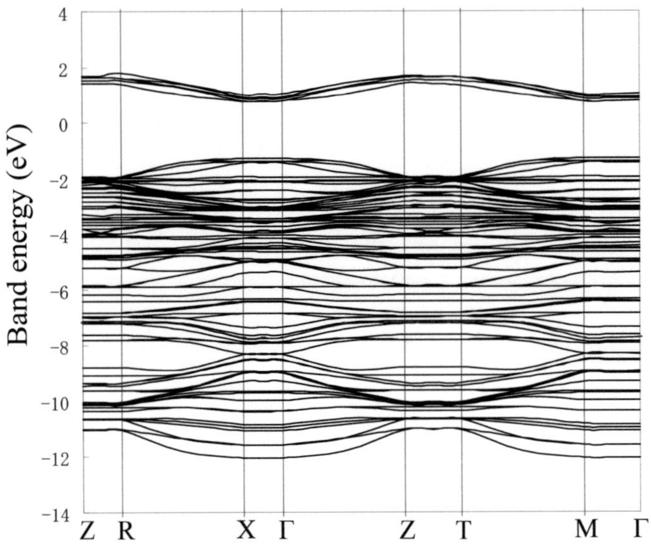

Fig. 6. Band energy diagram of $Si_{37}H_{28}$⟨001⟩ SiNW model: R–X, Γ–Z, and T–M sections correspond to the longitudinal paths, and other sections are the transverse ones

direct band gap at the Γ point. As the radius of the SiNW model increases for each orientation series, the band gap gets lower and asymptotically approaches 0.612 eV, the computational value of band gap in bulk Si obtained by our calculation, as shown in Fig. 7. The transport properties of SiNWs have been discussed in terms of the band structures for $Si_{89}H_{44}$⟨001⟩ (wire diameter $2R$ = 2.21 nm), $Si_{80}H_{28}$⟨110⟩ ($2R$ = 2.58 nm), and $Si_{74}H_{42}$⟨111⟩ ($2R$ = 1.62 nm) SiNW models, which respectively have the largest wire diameters in the ⟨001⟩, ⟨110⟩, and ⟨111⟩ series we have devised. The longitudinal tensile-strained model for $Si_{89}H_{44}$⟨001⟩ SiNW has 1% strain in the [001] direction, and the transverse tensile-strained model shows the same level in the [100] (or [010]) direction. The tensile-strained models for $Si_{80}H_{28}\langle 110\rangle$ SiNW have 1% strain in the [110] direction as the longitudinal tension, or in the $[1\bar{1}0]$ direction as the transverse tension. For $Si_{74}H_{42}\langle 111\rangle$ SiNW, 1% strain is adopted in the [111] direction as the longitudinal tension, or in the $[1\bar{1}0]$ direction as the transverse tension. We have observed the variations of the band structures with respect to strain on the SiNW models. Band diagrams in the vicinity of the valence-band (VB) top and the conduction-band (CB) bottom with/without tensile strain are shown in Figs. 8–13.

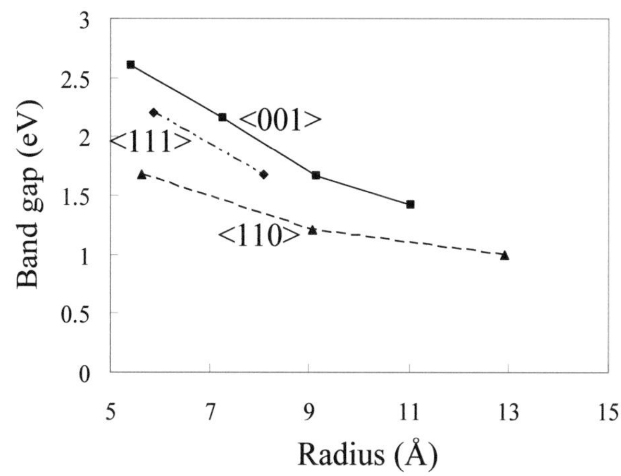

Fig. 7. Relationship between band gaps of SiNW models and wire radius

4. Theory of carrier conductivity

The electrical conductivity tensor G or the electrical resistivity tensor ρ can be represented in terms of carrier density and effective mass tensors (Kittel, 2005). Variations of band structure will exert an influence on them, and frequently contribute to a sudden turn of the conductivity. We have introduced the band carrier densities and their corresponding effective masses for each energy band, and the conductivity has been added up over all VB and CB subbands as follows,

$$G = \rho = e^2 \left(\sum_{j \in \mathrm{CB}} n_j \boldsymbol{m}_{\mathrm{e},j}{}^* \cdot \boldsymbol{\tau}_{\mathrm{e},j} + \sum_{j \in \mathrm{VB}} p_j \boldsymbol{m}_{\mathrm{h},j}{}^* \cdot \boldsymbol{\tau}_{\mathrm{h},j} \right), \quad (1)$$

where n_j is the jth CB carrier electron density, p_j is the jth VB hole density, $\boldsymbol{m}_{\mathrm{e},j}{}^*$ and $\boldsymbol{m}_{\mathrm{h},j}{}^*$ are the effective masse tensors, $\boldsymbol{\tau}_{\mathrm{e},j}$ and $\boldsymbol{\tau}_{\mathrm{h},j}$ are the relaxation time tensors, and e^2 is the square of the absolute value of the elementary electric charge. Subscripts e and h respectively denote electron and hole carriers. The band carrier densities are controlled by the Fermi energy E_{F} and temperature T,

$$n_j = \frac{2}{V} \sum_{k_z} w_{k_z} \left\{ \exp\left(\frac{E_{j,k_z} - E_{\mathrm{F}}}{k_{\mathrm{B}} T} \right) + 1 \right\}^{-1}; \quad (2)$$

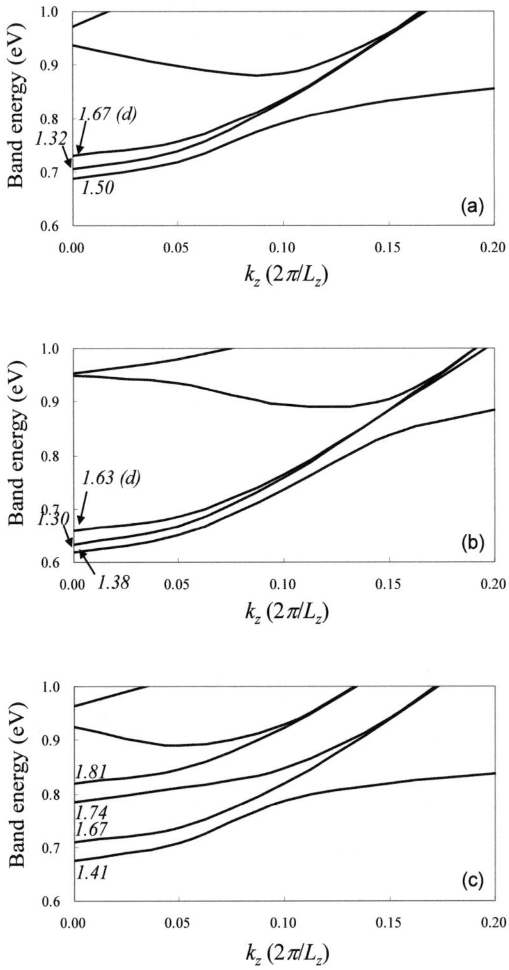

Fig. 8. Band diagram of $Si_{89}H_{44}$⟨001⟩ SiNW model in the vicinity of the conduction-band bottom (a) without stress, (b) with 1% longitudinal tensile stress, and (c) with 1% transverse tensile stress. Italic numbers denote the effective masses in units of m_0, the rest mass of an electron. *(d)* indicates degenerate subbands.

$$p_j = \frac{2}{V}\sum_{k_z} w_{k_z}\left\{\exp\left(-\frac{E_{j,k_z}-E_F}{k_B T}\right)+1\right\}^{-1}, \tag{3}$$

where $E_{j,kz}$ is the band energy of the jth subband at the k_z point, w_{kz} is the k-point weight for k_z, V is the volume of SiNW in the unit cell, and k_B is the Boltzmann constant. We have performed a sampling with 11 k_z points along the Γ–Z path. In the intrinsic semiconductor state, these carrier densities should satisfy

$$\sum_{j\in CB} n_j = \sum_{j\in VB} p_j \, . \tag{4}$$

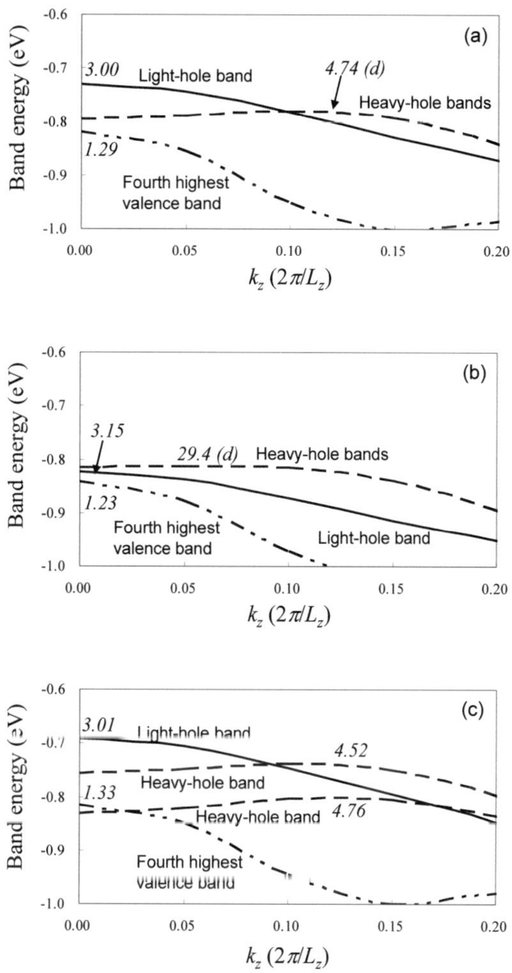

Fig. 9. Band diagram of $Si_{89}H_{44}$⟨001⟩ SiNW model in the vicinity of the valence-band top; (a)-(c) denote the same conditions as those in Fig. 8. Italic numbers denote the effective masses in units of m_0. *(d)* indicates degenerate subbands.

Intrinsic semiconductors have only an infinitesimal number of carriers, so that n- or p-type semiconductors are generally applied to piezoresistive sensors. However, an excess or lack of electrons as large as one particle per unit cell in order to carry out regular first-principles calculation should lead to an enormous overestimation of carrier density, about 5×10^{20} cm^{-3}, for our SiNW models. In actual n- or p-type SiNWs, the total number of carriers per unit cell, δ, must be less by a few orders than 1. Under the condition that a small amount of the carrier occupation does not cause significant change in the band structure, δ can be given by an appropriate shift of the Fermi energy as

$$\delta = \sum_{j \in \mathrm{CB}} n'_j V = 2 \sum_{j \in \mathrm{CB}} \sum_{k_z} w_{k_z} \left\{ \exp\left(\frac{E_{j,k_z} - E'_{\mathrm{F}}}{k_{\mathrm{B}} T} \right) + 1 \right\}^{-1} \qquad (5)$$

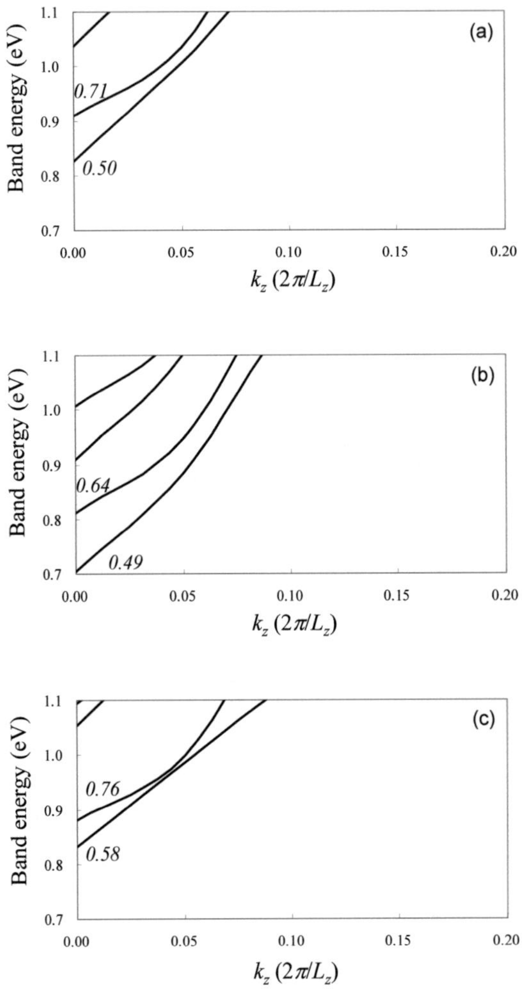

Fig. 10. Band diagram of $Si_{80}H_{28}\langle 110\rangle$ SiNW model in the vicinity of the conduction-band bottom: (a)–(c) denote the same conditions as those in Fig. 8. Italic numbers denote the effective masses in units of m_0.

in the n-type carrier occupation with an upward shift of the Fermi energy updated to E'_F, or

$$\delta = \sum_{j\in \mathrm{VB}} p'_j V = 2\sum_{j\in \mathrm{VB}}\sum_{k_z} w_{k_z}\left\{\exp\left(-\frac{E_{j,k_z}-E'_F}{k_B T}\right)+1\right\}^{-1} \tag{6}$$

in the p-type carrier occupation with a downward shift, where the set of $\{E_{j,kz}\}$ is identical to that in the intrinsic semiconductor state. Practically, we first have set δ to an appropriate constant such as 10^{-n} (n = 2, 3, and 4), and then E'_F in n- and p-type carrier occupations have been solved according to eqs. (5) and (6), respectively.

The effective mass is generally given by a 3 × 3 tensor. The reciprocal matrix of effective mass is defined as

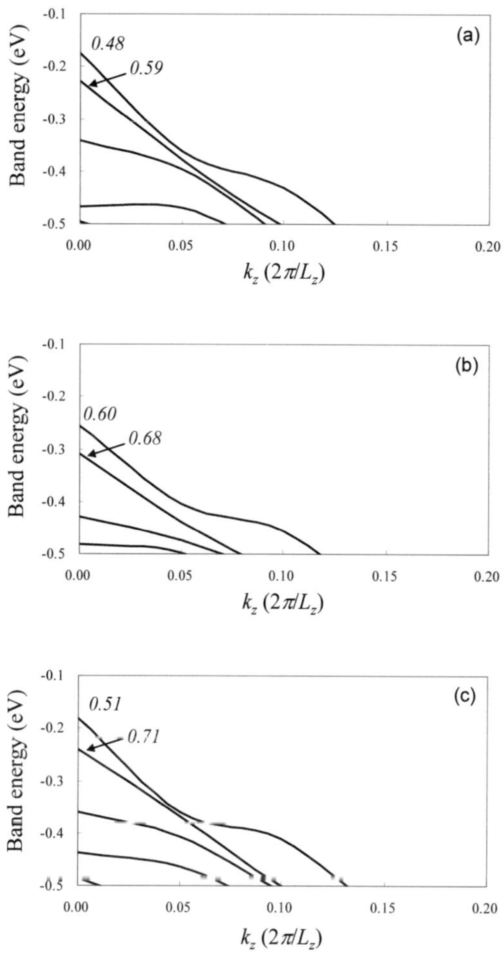

Fig. 11. Band diagram of $Si_{80}H_{28}\langle 110\rangle$ SiNW model in the vicinity of the valence-band top: (a)-(c) denote the same conditions as those in Fig. 8. Italic numbers denote the effective masses in units of m_0.

$$(m_j^*)^{-1} = \pm\frac{1}{\hbar^2}\begin{pmatrix} \dfrac{\partial^2 E_j}{\partial k_x^2} & \dfrac{\partial^2 E_j}{\partial k_x \partial k_y} & \dfrac{\partial^2 E_j}{\partial k_x \partial k_z} \\ \dfrac{\partial^2 E_j}{\partial k_y \partial k_x} & \dfrac{\partial^2 E_j}{\partial k_y^2} & \dfrac{\partial^2 E_j}{\partial k_y \partial k_z} \\ \dfrac{\partial^2 E_j}{\partial k_z \partial k_x} & \dfrac{\partial^2 E_j}{\partial k_z \partial k_y} & \dfrac{\partial^2 E_j}{\partial k_z^2} \end{pmatrix}, \tag{7}$$

where $\hbar$ is equal to Planck's constant divided by 2π. On the right hand, a positive sign is adopted for carrier electrons and a negative sign for holes. As mentioned above, the band energies of our SiNW models remain constant along the transverse directions, namely,

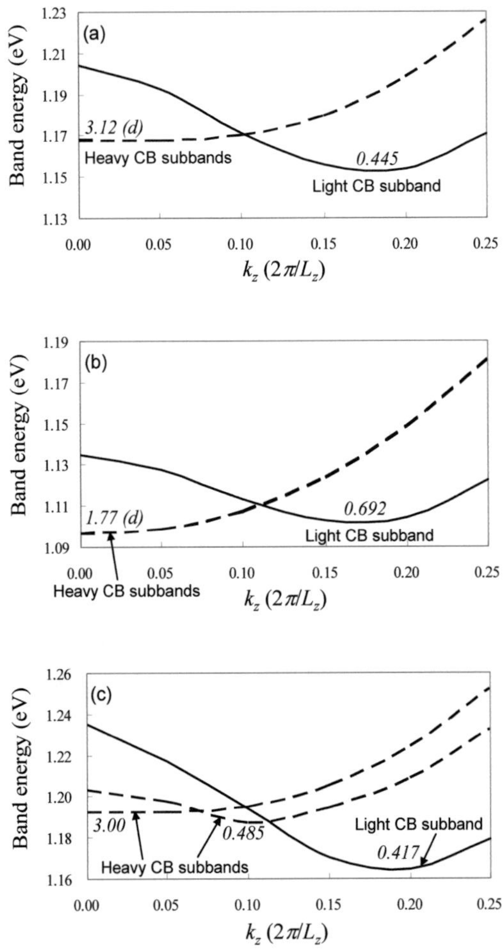

Fig. 12. Band diagram of $Si_{74}H_{42}\langle 111\rangle$ SiNW model in the vicinity of the conduction-band bottom: (a)–(c) denote the same conditions as those in Fig. 8. Italic numbers denote the effective masses in units of m_0. *(d)* indicates degenerate subbands.

$$\partial E_j / \partial k_x = \partial E_j / \partial k_y = 0, \tag{8}$$

and therefore, the effective mass of the *j*th band for the SiNW models can be defined simply as a scalar,

$$m_j{}^* = \pm\hbar^2 \left(\partial^2 E_j / \partial k_z^2 \right)^{-1} \tag{9}$$

at the maximum or minimum of E_j. The second derivative on the right hand of eq. (9) has been estimated numerically as

$$\frac{\partial^2 E_j[k_z]}{\partial k_z^2} = \frac{E_j[k_z + \Delta k_z] + E_j[k_z - \Delta k_z] - 2E_j[k_z]}{(\Delta k_z)^2} \tag{10}$$

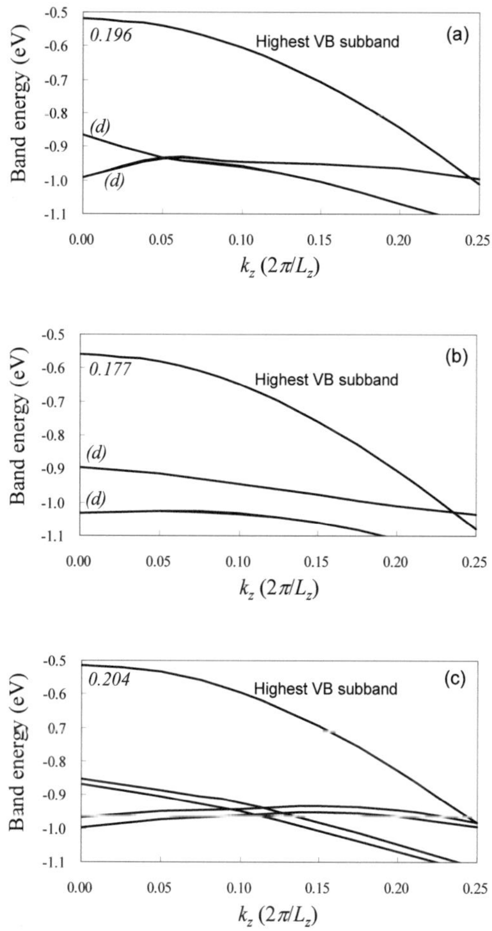

Fig. 13. Band diagram of $Si_{74}H_{42}\langle 111 \rangle$ SiNW model in the vicinity of the valence-band top: (a)–(c) denote the same conditions as those in Fig. 8. Italic numbers denote the effective masses in units of m_0. *(d)* indicates degenerate subbands.

with $\Delta k_z = 0.05 \times (2\pi L_z)$. For the relaxation time, we have introduced the approximation that all of the band relaxation times are equal and constant regardless of stress. This procedure seems to be rough to some extent, but the variation rate of carrier conductivity can be easily and adequately represented in consideration of the canceling of almost part of band relaxation times (Nakamura et al., 2009b).

5. Prediction of piezoresistance coefficients

The longitudinal piezoresistance coefficient π_l and the transverse one π_t are given by

$$\pi_l = \Delta\rho_l / \rho_0\sigma_l; \ \pi_t = \Delta\rho_t / \rho_0\sigma_t, \tag{11}$$

where $\sigma_{l,t}$ are the longitudinal and transverse tensile stresses, ρ_0 is the resistivity along the wire axis without stress, and $\Delta\rho_{l,t}$ are variations in ρ_0 due to $\sigma_{l,t}$. On the approximation of band relaxation times mentioned above, eq. (11) in the n- and p-type states is rewritten with respect to strain ε as follows,

$$(\pi_\alpha)^{\mathrm{n}} = \frac{1}{Y_\alpha \varepsilon_\alpha}\left(\sum_{j \in \mathrm{CB}} \frac{n'_{j,0}}{n'_{j,\alpha}} \frac{m_{\mathrm{e},j,\alpha}{}^*}{m_{\mathrm{e},j,0}{}^*} - 1 \right); \tag{12}$$

$$(\pi_\alpha)^{\mathrm{p}} = \frac{1}{Y_\alpha \varepsilon_\alpha}\left(\sum_{j \in \mathrm{VB}} \frac{p'_{j,0}}{p'_{j,\alpha}} \frac{m_{\mathrm{h},j,\alpha}{}^*}{m_{\mathrm{h},j,0}{}^*} - 1 \right), \tag{13}$$

where n'_j and p'_j are defined in eqs. (5) and (6), respectively, and Y is Young's modulus leading to normal stress with a linear response approximation according to the classical Hooke's law,

$$\sigma_{\mathrm{l}} = Y_{\mathrm{l}}\varepsilon_{\mathrm{l}};\ \sigma_{\mathrm{t}} = Y_{\mathrm{t}}\varepsilon_{\mathrm{t}}, \tag{14}$$

Subscript α in eqs. (12) and (13) denotes l or t, which means a property in relation to longitudinal or transverse tensile strain, and 0 denotes the strain-free condition. The value of ε_α is 0.01 strain for all cases in this chapter, and we assumed Y_α to be the experimental values of Young's modulus of bulk silicon (Wortman & Evans, 1965); 130.2 GPa for longitudinal and transverse tensile stresses in the $Si_{89}H_{44}\langle 001\rangle$ model, 168.9 GPa for both tensile stresses in the $Si_{80}H_{28}\langle 110\rangle$ model, and 187.5 GPa for longitudinal tensile stress and 168.9 GPa for transverse one in the $Si_{74}H_{42}\langle 111\rangle$ model.

Table 2 summarizes calculation results of the piezoresistance coefficients with respect to δ. Obviously, the values of $(\pi_{\mathrm{l}\langle 001\rangle})^{\mathrm{p}}$, longitudinal piezoresistance coefficients for the p-type $Si_{89}H_{44}\langle 001\rangle$ model, are by far the largest in Table 2. We have obtained 147×10^{-11} Pa^{-1} for $(\pi_{\mathrm{l}\langle 001\rangle})^{\mathrm{p}}$ with $\delta = 10^{-4}$ (Nakamura et al., 2008), and it is expected that p-type ⟨001⟩ SiNW will have giant longitudinal piezoresistivity, in consideration of the result that $(\pi_{\mathrm{l}\langle 001\rangle})^{\mathrm{p}}$ is about 10 times as large as $(\pi_{\mathrm{l}\langle 110\rangle})^{\mathrm{p}}$, which has been measured experimentally as 38.7×10^{-11} Pa^{-1} (Toriyama & Sugiyama, 2003), with a small carrier density.

δ	Carrier (cm^{-3})	n-doped piezoresistance constant (10^{-11} Pa^{-1})					
		$\pi_{\mathrm{l}\langle 001\rangle}$	$\pi_{\mathrm{t}\langle 001\rangle}$	$\pi_{\mathrm{l}\langle 110\rangle}$	$\pi_{\mathrm{l}\langle 110\rangle}$	$\pi_{\mathrm{l}\langle 111\rangle}$	$\pi_{\mathrm{t}\langle 111\rangle}$
10^{-4}	5×10^{16}	−3.37	0.84	−1.02	12.1	19.6	−19.2
10^{-3}	5×10^{17}	−3.66	0.63	−1.00	12.1	19.6	−19.1
10^{-2}	5×10^{18}	−3.66	0.77	−1.18	12.1	19.5	−18.8
δ	Carrier (cm^{-3})	p-doped piezoresistance constant (10^{-11} Pa^{-1})					
		$\pi_{\mathrm{l}\langle 001\rangle}$	$\pi_{\mathrm{t}\langle 001\rangle}$	$\pi_{\mathrm{l}\langle 110\rangle}$	$\pi_{\mathrm{l}\langle 110\rangle}$	$\pi_{\mathrm{l}\langle 111\rangle}$	$\pi_{\mathrm{t}\langle 111\rangle}$
10^{-4}	5×10^{16}	147	−1.44	14.2	4.71	−4.87	2.60
10^{-3}	5×10^{17}	146	−1.41	14.4	4.72	−4.87	2.60
10^{-2}	5×10^{18}	146	−1.41	14.1	4.53	−4.87	2.60

Table 2. Theoretical values of piezoresistance coefficients for H-terminated SiNW models

A sign of the giant longitudinal piezoresistivity in p-type H-terminated ⟨001⟩ SiNW can be observed in the variations of the VB structure. The $Si_{89}H_{44}\langle 001\rangle$ model has four VB subbands in the vicinity of the VB top as shown in Fig. 9. The highest VB subband of those without stress [Fig. 9(a)] has a small hole effective mass, called *the light-hole subband*, and by contrast, two of the second highest VB subbands in degeneracy have a larger hole effective mass, called *the heavy-hole subbands*. The uniaxial tensile stress in the [001] longitudinal direction [Fig. 9(b)] causes a sharp drop in the band energy of the light-hole subband, leading to the alternation of the order of band energy levels between the light-hole subband and the heavy-hole subbands, and then most of the holes will be redistributed to the heavy-hole subbands where hole effective mass is markedly raised due to the longitudinal tensile stress. This sudden change in the hole occupation with the increase in effective mass will bring a significant decrease in the hole conductivity in the p-type H-terminated ⟨001⟩ SiNW. The hole occupation numbers and effective masses of each band with/without stress are summarized in Table 3. $(\pi_{t\langle 001\rangle})^{p}$ should be very small because the hole occupation numbers and effective masses do not change so much due to the transverse tensile stress. Similarly, a drastic change due to tensile stresses in the carrier distribution or effective masses is not seen in Figs. 8, 10, and 11.

	strain-free		1% longitudial tensile-strained		1%transverse tensile-strained	
Subband	p_jV/δ	$m_{h,j}/m_0$	p_jV/δ	$m_{h,j}/m_0$	p_jV/δ	$m_{h,j}/m_0$
Light-hole	0.601	3.00	0.131	3.15	0.716	3.01
Heavy-hole	0.193	4.74	0.415	29.4	0.257	4.52
	0.193	4.74	0.415	29.4	0.023	4.76
4th highest	0.012	1.29	0.038	1.23	0.004	1.33

Table 3. Hole occupation numbers and effective masses of each band in the $Si_{89}H_{44}\langle 001\rangle$ model for $\delta = 10^{-3}$; m_0 denotes the rest mass of an electron

For the $Si_{74}H_{42}\langle 111\rangle$ model, the band energy of the highest VB subband shown in Fig. 13 is maximized at the Γ point. As seen in the band structure of the same model (Li & Freeman, 2006), the highest VB subband is located far from other lower subbands by more than 0.3 eV at the Γ point. Thus, it is easily found that most holes (more than 99.999% at 300 K) occupy the highest VB subband if there are no dangling bonds on the wire wall. The change in the hole occupation, as observed in the H-terminated ⟨001⟩ SiNW model, cannot take place, and accordingly, the conductivity depends only on the hole mobility or the band hole effective mass of the highest VB subband. Our calculation results of the band hole effective mass, 0.177–0.204, are not greatly affected by tensile strains, and as a result, no drastic change due to strain in the conductivity can be expected.

The longitudinal and transverse piezoresistance coefficients for the p-type H-terminated ⟨111⟩ SiNW model are very small and not dependent on temperature or hole concentration, because their values do not contain the contribution from the change in hole distribution controlled by temperature and concentration. The result that piezoresistance coefficients for p-type H-terminated ⟨111⟩ SiNW are small seems to be contradictory to the experimental result (He & Yang, 2006). However, in their experiment, SiNWs with a giant piezoresistancre coefficient could be obtained in a very low ratio. It is natural for SiNWs with a giant piezoresistance coefficient to be under an irregular wall-termination condition, not normal oxidation.

Contrary to the VB top, three subbands lie close together on the CB bottom, where the minima of their band energies are within a range less than 0.03 eV. In the band diagrams without strain, shown in Fig. 12(a), we can observe band crossing between a nondegenerate subband minimized on the way along the reciprocal path with a small carrier effective mass, called *the light CB subband*, and a pair of doubly degenerate bands minimized at the Γ point with large carrier effective masses, called *the heavy CB subbands*. As shown in Fig. 12(b), the degeneracy in the heavy CB subbands is not lifted by the longitudinal stress because the threefold rotational symmetry of the SiNW model is maintained. The longitudinal tensile stress can control which of the minima of band energies will be the lowest, leading to the redistribution of carrier electrons from the light CB subband to the heavy CB subbands with increasing effective mass. Therefore, it is expected that the conductivity of n-type ⟨111⟩ SiNW will be reduced under longitudinal uniaxial tensile stress.

On the other hand, the degeneracy in the heavy CB subbands should be lifted by transverse stress, and we found that one of the heavy CB subbands give rise to a new minimum of band energy with small effective mass, as shown in Fig. 12(c). This change in the effective mass for the heavy CB subband owing to the transverse uniaxial tensile stress will cause a significant increase in the carrier electron conductivity. Although a meaningful change in the carrier-electron conductivity owing to stresses can be observed in n-type H-terminated ⟨111⟩ SiNW, $(\pi_{l\langle 111\rangle})^n$ and $(\pi_{t\langle 111\rangle})^n$ are also much smaller than $(\pi_{l\langle 001\rangle})^p$. We can conclude that ⟨111⟩ SiNW, particularly in the p-type semiconductor state, is not a suitable candidate for nanoscale piezoresistors.

6. Conclusions

We have simulated the piezoresistance coefficients in H-terminated single-crystal SiNWs on the basis of first-principles calculations of model structures. The carrier conductivity along the wire axis has been calculated using band carrier densities and their corresponding effective masses derived from the one-dimensional band diagram by a novel approach for a small amount of carrier occupation. In the $Si_{89}H_{44}$⟨001⟩ SiNW model, the uniaxial tensile stress in the longitudinal direction causes a sharp drop in the band energy of the highest VB subband, called the light-hole subband, leading to the redistribution of holes to the VB subbands with a huge hole effective mass, called the heavy-hole subbands. The sudden change in the hole occupation with the increase in effective mass will bring a drastic decrease in the hole conductivity, and as a result, we have obtained a giant piezoresistance coefficient of 147×10^{-11} Pa^{-1} for the longitudinal piezoresistance coefficient for p-type H-terminated ⟨001⟩ SiNW with a small carrier density. It is well known that the bulk p-type ⟨001⟩ silicon is not useful for the piezoresistor due to its very low piezoresistive coefficient (Smith, 1954). However, it is found that p-type H-terminated ⟨001⟩ SiNW will be one of the most suitable candidates for nanoscale piezoresistors due to its giant piezoresistivity.

For p-type ⟨111⟩ SiNW, most holes occupy the highest VB subband if there are no dangling bonds on the wire wall, and therefore, the longitudinal and transverse piezoresistance coefficients for p-type H-terminated ⟨111⟩ SiNW are very small, in contrast to those in the case of p-type bare-walled ⟨111⟩ SiNW with dangling bonds (Cao et al., 2007). For n-type ⟨111⟩ SiNW, the longitudinal and transverse piezoresistance coefficients are also much smaller than the longitudinal one for the p-type ⟨001⟩ SiNW model. We predict that ⟨111⟩-oriented SiNW will not be a suitable candidate for nanoscale piezoresistors.

7. References

Bockstedte, M.; Kley, A.; Neugebauer, J. & Scheffler, M. (1997). Density-functional theory calculations for poly-atomic systems: electronic structure, static and elastic properties and ab initio molecular dynamics. *Computer Physics Communication,* Vol. 107, No. 1-3, 187-222, ISSN 0010-4655

Bui, T. T.; Dao, D. V.; Nakamura, K.; Toriyama T. & Sugiyama, S. (2009). Evaluation of the piezoresistive effect and temperature coefficient of resistance in single crystalline silicon nanowires, *Proceedings of 2009 International Symposium on Micro-Nano Mechatronics and Human Science,* pp. 462-466, ISBN 978-1-4244-5095-4, Nagoya, Japan, November 2009, IEEE

Cao, J. X.; Gong, X. G. & Wu R. Q. (2007). Giant piezoresistance and its origin in Si(111) nanowires: first-principles calculations. *Physical Review B,* Vol. 75, No. 23, 233302 (4 pages), ISSN 1098-0121

Hamann, D. R. (1989). Generalized norm-conserving pseudopotentials. *Physical Review B,* Vol. 40, No. 5, 2980-2987, ISSN 1098-0121

He, R. R. & Yang, P. D. (2006). Giant piezoresistance effect in silicon nanowires. *Nature Nanotechnology,* Vol. 1, No. 1, 42-46, ISSN 1748-3387

Hohenberg, P. & Kohn, W. (1964). Inhomogeneous electron gas. *Physical Review,* Vol. 136, No. 3B, B864-B871, ISSN 0031-899X

Kittel, C. (2005). *Introduction to Solid State Physics, 8th edition,* Wiley, ISBN 978-0-471-41526-8, New York

Kresse, G. & Hafner, J. (1993). Ab initio molecular dynamics for liquid metals. *Physical Review B,* Vol. 47, No. 1, 558-561, ISSN 1098-0121

Kresse, G. & Furthmüller, J. (1996). Efficient iterative schemes for ab initio total-energy calculations using a plane-wave basis set. *Physical Review B,* Vol. 54, No. 16, 11169-11186, ISSN 1098-0121

Li, J. & Freeman, A. J. (2006). First-principles determination of the electronic structures and optical properties of one-nanometer (001) and (111) Si nanowires. *Physical Review B,* Vol. 74, No. 7, 075333 (7 pages), ISSN 1098-0121

Nakamura, K.; Isono, Y. & Toriyama T. (2008). First-principles study on piezoresistance effect in silicon nanowires. *Japanese Journal of Applied Physics,* Vol. 47, No. 6, 5132-5138, ISSN 0021-4922

Nakamura, K.; Isono, Y.; Toriyama T. & Sugiyama S. (2009a). First-principles simulation on orientation dependence of piezoresistance properties in silicon nanowires. *Japanese Journal of Applied Physics,* Vol. 48, No. 6, 06FG09 (5 pages), ISSN 0021-4922

Nakamura, K.; Isono, Y.; Toriyama T. & Sugiyama S. (2009b). Simulation of piezoresistivity in n-type single-crystal silicon on the basis of first-principle band structure. *Physical Review B,* Vol. 80, No. 4, 045205 (11 pages), ISSN 1098-0121

Perdew, J. P.; Burke, K. & Ernzerhof, M. (1996). Generalized gradient approximation made simple. *Physical Review Letters,* Vol. 77, No. 18, 3865-3868, ISSN 0031-9007

Sasaski, S.; Seki, T.; Imanaka K.; Kimata M.; Toriyama T.; Miyano, T. & Sugiyama, S. (2007). Batteryless-wireless MEMS sensor system with a 3D loop antenna, *Proceedings of the 6th Annual IEEE Conference on Sensors,* pp. 454-457, ISBN 978-1-4244-1262-5, Atlanta, USA, October 2007, IEEE, Los Alamitos

Smith, C. S. (1954). Piezoresistance effect in germanium and silicon. *Physical Review,* Vol. 94, No. 1, 42-49, ISSN 0031-899X

Toriyama T. & Sugiyama, S. (2003). Single crystal silicon piezoresistive nano-wire bridge. *Sensors and Actuators A,* Vol. 108, No. 1-3, 244-249, ISSN 0924-4247

Tufte O. N. & Stelzer, E. L. (1963). Piezoresistive properties of silicon diffused layers. *Journal of Applied Physics,* Vol. 34, No. 2, 313-318, ISSN 0021-8979

Wortman, J. J. & Evans R. A. (1965). Young's modulus, shear modulus, and Poisson's ratio in silicon and germanium. *Journal of Applied Physics,* Vol. 36, No. 1, 153-156, ISSN 0021-8979

16

Quantum Confinement in Heterostructured Semiconductor Nanowires with Graded Interfaces

G. A. Farias, J. S. de Sousa and A. Chaves
Universidade Federal do Ceará
Brazil

1. Theoretical model for wires with graded interfaces

In a simple approximation, quantum wires can be modeled by cylinders composed by two semiconductor materials with different energy gaps, A and B, as shown in Fig. 1 (a) and (b), for block-by-block and core-shell wires, respectively. In both cases, the existence of smooth interfaces with thickness w between materials is considered, by assuming that the composition χ of the alloy $A_\chi B_{1-\chi}$ varies linearly in the interfacial region. The heterostructure potential depends on χ as $V^{het}(\vec{r}) = e_1\chi(\vec{r}) + e_2\chi^2(\vec{r})$, where e_1 and e_2 are interpolation parameters (Li, 2000), whereas the effective mass is given by a linear interpolation $m(\vec{r}) = m_A\chi(\vec{r}) + (1-\chi(\vec{r}))m_B$, with $m_{A(B)}$ as the effective mass of the material A (B). In Fig. 1 (a), the wire is *free-standing, i. e.,* not embedded in a matrix of another material, consequently, an infinite potential can be assumed for the external region.
In cylindrical coordinates, the inclusion of a magnetic-field $\vec{B} = B\hat{z}$ into the Hamiltonian is made through the symmetric gauge vector potential $\vec{A} = 1/2B\rho\hat{\theta}$. In addition, for all cases that will be discussed, our system presents circular symmetry, so that a solution $exp[il\theta]$, with $l = 0, \pm 1, \pm 2, \ldots$, is appropriate for the angular coordinate. Thus, for the general case where both the effective mass and the potential depend on the spatial coordinates, the Schrödinger equation, within the effective-mass approximation, describing the confinement of each carrier in QWR reads

$$\left\{-\frac{\hbar^2}{2}\left[\frac{1}{\rho}\frac{d}{d\rho}\frac{\rho}{m^{\parallel}(\vec{r})}\frac{d}{d\rho} + \frac{\partial}{\partial z}\frac{1}{m^{\perp}(\vec{r})}\frac{\partial}{\partial z}\right] + \frac{m^{\parallel}(\vec{r})\omega_c^2\rho^2}{8} + \frac{\hbar^2 l^2}{2m^{\parallel}(\vec{r})\rho^2} + \frac{l}{2}\hbar\omega_c + V(\vec{r})\right\}\Psi(\vec{r}) = E\Psi(\vec{r}) \quad (1)$$

Where $m^{\parallel}$ and $m^{\perp}$ are the in-plane and longitudinal effective masses, respectively, and $\omega_c = eB/m^{\parallel}$ is the cyclotron frequency. In core-shell wires, we consider a potential V which includes not only the heterostructure, but also the electron-hole Coulomb potential. For the zinc-blend materials we studied, the valence band presents heavy and light hole states and, since in block-by-block wires the charge carriers are confined in three dimensions, the valence band states cannot be discussed in terms of a single hole state and a multi-band k.p study is necessary in this case. Hence, in this work we study excitonic effects only for core-shell wires, and the investigation of the exciton energies in wires with longitudinal heterostructures is left as subject for future research.

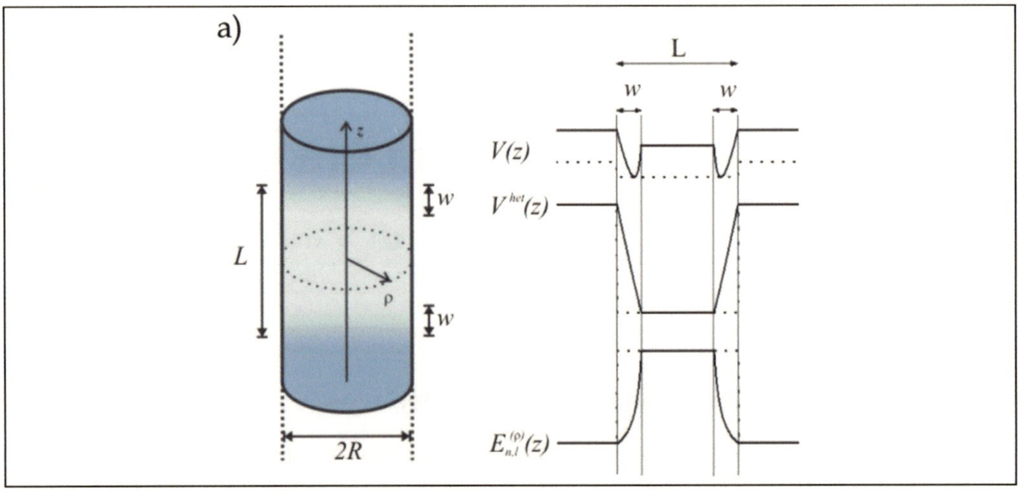

Fig. 1. (a) Sketch of our model of block-by-block quantum wires: the wire radius is R and a longitudinal heterostructured is considered in *z* direction; darker and lighter regions represent different materials and a smooth interface of thickness *w* between them is considered. A scheme of the effective potential in *z* is depicted for abrupt (dotted) and non-abrupt (solid) interfaces (see explanation in the text). (b) Sketch for the core-shell and core-multi-shell quantum wires considered in this work.

2. Longidudinal heterostructures

In this case, we consider a circular cylindrical quantum wire at an infinite potential region with a longitudinal block-by-block heterostructure. After a separation of variables $\Psi(\vec{r}) = \left(1/\sqrt{2\pi}\right)\exp[il\theta]R(r)Z(z)$, the Schrödinger equation for the radial confinement of one electron in this system, in the presence of an axial magnetic field, is obtained from Eq. (1) as

$$\left[-\frac{\hbar^2}{2m^{\|}(z)\rho}\frac{d}{d\rho}\rho\frac{d}{d\rho}+\frac{m^{\|}(z)\omega_c^2}{8}+\frac{\hbar^2 l^2}{2m^{\|}(z)\rho^2}+\frac{l}{2}\hbar\omega_c\right]R(\rho)=E^{(\rho)}R(\rho) \quad (2)$$

where the effective mass depends only on *z*, due to the existence of a heterostructure in this direction. In the absence of magnetic fields, it is straightforwardly verified that Eq. (2) becomes the Bessel equation, so that the eigenfunctions are given by the Bessel functions which do not diverge in the origin $J_l(k\rho)$, where $k=\sqrt{2m^{\|}(z)E_l^{(\rho)}/\hbar^2}$. The boundary condition at $\rho = R$, where an infinite potential is considered, implies that $J_l(kR)=0$, which leads to the quantization condition $k_{n,l} = x_{n,l}/R$, where $x_{n,l}$ is the *n*-th root of the Bessel funcion of order *l*. Hence, the radial confinement energies are easily analytically found as $E_{n,l}^{(\rho)}(z)=\hbar^2 x_{n,l}^2/2m^{\|}(z)R^2$.

In the presence of a magnetic field, analytical solutions for Eq. (2) can be found as well, not in terms of Bessel functions, but as Kummer functions instead. First, let us make the change of variables $\xi = \rho^2/2a_c$, where $a_c = \sqrt{\hbar/eB}$ is the cyclotron radius. Equation (2) now reads

$$\left[\xi\frac{d^2}{d\xi^2}+\frac{d}{d\xi}-\left(\frac{l}{2}+\frac{l^2}{4\xi}+\frac{\xi}{4}-\frac{E}{\hbar\omega_c}\right)\right]R(\xi)=0 \tag{3}$$

Taking account of the polynomial and assymptotic behaviors of $R(\xi)$, it is reasonable to try a solution of the form $R(\xi)=\xi^{|l|/2}e^{-\xi/2}F(\xi)$. With this solution, Eq. (3) is rewritten as

$$\xi\frac{d^2F(\xi)}{d\xi^2}+\left[(|l|+1)-\xi\right]\frac{dF(\xi)}{d\xi}-\left(\frac{l}{2}+\frac{|l|}{2}+\frac{1}{2}-\frac{E}{\hbar\omega_c}\right)F(\xi)=0 \tag{4}$$

Equation (4) is identified as a confluent hypergeometric differential equation, which can be solved in terms of the Kummer function of the first kind $F_1(\alpha,\beta,\xi)$, which remains finite in the origin. The eigenfunctions are then given by $R_{n,l}(\xi)=N\xi^{|l|/2}e^{-\xi/2}F_1(\alpha,|l|+1,\xi)$, where N is a normalization parameter and $\alpha = l/2+|l|/2+1/2-E/\hbar\omega_c$. The boundary condition at $\rho = R$ now leads to $F_1(\alpha,|l|+1,\xi_R)=0$, so that $\alpha = x_{n,|l|}$ must be the n-th zero of $F_1(\alpha,|l|+1,R^2/2a_c)=0$. Finally, the radial confinement energy in the presence of a magnetic field is written as

$$E_{n,l}^{(\rho)}(z)=\hbar\omega_c(z)\left(x_{n,|l|}+\frac{l}{2}+\frac{|l|}{2}+\frac{1}{2}\right) \tag{5}$$

Notice that this energy depends on z, for the effective mass is different for each material. Consequently, this energy can be viewed as an additional potential in this direction, which is summed up to the heterostructure potential V^{het}, leading to an effective potential $V(z)=V^{het}(z)+E_{n,l}^{(\rho)}(z)$. The effective potential $V(z)$ is then adjusted so that the energy referential is at the GaP (InP) layers.The shape of this effective potential is illustrated as dashed and solid curves in Fig. 1 (a), for abrupt and smooth interfaces, respectively. The radial confinement energy $E_{n,l}^{(\rho)}$ depends on the inverse of the effective mass and, since the electron effective masses at GaAs and InAs are lighter than those at GaP and InP, (Li, 2000) this energy exhibits a barrier-like profile as a function of z in GaP/GaAs/GaP or InP/InAs/InP heterostructures. For abrupt interfaces (dashed), when this energy is summed to the heterostructure potential, depending on the wire radius, the effective potential $V(z)$ can end up i) with a lower effective band offset, ii) with an inversion of the carriers localization, as in a type-II structure, or even iii) with a zero effective band offset, so that a carrier would behave like a free particle in z-direction, despite the presence of the longitudinal heterostructure (Voon & Willatzen, 2003). When graded interfaces are taken into account (solid), the effective potential at the interfacial region exhibits a peculiar shape, with cusps that are able to confine the carriers in this region. In the following results of this section, we will show the effects of this interfacial confinement on the energy states, as well as on the transport properties of longitudinally heterostructured QWR.

2.1 Single heterostructure

The parameters for the materials that compose the heterostructures are assumed as $m_{GaAs} = 0.063m_0$ ($m_{InAs} = 0.027m_0$) and $m_{GaP} = 0.33m_0$ ($m_{InP} = 0.077m_0$) for electron effective masses, $e_1 = 1.473$ eV and $e_2 = 0.146$ eV ($e_1 = 1.083$ eV and $e_2 = 0.091$ eV) for the interpolation parameters and the conduction band offset is assumed as $Q_e = 0.5$ ($Q_e = 0.68$), for GaP/GaAs (InP/InAs) heterostructures. (Li, 2000)

Figure 2 shows the confinement energies E_e of electrons in the absence of magnetic fields, for states with the set of quantum numbers (n, l, m) as (1, 0, 1) (symbols) and (1, 1, 1) (curves) in a GaP/GaAs QWR, as a function of the well width L, formed by the longitudinal heterostructure. The interface thickness between materials was considered as (a) w = 0 Å and (b) 20 Å, and two values for the wire radius are chosen, in order to observe the effect of enhancing the radial confinements on the energy levels.

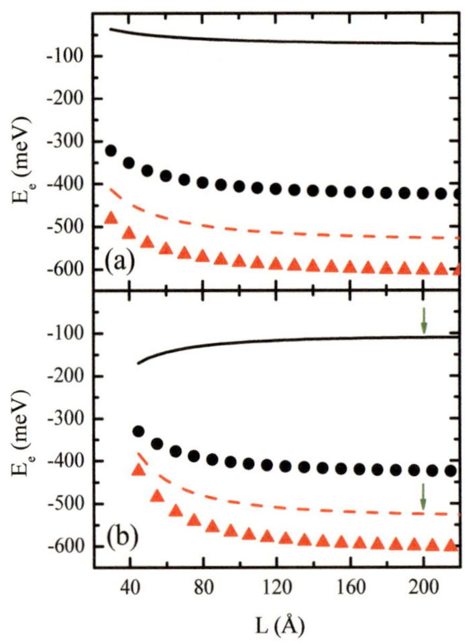

Fig. 2. Electron confinement energies in a cylindrical QWR with a longitudinal GaP/GaAs heterostructure, with interface thickness (a) 0 and (b) 20 Å , as a function of the well width L, for $l = 0$ (symbols) and $l = 1$ (curves). Two values of the wire radius R were considered: 35 Å (black, circles, solid) and 75 Å (red, triangles, dashed). The electron wavefunctions for the points indicated by green arrows (L = 200 Å) are plotted in Fig. 4(a).

For abrupt interfaces, a decrease in all energy levels is observed as the well width L is enlarged, similar to the case of quantum wells (Chaves et al., 2007). However, when graded interfaces are taken into account in a R = 35 Å wire, as shown in Fig. 2 (b), the energy of the electron $l = 1$ state increases with the well width, for small values of L. This indicates that in this system, such state is now confined at the interfacial regions, so that an enlargement on the well width L further separates the interfaces and increases the confinement energy of the $n = 1$ state, analogous to the case of the energy levels in double quantum wells. (Ferreira & Bastard, 1997) Conversely, in $l = 0$ states for $R = 35$ Å and in all the states shown for $R = 75$ Å, the same qualitative behavior is observed for abrupt and $w = 20$ Å interfaces, namely their energies decrease as increasing L, but quantitatively, the presence of graded interfaces still plays an important role, giving a significant blueshift on these energies, specially for small L and large R. For instance, if we consider a $L = 45$ Å QWR, the energy blueshift due to the graded interfaces is about $\Delta E_e = E_e(w) - E_e(0) = 30$ meV (105 meV) for $l = 0$ states in a $R = 35$ Å (75 Å) QWR. This blueshift in the energy states when non-abrupt interfaces are taken into account has also been predicted in quantum wells, *core-shell* QWR and quantum dots with graded interfaces. (Chaves et al., 2007; Costa e Silva et al. 2006; Caetano et al. 2003) These results are reasonable since, for large wire radius, the radial confinement energy $E_{n,l}^{(\rho)}$ is small, thus the presence of graded interfaces acts like in a quantum well, reducing the confinement region in the heterostructure potential $V^{het}(z)$, hence, enhancing the energy levels. Yet, for smaller values of wire radius, the presence of an interfacial region creates cusps that can confine carriers, as shown in Fig. 1 (b) (solid), which can i) reduce the energy blueshift due to interfaces in this case, or ii) lead to a redshift, as observed for $l = 1$ when $R = 35$ Å. Numerical results show that the electron confinement at the interfacial region for lower energy states, as well as the type-I (well) to type-II (barrier) transition in effective potential predicted in previous works (Voon & Willatzen, 2003; Chaves et al., 2008), occurs for small values of R, which can be troublesome for experimental verification of these features. Nevertheless, one could try to find another way to induce these variations on the carriers localization: The main effect produced by reducing the QWR radius is that the carriers wave functions are squeezed towards the wire axis. If a magnetic field is applied parallel to the wire axis, the same effect can be obtained, hence, type-I to type-II transitions and interfacial confinements, which were found for wires with small radii, are expected to be found for high magnetic fields as well.

In order to verify this alternative way of obtaining these effects, in Fig. 3 we study the influence of an applied magnetic field, parallel to the wire axis, on the confinement energies of electrons. The energies of $l = 1$ states in a GaP/GaAs QWR with $R = 42$ Å are shown as functions of the well width L, considering (a) $w = 0$ Å and (b) $w = 20$ Å. Three values of magnetic field are considered: $B = 0$ T (blue, dashed), 10 T (red, dotted) and 30 T (black, solid). The states with zero angular momentum l are not shown, for such states have negligible interaction with magnetic fields. From the Hamiltonian of Eq. (2), one can easily verify that $l = 1$ states are much more affected by increasing this field, since there is an additional term in this Hamiltonian involving the cyclotron frequency w_c and the angular momentum l. For abrupt interfaces (Fig. 3 (a)), the presence of a magnetic field enhances the confinement energies, but gives no appreciable change in qualitative behavior of E_e versus L curves. However, considering graded interfaces (Fig. 3 (b)), these curves are qualitatively different: the energy behavior $B = 30$ T (black, solid) is crescent as the well width L increases, whereas the opposite behavior is observed for $B = 0$ T (blue, dashed) and $B = 10$ T (red, dotted). This can be understood as an interfacial confinement of these states, not induced by reducing the wire radius, as in Fig. 2 (b), but due to the presence of a magnetic field parallel to the wire axis. It can be demonstrated by our model that, keeping the same interface thickness w = 20 Å, but considering a slightly smaller radius, R = 40 Å , the electron would be already confined at the interface, so that there is no need for applying a magnetic field to observe this kind of confinement if these values are chosen for the parameters of the system. However, with R = 40 Å and a smaller interface thickness w = 15 Å , the electron is confined at the GaAs layer and, in this case, interfacial confinements induced by magnetic fields were also found, but only for B = 30 T, whereas for R = 42 Å and w = 20 Å, we observed that an interfacial confinement for this system occurs already at a B = 20 T magnetic field. Our results also predict that interfacial confinement of l = 1 electrons due to magnetic fields can be found in InP/InAs QWR too, with a slightly larger radius R = 48 Å (not shown in this work). This can be an interesting feature of these systems since, once the QWR is grown its radius is fixed and an external parameter, namely the magnetic field, can just be tuned to obtain an electron confinement at the well or at the interfaces.

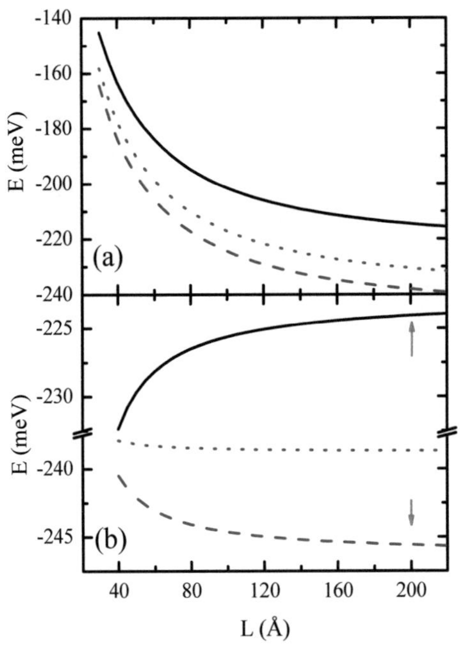

Fig. 3. Electron confinement energies as functions of the well width L for l = 1 states in a cylindrical GaP/GaAs QWR with wire radius R = 42 Å , considering interfaces thicknesses w = 0 (a) and 20 Å (b), under applied magnetic fields B = 0 T (blue, dashed), 10 T (red, dotted) and 30 T (black, solid). The electron wavefunctions for the points indicated by green arrows (L = 200 Å) are plotted in Fig. 4(b).

The interfacial confinement of states in GaP/GaAs QWR can be verified by analyzing the electron wave functions as a function of z in such systems, which is illustrated in Fig. 4, for (1, 1, 1) states for the results marked by arrows in Figs. 2 and 3. In Fig. 4 (a), two values of wire radius are considered, in the absence of magnetic fields, while in (b) the wire radius is

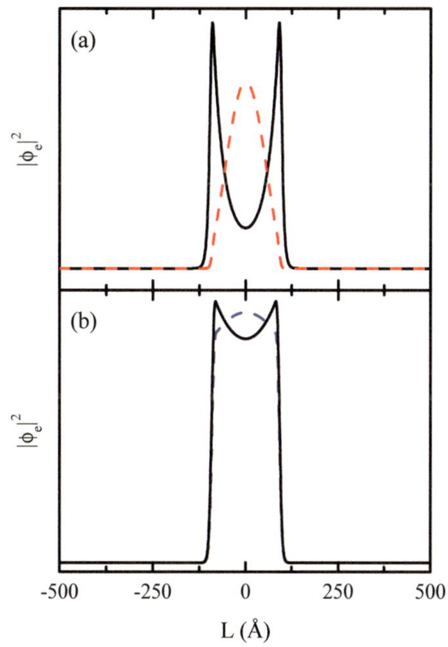

Fig. 4. Electron wave functions for $l = 1$ states in a cylindrical GaP/GaAs QWR with graded interfaces of w = 20 Å thickness and well width L = 200 Å, as a function of z, considering (a) wire radius R = 35 Å (black, solid) and R = 75 Å (red, dashed) in the absence of magnetic fields; and (b) magnetic fields B = 0 T (blue, dashed) and B = 30 T (black, solid) for a R = 42 Å wire radius.

kept and two values of the magnetic field B are considered. A confinement of l = 1 states is induced when the wire radius is reduced from R = 75 Å (red, dashed) to R = 35 Å (black, solid), as observed in Fig. 4 (a). When a magnetic field is applied, in Fig. 4 (b), the wave function of such states in a R = 42 Å QWR is also altered, and for B = 30 T (black, solid), two peaks at the interfacial regions can be seen. Therefore, we can conclude that, in the presence of graded interfaces, one can obtain confined states at the interfacial regions by reducing the wire radius as well as by increasing the magnetic field intensity.

In Fig. 5, the electron confinement energies are plotted as a function of the well width L, for l = 1 states in InP/InAs/InP QWR with R = 45 Å. For w = 0 Å (Fig. 5(a)), the confinement energies for such a state in the absence of magnetic fields (black, solid) decrease with increasing L, but for B = 10 T (red, dotted) the energy of the (1, 1, 1) is zero for all L, which shows that these states are not confined in z, despite the presence of a heterostructure. This indicates that the barrier like potential due to the radial confinement energy $E_{n,l}^{(\rho)}(z)$, for B = 10 T, is high enough to suppress the contribution of the heterostructure potential in $V(z)$ for this system, yielding a type-II effective potential in the case of abrupt interfaces. This suggests that, with a InP/InAs/InP QWR with abrupt interfaces, one could control the electron band offset or even change from a confinement to a scattering potential for non-zero l states, just by setting the external magnetic field. On the other hand, taking account of

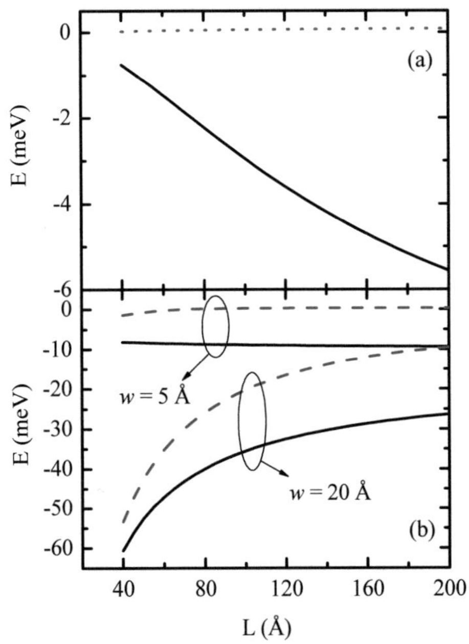

Fig. 5. (a) Electron confinement energies as functions of the well width L for $l = 1$ states in a cylindrical InP/InAs QWR with wire radius $R = 45$ Å and abrupt interfaces, under applied magnetic fields $B = 0$ T (black, solid) and 10 T (red, dotted). (b) The same of (a), but considering graded interfaces of $w = 5$ and 20 Å, and magnetic fields $B = 0$ T (black, solid) and 20 T (blue, dashed).

the existence of graded interfaces in real QWR, the results in Fig. 5 (b) show that for larger smooth interfaces ($w = 20$ Å), the magnetic field does not change the qualitative behavior of $l = 1$ electrons, which are confined at the interfaces for both $B = 0$ T (black, solid) and 20 T (blue, dashed). Considering a small interface thickness $w = 5$ Å, with $B = 0$ T (black, solid), the electron $l = 1$ state is confined at the InAs (well) layer and, due to the form of $V(z)$ in this case, it depends weakly on the well width: for L varying from 40 Å to 200 Å considering $w = 0$ Å, the energy of such a state decreases about 5 meV, whereas for $w = 5$ Å, it varies only about 1 meV. For a $B = 20$ T (blue, dashed) magnetic field, considering an interface thickness $w = 5$ Å, a weak interfacial confinement of (1, 1, 1) electrons is still observed for L lower than 80 Å, while for greater values of L, the barrier on the effective potential is large enough to avoid confinement at the interfaces, leading to a zero energy for such states and an electron localization at the InP layers, just like in a type-II system. Thus, for a perfect type-I to type-II transition induced by a magnetic field in InP/InAs/InP non-abrupt QWR, one would need a $R = 45$ Å wire with a large InAs layer and small interfaces (< 5 Å), *i. e.* with a very high quality of heterostructure modulation, in order to avoid interfacial confinements. The presence of magnetic fields has been shown to induce interfacial confinement for $l = 1$ states, because this field enhances the radial confinement energy $E_{n,l}^{(\rho)}(z)$ and then supports this change of localization. However, for $l = -1$ states, the magnetic field reduces $E_{n,l}^{(\rho)}(z)$, hence, it would never induce a change from well to interfacial localization of such states. Conversely, there are some cases where the l = -1 electron state is already confined at the interfaces, for instance, in a GaP/GaAs QWR with R = 39 Å and w = 15 Å, and a magnetic field B = 30 T can induce a transition from interfacial to well localization. Such a transition can also be obtained with R = 40 Å and w = 20 Å, but applying a higher magnetic field intensity, B = 35 T.

In present work, as one can verify in Sec. 2, the interfacial region is assumed to be a $XP_\chi As_{1-\chi}$ (X = Ga or In) alloy, with a P composition $\chi(z)$ varying linearly along z at the interfaces. The effective masses are then assumed to depend linearly on the composition, but other dependencies on $\chi(z)$, for instance, a linear dependence for the reciprocal effective mass, *i. e.*, $1/m(z) = (1/m_{XP})\,\chi(z) + (1/m_{XAs})[1 - \chi(z)]$ could be considered as well. Straightforward calculations with a simple algebra show that, even for this kind of variation, one can still tailor the system in order to find confining potentials at the interfaces. (nosso artigo)

The study of electron states in the conduction band for the systems we have considered so far can be done by the theoretical model we present in this work, as a good approximation. However, this simple one-band model fails when studying valence band states, because in cases where the wire diameter 2R and the heterostructure width L have almost the same dimensions, the lowest valence state is a combination of heavy-hole (HH) and light-hole (LH) states. Indeed, the problem of the confinement of holes in QWR with 2R close to L must be solved rather using a 4 x 4 Hamiltonian, which takes into account HH and LH states. Nevertheless, it can be inferred that the critical radii for type-I to type-II transitions, as well as those for interfacial confinements, are not expected to be the same for electrons and holes, since the effective masses of these carriers are completely different in each material and the changes in localization are strongly dependent on the differences between effective masses. In fact, a previous work (Voon et al., 2004) has shown that with a four-band *k.p* based theory one can obtain type-I to type-II transitions also for valence band states in abrupt InGaAs/InP and GaAs/AlAs QWR, and the critical radii they found are different for electrons and holes. Furthermore, considering a GaAs/GaP QWR, changes in valence band states localization would hardly be obtained, due to the small difference between the hole effective masses in GaAs and GaP. However, changes in the electrons localization are found in this system (see Fig. 2(b)). These electron hole separations might decrease the overlap between their ground state wave functions, which would reduce the probability of an interband transition for such states, a feature that is also commonly found in type-II systems. Thus, a low probability of interband transitions would be observed in cases where the hole remains at the well layer and the electron is confined at the interfaces or vice-versa.

2.2 Double heterostructure

Let us study a different case, where the longitudinal direction presents a double heterostructure, composed by InP and InAs layer. If these materials are disposed as InP/InAs/InP/InAs/InP, the electron will experience a double well structure in z direction in the case where the radial confinement energy is small, that is, when the wire radius is large: the InAs layers will act like confining wells and the InP layers, as barriers. However, as we have discussed earlier, as the wire radius is reduced, the radial confinement energy $E_{n,l}^{(\rho)}(z)$ becomes stronger and reduces the conduction band offset, leading even to a type-I to type-II transition for thin wires. In this section, we will study the double heterostructured system considering wire radii close to the critical radii that lead to the type-I to type-II transition, so that the InAs layers in this case can act like wells or barriers, depending on the wire radius. As we have the potential in z direction $V(z) = V^{het}(z) + E_{n,l}^{(\rho)}(z)$, we can easily calculate the transmission probabilities T for an electron passing through the double heterostructure, by means of the well known transfer matrix formalism. (Tsu & Esaki, 1973) In all cases studied in this section, we will deal only with ground state electrons for the radial confinement, *i. e.* considering $n = 1$ and $l = 0$.

Figure 6 shows the logarithm of the transmission probability T of the electron as a function of its energy, for three three values of the wire diameter $D = 2R$. The widths of the InP and InAs layers are chosen as L_{InP} = 70 Å and L_{InAs} = 100 Å, respectively. As one can se in the insets of Fig. 6 (a), for abrupt interfaces, the longitudinal effective potential *V(z)* for these wire diameters is a high double barrier, a very small double barrier and a double well, respectively. Since we have a double barrier for D = 40 Å (black, solid), the transmission probabilities in this case present some peaks which are related to the confined states of the quantum well formed in-between these barriers. When the electron energy reaches one of the eigenenergies of such quantum well, there is a complete transmission of the electron wavefunction, giving rise to the so called resonant tunneling effect. As the diameter of the

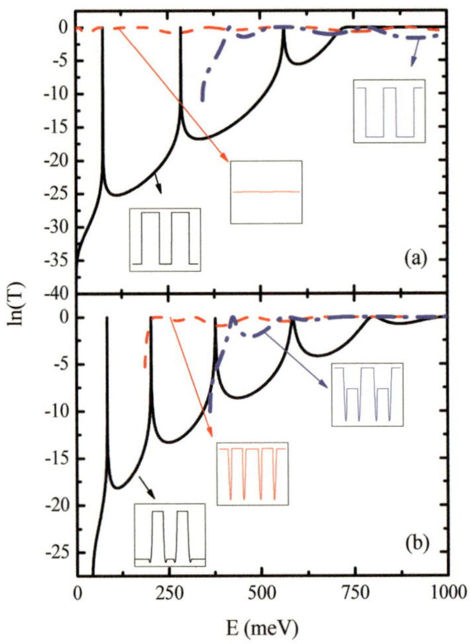

Fig. 6. Transmission probabilities (logarithm) as a function of the electron energy for a QWR with a longitudinal double heterostructure, three values of the wire diameter $D = 2R$: 40 Å (black, solid), 55 Å (red, dashed) and 80 Å (blue, dashed-dotted), considering abrupt (a) and w = 20 Å (b) interfaces. The widths of the InP and InAs layers are chosen as 70 Å and 100 Å, respectively. The insets illustrate a sketch of the longitudinal effective potential *V(z)* for each wire diameter.

ring increases, the barriers height is reduced and for D = 55 Å (red, dashed) we observe an almost flat conduction band for the electron. As a consequence, for this value of the wire diameter, the electron can be completely transmitted, with $T \simeq 1$, for any value of the electron energy E, as in a free electron case, despite the presence of the longitudinal heterostructure. For larger diameters, as in the D = 80 Å (blue, dashed-dotted) case, the InAs layers become wells and the transmission oscillations are only due to resonant states over the well region, in the continuum. The presence of w = 20 Å interfaces, as shown in Fig. 6 (b), gives rise to some relevant effects: i) the number of resonant tunneling peaks increases, due to the enlargement of the region between the two barriers, and ii) the oscillations in T for higher energies are altered, due to the new form of the interfacial potential.

The barriers width in double heterstructures plays an important role in the resonant tunnelling: as the barrier width becomes larger, the resonant transmission peak becomes narrower and the peak-to-valley difference increases, which is also reflected in the tunnelling time of the resonant states. (Singh, 1993) This is illustrated in Fig. 7 (a), for a QWR with diameter D = 50 Å, considering several values of the InAs layer width L_{InAs} and abrupt interfaces. This diameter is thin enough to make the effective potential present barriers for electrons at the InAs layers, hence, increasing the width L_{InAs} of these layers produces the effects we mentioned due to the enlargement of the barrier width. However, for an interface thickness w = 20 Å, this picture changes, because of the presence of the cusps in the

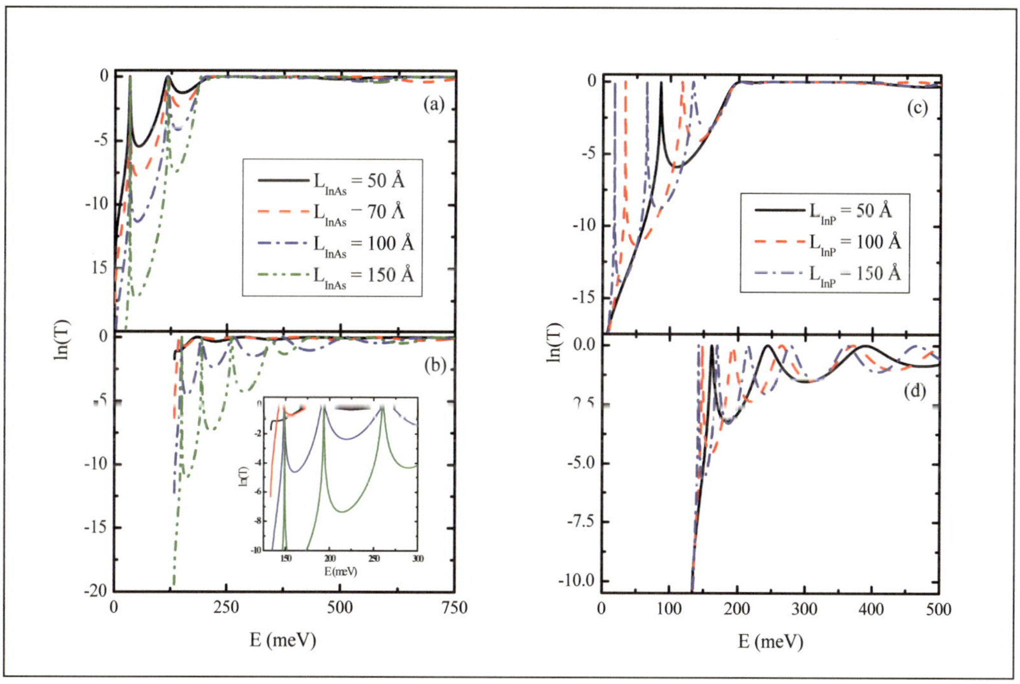

Fig. 7. Transmission probabilities (logarithm) for a 50 Å diameter wire with a double longitudinal heterostructure InP/InAs/InP/InAs/InP considering (a), (b) several values of width of the InAs layers L_{InAs}, while the width of the InP layer is kept the same L_{InP} = 70 Å, and (c), (d) several values of width of the InP layers L_{InP}, while the width of the InAs layer is kept the same L_{InAs} = 100 Å, for abrupt (a), (c) and w = 20 Å (b), (d) interfaces. The inset is a magnification of the results presented in (b).

interfacial region, as shown in Fig. 1. In this case, enlarging the InAs layer, we increase not only the barrier width, but also the separation between these cusps, hence, the effect of increasing L_{InAs} is not simply making the peaks narrower and changing peak-to-valley differences, as one can see in Fig. 7 (b). It can be seen that even the position of the peaks is slightly altered by enlarging L_{InAs} in the non-abrupt case, as highlighted by the inset.
By increasing the width of the InP layers L_{InP} in the D = 50 Å wire that we considered in Fig. 7, we are enhancing the well region and, consequently, more resonant tunnelling peaks are expected to occur, as one can verify in Fig. 7 (c), for abrupt interfaces: while the L_{InP} = 50 Å system (black, solid) has only one sharp peak, two and three peaks are observed for L_{InP} = 100 Å (red, dashed) and 150 Å (blue, dashed-dotted), respectivelly. Figure 7 (d) shows that, for these systems, the different effective potential $V(z)$ created by the presence of w = 20 Å graded interfaces is responsible for changes in the position of the peaks and leads to oscillations in the transmission probability for higher energies.

3. Core-shell quantum wires

As we will investigate the excitonic properties of core-shell quantum wires, we consider a potential in Eq. (1) that includes the electron-hole interaction, so that the Hamiltonian describing the confinement of these carriers now reads $H_{exc} = H_e + H_h + H_{e-h}$, where the single particle Hamiltonian H_i, with i = e (electron) or h (heavy hole), can be straightforwardly derived from Eq. (1) as

$$H_i = -\frac{\hbar^2}{2\rho_i}\frac{d}{d\rho_i}\frac{\rho_i}{m_i^{\parallel}(\rho_i)}\frac{d}{d\rho_i} + \frac{m_i^{\parallel}(\rho_i)\omega_c^2\rho_i^2}{8} + \frac{\hbar^2 l^2}{2m_i^{\parallel}(\rho_i)\rho_i^2} + \frac{l}{2}\hbar\omega_c + V_i^{het}(\rho_i) \tag{6}$$

and the electron hole interaction Hamiltonian H_{e-h} is given by

$$H_{e-h} = -\frac{\hbar^2}{2\mu^{\perp}}\frac{\partial^2}{\partial z^2} - \frac{e^2}{4\pi\varepsilon}\frac{1}{|\vec{r}_e - \vec{r}_h|} \tag{7}$$

where $\mu^{\perp} = 1/m_e^{\perp} + 1/m_h^{\perp}$ is the reduced mass in longitudinal direction and ε is the effective dielectric constant of the media. After a separation of variables, the single particle Hamiltonian H_i describing the confinement of each carrier in the radial direction is diagonalized numerically, by means of a finite differences scheme (Peeters & Schweigert, 1996), whereas a variational procedure is taken for the electron-hole interaction, by considering a Gaussian function

$$Z(z) = \frac{1}{\sqrt{\eta}}\left(\frac{2}{\pi}\right)^{\frac{1}{4}}\exp\left(-\frac{z^2}{\eta^2}\right) \tag{8}$$

as the solution for the z-direction, where η is a variational parameter that minimizes the exciton energy.
We have applied this theoretical model to study $Si/Si_{1-x}Ge_x$ core-shell quantum wires. It has been shown in previous papers that for lower Ge concentrations this heterostructure exhibits a type-I confinement for electrons, but a type-I to type-II transition occurs as the Ge concentration x increases (Kamenev et al., 2005). Then, in this work we study both types of

confinement, for the appropriate Ge concentrations in each case. The material parameters for Si and Ge can be easily found in litterature (Penn et al., 1999) and the parameters of the alloy were obtained by interpolation of those of the pure materials.

3.1 Type-I $Si/Si_{0.85}Ge_{0.15}$ wire

For a type-I system, we solve numerically the Schrödinger equation $H_i\psi_i(\rho_i)=E_i\psi_i(\rho_i)$ to obtain the energy and the radial wavefunctions of each charge carrier, so that the total exciton wavefunction reads $\Psi(\vec{r})=\left(1/\sqrt{2\pi}\right)\exp[il\theta]\psi_e(\rho_e)\psi_h(\rho_h)Z(z)$. Then we search for η so that $\langle\Psi|H_{exc}|\Psi\rangle=E_e+E_h+E_b(\eta)$ is minimized, with

$$E_b(\eta)=\langle\Psi|H_{e-h}|\Psi\rangle=-\frac{\hbar^2}{2\mu^{\perp}}\int_{-\infty}^{\infty}Z(z)\frac{\partial^2}{\partial z^2}Z(z)dz-\frac{e^2}{4\pi\varepsilon}\int_V\frac{|\psi_e(\rho_e)|^2|\psi_h(\rho_h)|^2Z^2(z)}{\sqrt{z^2+|\vec{\rho_e}-\vec{\rho_h}|^2}}dV \tag{9}$$

where the volume element is $dV=\rho_e d\rho_e\rho_h d\rho_h d\theta dz$. Using the Gaussian variational function of Eq. (8), the first integral in Eq. (9) is solved analytically, while the second integral can be simplified, yielding

$$E_b(\eta)=\frac{\hbar^2}{2\mu^{\perp}\eta^2}-\frac{e^2}{4\pi\varepsilon\eta}\sqrt{\frac{2}{\pi}}\int_0^{\infty}|\psi_e(\rho_e)|^2\rho_e\int_0^{\infty}|\psi_h(\rho_h)|^2\rho_h\int_0^{2\pi}\exp(a)K\left(\frac{a}{2}\right)d\rho_e d\rho_h d\theta \tag{10}$$

with $a=-2|\vec{\rho_e}-\vec{\rho_h}|/\eta$ and $K(x)$ is the modified Bessel function of the second kind (table of integrals). The remaining integral in Eq. (10) is then performed numerically and, after the minimization, we eventually obtain the total exciton energy as $E_{exc}=E_{gap}+E_e+E_h-E_b$, where E_{gap} is the gap of the core material.

The binding energy (black) and the ground state energy (red) of *e-hh* excitons in $Si/Si_{0.85}Ge_{0.15}$ QWR are shown in Fig. 8 as a function of the wire radius for interface thicknesses w = 0 Å (solid) and 15 Å (dashed). Notice that the binding energy increases as the wire radius increases up to a maximum at ρ_2 = 40 Å (50 Å), for abrupt (graded) interfaces and after this values it starts to decrease. This change of behavior as the wire radius increases is reasonable: reducing the wire radius makes the system seem like bulk Si, where the binding energies are naturally lower. (Ferreira et al., 2002; Costa e Silva et al., 2006) On the other hand, increasing the wire radius enlarges the region of confinement, which is responsible for reducing the binding energy. Hence, there must be some critical value of the radius ρ_2 where the behaviour of the binding energy as a function of the radius changes. For wire radii below this critical value, the inclusion of a graded interface shifts down the binding energies, while the opposite occurs for a larger radii. Numerical results show that for a 30 Å wire radius with interface thickness of w = 15 Å, for example, the binding energy is reduced about 25%, while for wire radii greater than 50 Å there is an average increase in the binding energies of about 5.5%, in relation to the abrupt case. The ground state exciton energy always decreases with the increase of the wire radius, and the inclusion of a graded interface shifts up this energy, especially for thin wires, where these shifts may reach about 30 meV in relation to the abrupt case. Indeed, for a wire radius as large as 120 Å, the inclusion of a 15 Å interface must affect weakly the exciton energy, whereas for a small wire radius, *e. g.* ρ_2 = 40 Å, the dimensions of the wire radius and the interface thickness are comparable, hence a stronger effect of the interface is expected for the exciton energy in this case.

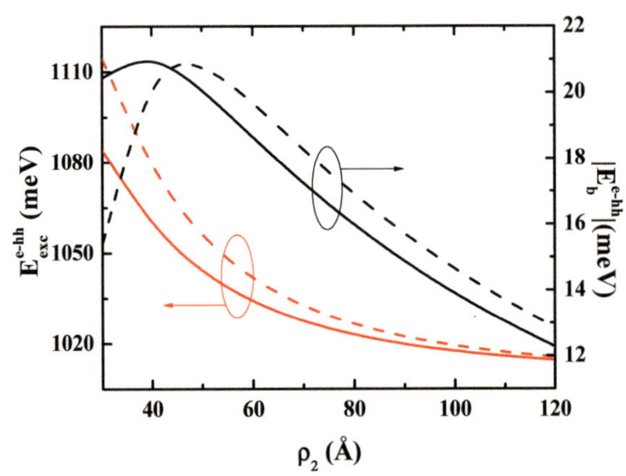

Fig. 8. Binding energy (black, right scale) and ground state exciton energy (red, left scale) of $Si/Si_{0.85}Ge_{0.15}$ type-I wires as a function of the wire radius ρ_2 for abrupt (solid) and w = 15 Å (dashed) interfaces.

The influence of applied magnetic fields on the binding (a) and total (b) ground state exciton energies is presented in Fig. 9 for two values of the wire radius, ρ_2 = 50 Å (black, solid, triangles) and 200 Å (red, dashed, squares), for an abrupt interface (lines) and for w = 15 Å (symbols). The magnetic field does not significantly affect these energies, giving them only small blueshifts of the order of 2 meV, for all values of wire radii considered. Wires with larger radii are more affected by this external field. These results are in good agreement with previous studies in such systems for other materials. (Branis et al., 1993)

3.2 Type-II $Si/Si_{0.70}Ge_{0.30}$ wire

In this case, since the conduction band forms a barrier for electrons, we cannot obtain the radial wavefunction for this carrier just by solving $H_e\psi_e(\rho_e) = E_e\psi_e(\rho_e)$, as we did previously for type-II, for the solution of this equation would lead to a free electron with $E_e = 0$. Hence, for a type-II structure, we must first solve the Schrödinger equation for the hole and use the obtained radial wavefunction $\psi_h(\rho_h)$ to construct an effective Coulombian potential, which is responsible for confining the electron in ρ-direction.

The Schrödinger equation for electrons in this system then reads

$$(E_h + H_e + I(\rho_e))|\psi_e\rangle = E_X|\psi_e\rangle \tag{11}$$

where the effective Coulomb potential $I(\rho_e) = \langle\psi_h Z|H_{e-h}|\psi_h Z\rangle$ is obtained from the hole wavefunction and from the Gaussian trial function of Eq. (8) as

$$I(\rho_e) = -\frac{\hbar^2}{2\mu^{\perp}}\int_{-\infty}^{\infty} Z(z)\frac{\partial^2}{\partial z^2}Z(z)dz - \frac{e^2}{4\pi\varepsilon}\int_{-\infty}^{\infty}\int_0^{2\pi}\int_0^{\infty}\frac{|\psi_h(\rho_h)|^2 Z^2(z)}{\sqrt{z^2 + |\vec{\rho_e} - \vec{\rho_h}|^2}}\rho_h d\rho_h d\theta dz \tag{12}$$

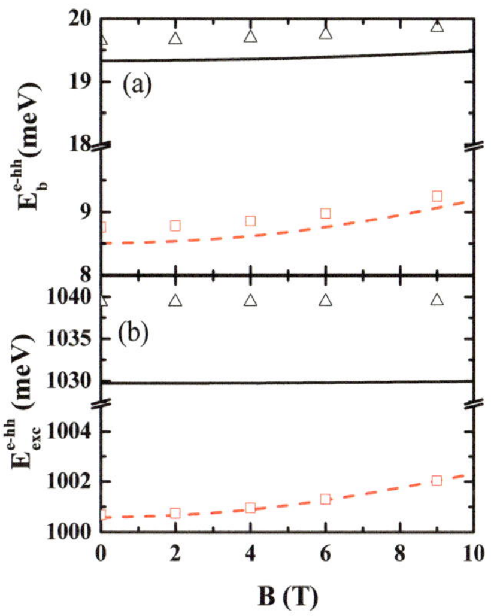

Fig. 9. Binding energy (a) and ground state exciton energies (b) of *e-hh* pairs in Si/SiGe type-I quantum wires as functions of the magnetic field, with w = 0 Å (lines) and w = 15 Å (symbols), for wire radii ρ_2 = 50 Å (black, solid, triangles) and 200 Å (red, dashed, squares).

As for Eq. (9) in the type-I case, the first integral in Eq. (12) can be solved analytically, while the second is simplified, leading to an expression that is similar to what we found in Eq. (10):

$$I(\rho_e) = \frac{\hbar^2}{2\mu^{\perp}\eta^2} - \frac{e^2}{4\pi\varepsilon\eta}\sqrt{\frac{2}{\pi}}\int_0^{\infty}\left|\psi_h(\rho_h)\right|^2 \rho_h \int_0^{2\pi} \exp(a) K\left(\frac{a}{2}\right) d\rho_h d\theta \tag{13}$$

Performing this integral numerically, we obtain the effective Coulomb potential for electrons and solve the Schrödinger equation given by Eq. (11). The parameter η is adjusted to minimize E_X and we obtain the total exciton energy by $E_{exc} = E_{gap} + E_X$.

Figure 10 shows the effective confining potential $V_{eff}(\rho_i) = V^{het}(\rho_i) + I(\rho_e)$ (black, solid) and the wave functions (red, dashed) for electrons (top) and holes (bottom) in type-II $Si/Si_{0.70}Ge_{0.30}$ QWR. A depression in the potential for the electron, due to the electron-hole Coulomb interaction, can be clearly seen. This depression is responsible for the electron bound state at the silicon layer near the $Si/Si_{0.70}Ge_{0.30}$ wire, despite the fact that the heterostructure forms a barrier for this carrier.

In Fig. 11, the binding energy (black) and the ground state energy (red) of *e-hh* excitons are plotted as a function of the QWR radius for $Si/Si_{0.70}Ge_{0.30}$ wires with abrupt (solid) and w = 15 Å (dashed) interfaces. Since in type-II systems the electron and hole are localized in different regions of space, the binding energies of type-II wires are lower than those of type I, as one can verify by comparing Figs. 8 and 11. Moreover, these energies always decrease when the wire radius is enlarged, and no change in the behavior of the binding energy versus wire radius curve is observed, contrary to the type-I case studied before. Actually, in the type-II case, in the limit of very thin wires, the system no more seems like bulk Si, but like a bulk system with a localized impurity, since the hole is confined within the thin wire, while the electron is bound to it on the Si layer. Considering a 40 Å wire radius and w = 15 Å interface thickness, the increase of binding energies is about 12% in relation to the abrupt case, while the total exciton energy is increased by about 20 meV.

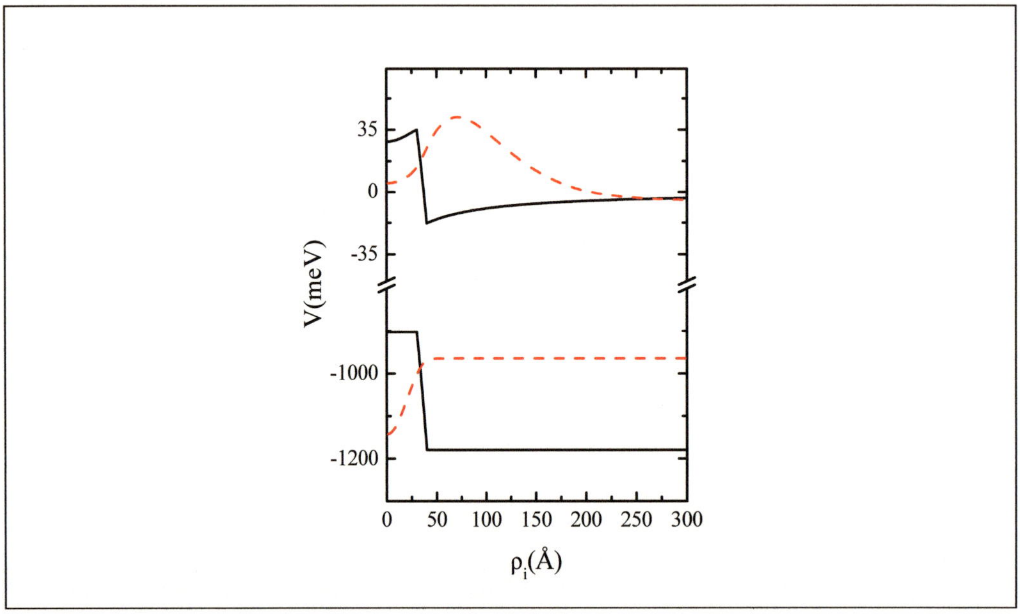

Fig. 10. Confining potentials (black, solid) for electrons (top) and holes (bottom) and their respective wave functions (red, dashed) in a ρ_2 = 40 Å type-II $Si/Si_{0.70}Ge_{0.30}$ QWR.

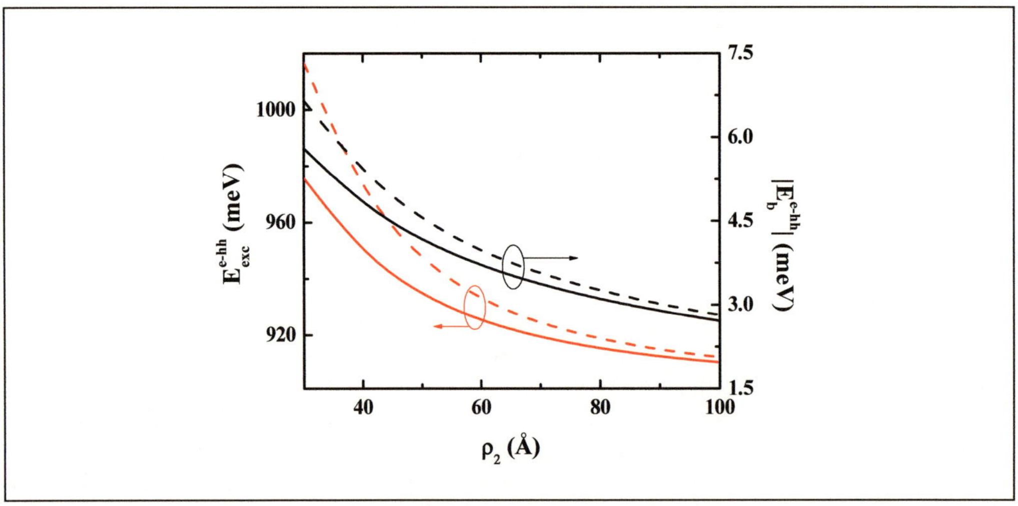

Fig. 11. Binding energy (black, right scale) and ground state energy (red, left scale) of *e-hh* excitons in $Si/Si_{0.70}Ge_{0.30}$ type-II quantum wires as a function of the wire radius, for interfaces thicknesses w of 0 Å (solid) and 15 Å (dashed).

The influence of a magnetic field parallel to the wire axis on the binding (a) and total exciton energies (b) of abrupt $Si/Si_{0.70}Ge_{0.30}$ type-II QWR is illustrated in Fig. 12, for the ground (solid) and first excited (dashed) states. The split between these two states is large for the wire with ρ_2 = 50 Å (black), but it is reduced as the wire radius increases, becoming very small for a 150 Å (red) wire radius, especially for the total exciton energy (b). As the magnetic field increases, the hole angular momentum remains zero in the ground state, but the angular momentum of the electron ground state changes almost periodically, leading to the energy oscillations observed in Fig. 12. Such oscillations can be compared to those observed in quantum rings, as a consequence of the Aharonov-Bohm (AB) effect, and the periodicity of these oscillations depends on the wire radius. (Chaves et al., 2008) The fact that, for the ground state, the hole angular momentum remains zero while the electron

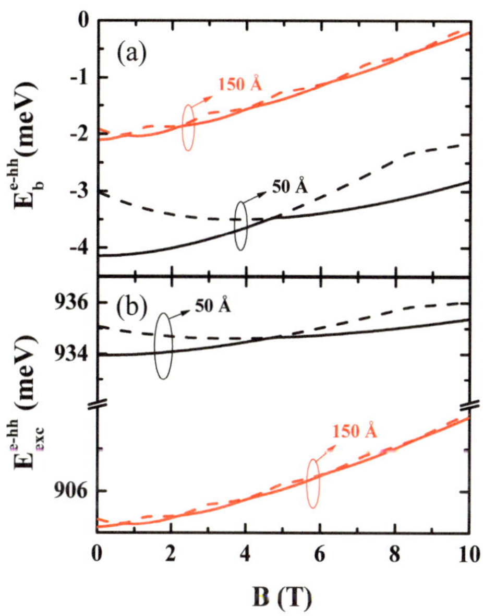

Fig. 12. Binding energies (a) and total exciton energies (b) of *e-hh* excitons in Si/$Si_{0.70}Ge_{0.30}$ type-II QWR as a function of the magnetic field for 50 Å and 150 Å wire radii, with abrupt interfaces. Solid lines are related to ground state excitons, while dotted lines are first excited states.

angular momentum changes as B increases is reasonable, since only the electrons in this system are localized around the wire, which causes the magnetic field to push the electron towards the barrier, giving rise to a change in l in each electron state, because this is energetically more favorable. Since the holes are localized inside the wire, they do not undergo any change in their angular momentum for the ground state. Aharonov-Bohm oscillations in the photoluminescence peaks of type-II stacked quantum dots have been reported recently in experimental works in literature. It has been suggested that the observation of this effect can be a useful tool for proving the radius of the system with a good resolution. (Sellers et al., 2008)

The periodicity of the Aharonov-Bohm oscillations in our results exhibits small changes as the magnetic field increases, in contrast to the oscillations observed in the energy spectrum of quantum-ring structures, where this periodicity is constant and well defined. (Szafran et al., 2004) As one can observe in Fig. 12, the periodicity clearly depends on the wire radius and, consequently, on the area enclosed by the ring-like potential, as usual for the Aharonov-Bohm effect. Hence, small changes in the periodicity of the electron energy oscillations are expected even for a constant wire radius, because increasing the magnetic field squeezes the hole wave function towards the wire axis, reducing the area enclosed by the effective Coulomb potential for electrons, implying in a larger periodicity for stronger magnetic fields.

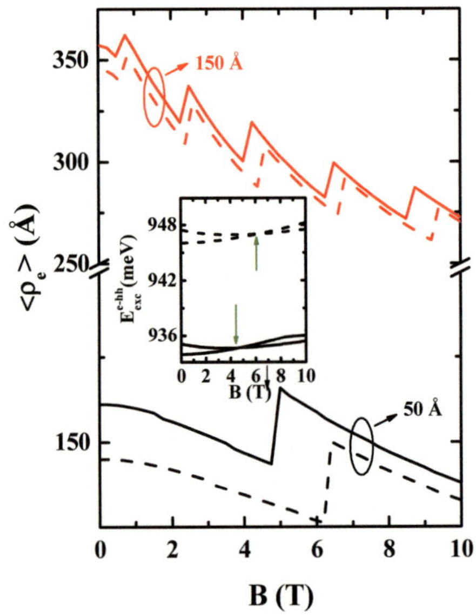

Fig. 13. Average radius of the electron ground state in Si/$Si_{0.70}Ge_{0.30}$ type-II QWR as a function of the magnetic field, for 50 Å (black) and 150 Å (red) wire radii, considering w = 0 Å (solid) and w = 15 Å (dashed). Inset: Exciton energy dependence with the applied magnetic field, for the ground and first excited states of a 50 Å QWR radius with w=0 Å (solid) and w = 15 Å (dashed). The green arrows highlight the angular momentum transition points.

The changes in the periodicity of the angular momentum transitions can be seen in Fig. 13, which illustrated the average radius of the electron ground state $<\rho_e>$ as a function of the applied magnetic field, for wire radii 50 Å (black) and 150 Å (red) with abrupt (solid) and w = 15 Å (dashed) interfaces. Since the electron angular momentum is changing almost periodically with the magnetic field, and due to the fact that wave functions for states with a larger modulus of angular momentum are more extensive than those for l = 0, one can expect that the electron average radius $<\rho_e>$ will also oscillate as a function of B. When the magnetic field pushes the electron towards the $Si_{1-x}Ge_x$ layer, its average radius decreases until a change of angular momentum occurs, when l assumes a higher modulus value, which implies a more spread electron wave function, as one can verify in Fig. 13. The inset shows the exciton energies related to a 50 Å QWR radius, for ground and first excited states, where a difference between the angular momentum transition points for abrupt (solid) and non-abrupt (dashed) interfaces, highlighted by the green arrows, can be easily seen. This difference, which is also reflected in the electron average radius, is expected to occur, since the inclusion of such an interface reduces the effective radius of the quantum wire potential.

The influence of graded interfaces on the transition points of AB oscillations can be observed even for larger radii. For a 150 Å wire radius with an abrupt interface, the fifth electron angular momentum transition occurs in a magnetic field B about 8.75 T, while for a w = 15 Å interface it occurs at B = 9.4 T. For a smaller radius, *e. g.* 50 Å, this effect is stronger: the first transition occurs at B = 5 T for an abrupt interface, whereas for w = 15 Å it occurs at B = 6.4 T. An exciton energy blueshift also appears due to the inclusion of graded interfaces, as one can observe in the inset of Fig. 13, which is consistent with earlier results of Fig. 11 (red).

3.3 Type-II Si/$Si_{0.70}Ge_{0.30}$ core-multi-shell structures

The growth of Si/SiGe core-multi-shell structures have been reported in literature (Lauhon et al., 2002) and the theoretical model presented in the last session can be straightforwardly adapted to describe such radial heterostructures. An interesting case to be studied is the system composed by a Si core wire, surrounded by a layer (internal shell) of $Si_{1-x}Ge_x$ and covered by a second layer (external shell) of Si. We consider that for a Ge concentration x = 0.30 this system exhibits a type-II potential for electrons in the conduction band, as in the case of core-shell wires we studied and Si/$Si_{1-x}Ge_x$ quantum wells in literature (Penn et al., 1999).
The type-II Si/$Si_{1-x}Ge_x$/Si core-multi-shell quantum wires have an interesting feature: the $Si_{1-x}Ge_x$ layer forms a well which confine the holes in the internal shell. However, for electrons, this layer is a barrier; thus, the electron must be confined either in the external Si shell, or in the internal Si core. These two situations are expected to lead to completely different situations of the system. For example, in Fig. 14 (a) we show the dependence of the electron-hole binding energy on the core radius ρ_1, considering w = 15 Å interfaces, for some values of the $Si_{1-x}Ge_x$ shell width S. It is observed that the energy increases with ρ_1 until a critical value of this parameter is reached. After this value, the energy starts to decrease. The results for abrupt interfaces (dotted lines) show the same qualitative behavior, although the critical radii in the abrupt case are slightly altered. The behavior of the binding energy as a function of ρ_1 in the first part (for smaller core radius) can be compared to the one of the confinement energy as a function of the average radius in a quantum ring with finite dimension (Song & Ulloa, 2001). In the second part (for larger core radius), these results resemble those found for the binding energy in a type-II wire in Fig. 11 (red). Indeed, if one analyzes the electron average radius $< \rho_e >$ in Fig. 14 (b), one observes that an abrupt transition occurs in these critical core radii, where the electron changes its radial position. A further analysis of this intriguing situation is made in Figs. 14 (c) and (d), where the electron (blue, solid) and hole (red, dashes) wave functions are presented, along with the effective electron confinement potential, for a shell width S = 100 Å and ρ_1 = 100Å (c) and 270 Å (d). Notice that the electron wave function jumps from the external Si shell towards the Si core when the core radius is increased, leading to the change we observed in the binding energies behavior.
The electron confinement in the external shell can lead to rather interesting features in the presence of an applied magnetic field parallel to the wire axis. As the magnetic field intensity increases, the hole confinement energy is expected to present AB oscillations, since this carrier is confined within the internal shell. As the electron is in the external shell for smaller core radii, one can expect that its energy will also exhibit AB oscillations. However, the magnetic field pushes the electron towards the center, which is also a Si layer, thus, for higher magnetic field intensities, the transition from the electron confinement at the external layer to the Si core is also observed and, once the electron is in the core for higher magnetic fields, its energy can not exhibit AB oscillations anymore. Our preliminary results show that

for a core radius ρ_1 = 200 Å and an internal layer width S = 100 Å, a magnetic field B = 0.8 T is enough to induce the electron confinement transition from the external shell to the Si core.

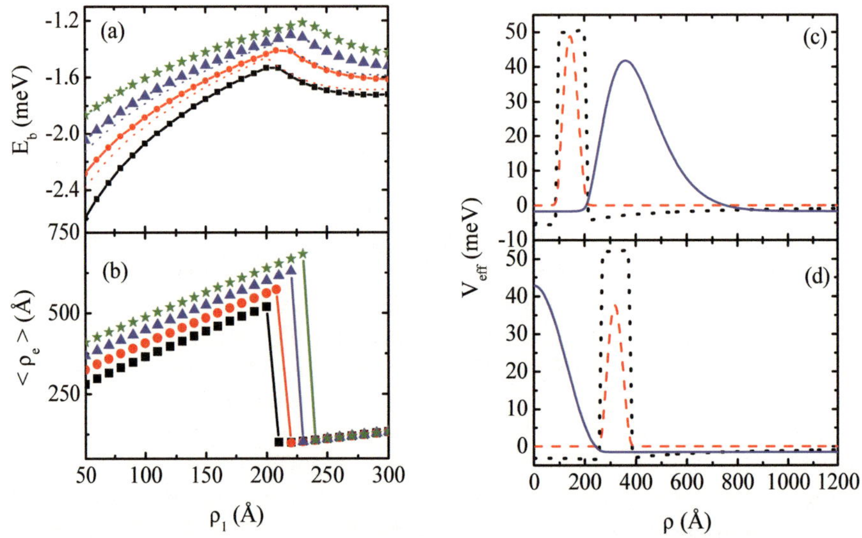

Fig. 14. Binding energy (a) and electron average radius (b) as a function of the core radius for a core-multi-shell structure with w = 15 Å interfaces, considering the internal shell width S as 50 Å (black squares), 75 Å (red circles), 100 Å (blue triangles) and 125 Å (green stars). Dotted lines are the results for abrupt interfaces in the 75 Å and 100 Å cases, while solid lines are just a guide for the eye. The electron (blue, solid) and hole (red, dashed) wave functions, as well as the electron effective potential (black, dotted) for S = 100 Å are shown for ρ_1 = 100Å (c) and 270 Å (d).

4. Conclusion

We have made a theoretical investigation of the electronic properties of two well known types of quantum wire: core-shell and longitudinally heterostructured quantum wires. In both cases, our results show that the existence of graded interfaces between materials can be responsible for significant fluctuations in the confinement energies, for a change in the qualitative behaviour of the wavefunctions and can even affect the excitonic properties of these systems.

For QWR with longitudinal heterostructures, we have shown that, due to the dependence of the radial confinement energy on the carriers effective masses, which are different for each material that compounds the heterostructure, this energy acts like an additional potential in the longitudinal direction, playing an important role on the carriers confinement in this direction. For abrupt interfaces, our results show that one can tune the conduction band offset by either growing a sample with a different radius or changing an applied magnetic field parallel to the wire axis. In this context, even a type-I to type-II transition for the electron longitudinal confinement can be observed by changing these parameters. In the presence of graded interfaces, we have shown that for wider diameters of the QWR, the electron confinement energies are blueshifted, while for smaller radii, an interfacial localization of the electron can be observed. The critical radii where the changes in the electron and hole localization occur are not expected to be the same; as a consequence, the overlap between the wave functions of these carriers can be small, leading to a low probability of interband transitions in these cases. In addition, significant qualitative changes on the electron transmission probabilities are observed for a longitudinal double heterostructure when non-abrupt interfaces are taken into account. Such changes can be reflected on the experimental observation of the electric current in longitudinally heterostructured QWR, which is left as a suggestion for future experimental research on these structures.

We have also studied Si/$Si_{1-x}Ge_x$ core-shell QWR, which can present type-I or type-II band alignments for electrons, depending on the Ge concentration x in the core. Our results show that the existence of graded interfaces affects significantly the binding and total exciton energies, especially for smaller wire radii, in both type-I and type-II cases. For the type-I structure, as electrons and holes are both confined in a thin quasi-one-dimensional wire, the excitonic properties are not greatly affected by the presence of an axially applied magnetic field. On the other hand, in type-II systems, the hole remains in the core region, whereas the electron is confined by Coulomb interaction with the hole at the shell region, surrounding a barrier like potential produced by the conduction bands mismatch between the core and shell materials. In this situation, as the intensity of an applied magnetic field parallel to the wire axis increases, AB oscillations are observed in the excitonic properties of the system. The period of such oscillations undergoes small changes for stronger magnetic fields, due to the shrinking of the hole in the radial direction induced by the field. Furthermore, although the exciton energy blueshift due to graded interfaces is very small for large wire radii, the presence of such interfaces in the type-II case is also responsible for a change in the angular momentum transition points of the AB oscillations, which can be observed even for wider diameters of the wire. A type-II core-multi-shell structure, where the hole is confined in the internal (first) shell, is also studied. We have shown that, depending on the core radius, the electron can be confined either on the external (second) shell or in the Si core, and the behaviour of the exciton energies as a function of the wire radius for each of these cases is qualitatively different. Such a transition between the electron confinement in the core or in the external shell regions can also be obtained by increasing the intensity of an applied magnetic field in the axial direction. The growth of core-multi-shell Si/SiGe QWR has been already reported in the literature, and our theoretical results show that such structures are very versatile, for one can obtain a system where the hole is in the internal shell, while the electron can be in another shell or in the core, depending on the core radius of the sample or simply by adjusting an external field.

5. References

Li, E. H. (2000). Material parameters of InGaAsP and InAlGaAs systems for use in quantum well structures at low and room temperatures. *Physica E*, Vol. 5, 215-273

Voon, L. C. L. Y. & Willatzen, M. (2003). Electron states in modulated nanowires. *Journal of Applied Physics*, Vol. 93, No. 12, (June 2003) 9997-10000

Chaves, A.; Costa e Silva, J.; Freire, J. A. K. & Farias, G. A. (2007). Excitonic properties of type-I and type-II Si/$Si_{1-x}Ge_x$ quantum wells. *Journal of Applied Physics*, Vol. 101, No. 11, (June 2007) 113703

Costa e Silva, J.; Chaves, A.; Freire, J. A. K.; Freire, V. N. & Farias, G. A. (2006). Theoretical investigation of excitons in type-I and type-II Si/$Si_{1-x}Ge_x$ quantum wires. *Physical Review B*, Vol. 74, No. 8, (August 2007) 085317

Caetano, E. W. S.; Mesquita, M. V.; Freire, V. N. & Farias, G. A. (2003). Strong graded interface related exciton energy blueshift in InGaN/GaN quantum dots. *Physica E*, Vol. 17, (April 2003) 22-23

Ferreira, R. & Bastard, G. (1997). Tunneling and relaxation in semiconductor double quantum wells. *Reports on Progress in Physics*, Vol. 60, No. 3, (March 1997) 345-387

Chaves, A.; Freire, J. A. K. & Farias, G. A. (2008). Grading effects in semiconductor nanowires with longitudinal heterostructures. *Physical Review B*, Vol. 78, No. 15, (October 2008) 155306

Voon, L. C. L. Y.; Lassen, B.; Melnik, R. & Willatzen, M. (2004). Prediction of barrier localization in modulated nanowires. *Journal of Applied Physics*, Vol. 96, No. 8, (October 2004) 4660-4662

Tsu, R. & Esaki, L. (1973). Tunneling in a finite superlattice. *Applied Physics Letters*, Vol. 22, No. 11, (June 1973) 562-564

Singh, J. (1993). *Physics of Semiconductors and Their Heterostructures*, McGraw-Hill Inc., ISBN: 0-07-057607-6

Peeters, F. M. & Schweigert, V. A. (1996). Two-electron quantum disks. *Physical Review B*, Vol. 53, No. 3, (January 1996) 1468-1474

Kamenev, B. V.; Tsybeskov, L.; Baribeau, J.-M. & Lockwood, D. J. (2005). Coexistence of fast and slow luminescence in three-dimensional Si/$Si_{1-x}Ge_x$ nanostructures. *Physical Review B*, Vol. 72, No. 19, (November 2005) 193306

Penn, C.; Schaffler, F.; Bauer, G. & Glutsch, S. (1999). Application of numerical exciton-wave-function calculations to the question of band alignment in Si/$Si_{1-x}Ge_x$ quantum wells. *Physical Review B*, Vol. 59, No. 20, (May 1999) 13314-13321

Ferreira, E. C.; da Costa, J. A. P.; Freire, J. A. K.; Farias, G. A. & Freire, V. N (2002). Interface related effects on confined excitons in GaAs/$Al_xGa_{1-x}As$ single quantum wells. *Applied Surface Science*, Vol. 190, No. 1-4, (May 2002) 191-194

Branis, S. V.; Li, G. & Bajaj, K. K. (1993). Hydrogenic impurities in quantum wires in the presence of a magnetic field. *Physical Review B*, Vol. 47, No. 3, (January 1993) 1316-1323

Chaves, A.; Costa e Silva, J.; Freire, J. A. K. & Farias, G. A. (2008). The role of surface roughness on the electron confinement in semiconductor quantum rings. *Microelectronics Journal*, Vol. 39, No. 3-4, (March-April 2007) 455-458

Sellers, I. R.; Whiteside, V. R.; Kuskovsky, I. L.; Govorov, A. O. & McCombe, B. D. (2008). Aharonov-Bohm excitons at elevated temperatures in type-II ZnTe/ZnSe quantum dots. *Physical Review Letters*, Vol. 100, No. 13, (April 2008) 136405

Szafran, B.; Peeters, F. M. & Bednarek, S. (2004). Electron spin and charge switching in a coupled quantum dot-quantum ring system. *Physical Review B*, Vol. 70, No. 12, (September 2004) 125310

Lauhon, L. J.; Gudiksen, M. S.; Wang, D. & Lieber, C. M. (2002). Epitaxial core-shell and core-multishell nanowire heterostructures. *Nature*, Vol. 420, (November 2002) 57-61

Song, J. & Ulloa, S. E. (2001). Magnetic field effects on quantum ring excitons. *Physical Review B*, Vol. 63, No. 12 (March 2001) 125302

17

Optical Modeling of Photoluminescence of Multilayered Semiconductor Nanostructures: Nanowires, Nanotubes and Nanocables

Xue-Wen Chen and Sailing He
Centre for Optical and Electromagnetic Research, Zhejiang University; Joint Research Centre of Photonics of the Royal Institute of Technology (Sweden) and Zhejiang University; China

1. Introduction

One-dimensional semiconductor nanostructures, such as nanowires (NWs), core-shell nanocables (NCs) and nanotubes (NTs), have great potential in building up nanometric elements for photonic applications, such as light-emitting diodes (*Yu et al., 1998; Könenkamp et al. 2004; Qian et al.,2005; Bao et al., 2006*), nanolasers (*Huang et al., 2001; Johnson et al.,2002; Duan et al.,2003*), optical interconnects (*Barrelet et al., 2004; Tong et al.,2005*), biolabeling (*Wang et al.,2005*), sensing and spectroscopy. For such applications, light emission and guiding properties of the nanostructures are critical and determine the performance of the photonic devices. While the waveguiding and cavtiy properties of nanowires can be readily obtained by following conventional optical waveguide theory (*Maslov & Ning,2003; Johnson et al.,2003*), light emission properties of a nanostructure, due to the pronounced Purcell effect (*Purcell, 1946; Lukosz, 1980*), change in relatively subtle way as the dimension of the structure scales down. For example, as observed experimentally (*Wang et al., 2001*), photoluminescence (PL) of NWs shows a strong dependence on the polarization state of the excitation light, which cannot be fully explained by just taking account of the effect of excitation since spontaneous emission becomes highly dependent on the orientation of the transition dipole moment of the excitons as the size of the NW decreases (*Chen et al., 2007a*). Therefore, a rigorous and efficient optical model of light emission from one-dimensional semiconductor nanostructure is desirable for understanding the physics and for engineering the emission and its interaction with the nanostructure.

In this chapter, we present a comprehensive and efficient model for analyzing PL of a single multilayered nanostructure based on classical electrodynamics and, as concrete examples, investigate the emissions from a single ZnO NW, NC and NT as functions of geometrical parameters and the polarization of the excitation. Here we assume the excitation intensity for PL is well below the lasing threshold of the nanostructure and consequently spontaneous emission process dominates. Our theoretical model consists of two parts, namely, the modeling of the spontaneous emission of a single emitter in the structure and the evaluation of the excitation effect. The chapter is organized as follows. In section 2, we present a rigorous optical model of the spontaneous emission of an emitter in an arbitrarily multilayered structure and introduce an efficient numerical implementation of the model.

Spontaneous emission properties, including the radiative decay rate, quantum efficiency and spatial distribution of the emission, are investigated for a single ZnO NW and NT. In Section 3, a full optical model of PL is given by including the effect of the excitation. We show the strong dependences of PL intensity on the polarization of the excitation and the geometrical parameters of ZnO-based nanostructures. Conclusion is given in Section 4.

2. Spontaneous emission in a cylindrically multilayered structure

2.1 Theory of spontaneous emission in nanostructures

The theoretical description of spontaneous emission in a cylindrically multilayered nanostructure is based on classical electromagnetics with the quantum emitter modeled as a classical electric dipole radiating with a constant dipole moment (*Lukosz, 1980; Sullivan & Hall, 1997*). As a consequence of Fermi's golden rule, the radiative decay rate of the quantum emitter in the nanostructure is modified according to

$$\Gamma_r = F \cdot \Gamma_r^0 \tag{1}$$

where Γ_r^0 and Γ_r are the radiative decay rate in the bulk medium and the nanostructure, respectively. F is the Purcell factor, which equals to the total radiation power of the dipole in the nanostruture normalized by the value in the bulk medium. Since the radiative decay and non-radiative decay are two competing processes, the quantum efficiency of the quantum emitter is modified in the nanostructure. Assuming that the non-radiative decay rate Γ_{nr} of the emitter does not change in the nanostructure, quantum efficiency η_m in the nanostructure is changed to

$$\eta_m \equiv \Gamma_r / (\Gamma_r + \Gamma_{nr}) = F\eta_0 / \left[F\eta_0 + (1-\eta_0)\right] \tag{2}$$

where $\eta_0 = \Gamma_r^0 / \left(\Gamma_r^0 + \Gamma_{nr}\right)$ is the quantum efficiency of the emitter in bulk. The spatial distribution of the emission, for example, the free space emission (emission goes to the far field) and the waveguide emission, can be obtained by evaluating the radiation property of the dipole source. Therefore, the radiation of a classical electric dipole in the cylindrically multilayered nanostructure basically reflects the emission properties of a quantum emitter in the structure. In principle, all the information can be obtained quite accurately by using a three-dimensional finite difference time domain (3D-FDTD) method (*Bermel et al., 2004*). However, the 3D-FDTD method is a brute force numerical method, which treats the problem as a black box and could not provide the physical insight. In addition, it is inefficient. Semi-analytical method based on cylindrical wave decomposition (*Lovell & Chew, 1987*) is applicable. This method involves the evaluation of the so-called Sommerfeld-like integral and is inefficient for a direct evaluation. Here we propose a simple technique to efficiently calculate the Purcell factor F and give an explicit formula to calculate the free space emission and the waveguide emission.

A cylindrically multilayered nanostructure is depicted in Fig. 1. An emitting layer is sandwiched between two stacks of shells, i.e., P outer shells and Q inner shells. The emitting medium and the outmost medium (e.g. air) are assumed to be non-absorbing at the emission wavelength while the other shells can be either transparent or absorptive. To study the spontaneous emission from the nanostructure, we assume the emitting layer consists of an ensemble of incoherent quantum emitters, each of which is modeled as an electric dipole radiating with a constant dipole moment. The orientation of the transition dipole moment of the emitter in a semiconductor nanostructure depends on the excitation. For the case of electroluminescence, the orientation is considered to be random while it is directed by the excitation for the case of PL. Here we investigate the emission of the emitters for three orthogonal orientations, namely, radially, azimuthally and z axially oriented. As displayed in Fig. 1, the emission from the emitter includes the waveguide emission W (emission coupled to the waveguide modes) and the free space emission U (emission coupled to the radiation modes). Here both of them are normalized by the total radiation power of the same dipole in the bulk medium. If all the materials are lossless, one has $F = U + W$.

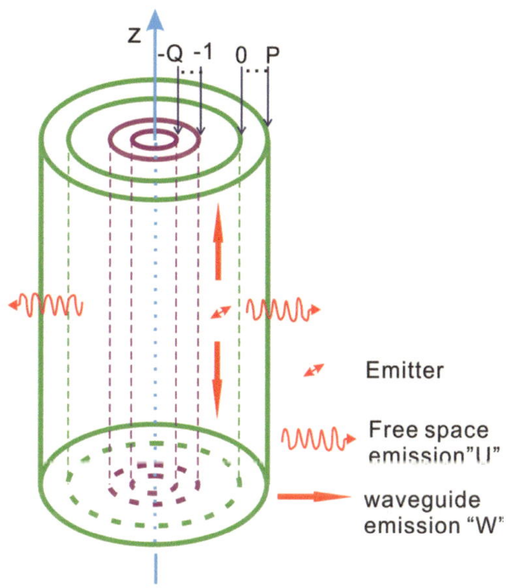

Fig. 1. Layout of a cylindrically multilayered structure. The spontaneous emission consists of free space emission and waveguide emission.

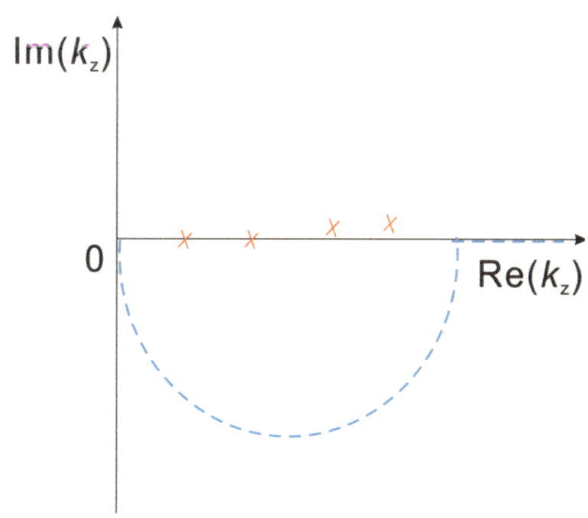

Fig. 2. Integration path in the complex plane of k. The crosses are the poles of the integral

In the emitting layer of the nanostructure, the electric field induced by an electric dipole can be written as

$$E = E^0 + E^r \tag{3}$$

where E^0 and E^r are the electric field of the dipole in the infinite medium and the reflected field due to the interfaces in the nanostructure, respectively. The normalized total emission power can be obtained as (*Lukosz, 1980*)

$$F_\alpha = 1 - \vec{\alpha} \cdot \mathrm{Im}\left(iE^r(\vec{r}\,',\vec{r}\,')\right) / (\omega\mu k\Theta / 6\pi) \tag{4}$$

where μ, ω, k, Θ, $\vec{\alpha}$, and $\vec{r}\,'$ are the permeability, angular frequency, wavenumber in the emitting medium, dipole moment, orientation vector and location of the dipole, respectively. Here Im () stands for the imaginary part of () and the subscript of F denotes the orientation. To calculate F, we expand the z component of the electric field E^0 and magnetic field H^0 in terms of cylindrical waves in polar coordinates (ρ, θ, z) (*Chew, 1995*)

$$\begin{bmatrix} E_z^0 \\ H_z^0 \end{bmatrix} = \begin{cases} \sum_{v=-\infty}^{\infty} e^{iv\theta} \int_{-\infty}^{\infty} dk_z \vec{a}_{00v} H_v^{(1)}(k_{\rho,e}\rho) e^{ik_z z} & (\rho > \rho') \\ \sum_{v=-\infty}^{\infty} e^{iv\theta} \int_{-\infty}^{\infty} dk_z \vec{b}_{00v} J_v(k_{\rho,e}\rho) e^{ik_z z} & (\rho < \rho') \end{cases} \tag{5}$$

where k_z and $k_{\rho,e}$ are the z component and radial component of the wavenumber in the emitting medium. $\vec{a}_{00v}$ and $\vec{b}_{00v}$ are the coefficient vectors of outgoing wave and standing wave, respectively. They are given by

$$\vec{a}_{00v} = -\frac{\Theta}{8\pi\varepsilon_e\omega} \begin{bmatrix} \left(\bar{z}k^2 + \frac{\partial}{\partial z'}\nabla'\right) \cdot \vec{\alpha} \\ i\omega\varepsilon_e \vec{\alpha} \cdot \bar{z} \times \nabla' \end{bmatrix} e^{-iv\theta' - ik_z z'} J_v(k_{\rho,e}\rho') \tag{6a}$$

$$\vec{b}_{00v} = -\frac{\Theta}{8\pi\varepsilon_e\omega} \begin{bmatrix} \left(\bar{z}k^2 + \frac{\partial}{\partial z'}\nabla'\right) \cdot \vec{\alpha} \\ i\omega\varepsilon_e \vec{\alpha} \cdot \bar{z} \times \nabla' \end{bmatrix} e^{-iv\theta' - ik_z z'} H_v^{(1)}(k_{\rho,e}\rho') \tag{6b}$$

Here ε_e is the permittivity of the emitting layer. J_v and $H_v^{(1)}$ denote the Bessel function and Hankel function of the first kind at order v. The operators with a prime act on the functions with primed coordinates. The coefficient vectors can be simplified for a specific dipole orientation. For the radially oriented dipole, the coefficient vectors read as

$$\bar{a}_{00v} = \begin{bmatrix} \frac{ik_z k_{\rho,e}\Theta}{8\pi\omega\varepsilon_e} J_v'(k_{\rho,e}\rho') \\ \frac{v\Theta}{8\pi\rho'} J_v(k_{\rho,e}\rho') \end{bmatrix} \tag{7a}$$

$$\bar{b}_{00v} = \begin{bmatrix} \frac{ik_z k_{\rho,e}}{8\pi\omega\varepsilon_e} H_v'(k_{\rho,e}\rho') \\ \frac{v}{8\pi\rho'} H_v(k_{\rho,e}\rho') \end{bmatrix} \tag{7b}$$

For the azimuthally oriented dipole, the coefficient vectors read as

$$\bar{a}_{00v} = \begin{bmatrix} \dfrac{vk_z}{8\pi\omega\varepsilon_e\rho'} J_v\left(k_{\rho,e}\rho'\right) \\ -\dfrac{ik_{\rho,e}}{8\pi} J_v'\left(k_{\rho,e}\rho'\right) \end{bmatrix} \tag{8a}$$

$$\bar{b}_{00v} = \begin{bmatrix} \dfrac{vk_z}{8\pi\omega\varepsilon_e\rho'} H_v\left(k_{\rho,e}\rho'\right) \\ -\dfrac{ik_{\rho,e}}{8\pi} H_v'\left(k_{\rho,e}\rho'\right) \end{bmatrix} \tag{8b}$$

For the z-axially oriented dipole, the coefficient vectors read as

$$\bar{a}_{00v} = \begin{bmatrix} -\dfrac{k^2 - k_z^2}{8\pi\omega\varepsilon_e} J_v\left(k_{\rho,e}\rho'\right) \\ 0 \end{bmatrix} \tag{9a}$$

$$\bar{b}_{00v} = \begin{bmatrix} -\dfrac{k^2 - k_z^2}{8\pi\omega\varepsilon_e} H_v\left(k_{\rho,e}\rho'\right) \\ 0 \end{bmatrix} \tag{9b}$$

The fields in the l-th layer of the multilayered structure are combinations of the outgoing wave and standing wave and their z components can be given by

$$\begin{bmatrix} E_z^l \\ H_z^l \end{bmatrix} = \sum_{v=-\infty}^{\infty} e^{iv\theta} \int_{-\infty}^{\infty} dk_z \left(\vec{a}_{lv} H_v^{(1)}\left(k_{\rho,l}\rho\right) + \vec{b}_{lv} J_v\left(k_{\rho,l}\rho\right)\right) e^{ik_z z} \tag{10}$$

The azimuthal components can be obtained from the z components and are expressed as

$$\begin{bmatrix} E_\theta^l \\ H_\theta^l \end{bmatrix} = \sum_{v=-\infty}^{\infty} e^{iv\theta} \int_{-\infty}^{\infty} dk_z \left(\vec{a}_{lv} \vec{H}_v^{(1)}\left(k_{\rho,l}\rho\right) + \vec{b}_{lv} \vec{J}_v\left(k_{\rho,l}\rho\right)\right) e^{ik_z z} \tag{11}$$

where $\vec{H}_v^{(1)}$ and $\vec{J}_v$ are defined as

$$\vec{H}_v^{(1)}\left(k_\rho\rho\right) = \frac{i}{2k_\rho}\left\{ \begin{bmatrix} ik_z & -\omega\mu \\ \omega\varepsilon & ik_z \end{bmatrix} H_{v-1}^{(1)}\left(k_\rho\rho\right) + \begin{bmatrix} ik_z & -\omega\mu \\ -\omega\varepsilon & ik_z \end{bmatrix} H_{v+1}^{(1)}\left(k_\rho\rho\right) \right\} \tag{12a}$$

$$\vec{J}_v\left(k_\rho\rho\right) = \frac{i}{2k_\rho}\left\{ \begin{bmatrix} ik_z & -\omega\mu \\ \omega\varepsilon & ik_z \end{bmatrix} J_{v-1}\left(k_\rho\rho\right) + \begin{bmatrix} ik_z & -\omega\mu \\ -\omega\varepsilon & ik_z \end{bmatrix} J_{v+1}\left(k_\rho\rho\right) \right\} \tag{12b}$$

The coefficient vectors of the neighbouring two layers are related according to the continuity of the tangentical field compoents.

$$\vec{a}_{lv} H_v^{(1)}\left(k_{\rho,l}R_l\right) + \vec{b}_{lv} J_v\left(k_{\rho,l}R_l\right) = \vec{a}_{l+1v} H_v^{(1)}\left(k_{\rho,l+1}R_l\right) + \vec{b}_{l+1v} J_v\left(k_{\rho,l+1}R_l\right) \tag{13a}$$

$$\vec{a}_{lv}\vec{H}_v^{(1)}\left(k_{\rho,l}R_l\right)+\vec{b}_{lv}\vec{J}_v\left(k_{\rho,l}R_l\right)=\vec{a}_{l+1v}\vec{H}_v^{(1)}\left(k_{\rho,l+1}R_l\right)+\vec{b}_{l+1v}\vec{J}_v\left(k_{\rho,l+1}R_l\right) \tag{13b}$$

where R_l is the radial coordinate of the l-th interface of the multilayered structure. For the fields in the emitting layer (l = 0), the two stacks of shells can be considered as two "black" shells characterized by the total downward reflection matrix N_{0v} and the total upward reflection matrix M_{0v}. The z components of the fields in the emitting layer due to the interfaces read as

$$\begin{bmatrix} E_z^r \\ H_z^r \end{bmatrix}=\sum_{v=-\infty}^{\infty} e^{iv\theta}\int_{-\infty}^{\infty} dk_z\left(\vec{a}_{0v}^{r}H_v^{(1)}\left(k_{\rho,e}\rho\right)+\vec{b}_{0v}^{r}J_v\left(k_{\rho,e}\rho\right)\right)e^{ik_z z} \tag{14}$$

where the coefficient vectors are

$$\vec{a}_{0v}^{r}=\left[\vec{I}-M_{0v}N_{0v}\right]^{-1}\left[M_{0v}\vec{b}_{00v}+\vec{a}_{00v}\right]-\vec{a}_{00v} \tag{15a}$$

$$\vec{b}_{0v}^{r}=\left[\vec{I}-N_{0v}M_{0v}\right]^{-1}\left[N_{0v}\vec{a}_{00v}+\vec{b}_{00v}\right]-\vec{b}_{00v} \tag{15b}$$

The total upward reflection coefficient matrix M_{0v} can be calculated recursively via (resulting from matching the boundary condition at each interface, Lovell & Chew,1987)

$$\vec{M}_{l+1,v}=\vec{\Gamma}_{l,v}^{d}+\vec{T}_{l,v}^{u}\vec{M}_{l,v}\left[I-\vec{\Gamma}_{l,v}^{u}\vec{M}_{l,v}\right]^{-1}\vec{T}_{l,v}^{d}\qquad (l=-Q,\ldots,-1) \tag{16}$$

$$\vec{M}_{-Q,v}=0 \tag{17}$$

where

$$\vec{T}_{l,v}^{d}=\frac{2\omega}{\pi k_{\rho,k+1}^{2}R_l}\vec{A}_{l,v}\begin{bmatrix}0 & -\mu\\ \varepsilon_{l+1} & 0\end{bmatrix} \tag{18a}$$

$$\vec{T}_{l,v}^{u}=\frac{2\omega}{\pi k_{\rho,l}^{2}R_l}\vec{A}_{l,v}\begin{bmatrix}0 & -\mu\\ \varepsilon_{l} & 0\end{bmatrix} \tag{18b}$$

$$\vec{\Gamma}_{l,v}^{d}=\vec{A}_{l,v}\left[J_v\left(k_{\rho,l}R_l\right)\vec{J}_v\left(k_{\rho,l+1}R_l\right)-J_v\left(k_{\rho,l+1}R_l\right)\vec{J}_v\left(k_{\rho,l}R_l\right)\right] \tag{18c}$$

$$\vec{\Gamma}_{l,v}^{u}=\vec{A}_{l,v}\left[H_v^{(1)}\left(k_{\rho,l}R_l\right)\vec{H}_v^{(1)}\left(k_{\rho,l+1}R_l\right)-H_v^{(1)}\left(k_{\rho,l+1}R_l\right)\vec{H}_v^{(1)}\left(k_{\rho,l}R_l\right)\right] \tag{18d}$$

$$\vec{A}_{l,v}=\left[H_v^{(1)}\left(k_{\rho,l+1}R_l\right)\vec{J}_v\left(k_{\rho,l}R_l\right)-J_v\left(k_{\rho,l}R_l\right)\vec{H}_v^{(1)}\left(k_{\rho,l+1}R_l\right)\right]^{-1} \tag{18e}$$

where I is the identity matrix. Similarly, the total downward reflection matrix N_{0v} can be obtained via

$$\vec{N}_{l,v}=\vec{\Gamma}_{l,v}^{u}+\vec{T}_{l,v}^{d}\vec{N}_{l+1,v}\left[I-\vec{\Gamma}_{l,v}^{d}\vec{N}_{l+1,v}\right]^{-1}\vec{T}_{l,v}^{u}\quad (l=P-1,P-2,..,0) \tag{19}$$

$$\vec{N}_{P,v}=\vec{0} \tag{20}$$

After determining the coeffcient vectors in the emitting layer, the coefficient vectors in the outmost layer are obtained recursively by using Eq. (13). The z and azimuthal components of the fields in the outmost layer read

$$\begin{bmatrix} E_z \\ H_z \end{bmatrix} = \sum_{\upsilon=-\infty}^{\infty} e^{i\upsilon\theta} \int_{-\infty}^{\infty} dk_z \vec{a}_{P\upsilon} H_{\upsilon}^{(1)}\left(k_{\rho,P}\rho\right) e^{ik_z z} \tag{21}$$

$$\begin{bmatrix} E_\theta \\ H_\theta \end{bmatrix} = \sum_{\upsilon=-\infty}^{\infty} e^{i\upsilon\theta} \int_{-\infty}^{\infty} dk_z \left(\vec{a}_{P\upsilon} \vec{H}_{\upsilon}^{(1)}\left(k_{\rho,P}\rho\right)\right) e^{ik_z z} \tag{22}$$

2.2 Efficient evaluation of the Purcell factor and free space emission

As seen from Eq. (4) and Eq. (14), the calculation of Purcell factor F involves the evaluation of an integral with pole singularities, which physically correspond to the guided modes in the cylindrically layered structure. This is similar to the case of a planar multilayered structure (*Gay-Balmaz & Mosig, 1997; Chen et al, 2007b*) and a direct evaluation of the integral in Eq. (14) tends to fail. These poles are located in the real axis of k_z or in the first quadrant of the complex plane of k_z. Here, to get round the problem, we change the 1D integral of Eq. (14) in the real space to a line integral in the complex plane of k_z and use the Cauchy integral theorem to perform the integration. Since the poles of the integrand in Eq. (14) are located in the first quadrant of the complex plane of k_z, we select an integration path in the fourth quadrant such that the domain enclosed by the new path and the real axis of k_z does not contain any pole. The new path shown by the dashed line in Fig. 2 consists of a semicircle in the fourth quadrant of the complex plane and a straight line along the real axis. The integral in Eq. (14) can be rewritten as

$$\begin{bmatrix} E_z^r \\ H_z^r \end{bmatrix} = 2\sum_{\upsilon=-\infty}^{\infty} e^{i\upsilon\theta} \left(\int_{C_R} + \int_{2k_r}^{\infty} \right) \left(\vec{a}_{0\upsilon} H_{\upsilon}^{(1)}\left(k_{\rho,e}\rho\right) + b_{0\upsilon} J_{\upsilon}\left(k_{\rho,e}\rho\right)\right) e^{ik_z z} dk_z \tag{23}$$

where k_r is the radius of the semicircle C_R. k_r should be large enough so that the semicircle can bypass all the poles of the integrand. In this way, the integration can be efficiently evaluated by using conventional numerical integration routines such as the Gaussian quadrature technique (*Chen et al, 2007b*).

The free space emission power U is the power leaving the cylindrically layered structure from the lateral surface to far field. Thus, U can be expressed as

$$U \equiv \int_0^{2\pi} \rho d\theta \int_{-\infty}^{\infty} dz \frac{1}{2} \mathrm{Re}\left(E_\theta \times H_z^* - E_z \times H_\theta^*\right) / P_0 \tag{24}$$

where $P_0 = \sqrt{\mu / \varepsilon_e} k_e^2 / \left(48\pi^3\right)$ is the normalization constant and $Re()$ stands for the real part of (). By using Eqs. (21) and (22), Eq.(24) can be written as

$$U = \frac{\omega\rho}{P_0} \sum_{\upsilon=-\infty}^{\infty} \int_0^{k_{out}} dk_z \mathrm{Re}\left[i\varepsilon_P H_{\upsilon}^{(1)}\left(k_{\rho,P}\rho\right) H_{\upsilon}^{'(1)*}\left(k_{\rho,P}\rho\right) \left|a_{P\upsilon}(1)\right|^2 - i\mu H_{\upsilon}^{(1)*}\left(k_{\rho,P}\rho\right) H_{\upsilon}^{'(1)}\left(k_{\rho,P}\rho\right) \left|a_{P\upsilon}(2)\right|^2 \right] / k_{\rho,P} \tag{25}$$

Here ε_P and k_P are the permittivity and wavenumber of the outmost medium, respectively. If the outermost medium is lossless, U does not depend on ρ. If the whole structure is lossless, the waveguide emission W can be readily obtained by $W = F-U$.

2.3 Examples: Spontaneous emission from a single ZnO NW and a ZnO NT

In this section, we employ the optical model developed above to study the spontaneous emission from a single ZnO NW and a ZnO NT. Light emission properties, including the radiative decay rate, the quantum efficiency and the fraction of the waveguide emission to the total emission, are investigated as a function of the size of the NW and NT at an emission wavelength of 387 nm. The cross-section of the ZnO NW is treated approximately as circular here and an initial quantum efficiency $\eta_0 = 0.85$ is used (Zhang & Russo, 2005). A refractive index of 2.5 is used for the ZnO material here.

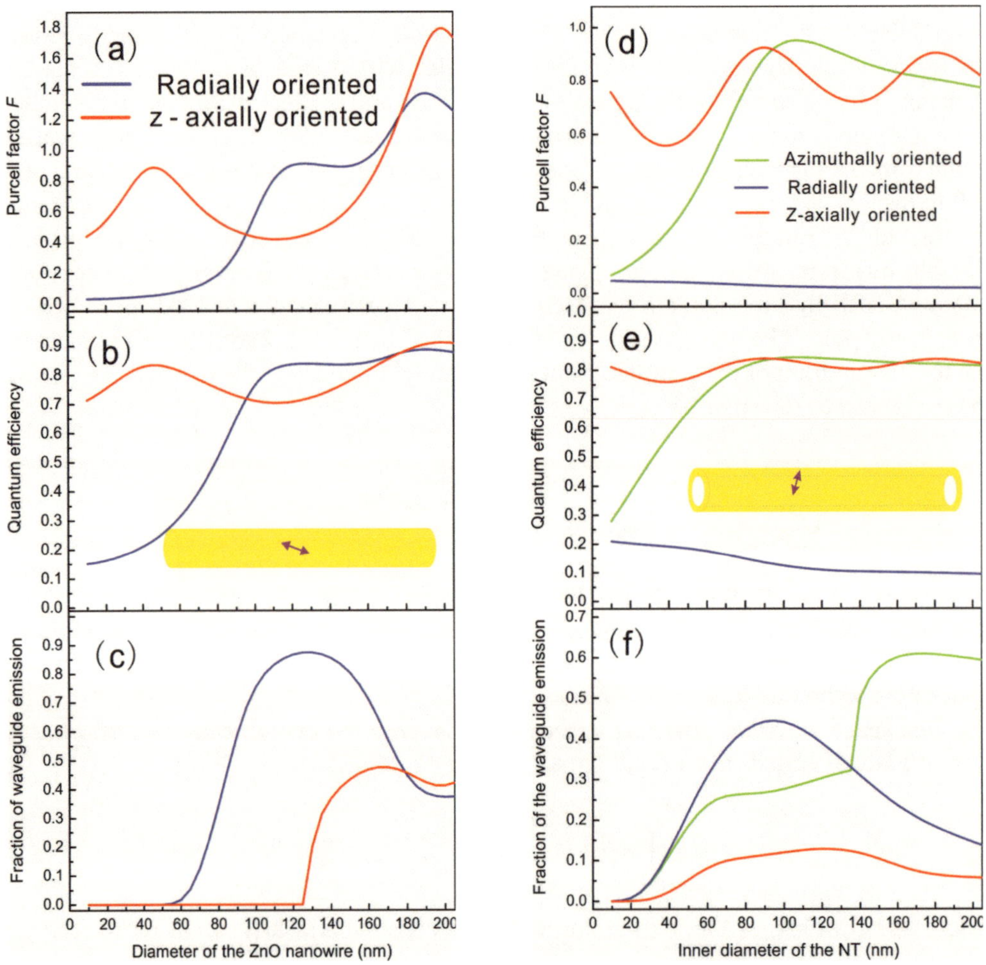

Fig. 3. Spontaneous emission properties of a single ZnO NW (a-c) and a single ZnO NT (d-f)

Figures 3(a)-(c) show the spontaneous emission properties of an emitter as a function of the ZnO NW diameter. The emitter is on the z axis and two kinds of orientations, namely the radial and z-axial orientations are considered in the simulation. Several remarkable features can be observed in these figures. Firstly, the Purcell factor of the radially oriented emitter is almost one order of magnitude smaller than the value of the z-axially oriented emitter for ZnO NWs with a small diameter. As the diameter of the NW decreases further, F_ρ has a limiting value of 0.03, which solely depends on the refractive index of the ZnO material. Since the radiative decay rate is proportional to F (c.f. Eq. (1)) and radiative transition and non-radiative transition are two competing processes, the quantum efficiency of the radially oriented emitter is significantly lower than the value of the z axially oriented emitter, as shown in Fig. 3(b). As the NW diameter increases, F_ρ goes up quickly and reaches a value close to F_z and then starts to oscillate, similar to F_z. Secondly, there are two critical diameters of the NW, i.e. 60 nm and 125 nm for the radially oriented emitter and z-axially oriented emitter, respectively. For NWs thinner than 60 nm, as seen in Fig. 3(c), there is no waveguide emission for both kinds of emitters. But for NWs thicker than 60 nm but thinner than 125 nm, the z-axially oriented emitter has almost zero waveguide emission. Here a larger critical diameter for the z axially oriented emitter is due to the fact that the emission from the z-axially oriented emitter located on the z axis can only couple to TM_{01} waveguide mode because the coefficient vectors in Eq.(9a) are nonzero only for n = 0. TM_{01} mode is not the fundamental mode and is cut-off until the diameter is larger than 125 nm. However, the radially oriented emitter has a good coupling to the waveguide modes starting from the fundamental mode, i.e., HE_{11} mode. For the NW thinner than 60 nm, the emission is purely free space emission and the z-axially oriented emitter is much more efficient than the radially oriented emitter (see Fig. 3(b)). As the diameter increases from 60 nm to 125 nm, the emission efficiency of the radially oriented emitter steadily increases and the waveguide emission gradually becomes dominant.

Figures 3(d)-(f) show the Purcell factor, quantum efficiency and fraction of waveguide emission as a function of the inner diameter of a ZnO NT with a shell thickness of 20 nm. The emitter sits in the middle of the ZnO shell and can have three orthogonal orientations, namely, the radial, azimuthal and z axial orientations. The emission properties are distinct for different orientations. For the ZnO NT studied here, F_ρ is always the smallest and decreases as the inner diameter of the NT increases. Thus the radially oriented emitter is the most inefficient one. F_z oscillates as a function of the inner diameter and is much larger than F_θ and F_ρ for the NTs with small inner diameters. As the inner diameter increases, F_θ goes up quickly to a maximum value of 0.95 and then decreases slowly. Figure 3(e) shows the quantum efficiencies of the three kinds of emitters. As the inner diameter increases from 10 nm to 80 nm, the z axially oriented emitter remains the most efficient one while the efficiency of the azimuthally oriented emitter increases from 28% to 81%. As the diameter further increases, the quantum efficiencies of the above two emitters become similar and remain above 80%. For the waveguide emission, different from the case of NW, there is no critical inner diameter below which no waveguide emission exists.

In summary, we have presented an efficient and accurate optical model of spontaneous emission from a single cylindrically multilayered nanostructure. The Purcell factor is represented as an integral and can be efficiently evaluated by selecting a new integration path in the complex plane. An explicit formula has been derived to calculate the free space emission. As examples of practical interest, the emission properties of a single ZnO NW and a ZnO NT are comprehensively studied and discussed. We found that the emission properties depend strongly on the orientation of the emitter and consequently the PL should show strong dependence on the polarization state of the excitation light.

3. Photoluminescence of cylindrically multilayered nanostructure

In section 2, we studied the spontaneous emission of a single emitter with a given orientation in a cylindrically multilayered structure without considering the effect of excitation. For PL of the semiconductor nanostructure, the excitation light determines the spatial distribution of the density of the excitons and the orientation of the transition dipole moment, and therefore governs the emission of the nanostructure. The exciton density in the semiconductor nanostructure is proportional to the local intensity of excitation light I_{ex} and the orientation of the dipole moment is directed by the local electric field. Therefore, the total emission intensity I_t of a cylindrically multilayered nanostructure can be expressed as

$$I_t = \int_{r_n}^{r_{n+1}} \rho d\rho \int_0^{2\pi} d\theta \left[I_{ex,\rho}\eta_\rho + I_{ex,\theta}\eta_\theta + I_{ex,z}\eta_z \right] / \left(\pi r_{n+1}^2 - \pi r_n^2 \right) \tag{26}$$

where the subscript n denotes the emitting layer which is between two interfaces labeled by r_n and r_{n+1} respectively. η_x ($x = \rho, \theta, z$) is the quantum efficiency of the emitter oriented along x direction while $I_{ex,x}$ is the intensity of the excitation along x direction at the position of the emitter. The free space emission is given by

$$I_u = \int_{r_n}^{r_{n+1}} \rho d\rho \int_0^{2\pi} d\theta \left[I_{ex,\rho}\eta_\rho \frac{U_\rho}{F_\rho} + I_{ex,\theta}\eta_\theta \frac{U_\theta}{F_\theta} + I_{ex,z}\eta_z \frac{U_z}{F_z} \right] / \left(\pi r_{n+1}^2 - \pi r_n^2 \right) \tag{27}$$

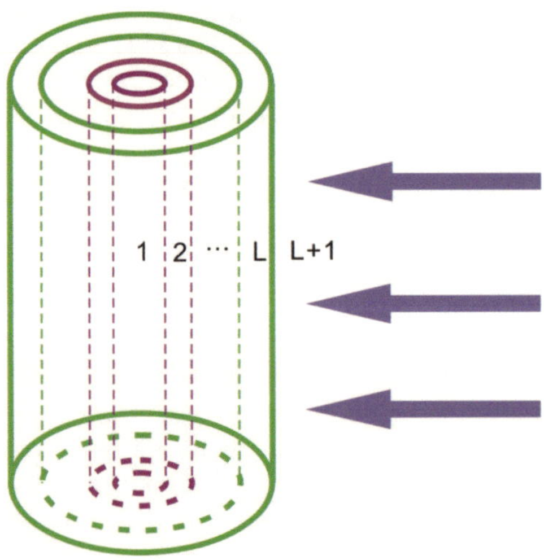

Fig. 4. Schematic diagram of a cylindrically multilayered structure illuminated by a plane wave.

By following the model in Section 2, one could easily evaluate η_x , U_x and F_x at arbitrary position. For a lossless structure, the waveguide emission is I_t - I_u. In this section, we first introduce how to calculate the local excitation field by a plane wave excitation and then simulate the PL from a single ZnO NW, NC and NT.

3.1 Modeling of the excitation

In this subsection, we consider a single cylindrically multilayered nanostructure illuminated by a polarized plane wave and formulate the problem to calculate the electric field at an arbitrary position in the structure. In polar coordinates, an arbitrary polarized plane wave can be decomposed into two polarizations, namely the vertical polarization with electric field parallel to the z axis and the horizontal polarization with electric field perpendicular to the z axis, each of which can be expanded in terms of cylindrical waves with coefficients determined by the wave vector and amplitude of the plane wave. By summing up the response of the nanostructure from each cylindrical wave, one obtains the response of the nanostructure by the plane wave. A plane wave E_i is decomposed as the following (Tsang et al., 2001)

$$E_i(\vec{r}) = \left(\hat{v}_i E_{vi} + \hat{h}_i E_{hi}\right) e^{i\vec{k}\cdot\vec{r}} = \sum_{n=-\infty}^{\infty} \frac{-i^n e^{-in\theta_k}}{k_\rho}\left[E_{vi}\vec{N}_n^J(k_\rho, k_z, \vec{r}) - iE_{hi}\vec{M}_n^J(k_\rho, k_z, \vec{r})\right] \quad (28)$$

where θ_k is the azimuthal angle of the incident wave vector $\vec{k}$, subscript v and h denote the vertical and horizontal components, respectively. k_ρ and k_z are respectively the radial and z axial components of the wave vector of the incident plane wave. The cylindrical vector wave functions are defined as

$$\vec{M}_n^X(k_\rho, k_z, \vec{r}) = \left(\hat{\rho}\frac{in}{\rho} X_n(k_\rho \rho) - \hat{\theta} k_\rho X_n'(k_\rho \rho)\right) e^{ik_z z + in\theta} \quad (29)$$

$$\vec{N}_n^X(k_\rho, k_z, \vec{r}) = \left(\hat{\rho}\frac{ik_\rho k_z}{k} X_n'(k_\rho \rho) - \hat{\theta}\frac{nk_z}{k\rho} X_n(k_\rho \rho) + \hat{z}\frac{k_\rho^2}{k} X_n(k_\rho \rho)\right) e^{ik_z z + in\theta} \quad (30)$$

where X_n denotes the bessel function (if $X = J$) or the Hankel function of first kind (if $X = H$) at order n. These cylindrical vector wave functions satisfy the following relations

$$\nabla \times \vec{M}_n^X(k_\rho, k_z, \vec{r}) = k\vec{N}_n^X(k_\rho, k_z, \vec{r}) \quad (31)$$

$$\nabla \times \vec{N}_n^X(k_\rho, k_z, \vec{r}) = k\vec{M}_n^X(k_\rho, k_z, \vec{r}) \quad (32)$$

In the l-th layer of the nanostructure shown in Fig. 4, the electric field and magnetic field are given by

$$\vec{E}_l(\vec{r}) = \sum_{n=-\infty}^{\infty}\begin{bmatrix} A_{l,n}^M \vec{M}_n^J(k_{l,\rho}, k_z, \vec{r}) + A_{l,n}^N \vec{N}_n^J(k_{l,\rho}, k_z, \vec{r}) + \\ B_{l,n}^M \vec{M}_n^H(k_{l,\rho}, k_{iz}, \vec{r}) + B_{l,n}^N \vec{N}_n^H(k_{l,\rho}, k_z, \vec{r}) \end{bmatrix} \quad (33)$$

$$H_l = \frac{i}{\omega\mu}\nabla \times E_l = \frac{ik_l}{\omega\mu}\sum_{n=-\infty}^{\infty}\begin{bmatrix} A_{l,n}^M \vec{N}_n^J(k_{l,\rho}, k_z, \vec{r}) + B_{l,n}^M \vec{N}_n^H(k_{l,\rho}, k_z, \vec{r}) \\ +A_{l,n}^N \vec{M}_n^J(k_{l,\rho}, k_z, \vec{r}) + B_{l,n}^N \vec{M}_n^H(k_{l,\rho}, k_z, \vec{r}) \end{bmatrix} \quad (34)$$

where $A_{l,n}^M$, $A_{l,n}^N$, $B_{l,n}^M$ and $B_{l,n}^N$ are the coefficients of the cylindrical vector wave functions. The coefficients between the l-th layer and $l+1$-th layer are related through the boundary condition, i.e. the continuity of the tangential field components, which are given by

$$A_{l+1}^{M}k_{l+1,\rho}J_n'(k_{l+1,\rho}r_l)+B_{l+1}^{M}k_{l+1,\rho}H_n^{(1)}{}'(k_{l+1,\rho}r_l)+A_{l+1}^{N}\frac{nk_z}{k_{l+1}r_l}J_n(k_{l+1,\rho}r_l)+B_{l+1}^{N}\frac{nk_z}{k_{l+1}r_l}H_n^{(1)}(k_{l+1,\rho}r_l)$$
$$=A_l^{M}k_{l,\rho}J_n'(k_{l,\rho}r_l)+B_l^{M}k_{l,\rho}H_n^{(1)}{}'(k_{l,\rho}r_l)+A_l^{N}\frac{nk_z}{k_lr_l}J_n(k_{l,\rho}r_l)+B_l^{N}\frac{nk_z}{k_lr_l}H_n^{(1)}(k_{l,\rho}r_l) \quad (35)$$

$$A_{l+1}^{N}\frac{k_{l+1,\rho}^2}{k_{l+1}}J_n(k_{l+1,\rho}r_l)+B_{l+1}^{N}\frac{k_{l+1,\rho}^2}{k_{l+1}}H_n^{(1)}(k_{l+1,\rho}r_l)=A_l^{N}\frac{k_{l,\rho}^2}{k_l}J_n(k_{l,\rho}r_l)+B_l^{N}\frac{k_{l,\rho}^2}{k_l}H_n^{(1)}(k_{l,\rho}r_l) \quad (36)$$

$$A_{l+1}^{M}k_{l+1,\rho}^2J_n(k_{l+1,\rho}r_l)+B_{l+1}^{M}k_{l+1,\rho}^2H_n^{(1)}(k_{l+1,\rho}r_l)=A_l^{M}k_{l,\rho}^2J_n(k_{l,\rho}r_l)+B_l^{M}k_{l,\rho}^2H_n^{(1)}(k_{l,\rho}r_l) \quad (37)$$

$$k_{l+1}\left[-A_{l+1}^{N}k_{l+1,\rho}J_n'(k_{l+1,\rho}r_l)-B_{l+1}^{N}k_{l+1,\rho}H_n^{(1)}{}'(k_{l+1,\rho}r_l)-A_{l+1}^{M}\frac{nk_z}{k_{l+1}r_l}J_n(k_{l+1,\rho}r_l)-B_{l+1}^{M}\frac{nk_z}{k_{l+1}r_l}H_n^{(1)}(k_{l+1,\rho}r_l)\right]$$
$$=k_l\left[-A_l^{N}k_{l,\rho}J_n'(k_{l,\rho}r_l)-B_l^{N}k_{l,\rho}H_n^{(1)}{}'(k_{l,\rho}r_l)-A_l^{M}\frac{nk_z}{k_lr_l}J_n(k_{l,\rho}r_l)-B_l^{M}\frac{nk_z}{k_lr_l}H_n^{(1)}(k_{l,\rho}r_l)\right] \quad (38)$$

where the subscript n of the coefficients is droped for simplicity. The coefficients in the outmost and innermost layers satisfy the following the conditions.

$$A_{L+1,n}^{M}=\frac{i^ne^{-in\phi_k}}{k_{L+1,\rho}}iE_{hi} \quad A_{L+1,n}^{N}=-\frac{i^ne^{-in\phi_k}}{k_{L+1,\rho}}E_{vi} \quad (39)$$

$$B_{1,n}^{M}=0 \quad B_{1,n}^{N}=0 \quad (40)$$

By using Eq.(35) – (40), all the coefficients can be readily determined and consequently the electric field at any position in the nanostructure can be computed according to Eq. (33).

3.2 Results and discussion

In this subsection, we employ the approach introduced in Section 3.1 and the spontaneous emission model developed in Section 2 to respectively investigate the PL of a single ZnO NW, ZnO NC and ZnO NT illuminated by a plane wave at a wavelength of 325 nm. The wave vector is in the horizontal plane (having no z component) and two kinds of polarizations, namely, the vertical and horizontal polarizations, are considered in the simulation. The ZnO nanostructures emit light spontaneously at 387 nm.

Figures 5 (a) and (b) show the PL intensity of the ZnO NW as a function of the diameter for two kinds of polarizations. In the figures, one critical diameter, i.e. 70 nm, is observed. For the NWs thinner than 70 nm, the waveguide emission is negligible for both kinds of polarizations. The critical diameter obtained previously from the spontaneous emission model is larger than 60 nm because the PL is the product of the quantum efficiency and excitation field (c.f. Eq. (26)). The latter is weak for the horizontally polarized illumination as seen in Fig. 5(b). The second critical diameter does not show up in Fig. 5(a) because the emission by the emitters sitting away from z axis has nonzero coupling to the HE_{11} mode.

For thin NWs, PL by the excitation with a vertical polarization is by tens of times larger than the emission from the excitation with a horizontal excitation. As the diameter of the NW increases, PL from two kinds of excitation becomes comparable and considerable amount of waveguide emissions appear. These properties should be interesting for the applications of nanowire LEDs and nanolasers.

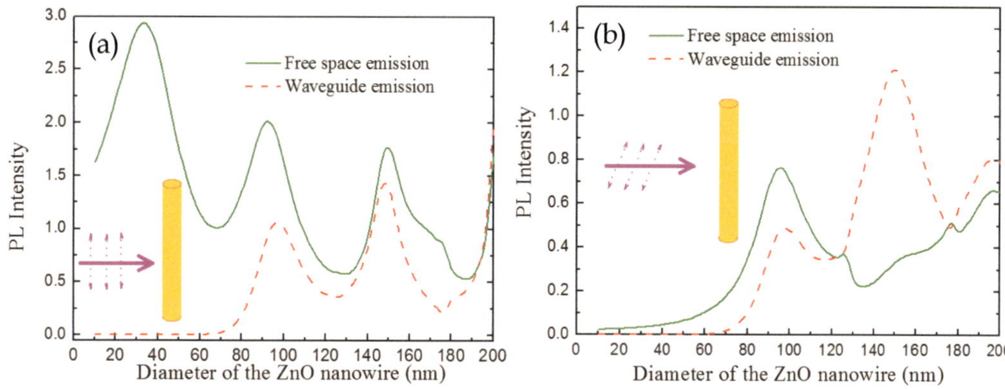

Fig. 5. PL as a function of the ZnO NW diameter (a) excitation field polarized vertically (b) excitation field polarized horizontally

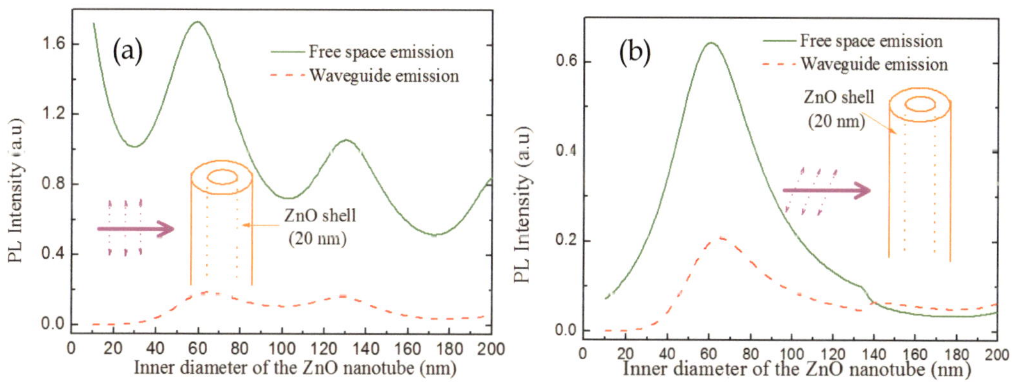

Fig. 6. PL as a function of the inner diameter of the ZnO NT diameter (a) excitation field polarized vertically (b) excitation field polarized horizontally

Figures 6(a) and (b) depict the PL of a single ZnO NT for two kinds of polarizations. For thin NTs, the free space emission is always larger than the waveguide emission. The PL by the excitation with a horizontal polarization is much weaker than the emission by the vertically polarized excitation. At an inner diameter of 65 nm, PLs, including free space emission and waveguide emission, show a maximum for both kinds of excitation.

Figures 7(a) and (b) display the PL of a single ZnO NC (i.e. a 20nm ZnO core with a silica shell) as a function of the thickness of the silica shell. As observed from the Figures, the PL in (a) is larger than the value in (b) by more than an order of magnitude. For vertically polarized excitation in Fig. 7(a), the free space emission always dominates. Since the excitons are confined in the 20nm ZnO core, there are two different critical values of the shell thickness, below which no waveguide emission exists, for excitations with different polarizations. For the excitation with vertical polarization, the critical shell thickness is 120 nm while it is 40 nm for the excitation with horizontal polarization.

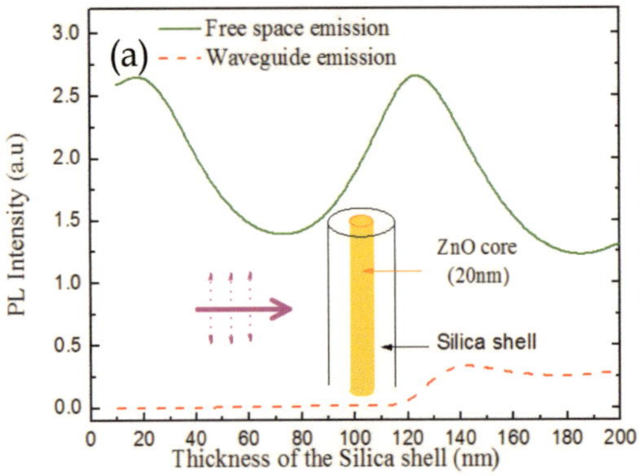

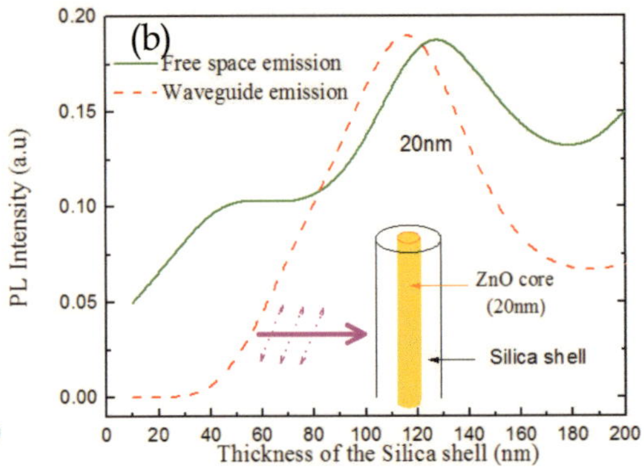

Fig. 7. PL as a function of the thickness of the silica shell of the ZnO NC (a) excitation field polarized vertically (b) excitation field polarized horizontally

4. Conclusion

In conclusion, we have presented an efficient and accurate optical model for studying the PL of a single cylindrically multilayered nanostructure. The model consists of two parts, namely, the modeling of the spontaneous emission of a single emitter in the nanostructure at the emission wavelength and the evaluation of the excitation effect at the illumination wavelength. As examples of practical interest, we have studied the PL of a single ZnO nanowire, ZnO nanotube and ZnO/Silica nanocable under plane wave illumination with two orthogonal polarization states. We found that the PL is highly excitation polarization dependent and the polarization dependence is stronger for thinner nanowires, nanotubes and nanocables. The free space emission and waveguide emission have been calculated as a function of the size of the nanostructure and the polarization state of the excitation. The optical model and the results should be useful and interesting for designing one-dimensional semiconductor active nanostructure for various applications.

5. Acknowledgement

This work is supported partially by the National Natural Science Foundation of China under grant number 60688401. One of the authors (X.-W. Chen) would like to thank Wallace C. H. Choy for helpful discussions and encouragement during the author's stay at the University of Hong Kong.

6. References

Bao J.M., Zimmler M.A., Capasso F., Wang X.W., Ren Z. F. (2006). Broadband ZnO single-nanowire light-emitting diode, *Nano Lett.* Vol.6, No. 8, 1719-1722, *ISSN* 1530-6984.

Barrelet C.J., Greytak A.B., Lieber C.M. (2004). Nanowire photonic circuit elements. *Nano Lett.* Vol. 4, No. 10, 1981-1985, ISSN 1530-6984.

Bermel P., Joannopoulos J.D., Fink Y. (2004) Properties of radiating pointlike sources in cylindrically omnidirectionally reflecting waveguides. *Phys. Rev. B* Vol. 69, No. 3, 035316, ISSN 0163-1829.

Chen X.W., Choy W.C.H., He S.L. (2007a) Efficient optical modeling of spontaneous emission in a cylindrically layered nanostructure. *Opt. Express* Vol. 15, No. 16, 10356-10361, ISSN 1094-4087.

Chen X.W., Choy W.C.H., He S.L., Chui P.C. (2007b) Efficient and Rigorous Modeling of Light Emission in Planar Multilayer Organic Light-Emitting Diodes. IEEE/OSA, *Journal of Display Technolog*, Vol. 3, No. 2, 110-117, ISSN 1551-319X.

Chew W.C. (1995). Waves and Fields in Inhomogeneous Media. IEEE Press, ISBN 0780311167, New York.

Duan X.F., Huang Y., Agarwall R., Lieber C.M. (2003). Single-nanowire electrically driven lasers. *Nature* Vol. 421, (16 January 2003), 241-245, ISSN 0028-0836.

Gay-Balmaz P., Mosig J.R.(1997). Three-Dimensional planar radiating structures in stratified media. *Int. J. Microwave Millimeter Wave Computer-Aided Eng.* Vol. 37, 330-343, ISSN. 1050-1827.

Huang M. H., Mao S., Feick H., Yan H.Q., Wu Y.Y., Kind H., Weber E., Russo R., Yang P.D. (2001). Room-temperature ultraviolet nanowire nanolasers. *Science* Vol. 292. No. 5523, 1897 - 1899 *ISSN* 0036-8075.

Johnson J.C., Choi H.J., Knutsen K.P., Schaller R.D., Yang P.D., Saykally R.J. (2002). Single gallium nitride nanowire lasers. *Nature Materials* Vol. 1, (15 September 2002), 106–110, ISSN 1476-1122.

Johnson J.C., Yan H.Q., Yang P.D., Saykally R.J. (2003). Optical cavity effects in ZnO nanowire lasers and waveguides. *J. Phys. Chem. B,* Vol. 107, No. 34, 8816-8828, ISSN 1089-5647.

Könenkamp R., Word R. C., and Schlegel C. (2004). Vertical nanowire light-emitting diode. *Appl. Phys. Lett.* Vol. 85, No.24 , 6004-6006, *ISSN* 0003-6951.

Lovell J. R., Chew W.C. (1987) Response of a point source in a multicylindrically layered medium. *IEEE Trans. Geosci. Remote Sensing,*Vol. GE-25,No.6, 850-858, ISSN 0196-2892.

Lukosz W. (1980). Theory of optical-environment-dependent spontaneous emission rates for emitters in thin layers. *Phys. Rev. B* Vol. 22, No. 6, 3030-3038, ISSN 1098-0121.

Maslov, A. V., Ning, C. Z. (2003). Reflection of guided modes in a semiconductor nanowire laser. *Appl. Phys. Lett.*, Vol. 83, No. 6, 1237-1239, ISSN 0003-6951.

Purcell E.M. (1946), Spontaneous emission probabilities at radio frequencies. *Phys. Rev.* Vol. 69, 681.

Qian F., Gradecak S., Li Y., Wen C.Y., Lieber C.M. (2005) Core/multishell nanowire heterostructures as multicolor, high-efficiency light-emitting diodes. *Nano Lett.* Vol. 5, No. 11, 2287-2291, *ISSN* 1530-6984.

Sullivan K.G. & Hall D.G.(1997). Enhancement and inhibition of electromagnetic radiation in plane-layered media. I. Plane-wave spectrum approach to modeling classical effects. *J. Opt. Soc. Am. B*, Vol. 14, No. 5, 1149-1159, ISSN 0740-3224.

Tong L.M., Lou J.Y., Gattass R.R., He S.L., Chen X.W., Liu L. Mazur E. (2005). Assembly of silica nanowires on silica aerogels for microphotonic devices, *Nano Lett.* Vol. 5, No. 2, 259-262, ISSN 1530-6984.

Tsang L., Kong J.A., Ding K.H. (2001), *Scattering of Electromagnetic Waves: Theories and Applications*, Wiley-Interscience, ISBN 0471387991, Boston.

Wang Y., Tang Z.Y., Kotov N.A. (2005). Bioapplication of nanosemiconductors, *Materials Today* Vol. 8, No. 5 (Supplement 1, May 2005), 20-31, ISSN 1369-7021.

Wang J.F., Gudiksen M.S., Duan X.F., Cui Y., Lieber C.M. (2001). Highly Polarized Photoluminescence and Photodetection from Single Indium Phosphide Nanowires. *Science* Vol. 293, No. 5534 1455-1457, ISSN 0036-8075.

Yu D. P., Hang Q. L., Ding Y., Zhang H. Z., Bai Z. G., Wang J. J., Zou Y. H., Qian W., Xiong G. C., and Feng S. Q. (1998). Amorphous silica nanowires: Intensive blue light emitters. *Appl. Phys. Lett.* Vol. 73, No. 21, 3076-3078, *ISSN* 0003-6951.

Zhang Y. Russo R.E. (2005). Quantum efficiency of ZnO nanowire nanolasers. *Appl. Phys. Lett.* Vol. 87, No.4, 043106-043108, *ISSN* 0003-6951.

18

Multi-physics Properties in Ferroelectric Nanowires and Related Structures from First-principles

Takahiro Shimada and Takayuki Kitamura
Department of Mechanical Engineering and Science, Kyoto University
Japan

1. Introduction

Ferroelectric (FE) nanowires (Spanier et al., 2006) have drawn much attention as one-dimensional (1D) multifunctional materials for their technological applications, e.g., ferroelectric random access memories (FeRAM) (Scott, 2000). In recent years, single-crystalline $PbTiO_3$ nanowires with a diameter of 5-12 nm (Gu et al., 2007) have been manufactured, and smaller nanowires with several lattice spacings are anticipated in the near future. The ferroelectric nanowires are surrounded by both surfaces and atomically sharp edges consisting of (100) and (010) surfaces (Yamashita et al., 2002). The edge structure is commonly observed in nanostructured perovskite-type oxides, e.g., the step on the $SrTiO_3$ (001) surface (Cho et al., 2004), the zigzag (110) surface structure (Kotomin et al., 2004), and the edge of a $PbTiO_3$ nano-island (Chu et al., 2004). Thus, the edges are among the characteristic nanostructures in perovskite oxides.
Ferroelectricity is defined as a property that materials exhibit non-vanishing spontaneous polarization, which is a dipole moment per unit volume, even in the absence of external electric fields. Figure 1 shows the crystal structures of perovskite oxides ABO_3 in the paraelectric (PE) and ferroelectric (FE) phases. In the FE phase where atoms are displaced in a specific direction from their ideal lattice sites, spontaneous polarization emerges due to its non-symmetrical atomic structure, while there is no polarization in the PE phase where the structure is centro-symmetric. The FE-phase structure is energetically favorable, and it has, in general, two equivalent states with upward and downward spontaneous polarization, $\boldsymbol{P}_{UP}$ and $\boldsymbol{P}_{DOWN}$, respectively. They are switchable each other by applied external electric fields.
The rapid change in coordination number at the surface and edge can significantly affect the ferroelectric distortions in the nanowires because ferroelectricity originates from the delicate balance between the short-range covalent and long-range Coulomb interactions. In addition, nanowires are normally subjected to axial tension or compression, which sometimes enhances or destabilizes the ferroelectric distortions. Such a coupling effect between mechanical deformation and electric properties is known as "multi-physics properties" and it is of importance for the design of nanoscale materials. For example, strained silicon (Chakrabarti et al., 2001) is one of the well-known representatives leveraging the multi-physics property: Applied strain changes the band structure of silicon, and at the critical

strain silicon undergoes the semiconducting - metallic transition (Umeno et al., 2002). Such multi-physics properties have been also intensively studied in the ferroelectric or ferromagnetic materials. The mechanical deformation not only changes the electronic polarization but also induces the ferroelectric-paraelectric phase transition.
First-principles (*ab initio*) density functional theory calculations (Hohenberg & Kohn, 1964) (Kohn & Shan, 1965), which can analyze the electronic ground-state as well as atomic structure by solving the Schrödinger equation within the one-electron approximation, are the powerful tool for the study of multi-physics properties. The method has been applied to various kinds of materials, for example, the calculations have revealed the characteristic change in ferroelectric properties (including its disappearance) of nano-structured ferroelectrics, such as nanofilms induced by applied strain.
In this chapter, we demonstrate the characteristic ferroelectricity at the surfaces and edges in $PbTiO_3$ nanowires, and the crucial role of mechanical strain, namely, multi-physics properties from the atomistic and electronic points of view using first-principles (*ab initio*) density-functional theory calculations.

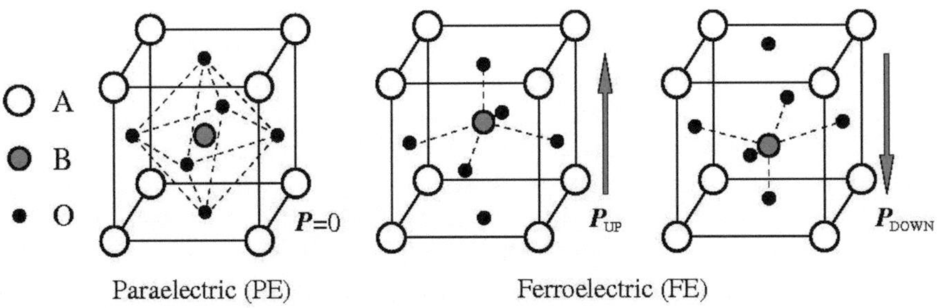

Fig. 1. Atomic structures of perovskite oxides ABO_3 in the paraelectric (PE) phase and ferroelectric (FE) phase with upward and downward polarization.

2. First-principles (*ab initio*) density functional theory calculations

The description of macroscopic solids from first-principles is based on the determination of the quantum mechanical ground-state associated to their constituting electrons and nuclei. It consists in calculating the quantum mechanical total energy of the system and in the subsequent minimization of that energy with respect to the electronic and nuclear coordinates (variational principle). This defines a complex many-body problem of interacting particles. Density Functional Theory (DFT), proposed in the 1960's by Hohenberg and Kohn (Hohenberg & Kohn, 1964) and Kohn and Sham (Kohn & Shan, 1965), provided a simple method for describing the effect of electron-electron interactions. Hohenberg and Kohn first proved that the total energy of an electron gas is a unique functional of the electron density. This means that instead of seeking directly for the complex many-body wave function of the system, we can adopt an intrinsically different point of view and consider the electronic density as the fundamental quantity of the problem. The minimum value of the total energy density functional is the ground-state energy of the system and the density yielding this minimum value is the exact ground-state density.
According to the density functional theory (DFT) (Hohenberg & Kohn, 1964) (Kohn & Shan, 1965), the Kohn-Sham total-energy functional in the presence of a static external potential, $v(\boldsymbol{r})$, can be written as a functional of the electronic density, $\rho(\boldsymbol{r})$,

$$E_{tot} = \int v(\mathbf{r})\rho(\mathbf{r})d\mathbf{r} + T[\rho(\mathbf{r})] + \frac{1}{2}\iint \frac{\rho(\mathbf{r})\rho(\mathbf{r}')}{|\mathbf{r}-\mathbf{r}'|}d\mathbf{r}'d\mathbf{r} + E_{xc}[\rho(\mathbf{r})], \tag{1}$$

where the first, second, third and fourth terms denote the static total external (or electron-ion) potential energy, the kinetic energy of the electrons, the Coulomb energy associated with interactions among the electrons and the exchange-correlation energy, respectively.
It is necessary to determine the set of wave functions, $\psi_i(\mathbf{r})$, that minimize the Kohn-Sham energy functional. These are given by the self-consistent solutions to the Kohn-Sham equations,

$$\left[-\frac{1}{2}\nabla^2 + v_{eff}(\mathbf{r})\right]\psi_i(\mathbf{r}) = \varepsilon_i\psi_i(\mathbf{r}), \tag{2}$$

where $\psi_i(\mathbf{r})$ is the wave function of electronic state i, ε_i is the Kohn-Sham eigenvalue, and $v_{\mathrm{eff}}(\mathbf{r})$ is the effective one-electron potential which is given by

$$v_{eff}(\mathbf{r}) = v(\mathbf{r}) + V_{coul}(\mathbf{r}) + \varepsilon_{xc}(\mathbf{r}) \tag{3}$$

where $V_{\mathrm{coul}}(\mathbf{r})$ and $\varepsilon_{\mathrm{xc}}(\mathbf{r})$ denote the Coulomb potential of electrons and the exchange-correlation potential, respectively. They can be written

$$V_{coul}(\mathbf{r}) = \int \frac{\rho(\mathbf{r})}{|\mathbf{r}-\mathbf{r}'|}d\mathbf{r}' \tag{4}$$

and

$$\varepsilon_{xc}(\mathbf{r}) = \frac{\partial E_{xc}[\rho(\mathbf{r})]}{\partial\rho(\mathbf{r})} \tag{5}$$

The electronic density, $\rho(\mathbf{r})$, is given by

$$\rho(\mathbf{r}) = 2\sum_{i}^{occ}|\psi_i(\mathbf{r})|^2 \tag{6}$$

The Kohn-Sham equations must be solved self-consistently so that the occupied electronic states generate a charge density that produces the electronic potential that was used to construct the equations.
It has been validated that first-principles calculations can accurately predict not only atomic structures in solids, at surfaces, domain walls and grain boundaries but also the electronic properties (band structure, magnetic moment, spontaneous polarization and conductivity). For ferroelectrics, atomic structures, ferroelectric characters, mechanical and vibrational properties are also well reproduced by the first-principles calculations.

3. Ferroelectricity and multi-physics properties at surfaces

3.1 Simulation procedure and models

Figure 2 shows simulation models for the study of $PbTiO_3$ (001) surface with both TiO_2- and PbO-terminations with a (1×1) periodicity. Each slab model has 9 atomic layers. The three dimensional periodic boundary conditions are applied for all the models. The thickness of the vacuum regions are set to 12 Å so that undesirable interaction from imaginary films does not appear.

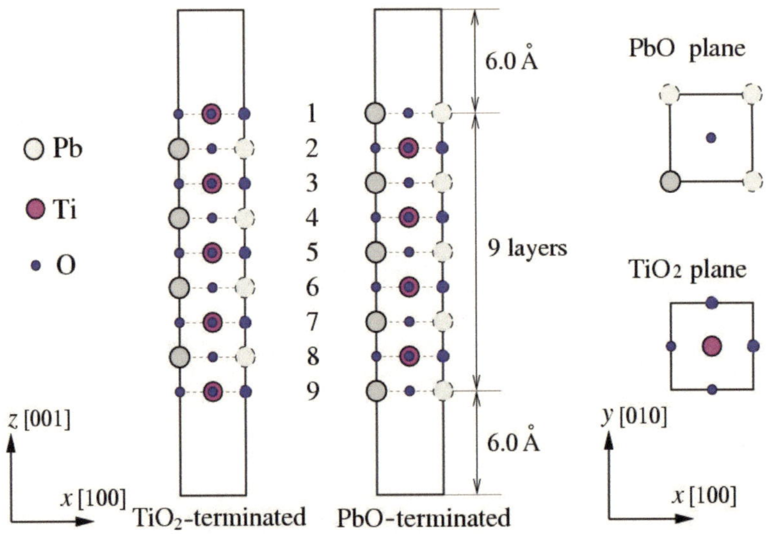

Fig. 2. Simulation models of $PbTiO_3$ (001) surface of TiO_2- and PbO-termination with a (1×1) periodicity. Solid boxes represent simulation cells.

Ab initio calculations based on the projector augmented wave (PAW) method (Blochl, 1994) are conducted using the VASP (Vienna Ab-initio Simulation Package) code (Kresse & Hafner, 1993) (Kresse & Furthmuller, 1996). The plane-wave cutoff energy is set to 500 eV. The local density approximation (LDA) of the Ceperley-Alder form (Ceperley & Alder, 1980) is used for the evaluation of the exchange correlation energy. The pseudopotentials include the O 2s and 2p, the Ti 3s, 3p, 3d and 4s, the Pb 5d, 6s and 6p electrons in the valence states. The Brillouin-zone integrations are carried out with a 4×4×2 Monkhorst-Pack (Monkhorst & Pack, 1976) *k*-point mesh. Structural relaxation is performed by means of the conjugate gradient method until all the forces acting on atoms are less than 0.005 eV/Å.
Initially the atoms are arranged on the lattice points of the cubic perovskite structure with its theoretical lattice constant of a_0 = 3.891 Å. Then, the atomic coordinates are fully relaxed with a centrosymmetry in the z-direction to confine the ferroelectricity only in the lateral directions. In this study, (a) in-plane isotropic ($\varepsilon_{xx} = \varepsilon_{yy} \neq 0$) and (b) [100]-uniaxial ($\varepsilon_{xx} \neq 0$, $\varepsilon_{yy} = 0$) strain conditions are simulated (see also Fig. 3). On both cases, a small strain is applied stepwise and atomic coordinates are fully relaxed at each straining step.

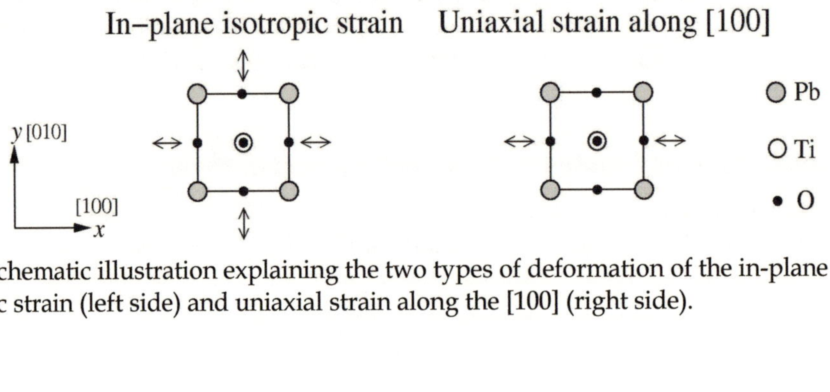

Fig. 3. Schematic illustration explaining the two types of deformation of the in-plane isotropic strain (left side) and uniaxial strain along the [100] (right side).

3.2 Ferroelectricity of unstrained surface structures

We introduce δ defined as follows to represent strength of polarization distortion of each layer,

$$\delta_i = \begin{cases} [\overline{\delta_i}(\mathrm{Pb}) - \overline{\delta_i}(\mathrm{O})] & (\mathrm{PbO\ \ layer}) \\ [\overline{\delta_i}(\mathrm{Ti}) - \overline{\delta_i}(\mathrm{O})] & (\mathrm{TiO_2\ \ layer}) \end{cases} \quad (i = x, y) \tag{7}$$

$$\delta = \sqrt{\delta_x^2 + \delta_y^2} \tag{8}$$

where δ_i is the layer-averaged atomic displacement relative to ideal lattice sites (see also Fig. 4). We simulated the [100]- and [110]-polarized films and found that the polarization in [110] direction is preferred with a small energy advantage of 2.96 mJ/m^2 over [100] polarization in both TiO_2- and PbO-terminations. The layer-by-layer δ of both the [100]- and [110]-polarized films is shown in Table 1 although the [110]-polarized case is more stable. The δ gradually decreases toward the surface layer for the TiO_2-termination, while the opposite trend is found in the PbO-termination. The magnitude of δ at Layer 5 is close to that of the bulk value for both terminations indicating that our models have sufficient thickness to avoid the effect from another surface.

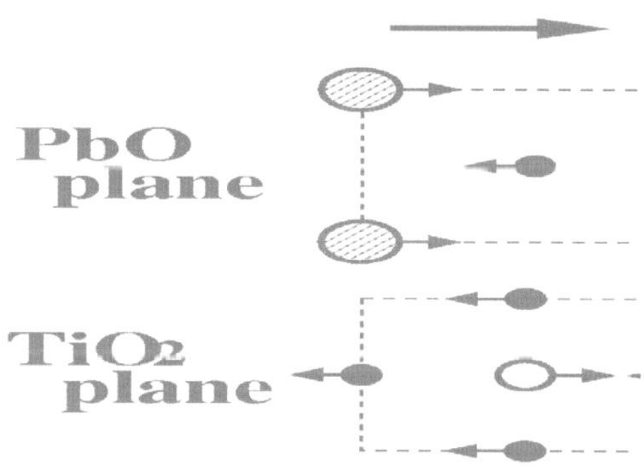

Fig. 4. Schematic illustration explaining the structural parameter of polarization distortion, δ.

3.2 Influence of in-plane isotropic tension/compression

The polarization distortion per area must be the relevant value to represent the response of polarization to strain because polarization of bulk material is defined as dipole moment per volume. Change in the polarization distortion per area, $\delta' = \delta / A$ (A is the cross-sectional area of simulation cell in the xy direction), with respect to in-plane isotropic strain is shown in Fig. 5. Here, the polarization direction is always in [110] direction because we confirmed that [100]-polarized structure has larger energy under nonzero isotropic strain as well. In the case of TiO_2-termination, δ increases smoothly with increasing tensile strain, showing the enhancement of ferroelectricity by tangential stretch. On the other hand, ferroelectricity is suppressed under compression and vanishes at $\varepsilon = -0.06$, indicating the existence of the critical compressive strain for ferroelectricity. When the surface is further compressed the ferroelectricity emerges again. The response of δ to the tangential strain in the PbO-terminated surface is similar. However, under compression, δ does not vanish because of the enhancement of FE distortion by PbO-termination.

	TiO_2 termination				PbO termination			
	P[100]		P[110]		P[100]		P[110]	
Layer	TiO_2	PbO	TiO_2	PbO	TiO_2	PbO	TiO_2	PbO
1	2.29		2.58			13.48		12.10
2		3.40		3.64	4.75		5.26	
3	3.48		3.82			8.77		8.70
4		6.77		6.62	4.09		4.34	
5	3.81		4.04			8.05		7.93
Bulk	3.95	7.67	4.14	7.47	3.95	7.67	4.14	7.47

Table 1. Polarization distortion, δ, in percentage of the lattice parameter (a_0 = 3.891 Å) at the TiO_2- and PbO-terminated surfaces with a (1×1) periodicity.

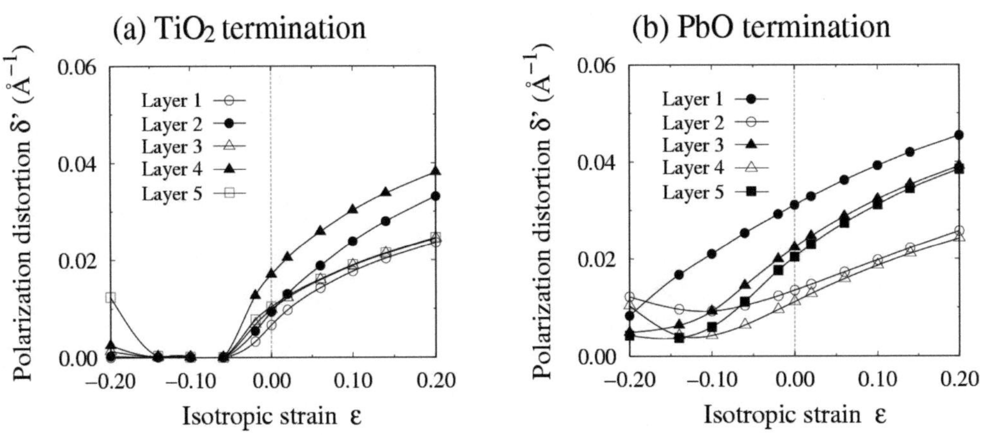

Fig. 5. Polarization distortion per area, δ', as a function of in-plane isotropic strain, ε, in the (1×1) model with (a) TiO_2 and (b) PbO termination.

3.3 Effect of uniaxial tension/compression

Tensile and compressive uniaxial strain in [100] changes the tetragonality of the $PbTiO_3$ lattices, which can cause variation in the polarization direction as well as the magnitude. Here, we introduced the polarization direction angle,

$$\varphi = \tan^{-1}(\delta_y / \delta_x). \tag{9}$$

(φ = 0° in P[100] and φ = 45° in P[110]). Figure 6 shows changes in the polarization distortion per area (δ′) and direction (φ) at the (1×1) surfaces with respect to ε_{xx}. In both terminations, the polarization direction rotates toward [100] under tension and [010] under compression due to the preference of the polarization to the longitudinal direction of rectangular lattices. Although the polarization distortions differ with the layer, their directions are identical in all the layers. While the evolution of δ with growing tensile strain is similar to that in the case of isotropic tension, the distortion is less suppressed under compression. No critical strain for ferroelectricity is found regardless of the termination layer.

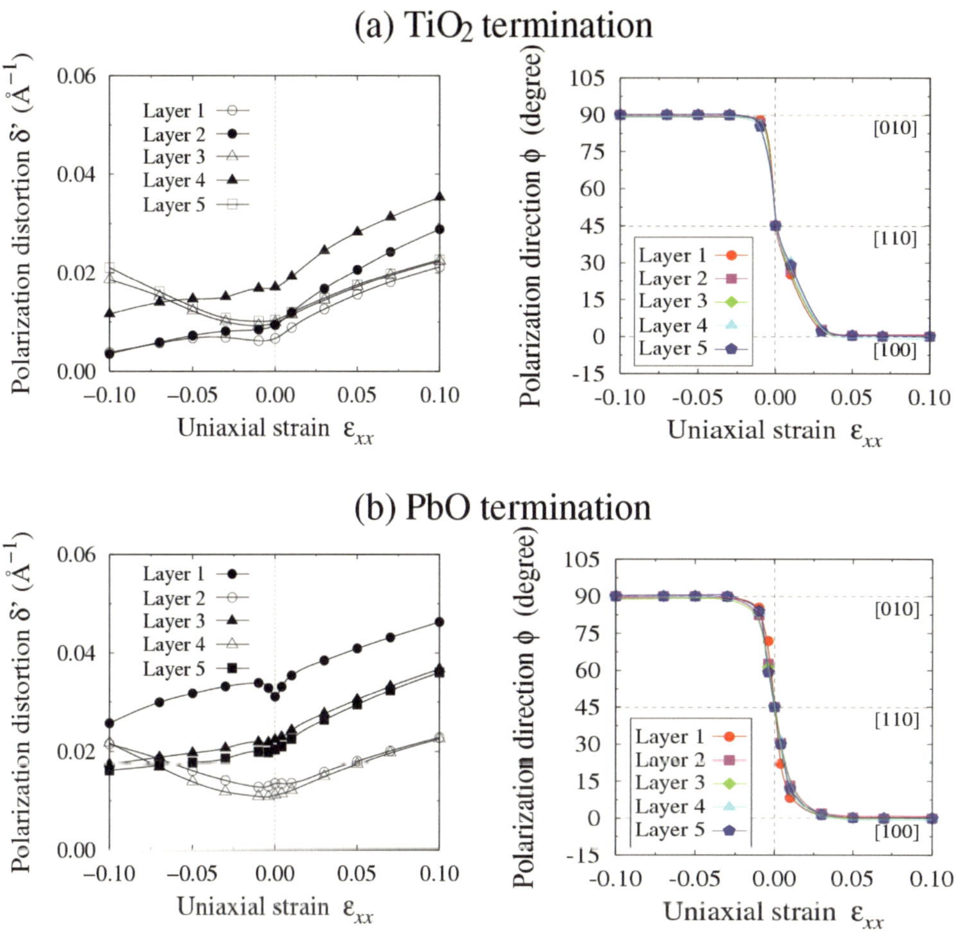

Fig. 6. Polarization distortion per area, δ', and polarization direction angle, φ, as a function of uniaxial strain, ε_{xx}. (a) TiO_2- and (b) PbO-terminations.

3.4 Effect of c(2×2) reconstruction on ferroelectricity at surfaces

For the PbO-terminated surface, there exists the c(2×2) surface reconstruction incorporated with the rotation of the TiO_4 square, which is denoted as the antiferrodistortive (AFD). The atomic displacement of AFD is depicted in Fig. 7.

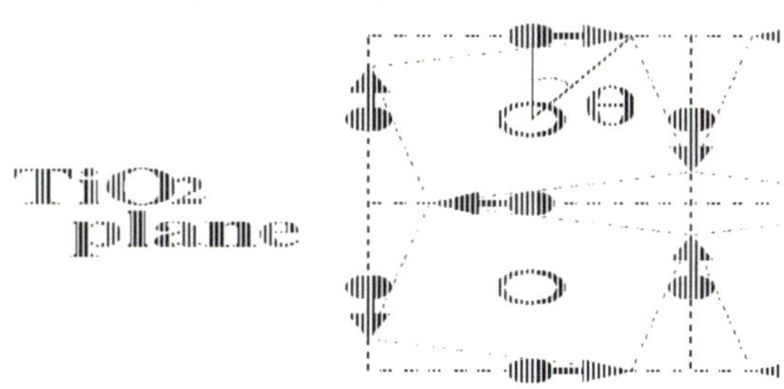

Fig. 7. Schematic illustration explaining the structural parameter of antiferrodistortive (AFD) rotation angle, θ.

Figure 8 shows the simulation model for the c(2×2) reconstructed surface. The simulation procedure is the same as the (1×1) case in the previous section.

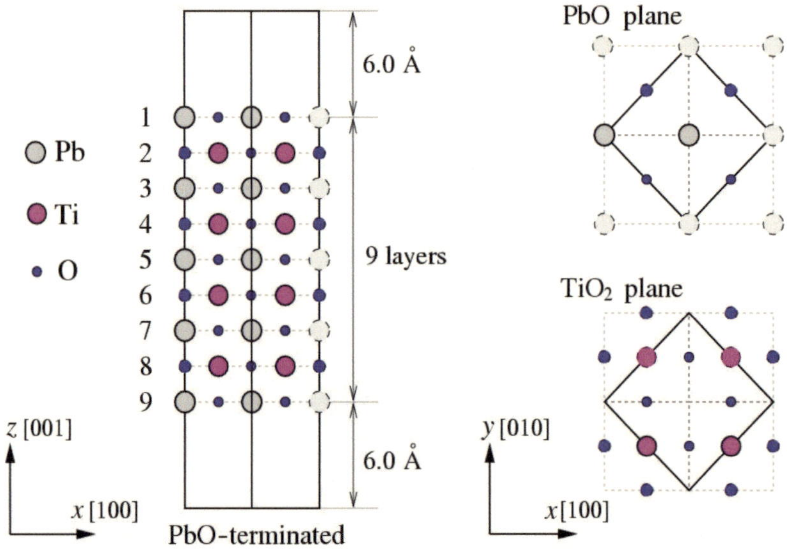

Fig. 8. Simulation models of $PbTiO_3$ (001) surface of the PbO-termination with the c(2×2) reconstruction. Solid boxes represent simulation cell.

Figure 9 shows atomic structures of the PbO-terminated c(2×2) surface, where FE and AFD coexist, and schematic illustrations of FE and AFD distortions. In this phase, the polarization directions of cells A and B in Fig. 9 are not identical. The polarizations are deviated from their average directions ([100] or [110]) clockwise or anticlockwise, which is clearly due to the contribution of AFD. This phase will be denoted as "FE+AFD (coexisted) phase" hereafter. The energy, the polarization distortion and the AFD rotation in this phase are listed in Table 2. The φ_{ave} is the averaged polarization direction (0° in P[100] and 45° in

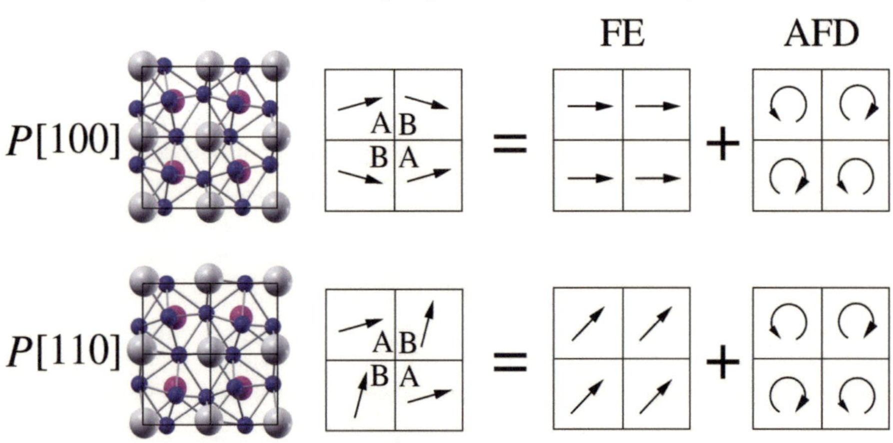

Fig. 9. Atomic structures of c(2×2) PbO-terminated surfaces. Only the first two layers are shown. Arrows indicate FE and AFD distortions of the first PbO layer and AFD rotation of the adjacent TiO_2 layer.

P[110]), thus φ - φ_{ave} denotes the strength of polarization rotation induced by AFD. As for θ, we take the rotation angle of the oxygen squares in the TiO_2 planes (see also Fig. 7). An important finding here is that the [110]-polarized structure is energetically the most favored with the nontrivial energy advantage of about 0.075 eV (39.7 mJ/m^2 for one surface side) with respect to the [100]-polarized case. Although the magnitude of the AFD rotation is almost comparable among the two cases, the magnitude of φ - φ_{ave} of PbO-layers in P[110] is more than twice larger than that in P[100].

Figure 10 represents change in the ferroelectric and antiferrodistortive distortions of P[110] phase of the c(2×2) PbO-terminated surface. The response of the polarization distortion (δ') to the variation of the isotropic strain is similar to that in the (1×1) model. The AFD rotation

	Layer	P[100] (φ_{ave}=0°)	P[110] (φ_{ave}=45°)
δ	1	12.22	10.68
	2	4.18	4.36
	3	6.70	6.35
	4	3.88	4.10
	5	7.69	7.29
φ-φ_{ave}	1	13.9	-36.1
	2	-6.8	1.1
	3	-5.4	24.7
	4	1.0	-0.8
	5	5.0	-18.2
θ	2	11.0	10.6
	4	-0.3	-0.4

Table 2. Polarization distortion, δ, in percentage of the lattice parameter (a_0 = 3.891 Å), polarization direction angle, φ, and AFD rotation angle, θ, of the cell A in Fig. 9 at PbO-terminated c(2×2) surfaces.

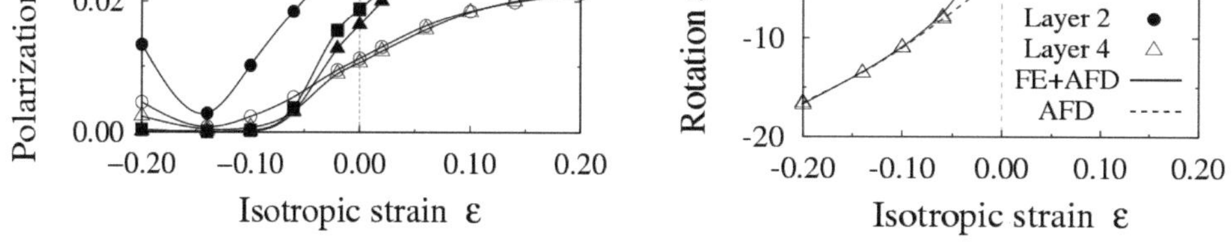

Fig. 10. Polarization distortion per area, δ', and AFD rotation, θ, as a function of isotropic strain, ε, in the c(2×2) PbO-terminated surface model. Dashed lines in the right figure represent the results of AFD-only phase for comparison.

has a nature of the opposite trend to FE case (see dashed lines in Fig. 10, which are the results of AFD-only phase). In the FE+AFD coexisted phase, FE and AFD compete with each other, resulting in the AFD rotation being more suppressed by the FE distortion under tension and the FE being more weakened by the AFD under compression.

Figure 11 shows the polarization direction and distortion in the FE+AFD coexisted phase of the c(2×2) PbO-terminated model. Although the polarization distortion as a function of [100] strain is quite similar to that of (1×1) model, the polarization direction behaves differently. The direction φ differs among the layers. The surface layer prefers to have polarization in [110] direction and the adjacent layer is slightly affected. This result suggests that different polarization directions depending on the distance from the surface can be observed at the $PbTiO_3$ surface subjected to a high uniaxial strain.

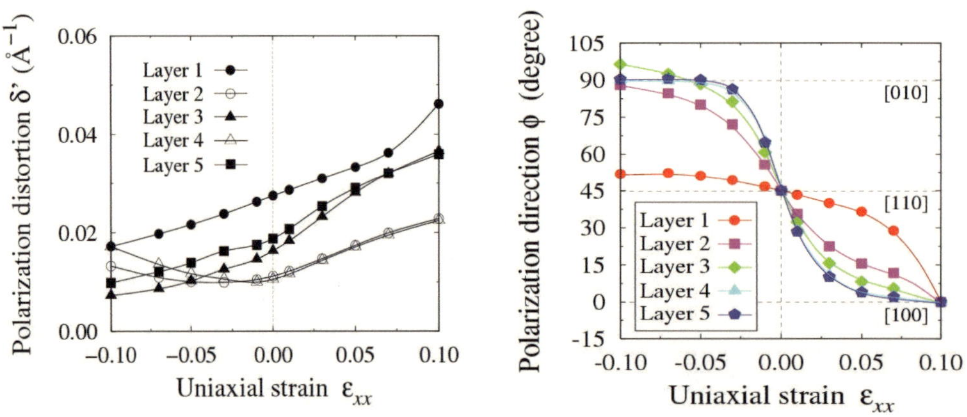

Fig. 11. Polarization distortion per area, δ′, and polarization direction angle, φ, as a function of uniaxial strain, ε_{xx}, in the c(2×2) PbO-terminated surface model.

Figure 12 depicts change in the atomistic and electronic configurations at the surface layer of the PbO-terminated models. In the (1×1) model, the central oxygen atom is shifted toward a corner Pb with a strong covalent bond (α) and two equivalent weak bonds (β and γ) at ε_{xx} = 0.0. When a strain is applied in the x direction, the oxygen immediately moves towards either of the equivalent bonds to form two strong bonds; α and β under tension or α and γ under compression. This bond switching process, which is also found in the TiO_2-terminated surface, explains the polarization rotation very sensitive to the uniaxial strain in the (1×1) periodicity. On the other hand, the c(2×2) PbO-terminated surface has the different bond structure. Oxygen atoms are shifted in different directions in the lattices A ([010]) and B ([100]), forming two strong covalent bonds with the nearest Pb atoms. The bond structure in the lattice A does not immediately change when a small uniaxial tension is applied. At relatively high strain (ε_{xx} > 0.07) the shift of the oxygen occurs with the break of γ bond followed by the construction of β bond, consequently forming [100] polarization in both lattices. It is worth noting that the [100]-polarized structure becomes energetically almost identical to the [100]/[010] mixed polarization at a lower strain, ε_{xx} ~ 0.05, which indicates the existence of small energy barrier between those structures for the bond reconstruction of β and γ bonds. In contrast, the mixed polarization is always more favored than [010] under compression despite the increase in the tetragonal ratio. This is because compression is applied keeping L_y (lattice parameter in the y direction) in our calculation, which does not lead to bond breaking required for the reconstruction of polarization structure.

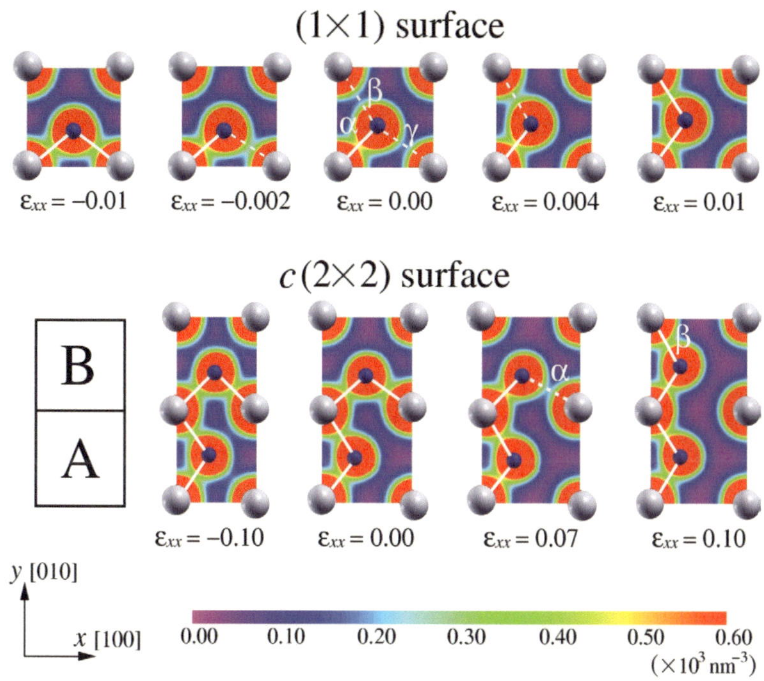

Fig. 12. Change in atomistic and electronic configurations at the first surface layer of the (1×1) and c(2×2) PbO-terminated surface models under uniaxial strain, ε_{xx},.

4. Ferroelectricity and multi-physics properties in edged nanowires

4.1 Simulation models and procedure

We study $PbTiO_3$ nanowires with edges consisting of (100) and (010) surfaces, where there are two possible terminations of the PbO and TiO_2 atomic layers. Figures 10(a) and (b) show the simulation models of the PbO- and TiO_2-terminated nanowires, respectively. Each nanowire has a cross-section in which four perovskite unit cells are arranged in the x and y directions, namely 4×4 unit cells. Spontaneous polarization aligns along the z direction (the axial direction in the nanowire). Since a three-dimensional periodic boundary condition is applied in the calculations, a vacuum region of $l_v = 12$ Å thickness is introduced in both the x and y directions so that undesirable interaction between the neighboring nanowires is sufficiently avoided. The simulation cell dimensions in the x, y and z directions are initially set to $4a+l_v$, $4a+l_v$ and c using the theoretical lattice constants of the tetragonal bulk, $a = 3.867$ Å and $c = 4.034$ Å ($c/a = 1.043$). The Brillouin zone (BZ) integration is carried out with a 2×2×6 k-point mesh generated by the Monkhorst-Pack scheme. The condition of *ab initio* calculations are the same as the case of thin films in the previous section.

As shown in Fig. 10, the unit cells at an edge, on a surface and inside the nanowire are classified into "edge cell", "surface cell" and "inside cell", respectively. The terms "edge bond", "surface bond" and "inside bond" denote the Pb-O covalent bonds in the corresponding cells.

To obtain an equilibrium structure, atomic positions and a cell size in the z direction are fully relaxed using the conjugate gradient method until all the Hellmann-Feynman forces and the stress component σ_{zz} are less than 1.0×10^{-3} eV/Å and 1.0×10^{-2} GPa, respectively.

Smaller nanowires with a cross-sections of 1×1, 2×2 and 3×3 unit cells are investigated in the same manner in order to elucidate the effects of nanowire size and axial strain on ferroelectricity. For tensile simulations, a small axial strain, $\Delta\varepsilon_{zz}$, is applied to the simulation cell step by step. At each strain, internal atoms are fully relaxed under the fixed cell dimensions.

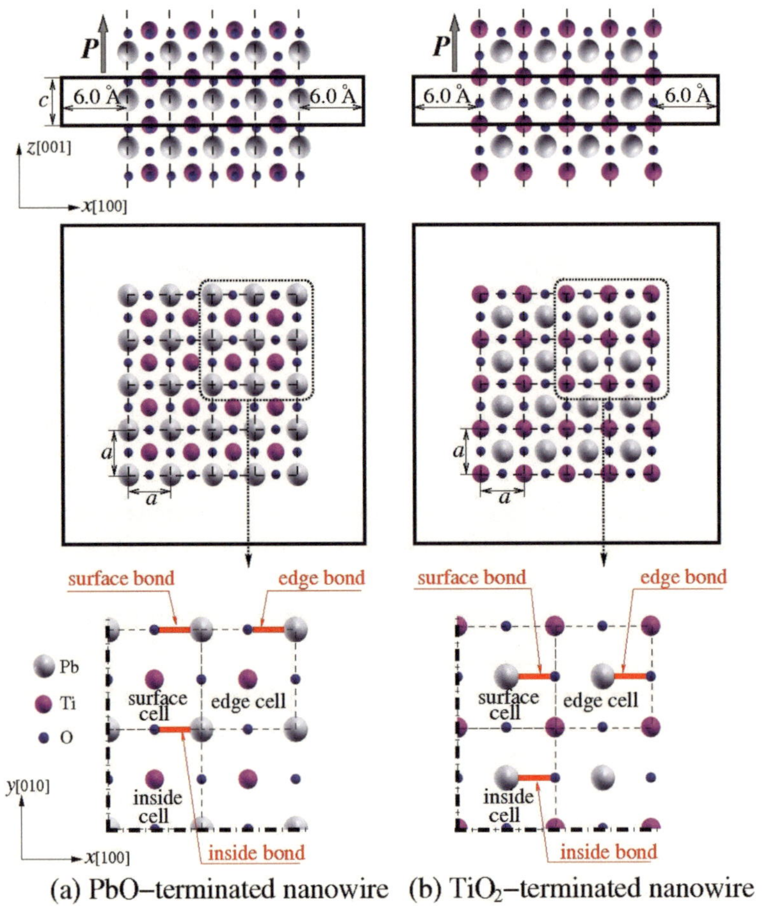

Fig. 13. Simulation models of (a) PbO-terminated and (b) TiO_2-terminated nanowires with a cross-section of 4×4 unit cells. The solid boxes represent the simulation cells. Polarization ***P*** lies along the axial *z* direction.

4.2 Ferroelectricity in unstrained nanowires

The local polarization in the cell, P, is evaluated by

$$\mathbf{P} = \frac{e}{\Omega_c} \sum_j w_j \mathbf{Z}_j^* \mathbf{u}_j, \tag{10}$$

where Ω_c, e and $\boldsymbol{u}_j$ denote the volume of the unit cell, the electron charge and the atomic displacement vector from the ideal lattice site of atom j, respectively. Index j runs over all atoms in the unit cell i. $\mathbf{Z}_j^*$ is the Born effective charge tensor of the cubic bulk $PbTiO_3$. In this study, we use the theoretical values of the Born effective charge tensors calculated by Zhong et al. (Zhong et al., 1994). Weights are set to w_{Pb}=1/8, w_{Ti}=1 and w_{O}=1/2 for the Pb-edged cell, and w_{Pb}=1, w_{Ti}=1/8 and w_{O}=1/2 for the Ti-edged cell.

Table 3 shows local polarization of the edge, surface and inside cells in the PbO-terminated and TiO_2-terminated nanowires with the 4×4-cells cross-section. The results of the [100]-polarized nine-layered $PbTiO_3$ film with the (001) surfaces (Umeno et al., 2006) and the bulk are shown for comparison. For the film, polarization is taken from the magnitude in the unit cell on the surface. In the PbO-terminated nanowire, the edge cell exhibits the highest local polarization. The local polarization of the surface cell, which is second highest, is nearly equal to that of the film. Polarization in the inside cell is also comparable to that of the bulk. On the other hand, little polarization is found in the TiO_2-terminated nanowire. More specifically, the inner cell in the nanowire exhibits higher polarization, which is the opposite of the trend found in the PbO-terminated nanowire. Thus, ferroelectricity in the vicinity of the edge is enhanced in the PbO-terminated nanowire, while it is considerably suppressed in the TiO_2-terminated nanowire. Note that the TiO_2-terminated nanowire shrinks in the z (axial) direction at equilibrium, resulting in considerable suppression of tetragonality, $c/\bar{a}$=0.973 (c/a=1.043 for bulk), where $\bar{a}$ denotes the averaged lateral lattice parameter of the wire. It was reported that ferroelectricity in $PbTiO_3$ is suppressed as uniaxial compression is applied along the c-axis, meaning the decrease of tetragonality (Shimada et al., 2008). Hence, the tiny magnitude of polarization of the TiO_2-termination even in the inside cell is due to not only the effect of TiO_2-terminated surface but also the highly reduced tetragonality.

	Nanowire			Film	Bulk
	Edge cell	Surface cell	Inside cell		
PbO termination	103.2	96.5	81.0	93.8	85.8
TiO_2 termination	0.55	0.70	1.61	47.0	85.8

Table 3. Local polarization evaluated via Eq. (10), P (in μC/cm²), of the edge, surface, and inside cells in the PbO-terminated and TiO_2-terminated nanowires with a 4×4-cell cross section. The results for [100]-polarized film with (001) surfaces (Umeno et al., 2006) and the bulk are shown for comparison.

It was both theoretically (Cohen, 1992) and experimentally (Kuroiwa et al., 2001) revealed that the prominent ferroelectricity and large lattice tetragonality in $PbTiO_3$ stems from the formation of the strong covalent Pb-O bond through the hybridization of the Pb 6s and O 2p orbitals, unlike $BaTiO_3$ where the Ba-O interaction is almost ionic. In addition, the Pb-O bond often plays a significant role in determining characteristic structures, such as a complex c(2×2) surface reconstruction (Shimada et al., 2006) and an atomically sharp 90° domain wall structure (Shimada et al., 2008). Thus, the Pb-O bond characterizes ferroelectric instability in $PbTiO_3$. Table 4 lists the bond length, d, and minimum charge density, ρ_{min}, of the edge, surface and inside Pb-O bonds in both the PbO- and TiO_2-terminated nanowires (see Fig. 10). The minimum charge density is calculated on a line between the bonded Pb and O atoms. For comparison, we also show those of the bonds in the bulk and in the film as mentioned above, whose sites correspond to the surface and inside bonds in the nanowire.

In the PbO-terminated nanowire, the bond length is in order, $d_{edge} < d_{surface}$ ($< d_{film}$) $< d_{inside}$ ($\approx d_{bulk}$), and the minimum charge density is in order, $\rho_{min}^{edge} > \rho_{min}^{surface} >$ ($\rho_{min}^{film} >$) ρ_{min}^{inside} ($\approx \rho_{min}^{bulk}$). These findings indicate that a highly strengthened covalent bond is formed at the edge. On the other hand, d and ρ_{min} are longer and lower than those of the bulk in the TiO_2-terminated nanowire, suggesting that the Pb-O bond is weakened. Therefore, the large (respectively, small) ferroelectric distortion in the edge cell arises from the strengthened (respectively, weakened) covalent Pb-O bond in the PbO- (respectively, TiO_2-) terminated nanowire.

	Nanowire			Film	Bulk
	Edge bond	Surface bond	Inside bond		
PbO termination					
d	2.243	2.299	2.464	2.386	2.506
ρ_{min}	0.534	0.473	0.332	0.392	0.312
TiO_2 termination					
d	2.675	2.722	2.697	2.631	2.506
ρ_{min}	0.230	0.203	0.216	0.240	0.312

Table 4. Bond length d (in Å), and minimum charge density ρ_{min} (in $Å^{-3}$), of the Pb-O covalent bond in the edge, surface, and inside cells. The corresponding Pb-O bonds in the bulk and on the (001) surface of the nine-layered film (Umeno et al., 2006) are shown for comparison.

Figure 11 depicts the difference in site-by-site minimum charge density between the corresponding site of the bulk and those in the PbO- and TiO_2-terminated nanowires. In the PbO-terminated nanowire, the charge density at the edge and surface Pb-O bonds increases, while that of the Ti-O bond decreases. This indicates that the electrons transfer from the Ti-O site to the Pb-O site. Because the coordination number of the Pb atom at the edge is smaller than that of the bulk, the number of electrons contributing to the bond increases in the PbO-terminated nanowire. In fact, the Pb atoms associated with the edge, surface and

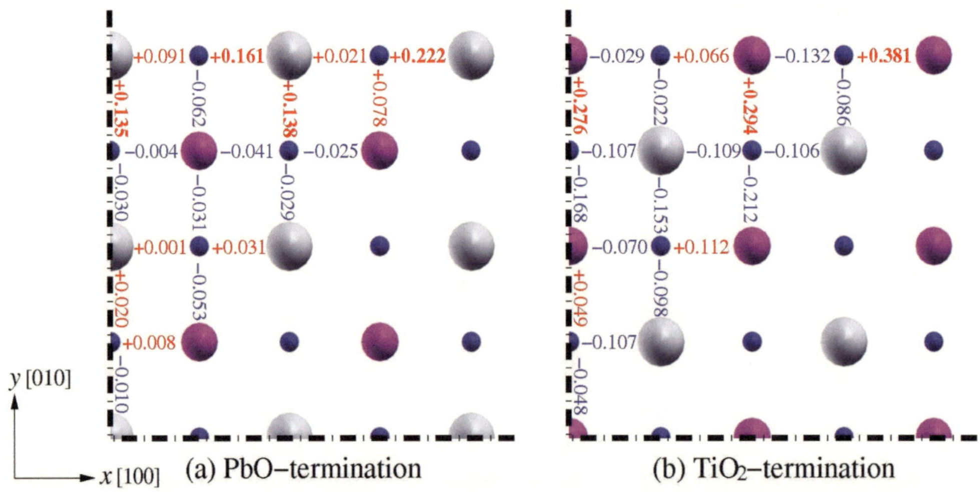

Fig. 14. Difference in site-by-site minimum charge density from the corresponding site in the bulk in (a) PbO-terminated and (b) TiO_2-terminated nanowires with a cross-section of 4×4 unit cells (in $Å^{-3}$). Only the top right quarter of the cross-section is shown by symmetry. The number in red and blue indicate the increase and decrease, respectively.

inside bonds form 2, 3 and 4 Pb-O bonds, respectively, which correspond well to the order in the minimum charge density although the relationship is not simply linear (see Table 4). Thus, the electron concentration on the edge bond due to the reduced coordination number results in locally enhanced ferroelectricity. On the other hand, charge density increases at the Ti-O bond, especially, at the edge in the TiO_2-terminated nanowire, while it decreases at the Pb-O bond. The enhancement of the Ti-O bond can be explained in the same manner as that of the Pb-O bond in the PbO-terminated nanowire. As a result of the charge redistribution from the Pb-O site to Ti-O site in the nanowire, the covalent Pb-O bond is relatively weakened, leading to strong suppression of ferroelectricity. Thus, ferroelectricity in the $PbTiO_3$ nanowire depends significantly on the terminations.

4.2 Finite size effect of nanowires on ferroelectricity

Smaller nanowires with a cross-sections of 1×1, 2×2 and 3×3 unit cells are investigated in the same manner in order to elucidate the effects of nanowire size on ferroelectricity. Table 5 lists the averaged lateral lattice parameter and the axial lattice parameter, $\bar{a}$ and c, and the averaged polarization, P, in both the PbO- and TiO_2-terminated nanowires as a function of cross-sectional size. As the size of the PbO-terminated nanowire decreases, both the lateral and axial lattice parameters shrink. However, the tetragonality of lattice, $c/\bar{a}$, which plays an important role in stabilizing ferroelectric distortion, increases in the smaller nanowire. Surprisingly, even the smallest PbO-terminated nanowire with a 1×1 unit cell size exhibits spontaneous polarization, which has the highest magnitude among the nanowires and bulk. This is because the ratio of the edge structure that enhances ferroelectricity with respect to the entire wire volume becomes more dominant with decreasing nanowire size. The magnitude of polarization approaches that of the bulk as the size increases. On the other hand, the lattice parameters of the smaller TiO_2-terminated nanowire also shrink, but the tetragonality decreases because the axial lattice parameter is highly suppressed. Moreover, nanowires smaller than 4×4 cells in cross-section, which corresponds to a diameter of about 17 Å, do not exhibit spontaneous polarization, indicating that there is the critical size for ferroelectricity. This indicates that critical size for ferroelectricity in perovskite nanowires is much sensitive to surface terminations.

Nanowire size	1×1	2×2	3×3	4×4	Bulk
		PbO termination			
$\bar{a}$	3.60	3.64	3.70	3.73	3.87
c	3.92	3.97	3.99	4.01	4.03
$c/\bar{a}$	1.088	1.090	1.081	1.074	1.043
P	103.7	99.9	95.7	94.3	85.8
		TiO_2 termination			
$\bar{a}$	3.77	3.81	3.84	3.85	3.87
c	3.39	3.57	3.67	3.74	4.03
$c/\bar{a}$	0.900	0.935	0.956	0.973	1.043
P	0.00	0.00	0.00	0.89	85.8

Table 5. Averaged lateral and axial lattice parameters, $\bar{a}$ and c (in Å), and averaged polarization P (in $\mu C/cm^2$), in both the PbO-terminated and TiO_2-terminated nanowires as a function of cross-sectional size.

4.3 Role of axial tensile strain

Figure 12 plots the averaged polarization, P, in the PbO-terminated and TiO_2-terminated nanowires as a function of the tensile strain, ε_{zz}. Averaged polarization increases almost linearly with respect to tensile strain for all the PbO-terminated nanowires. This suggests that the axial tensile strain enhances the ferroelectricity. For the TiO_2-terminated nanowires which are initially paraelectric, ferroelectricity appears under axial tension. The critical tensile strain to recover ferroelectricity is listed in Table 6. Smaller nanowires require larger critical strain because the TiO_2-terminated edge structure, which suppresses ferroelectricity, strongly affects thinner nanowires. After the critical strain, averaged polarization increases smoothly with tensile strain.

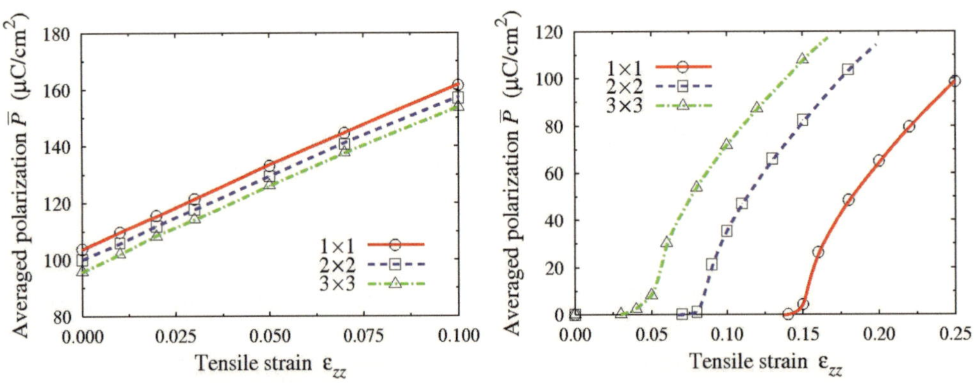

(a) PbO-terminated nanowires (b) TiO_2-terminated nanowires

Fig. 15. Average polarization P in (a) PbO-terminated and (b) TiO_2-terminated nanowires with a cross-section of 1×1, 2×2 and 3×3 unit cells as a function of tensile strain ε_{zz}.

Nanowire size	1×1	2×2	3×3
ε_c	0.15	0.08	0.04

Table 6. Critical tensile strain to recover ferroelectricity ε_c as a function of the TiO_2-terminated nanowire size.

Figure 13 shows the change in atomistic and electronic configurations on the PbO (010) planes in the PbO-terminated and TiO_2-terminated nanowires with a cross-section of 3×3 unit cells under axial tension. At equilibrium in the PbO-terminated nanowire, O atoms are displaced in the $-z$ direction from their ideal lattice sites and the covalent Pb-O bonds emphasized by white lines are formed, indicating ferroelectric distortion. During axial tension, all the Pb-O covalent bonds are sustained and the charge density at these sites increases. On the other hand, the upper and lower Pb-O bonds are equivalent to each other and there is no ferroelectric distortion in the unstrained TiO_2-terminated nanowire. After the critical strain of ε_c =0.04, the upper bond is strengthened while the lower one is weakened for all the PbO planes accompanying the upward displacement of the Pb atom. As a result, the nanowire undergoes a paraelectric-to-ferroelectric (PE-FE) phase transition.

For more detailed discussion, we define the "edge", "surface" and "inside" bonds in the 3×3 nanowire in the same manner as in the 4×4 nanowires (see Fig. 10). Figure 14 plots the minimum charge density, ρ_{min}, at the edge, surface and inside bonds in the PbO-terminated and TiO_2-terminated nanowires with a cross-section of 3×3 cells as a function of axial tensile

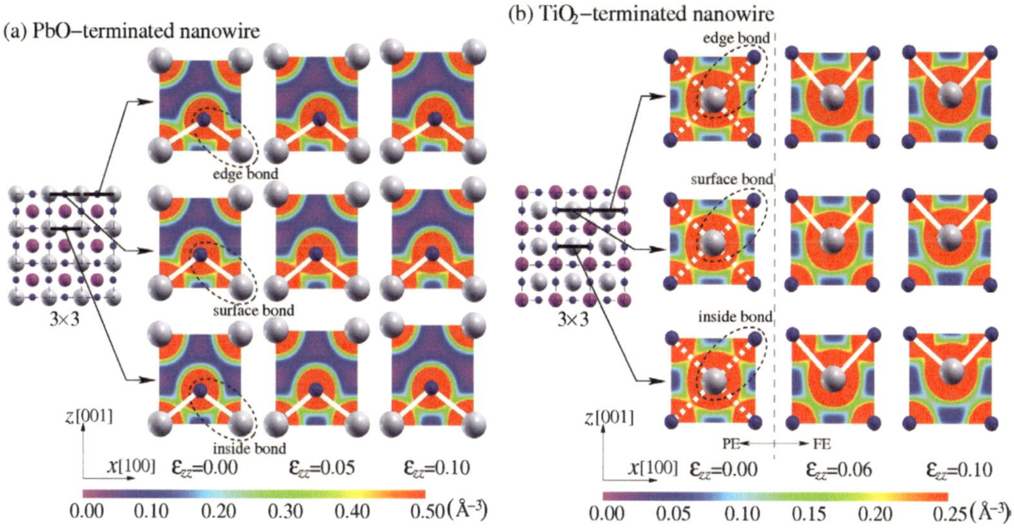

Fig. 16. Change in atomic structures and charge density distribution on the PbO (010) planes in (a) PbO-terminated and (b) TiO_2-terminated nanowires with a cross-section of 3×3 unit cells under axial tensile strain ε_{zz}. The gray and blue spheres indicate Pb and O atoms, respectively. The covalent Pb-O bonds are shown by white lines.

strain, ε_{zz}. The minimum charge density at each bond increases monotonically with respect to the tensile strain in the PbO-terminated nanowire. This corresponds well to the linear increase in the averaged polarization. Under high axial tension, the edge bond keeps the highest charge density of the three. In the TiO_2-terminated nanowire, the minimum charge density decreases before the critical strain. This indicates that the equivalent bonds at the upper and lower sites of the PbO plane are uniformly stretched because the structure is paraelectric (symmetric in the *z* direction). At the critical strain, the trend for all the bonds concurrently changes from decreasing to increasing. This indicates that ferroelectricity

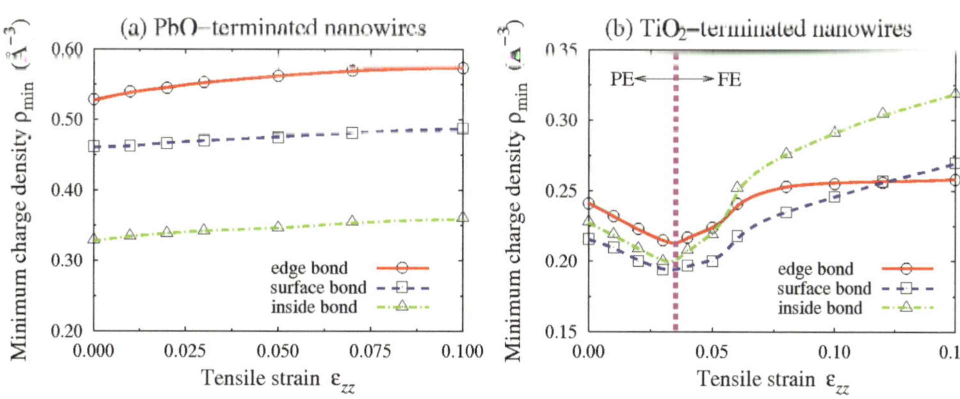

Fig. 17. Minimum charge density ρ_{min} at the edge, surface and inside bonds (see also Fig. 11) in (a) PbO-terminated and (b) TiO_2-terminated nanowires with a cross-section of 3×3 unit cells as a function of axial tensile strain ε_{zz}. The dashed line indicates the critical strain of the PE-FE phase transition.

appears in all the sites of the nanowire at the same strain. After that, all the upper bonds are partially strengthened. Note that the increase in charge density at the edge bond is somewhat moderate, which leads to the suppression of ferroelectricity at the edge with respect to the other sites.

5. Conclusion

This chapter highlights the prominent ferroelectricity in nanowires focusing on their surface and edge structures from the atomistic and electronic points of view by means of first-principles (*ab initio*) density functional theory calculations. In addition, the coupling effect between the mechanical deformation and ferroelectrictricity, namely, the multi-physics property is reviewed, as well.

At a surface, ferroelectricity, which depends on the terminations, is suppressed by TiO_2-termination and is enhanced by PbO-termination. We demonstrate the strong influence of lateral isotropic strain on the FE and AFD distortions. The ferroelectricity is enhanced by lateral tensile strain while it is suppressed by compression. At the TiO_2-terminated surface, ferroelectricity disappears when the compressive strain reaches a critical value. On the other hand, in the PbO termination, ferroelectricity is not suppressed by the surface and thus polarization distortion does not vanish under high compressive strain. These results indicate the surface FE and AFD distortions can be tuned by applied load. Furthermore we examine the effect of uniaxial [100] strain and find the rotation of ferroelectric polarization direction, which prefers to be along the longitudinal direction of rectangular lattices. While the response is sharp at the (1×1) surfaces, the surface layer exhibit [110]-oriented FE polarization up to relatively high tensile/compressive strains at the PbO-terminated c(2×2) surface. In addition, polarization direction differs among the layers (Shimada et al., 2006).

For the nanowires with atomically sharp edges consisting of (100) and (010) surfaces, the ferroelectric distortions at the edge of the PbO-terminated nanowire are enhanced, while they are entirely suppressed in the TiO_2-terminated nanowire. A precise investigation from the electronic state has shown that ferroelectricity is enhanced as a result of the formation of strong Pb-O covalent bonds at the edge of the PbO-terminated nanowire. This arises from the increase of electrons contributing to the bond due to the relative decrease in the coordination number of the Pb atom. In the TiO_2-terminated nanowire, the Pb-O bonds are weakened by the charge-transfer from the Pb-O site to the Ti-O site. Surprisingly, the smallest PbO-terminated nanowire with a cross-section of only one unit cell exhibits spontaneous polarization, which has a higher magnitude than the bulk. By contrast, the TiO_2-terminated nanowires with a cross-section smaller than 4×4 cells (about 17 Å in a diameter) cannot sustain ferroelectricity, indicating the existence of critical size. Spontaneous polarization in the PbO-terminated nanowires increases almost proportionally to the applied tensile strain accompanying the homogeneous enhancement of the Pb-O bond. In the TiO_2-terminated nanowires, which are initially paraelectric, the ferroelectricity emerges under axial tension. A larger critical strain is required for smaller nanowires, where the suppressing effect of the edge on ferroelectricity becomes predominant (Shimada et al., 2009).

There are further challenging topics on the ferroelectric nanowires remained. For example, a magnetic-like vortex ferroelectric structure was predicted in a nanowire by macroscopic conventional theory, but such vortex structure has not yet been investigated from the atomistic points of view. Besides, a recent experimental investigation revealed that the characteristic domain structure is incorporated with the nanowires, and the domain period is highly depends on the size of nanowires. It is expected that the atomic and electronic structure at the junction of atomically sharp edge (or surface) in nanowires and the domain wall significantly differs from those of bulk. It would be also interesting to study the influence of the junction in the nanowires. These issues will be addressed in the near future.

6. Acknowledgements

We acknowledge financial support by a Grant-in-Aid for Scientific Research (S) (Grant No. 21226005) and a Grant-in-Aid for Young Scientists (B) (Grant No. 21760073), of the Japan Society of the Promotion of Science (JSPS). The authors would also like to acknowledge fruitful discussions and kind support with Mr. Shogo Tomoda.

7. References

Spanier, J.E. ; Kolpak, A.M. ; Urban, J.J. ; Grinberg, I. ; Ouyang, L. ; Yun, W.S. ; Rappe, A.M. & Park, H. (2006). Ferroelectric phase transition in individual single-crystalline $BaTiO_3$ nanowires, *Nano Letters*, Vol. 6, No. 4 (April 2006) pp. 735-739, ISSN 1530-6984

Scott, J.F. (2000). *Ferroelectric Memories*, Springer, Berlin

Gu, H. ; Hu, Y. ; You, J. ; Hu, Z. ; Yuan, Y. & Zhang, T. (2007). Characterization of single crystalline $PbTiO_3$ nanowire growth via surfactant-free hydrothermal method, Journal of Applied Physics, Vol. 101, No. 2 (January 2007) 024319, ISSN 0021-8979

Yamashita, Y. ; Mukai, K. ; Yoshinobu, J. ; Lippmaa, M. ; Kinoshita, T. & Kawasaki, M. (2002). Chemical nature of nanostructures of $La_{0.6}Sr_{0.4}MnO_3$ on $SrTiO_3$ (100), *Surface Science*, Vol. 514, No. 1-3 (August 2002) pp. 54-59, ISSN 0039-6028

Cho, G.-B. ; Yamamoto, M. & Endo, Y. (2004). Surface features of self-organized $SrTiO_3$ (001) substrates inclined in [100] and [110] directions, *Thin Solid Films*, Vol. 464-465, (October 2004) pp. 80-84, ISSN 0040-6090

Kotomin, E.A. ; Heifets, E. ; Dorfman, S. ; Fuks, D. ; Gordon, A. & Maier, J. (2004). Comparative study of polar perovskite surfaces, *Surface Science*, Vol. 566-568, (September 2004) pp. 231-235, ISSN 0039-6028

Chu, M.W. ; Szafraniak, I. ; Scholz, R. ; Harnagea, C. ; Hesse, D. ; Alexe, M. & Gosele, U. (2004). Impact of misfit dislocations on the polarization instability of epitaxial nanostructured ferroelectric perovskites, *Nature Materials*, Vol. 3, No. 2 (February 2004) pp. 87-90, ISSN 1476-1122

Chakrabarti, N.B. ; Ray, S.K. & Maiti, C.M. (2001). *Strained Silicon Heterostructures : Materials and Devices*, Inspec/Iee, England

Umeno, Y. & Kitamura, T. (2002). Ab initio simulation on ideal shear strength of silicon, *Materials Science and Engineering B*, Vol. 88, No. 1 (January 2002) pp. 79-84, ISSN 0921-5107

Hohenberg, P. & Kohn, W. (1964). Inhomogeneous electron gas, *Pysical Review*, Vol. 136, No. 3B, B864, ISSN 0163-1829

Kohn, W. & Sham, L. (1965). Self-consistent equations including exchange and correlation effects, *Physical Review*, Vol. 140, No. 4A, 1133, ISSN 0031-899X

Blochl, P.E. (1994). Projector augmented-wave method, *Physical Review B*, Vol. 50, No. 24 (December 1994) pp. 17953-17979, ISSN 0163-1829

Kresse, G. & Hafner, J. (1993). Ab initio molecular dynamics for liquid metals, *Physical Review B*, Vol. 47, No. 1 (January 1993) pp. 558-561, ISSN 0163-1829

Kresse, G. & Furthmuller, J. (1996). Efficient iterative schemes for ab initio total-energy calculations using a plane-wave basis set, *Physical Review B*, Vol. 54, No. 16 (October 1996) pp. 11169-11186, ISSN 0163-1829

Ceperley, D.M. & Alder, B.J. (1980). Ground state of electron gas by a stochastic method, *Physical Review Letters*, Vol. 45, No. 7, pp. 566-569, ISSN 0031-9007

Monkhorst, H.J. & Pack, J.D. (1976). Special points for Brillouin-zone integrations, *Physical Review B*, Vol. 45, No. 12, pp. 5188-5192, ISSN 0163-1829

Zhong, W. ; King-Smith, R.D. & Vanderbilt, D. (1994). Giant LO-TO splitting in perovskite ferroelectrics, *Physical Review Letters*, Vol. 72, No. 22 (May 1994) pp. 3618-3621, ISSN 0031-9007

Umeno, Y. ; Shimada, T. ; Kitamura, T. & Elsaesser, C. (2006). Ab initio density functional theory study of strain effects on ferroelectricity at $PbTiO_3$ surfaces, *Physical Review B*, Vol. 74, No. 17 (November 2006) 174111, ISSN 1098-0121

Shimada, T. ; Wakahara, K. ; Umeno, Y. & Kitamura, T. (2008). Shell model potential for $PbTiO_3$ and its applicability to surfaces and domain walls, *Journal of Physics : Condensed Matter*, Vol. 20, No. 32 (Augst 2008) 325225, ISSN 0953-8984

Cohen, R.E. (1992). Origin of ferroelectricity in perovskite oxides, *Nature*, Vol. 358 (July 1992) pp. 136-138, ISSN 0028-0836

Kuroiwa, Y. ; Aoyagi, S. ; Sawada, A. ; Harada, J. ; Nishibori, E. ; Tanaka, M. & Sakata, M. (2001). Evidence for Pb-O covalency in tetragonal $PbTiO_3$, *Physical Review Letters*, Vol. 87, No. 21 (November 2001) 217601, ISSN 0031-9007

Shimada, T. ; Umeno, Y. & Kitamura, T. (2008). Ab initio study of stress-induced domain switching in $PbTiO_3$, *Physical Review B*, Vol. 77, No. 9 (March 2008) 094105, ISSN 1098-0121

Shimada, T. ; Tomoda, S. & Kitamura, T. (2009). Ab initio study of ferroelectricity in edged $PbTiO_3$ nanowires under axial tension, *Physical Review B*, Vol. 79, No. 2 (January 2009) 024102, ISSN 1098-0121

19

Progress Toward Nanowire Device Assembly Technology

Yanbo Li and Jean-Jacques Delaunay
The University of Tokyo
Japan

1. Introduction

The advancement of integrated circuits (ICs) has been following Moore's Law well since 1960s. For the sustaining of Moore's Law, technologists in the microelectronics industry are, on one hand, trying to push lithography technology to the limit for making devices with smaller length scales. Extreme ultraviolet, e-beam, nanoimprint or other lithography technologies have been developed as candidate replacement technologies for the conventional optical lithography (Gwyn et al., 1998; Vieu et al., 2000; Chou et al., 1996). On the other hand, technologists are also exploring the third dimension for the 3D integration of chips (Baliga, 2004). Although the advancement of lithography technologies and 3D integration technology can keep the IC industry abreast of Moore's Law for the next decade, the problems we will face at the end of that period are becoming visible. The emerging of nanowires/nanotubes as building blocks of ICs will bring fundamental changes to the future IC industry and offer continuance of Moore's Law. Besides the applications in logic circuits, nanowires have very promising applications as sensing elements in highly sensitive bio/chemical/photon sensors and detectors.

Nanowires are commonly grown by vapor-liquid-solid (VLS) process (Wagner & Ellis, 1964), vapor-solid (VS) process (Zhang et al., 1999), electrochemical deposition into nanoporous templates (Sander et al., 2002), and solution growth (Govender et al., 2002). In the past 20 years, nanowires of a diverse range of compositions have been produced at a relatively low cost with precisely controlled parameters including structure, size, defect, and doping. Nanowire devices such as field effect transistors (FETs) (Ju et al., 2007), single virus detector (Patolsky et al., 2004), pH sensor (Cui et al., 2001), gas sensors (Zhang et al., 2004), and photodetectors (Soci et al., 2007) have been demonstrated to show superior performance than their thin-film counterparts or even exhibit novel properties that have never been achieved by thin-film technology. However, most of the nanowire devices are limited to the demonstration of single device, not adequate for production on a large scale at low cost. Ultimately, cost and yield will decide whether nanowire devices find their way into market. Developing cost-effective means to integrate nanowires into working devices on large scales is essential for the prosperity of nanotechnology. In this chapter, we focus on progress toward nanowire device assembly technologies, which may benefit for the mass production of nanowire devices in the future. Generally, two strategies exist for the fabrication of devices from nanowires, namely, transfer pre-grown nanowires onto a surface with alignment and direct growth of nanowires onto a substrate at desired locations. Examples of nanowire devices demonstrated by these two strategies are given. It should be noted that although the assembly techniques described here are for nanowires, in most of the cases, they can also be applied to other 1D nanostructures, such as nanorods, nanotubes, and nanobelts.

2. Transfer with alignment

While manipulating atoms is still a dream to be realized in future nanotechnology, manipulating an assembly of them, such as nanowires, is a tough but possible task at current stage. At the early stage of nanowire research, nanowire devices are commonly made by "pick and place" method (Cui & Lieber, 2001). Nanowires grown by bottom-up process are removed from their substrates and suspended into a solution. Then, nanowires are dispersed randomly onto another substrate. Before nanowire dispersion, markers are made on the substrate, so that the position and angle of the nanowires can be noted under SEM observation. Next, a lift-off process is applied to pattern metallic contacts to the nanowires. This method is suitable for studying fundamental properties of the nanowires because the structure is well defined (Keem et al., 2006). However, it is complicated and costly and, therefore, not suitable for production. In some other cases, the metal electrodes are made before nanowire dispersion (Kind et al., 2002). Successful placement of nanowires between the pre-fabricated electrodes requires luck and the contact barriers between the electrodes and the nanowires are usually very high, which make the technique non-reproducible and not suitable for many device applications. The assembly of many individual nanowires over large areas with controlled directions and interspacing is desired for the fabrication of complex circuits of nanowires with logic functions. In this section, we concentrate on techniques that have been developed to transfer pre-fabricated nanowires and align them parallel with each other on substrates where standard lithographic processes can be applied to fabricate devices.

2.1 Alignment with fluid flow in microchannels

The shear force created by the motion of a fluid against a solid boundary can be used to align nanowires that are suspended in a solution. The nanowires will reorient to the direction of the fluid flow to minimize the fluid drag forces. The shear forces from the evaporation of a droplet can align nanowires, but the resulting pattern is a ring because the nanowires dispersed in the drop are advected to the contact line (Deegan, 2000). By flowing a stream of fluid across a substrate surface, nanowires can be reoriented towards the flow direction and become quasi-aligned (Wang et al., 2005a). A better technique to align nanowires with fluid flow is to confine the fluid flow to a microfluidic channel (Huang et al., 2001a). The microchannel can establish a shear force that is more uniform than the previous techniques. In this flow assembly technique, a poly(dimethylsiloxane) (PDMS) mold with a microchannel with width ranging from 50 to 500 μm and length from 6 to 20 mm is brought into contact with a flat substrate. Parallel nanowire arrays are achieved by flowing a nanowire suspension inside the microchannel with a controlled flow rate for a set duration. The degree of the alignment can be controlled by the flow rate: the angular distribution of the nanowires narrows substantially with increasing flow rates. The nanowire density increases systematically with the flow duration. High-performance p-Si nanowire transistors have been demonstrated using this fluid flow alignment technique (Duan et al., 2003).

The ability to assemble nanowires into complex crossed structures makes this technique more interesting for building dense nanodevice arrays. Crossed nanowire arrays can be obtained by alternating the flow in orthogonal directions in a two-step flow assembly process. An equilateral triangle of nanowires can be obtained in a three-step assembly process, with 60° angles between flow directions. The important feature of this layer-by-layer assembly scheme is that each layer is independent of the others and, therefore, a variety of homo- and hetero-junctions can be obtained at the crossed points. By using nanowires with different conduction types (e.g., p-Si and n-GaN nanowires) in each step, the authors have demonstrated logic gates with computational functions from the assembled crossbar nanowire structures (Huang et al., 2001b). The weakness of this technique is that the area for nanowire alignment is limited by the size of the fluidic microchannels. It will be more difficult to establish a uniform shear force in a large channel.

2.2 Alignment by interactions with chemically patterned surfaces

Nanowires deposited on a substrate from a suspension have random orientations. The deposition site of the nanowires strongly depends on the surface chemical functionality. The deposition sites can be controlled through van der Waals and hydrogen bonding interactions between the nanowires and the chemically functionalized substrate. Selective deposition of nanowires onto the chemically patterned area can be achieved under proper deposition conditions. One common technique to achieve such interactions is through the hybridization of complementary DNAs (Mbindyo et al., 2001; Lee et al., 2007b), which is well established in biology. A suspension of nanowires whose surface is modified with single-stranded DNA (ssDNA) is cast onto a substrate patterned with the complementary DNA (cDNA) strands and polyethylene glycol (PEG) (Wang & Gates, 2009). Nanowires deposit on the areas that are patterned with cDNA through hybridization of ssDNA with cDNA, but not on the surfaces that are passivated by PEG. This assembly technique has also been exploited for bridging nanowires across two electrodes by selective DNA hybridization (Lee et al., 2007b). The assembled nanowires show ohmic contact with minimum contact resistance, which proves this is an effective way for nanowire assembly. Similar approaches to enhance the interaction between the nanowires and the substrate include using biotin-avidin linkages (Chen et al., 2006), block-copolymer modifications (Nie et al., 2007), and hydrophobic/hydrophilic surface modifications (Ou et al., 2008). Future works should focus on improving the efficiency of the nanowire-substrate interaction and on using this technique to make ordered and patterned nanowire arrays.

Although selective deposition of nanowires can be achieved by this surface modification technique, the nanowires are not aligned after the assembly process. The alignment is driven largely by the shear force during solvent evaporation. To solve this problem, Huang et al. have combined the surface modification technique with the flow assembly technique introduced above to obtain periodic aligned nanowire arrays (Huang et al., 2001a). The SiO_2/Si substrate is patterned with NH_2-terminated monolayers in the shape of parallel stripes with a separation of a few micrometers. During the flow assembly process, the nanowires are preferentially attracted to the NH_2-terminated regions of the surface. The orientation of the nanowires is controlled by the shear force generated from the fluidic flow in the microchannels. Controlling both the location and orientation of the nanowires is therefore realized.

2.3 Alignment by Langmuir-Blodgett technique

When a solid surface is vertically dipped into a liquid containing a Langmuir monolayer and then pull out properly, the monolayer will deposit homogeneously onto the surface. This process creates Langmuir-Blodgett (LB) films. The LB technique is usually used to transfer organic monolayer from water onto a solid substrate to form extremely thin films with high degree of structure order. This technique can also be applied to prepare LB films of nanowires if the materials for assembly meet the following requirements: 1) soluble in water-immiscible solvents; 2) formation of stable floating monolayers at the surface of the subphase with internally oriented, cohesive, and compact structure that are sheer resistant. To meet these requirements, the nanowires used to form LB films are usually functionalized by surfactants. Without functionalization, the nanowires do not form stable suspensions in the organic solvents and sink into water. The process of the LB technique is illustrated in Fig. 1a. A nanowire-surfactant monolayer is initially formed on a liquid (usually water) surface in an LB trough. The monolayer is then compressed using the barrier under an appropriate level of compression. The nanowires are close-packed as parallel arrays with their longitudinal axes aligned perpendicular to the compression direction to minimize the surface energy of the liquid. The formed nanowire monolayers resemble a microscopic version of "logs-on-a-river". The monolayer of the aligned nanowires is then transferred onto a substrate through vertical-dipping (i.e., LB) or horizontal-lifting (i.e., Langmuir-Schaefer (LS)) techniques. The spacing between the parallel nanowires can be adjusted by the lifting speed and by the pressure of the compression.

The LB technique has so far been adopted by many researchers to assemble various nanowires in large scales. Tao and coworkers used the LB technique to assemble aligned monolayers of silver nanowires that are ~50 nm in diameter and 2-3 μm in length over areas as large as 20 cm^2 (Tao et al., 2003). The SEM images of the resulting Ag nanowire arrays on a Si wafer are shown in Fig. 1b-c. The Ag nanowire monolayers serve as excellent substrates for surface-enhanced Raman spectroscopy with large electromagnetic field enhancement factors.

Other aligned nanowire monolayers are also realized using the LB technique, including Ge nanowires (Wang et al., 2005b), ultrathin (~1.3 nm in diameter) ZnSe nanowires (Acharya et al., 2006), V_2O_5 nanowires (Park et al., 2008), and VO_2 nanowires (Mai et al., 2009). Whang et al. used the LS technique to transfer Si nanowire monolayers on a 1 cm × 3 cm substrate (Whang et al., 2003). The spacing of the transferred nanowires is controlled from micrometer scale to well-ordered and close-packed structures by the compression process, as shown in Fig. 1d-f. Using the LS assembly of Si nanowire arrays, they have also fabricated FET arrays over large areas without the need to register individual nanowire-electrode interconnects (Jin et al., 2004). The non-registration integration method is very useful because it can also be applied to nanowire arrays assembled by other methods. The ability to assemble hierarchical nanowire structures makes the LB and LS techniques more interesting. Hierarchical structures are produced by repeating the assembly process after changing the orientation of the substrate (Whang et al. 2003; Acharya et al., 2006). The challenges of the LB and LS techniques are the aggregation of nanowires in the Langmuir monolayer and the reorientation of nanowires during the post processes. The applications of LB techniques for the assembly of nanomaterials including nanoparticles, nanorods, nanowires, nanotubes, and nanosheets have been well summarized in a recent review (Acharya et al., 2009).

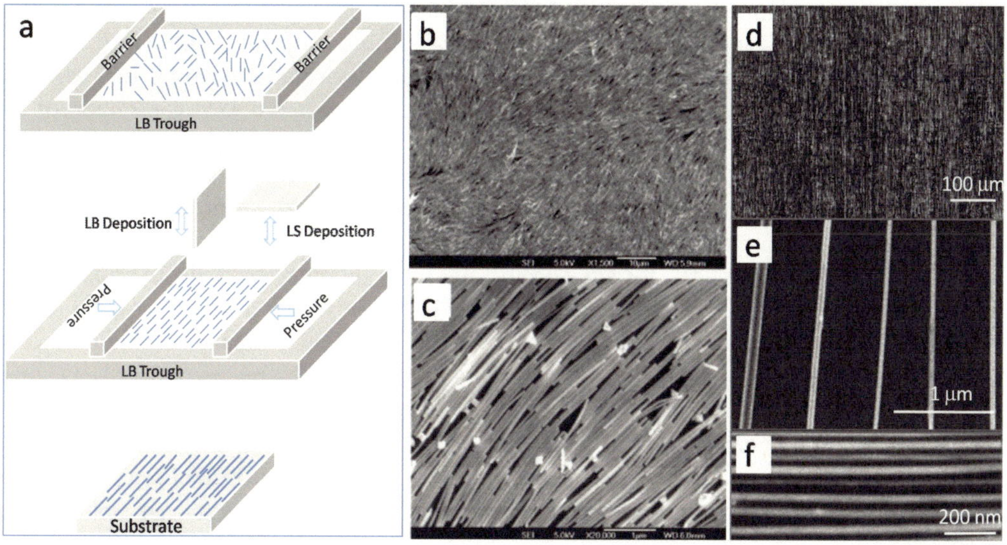

Fig. 1. Alignment of nanowires by LB and LS techniques. (a) Schematic processes of the LB and LS techniques. (b & c) SEM images of an LB assembly of Ag nanowires. Reprinted with permission from (Tao et al., 2003), © 2003 American Chemical Society. (d-f) SEM images of an LS assembly of Si nanowires. Reprinted with permission from (Whang et al., 2003), © 2003 American Chemical Society.

2.4 Electric and magnetic fields assisted orientation

For nanowires with an inherent charge or modified to adopt a specific charge, electrostatic interactions between the nanowires and patterned substrates can be used for nanowire assembly (Myung et al., 2005). The process is similar to the nanowire assembly using chemically patterned surfaces. Nanowires can also be aligned with the assistance of external electric and magnetic fields. The process for the alignment of nanowires in a fluid under electric field is called dielectrophoresis (DEP) assembly of nanowires. The dielectrophoretic forces which arise from induced dipole moments embedded in nonuniform electric fields are widely used to manipulate micro and nanoscale particles. The structures align with a minimum energy configuration within the applied field, such as along the field lines. In practice, electrodes with proper gap distances are fabricated by lithography on the substrate for DEP. Nanowire dispersion is dropped onto the substrate and a direct current (dc) or an alternating current (ac) electric field is applied to the electrodes. The density of the assembled nanowires can be controlled from a dense array of nanowires (Zhou et al., 2007) to a single nanowire (Duan et al., 2001). The density of the nanowires is generally increased with the increase in the applied voltage or with the decrease in the gap distance, that is, with the increase in the applied electric field. In the cases of ac electric field, the nanowire assembly becomes tighter, straighter, and more uniformly aligned as the frequency increases (Wang et al., 2007). A gate voltage is also used to control the deposition position of the nanowires (Wang et al., 2007). Due to the large geometric aspect ratio of nanowires, the induced dipole of the nanowires is proportional to their conductivity. Therefore, the DEP assembly works well with metallic nanowires (Smith et al., 2000). For semiconducting nanowires, super-band-gap illumination has been used to increase their conductivity, which results in an increase in the wire DEP mobility (Zhou et al., 2007). Theoretical understanding of the DEP assembly has also been done for further improving this technique (Liu et al., 2006). The weakness of the technique is the need of prefabricated electrodes for nanowire assembly and the reorientation and aggregation of the nanowires upon evaporation of the solvent.

For magnetic nanowires (e.g., Fe, Co, and Ni), a magnetic field can be applied to align them in a liquid. The nanowires align along the magnetic field lines. Pre-fabricated ferromagnetic microelectrodes on substrates provide an additional degree of control, dominating dipole interactions among nanowires, for site-specific assembly (Hangarter et al., 2007; Yoo et al., 2006). Magnetic nanowires can be effectively trapped on templates with nanomagnet arrays under a low external magnetic field if magnetic charge and dimension are matched between the magnetic nanowires to be assembled and the gaps between the two nanomagnets (Liu et al., 2007). Nonmagnetic nanowires can be capped with magnetic ends and assembled using magnetic field (Hangarter & Myung, 2005). The field gradients can also control the alignment of nonmagnetic nanowires immersed inside magnetic fluids (Ooi & Yellen, 2008). Due to the competition between a preference to align with the external field and a preference to move into regions of minimum magnetic field, the nanowires align perpendicular to the external field at low field strengths, but parallel to the external field at high field strengths. Magnetic nanowires assembled in conjunction with micropatterned magnet arrays have been demonstrated to be a flexible tool for manipulation and positioning of mammalian cells (Tanase et al., 2005). One major problem for the magnetic assembly is the lateral aggregation and edge accumulation of the nanowires. The concentration of nanowires in the solution and the applied magnetic field are very important parameters in reducing these problems (Hangarter et al., 2007).

2.5 Alignment in blown bubble films

Another technique for large-scale assembly of nanowires is using shear force created by the expansion of a blown-bubble film (Yu et al., 2007). Blown film extrusion is a well-developed process for the manufacture of plastic films in large quantities. Yu et al. applied this technique to the formation of nanocomposite films where the density and orientation of the nanowires were controlled within the films. The basic steps in their approach consist of (1) preparation of a homogenous, stable, and controlled concentration polymer suspension of nanowires; (2) expansion of the polymer suspension using a circular die to form a bubble at controlled pressure and expansion rate; (3) transfer of the bubble film to substrates or open frame structures. The nanowires within the film align along the shear force created by the expansion of the film. More than 85% of the nanowires are aligned within $\pm 6°$ of the upward expansion direction. Si nanowires are transferred conformally to single-crystal wafers up to 200 mm in diameter, flexible plastic sheets up to 225 mm × 300 mm, highly curved surfaces, and also suspended across open frames. The nanowire density within the film can be controlled by the concentration of the nanowire in the polymer suspension. Large nanowire FET arrays were also fabricated using transferred Si nanowire blown-bubble films. The limitation of this method lies in the necessity to embed nanowires in the bubble films which result in contamination of the nanowires and degradation of their performance. Also, excess epoxy matrix should be removed using processes such as reactive ion etching before the fabrication of nanodevices.

3. Integration by direct growth

Even though the transfer techniques allow one to align nanowires and control the position to a certain degree, their need for transfer media in most cases liquid, poses potential harm to the unique properties of the nanowires. Especially when the surface cleanness of the nanowires are critical to the performance of devices (such as detectors and sensors), reducing or eliminating post-processing of the nanowires is necessary. In this respect, integration of nanowires into devices on desired locations by direct growth possesses apparent advantage over the transfer methods. The synthesis methods can be modified to achieve selective growth of nanowires on desired locations by patterning of catalysts/seeds. Both vertical and planar integration of nanowires with respect to the substrate are possible. In this section, examples are given on the direct growth strategy for nanowire device integration.

3.1 Direct growth of vertical nanowire arrays

The simplest example of this strategy that can be given is the growth of vertically aligned nanowire arrays for application in devices such as nanolasers (Huang et al., 2001), light-emitting diodes (LEDs) (Könenkamp et al., 2004; Lai et al., 2008), and solar cells (Law et al., 2005). Starting with a substrate covered by nanosized catalysts or seeds, vertically aligned nanowires are grown on top of it by either vapor phase or solution growth methods. The substrate is chosen to have a good lattice match with the nanowires so that the nanowires can be epitaxially grown from the substrate to achieve better alignment. Depending on the purpose, post-treatments of the nanowire array may be necessary for the final device.

For laser applications, the nanowires are optically pumped and laser emission is observed when the excitation intensity exceeds a threshold. The lasing threshold is much lower than those for random lasing obtained in disordered particles or thin films. The vertically aligned nanowires serve as natural resonance cavities, so that lasing action is observed in the nanowire array without any fabricated mirrors.

For LED applications, the nanowires should be grown from a thin film with different conduction type to form p-n junction. To protect the nanowires and to form a buffer layer between the thin film, the nanowire array is buried in spin-on glass (Luo et al., 2006) or in high-molecular-weight polymers such as polystyrene (PS) and poly(methyl methacrylate) (PMMA). After etching the insulating layer on top of the nanowire array, ohmic contacts are formed on the thin film and on the exposed nanowire tips. Similar process has been proposed recently as a generic approach for vertical integration of nanowires (Latu-Romain et al., 2008). Electroluminescence is observed under a forward bias, which originates from the recombination of minority carriers that are injected across the junction between the thin film and the nanowire array. The vertical nanowire architecture of the device leads to waveguided emission, which is one of the advantages of using a vertical nanowire array over a thin film. Similar device architecture has been realized recently using InAs vertical nanowires on Si substrate for photovoltaic and photodetection applications (Wei et al., 2009).

Vertical nanowire arrays are also used to construct the anodes of dye-sensitized solar cells (DSCs) to replace anodes made by thick films and nanoparticles. Vertical nanowire arrays are grown on a transparent conductive substrate, dye-coated, sandwiched together and then bonded. The internal space of the cell is filled with a liquid electrolyte by capillary action. The nanowire anode features a large surface area as well as direct electrical pathways which ensure rapid collection of carriers generated throughout the device, which promotes high device efficiency. Although nanoparticle DSCs offer larger surface area, they rely on trap-limited diffusion for electron transport, a slow mechanism that can limit the device efficiency.

3.2 Patterned growth of vertical nanowires

Above devices are made using dense nanowire arrays without precisely controlling the locations of the nanowires. Although this technique offers the advantage of simplicity and low cost in the fabrication process, the alignment of the nanowires, far from ideal, presents a certain degree of randomness. Controlling the density, position, and arrangement of the nanowires is of great interest for most applications in nanoelectronics, nanophotonics, and optoelectronics. The patterned growth of nanowires in a periodic fashion over a large scale (square-centimeter and above) by inexpensive methods is highly desired. This is realized using patterned arrays of metal nanodots as catalysts made by nanopatterning techniques. The nanowires would copy the pattern of the metal nanodots due to the selective growth of nanowires via the VLS process. Existing nanopatterning techniques include photolithography (Greyson et al., 2004), e-beam lithography (Ng et al., 2004), nanosphere lithography (Fuhrmann et al., 2005; Zhou et al., 2008), nanoimprint lithography (Martensson et al., 2004), and nanoporous mask patterning (Fan et al., 2005). The nanowires are usually grown by CVD, MOVPE, MOCVD, and MBE. The patterned growth of vertical nanowires has been realized with various semiconductors, such as Si, InP, InAs, ZnO, and GaN. The key advantages of this technique are that the diameter, height, orientation, and location of the nanowires can all be controlled. For more details about the patterned growth of semiconductor nanowires, readers can refer to an extensive review given by (Fan et al., 2006) and the references therein. Here, we will focus on the applications of these patterned nanowires for device assembly.

One demonstrative device using highly ordered vertical nanowires is the nanowire vertical surround-gate field-effect transistor (VSG-FET). This device was first realized by (Ng et al., 2004) using an array of individual ZnO nanowires. By patterning Au catalyst pads with diameter of ~180 nm and thickness of ~1.5 nm using e-beam lithography, a single ZnO nanowire with an average diameter of ~35 nm was grown at each catalyst spot in the CVD process. A heavily doped SiC was used as the device substrate because the SiC (0001) plane has very small lattice mismatch (~5.5%) between the ZnO (0001) which facilitates epitaxial growth of vertically aligned ZnO nanowires and its high conductivity offers bottom electrical contacts to the nanowires. Fig. 2a-f show a generic process flow for the fabrication of a functional VSG-FET. A 3D schematic cartoon and a FE-SEM cross-sectional image are shown in Fig. 2g-h. In principle, similar fabrication schemes can be used with other nanowire/substrate combinations. Similar strategy has been employed for the realization of a VSG-FET using Si nanowire grown epitaxially by CVD on a (111)-oriented p-type Si substrate (Schmidt et al., 2006). The surround gate allows better electrostatic gate control of the conducting channel and offers the potential to drive more current per device area than is possible in a conventional planar architecture (Wang et al., 2004). 1D nanowires obtained using the bottom-up approach completely eliminate the lithography and etching processes typically employed in the top-down approach to obtain nanopillars (Endoh et al., 2003). Direct integration of the vertically aligned nanowires using current semiconductor processing technology bridges the gap between microtechnology and nanotechnology.

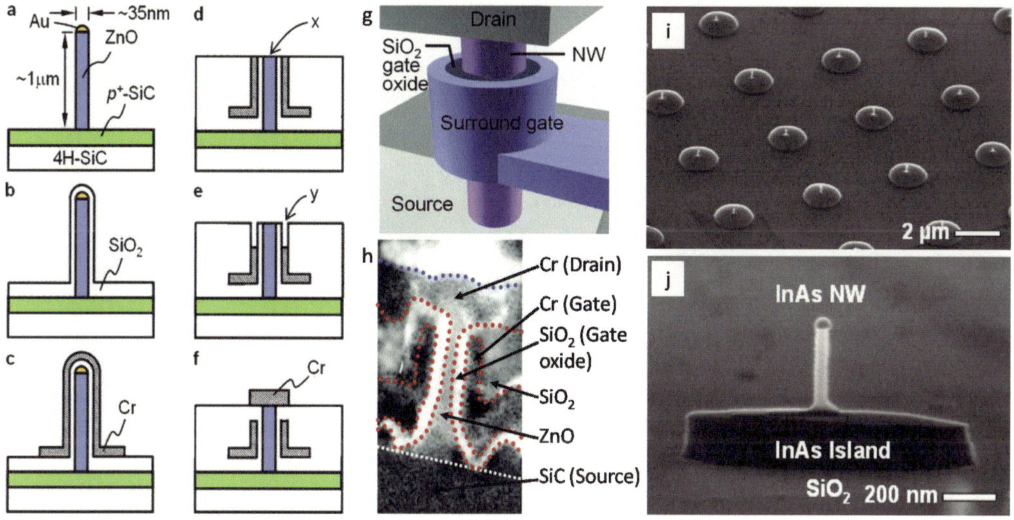

Fig. 2. (a-f) Schematics of a process flow showing the major steps to fabricate a ZnO nanowire VSG-FET. (g) A 3D schematic illustrating the critical components of the device. (h) SEM cross-sectional image of the vertical surround-gate FET. Reprinted with permission from (Ng et al., 2004), © 2004 American Chemical Society. (i-j) SEM images of vertical and electrically isolated InAs nanowires on SiO_2/Si. Reprinted with permission from (Dayeh et al., 2008), © 2008 American Institute of Physics.

The processes in the two examples given above are designed only for single device demonstration. The presence of the underlying semiconducting substrate precludes electrical isolation and individual addressability of single nanowire. For practical applications, addressable patterned underlying electrical contacts are needed. Dayeh and co-workers (Dayeh et al., 2008) tried to address this problem by a layer transfer technique that combines hydrogen ion implantation and wafer bonding, known as ion-cut or Smart-cut® process (Bruel, 1995). A thin InAs layer is transferred onto SiO_2/Si by Smart-cut® technique and ordered InAs nanowires are epitaxially grown on the layer by MOVPE. After nanowire growth, the InAs layer in the regions between the nanowires is etched resulting in an ordered, vertical, and electrically isolated InAs nanowire array as shown in Fig. 2i-j. Combining this technique with the VSG-FET technique would result in the realization of individually addressable, high density VSG-FET arrays suitable for 3D circuit applications.

Nanowires and nanotubes have been proved to exhibit excellent field emission properties due to their high aspect ratio and tip-like shape which maximize the geometrical field enhancement (Au et al., 1999; Wong et al., 1999; Nilsson et al., 2000). Another important device demonstrated using site-determined vertical nanowires is the self-aligned, gated arrays of individual nanotubes/nanowire emitters (Gangloff et al., 2004). The fabrication process of the device is shown in Fig. 3a-d. A resist hole is first patterned on a gate electrode/insulator/emitter electrode sandwich. The gate and insulator are then isotropically etched. A thin film of catalyst and a barrier layer are deposited on the structure. A lift-off process is performed to remove the unwanted catalyst on top of the gate followed by the growth of nanowire/nanowire inside the cavity. The gated nanotube cathode array has a low turn-on voltage of 25 V and a peak current of 5 μA at 46 V, with a gate current of 10 nA, which corresponds to a gate transparency of ~99%. These low operating voltage cathodes are potentially useful as electron sources for field emission displays or miniaturizing electron-based instrument.

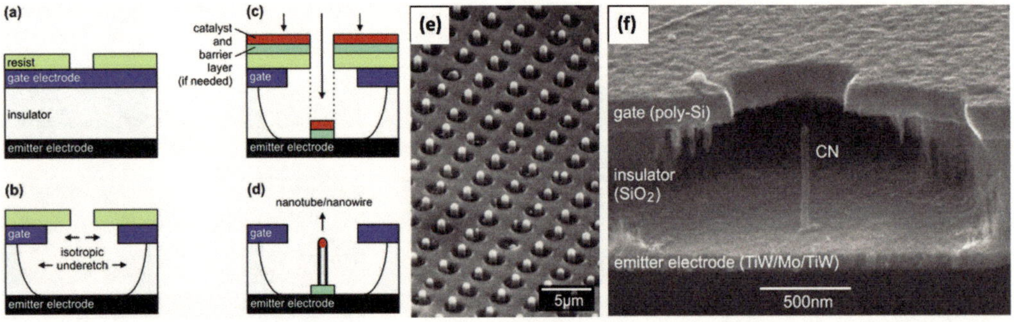

Fig. 3. (a-d) Schematics of the self-aligned process for fabricating individual nanowire/nanotubes emitters with integrated gates. (e) An array of integrated gate Si nanowire cathodes. (f) Cross-sectional view of an integrated gate carbon nanotube cathode. Reprinted with permission from (Gangloff et al., 2004), © 2004 American Chemical Society.

3.3 Direct growth of horizontally aligned nanowires

The growth of vertically aligned nanowires can be easily achieved if the substrate is terminated with a surface which allows epitaxial growth of the material and the crystal orientation of the surface matches the fast growth direction of the nanowires. The growth of horizontally aligned nanowires is, however, not easily achieved and less studied. Horizontally aligned nanowires offer a benefit of fabricating integrated nanodevices. As we discussed before, most of the works focus on the transfer of nanowires with alignment to a different substrate. It would be better if nanowires could be grown horizontally on desired locations of the substrates.

Nikoobakht has proposed a method to grow horizontally aligned nanowires on desired locations and directions (Nikoobakht, 2007). In this architecture, nanowires are grown where the nanodevices will later be fabricated on. Gold nanodroplets are first patterned on an α-plane ($11\bar{2}0$) sapphire substrate. ZnO nanowires with diameter of ~10 nm are then grown selectively on the predefined gold sites. The growth direction of the nanowires is controlled using the anisotropic crystal match between ZnO and the underlying substrate. Subsequently, metal electrodes are deposited on nanowires at once and in a parallel fashion. Large numbers of top-gated ZnO nanowire field-effect transistors are fabricated using only three photolithographic steps. The advantages of this technique are: 1) the starting coordinates of the horizontally grown nanowires are defined; 2) the number of nanowires in each device is controlled; and 3) the technique is scalable and therefore capable of industrial production. The horizontally grown nanowires have also been utilized to integrate horizontal nanochannels with known registries to microchannels (Nikoobakht, 2009).

Horizontally aligned GaAs nanowires have also been realized on GaAs (100) substrates using atmospheric pressure MOCVD with Au as catalyst (Fortuna et al., 2008). GaAs nanowires with diameter of ~30 nm and length of several microns are grown in plane in either the [$1\bar{1}0$] or [$\bar{1}10$] direction axially at 460-475 °C. The spacing between the adjacent planar nanowires can be controlled by the density of the gold catalysts. The drawback of this method is that the nanowires are grown on conducting substrates. Therefore they cannot be directly integrated into nanodevices. To solve this problem, the authors have applied a direct transfer process to the nanowires. The position and alignment of the nanowires are maintained after the transfer process, which makes possible the nanowire integration afterwards.

A general method for growing laterally aligned and patterned ZnO nanowire arrays on any flat substrate is also proposed (Qin et al., 2008). The nanowires are grown by a solution based growth process, in which the orientation control is achieved using the combined effects from a ZnO seed layer and a catalytically inactive Cr (or Sn) layer for nanowire growth. Because the growth temperature is low (<100 °C), the method can be applied to any substrate. However, the alignment of the nanowires should be improved for device applications such as FETs. Xu et al. recently achieved highly ordered horizontal ZnO nanowire arrays using a hydrothermal decomposition method onto the $[2\bar{1}\bar{1}0]$ surface of a single crystal ZnO substrate (Xu et al., 2009). However, the diameter of the nanowires is quite large (> 400 nm) and increases with the length of the nanowire. It is apparent that there is much work to do in this area before the techniques can benefit to practical applications.

3.4 Nanowire integration by bridging method

For horizontally aligned nanowires obtained either by a transferring method or by a direct growth method, electrodes are made on top of the nanowires afterwards. The post processing usually introduces contamination to the nanowires, which may deteriorate the performance of the nanodevices. Especially, keeping the nanowire surface clean is of significant importance for nanowire sensors and detectors. A solution for this would be to grow nanowires from one desired location directly to another desired location, such as between two electrodes. Nanowires bridge two desired locations in the "bottom-up" process, therefore this technique is called "bridging method". In this method, a substrate is etched to form two electrode posts using microfabrication process. Catalysts or seed layers are then deposited on the side walls of the posts. Nanowires are grown across the trench between the electrode posts and form bridges in a VLS process. Since the electrode posts are fabricated prior to the growth of nanowires, surface contamination is minimized. The nanowire integration process is also more efficient than the former methods. The bridging method was first demonstrated by Haraguchi and coworkers by growing GaAs nanowhiskers across a trench between GaAs posts (Haraguchi et al., 1996; Haraguchi et al., 1997). Although the bridging nanowhiskers in this architecture cannot be used for electrical characterization because the substrate between the two posts is conductive, it shows the potential to directly integrate an ensemble of nanowires on-chip. In recent years, this technique has been used to fabricate bridging nanowire devices such as gas sensors, photodetectors, and transistors with Si, GaN, and ZnO. Some representative works are given below.

Because of silicon's compatibility with existing IC process, Si nanowires are especially attractive. Using the vast knowledge of Si technology, the Hewlett-Packard (HP) group first demonstrated ultrahigh-density Si bridging nanowires across a trench etched into a (110)-oriented Si wafer (Islam et al., 2004). The sidewalls of the trench are (111) planes, on which Si nanowires grow vertically to the surface. The bottom of the trench is still conductive in this case. However, this problem can be easily solved by using a silicon-on-insulator (SOI) wafer. The trench is etched into the SOI until reaching the buried insulator layer. By using Si bridging nanowires grown on an SOI wafer, the HP group has realized a gas sensor and proposed a concept for using bridging nanowires to build a sensor system (Kamins et al., 2006).

A typical process for the fabrication of Si bridging nanowires on an SOI is as follows. SOI wafers consisting of a 20-80 μm thick Si(110) layer, a 0.5-2 μm thick thermally grown SiO_2 layer, and a ~400 μm thick Si(100) handle layer are used as the substrates. A 0.5-1 μm-thick thermal SiO_2 layer is first grown on the Si(110) surface and patterns designed for trenches are made by photolithography and transferred onto the SiO_2 layer by plasma etching. The trenches are made by DRIE process using the patterned SiO_2 layer as a mask. The SiO_2 mask is removed by wet etching afterwards. Gold colloids are dispersed on the substrate at catalysts for VLS growth of nanowires. Si bridging nanowires are then grown by CVD. In this process, the lengths, diameters, and densities of the bridging nanowires can be controlled. The lengths of the nanowires can be tailored to fit in trenches of varying widths by controlling the growth time. The diameters of the nanowires can be defined by the sizes of the Au colloids. The density of the nanowires in the trenches can be controlled by the surface density of the Au colloids. Using a single Si bridging nanowire fabricated by this process, giant piezoresistance effect was demonstrated (He & Yang, 2006).

Apart from Si bridging nanowires, other compound semiconductors can also be fabricated into a bridging architecture. GaN bridging nanowires are fabricated using a similar process (Chen et al., 2008). A wafer with a ~2-μm-thick layer of highly n^+-doped c-plane GaN on sapphire is used as the substrate for fabrication of the bridging nanowire device. Ni electrodes with thickness of 0.2-0.4 μm are patterned on the substrate by photolithography and lift-off. The trenches are then made by RIE using the patterned Ni electrodes as a mask. To ensure the electrical isolation between two electrodes, the n^+- GaN layer is over-etched down to the sapphire surface. After sputtering a thin (< 10 nm) layer of Au catalyst, GaN nanowires are grown by CVD to bridge the electrodes. The formed nanobridge device shows a linear *I-V* characteristic in dark, which suggests that there is no contact barrier between the electrodes and the bridging nanowires. The device shows ultrahigh (~10^5 A W^{-1}) photocurrent responsivity to UV light, which could be used as a visible-blind photodetector. The problem with this device is that the dark current is too large compared to the photocurrent, which should be addressed before using as a photodetector. One possible solution is to improve the crystal quality and lower the defect density of the GaN nanowires. The horizontally aligned GaN bridging nanowires can also be used as nanoelectromechanical resonators (Henry et al., 2007).

The formation of bridging nanowires with other material is also possible using Si trenches on a SOI wafer as substrate. This allows the integration of other functional elements into Si microelectronics. Conley et al. have demonstrated a gas and UV sensor using ZnO nanowires bridging between n^+-Si electrodes (Conley et al., 2005). Lee et al. also have successfully achieved ZnO bridging nanowires between Si electrode posts by a single-step thermal evaporation method (Lee et al., 2006). The fabricated ZnO nanobridge device shows very fast response upon turning on/off UV (Lee et al., 2007a). This kind of heterostructures have contact barrier between the electrode posts and the bridging nanowires, which may not be favored by some devices. On the other hand, the heterojunction can be beneficial for some devices if it is properly designed.

Recently, the HP group has advanced this technique to realize a top-gated MOSFET (Quitoriano & Kamins, 2008). Fig. 4a-f shows the schematic process flow for fabricating such a Si bridging nanowire MOSFET. An SOI wafer consisting of a 100 nm thick, (100)-oriented, n^+-Si layer on a 200 nm thick buried SiO_2 layer on a p-Si handle layer is used as the starting substrate. A 70 nm Si_3N_4 layer is deposited on the n^+-Si, patterned, and used as hard mask to etch the exposed Si to form electrically isolated electrodes. The SiO_2 layer is then etched to undercut the top n^+-Si layer. Colloidal Au nanoparticles are deposited only on the exposed Si surfaces by a selective placement technique, as schematically shown in Figure 4c. Si bridging nanowires are grown across the gap between the source and drain in a CVD process. The growth direction of the Si nanowires is guided by using the SiO_2 surfaces. The mechanism of the guided VLS nanowire growth using SiO_2 is also studied by the same group (Quitoriano et al., 2009). The nanowires are thermally oxidized to form a 13-21 nm oxide. Al (200 nm in thickness) is deposited to make contacts to the source and drain and Ti (170 nm in thickness) is deposited as the top gate. The SEM and TEM images of the fabricated MOSFET are shown in Fig. 4g-i. The measured drain characteristics with gate voltages from 0 to -1.35 V show good saturation. The gate characteristics measured at a drain voltage of 0.1 V show high I_{on}/I_{off} ratio of ~10^4 and inverse sub-threshold slope of ~155 mV/decade. The integration of nanowires into MOSFETs by bridging method can help realize the full promise of semiconducting nanowires since practical applications of nanowires are likely to use a combination of top-down patterning with self-assembly to integrate nanowires with conventionally formed microstructures.

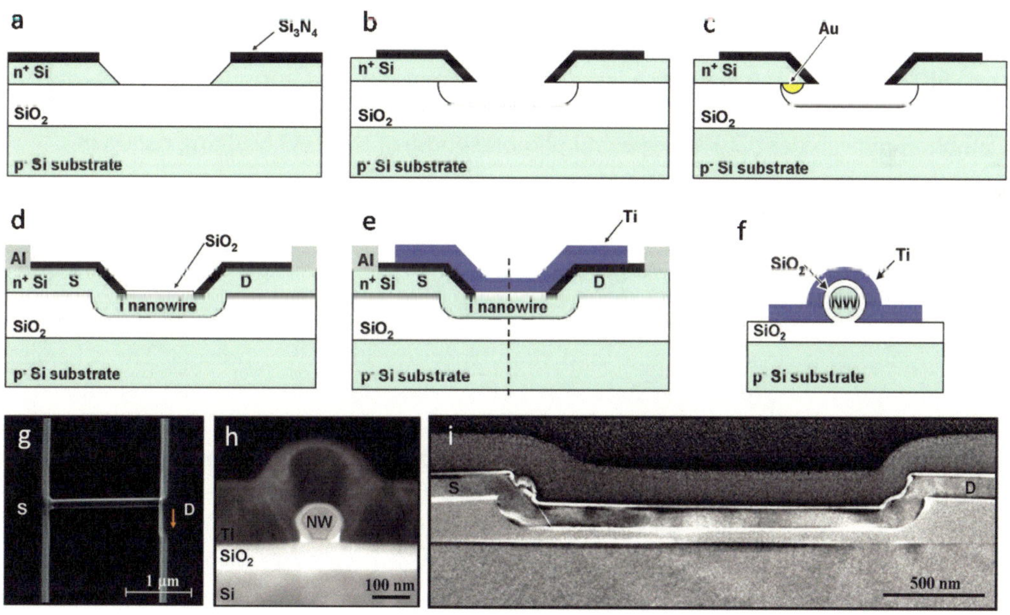

Fig. 4. (a-f) Process flow and schematic architecture of the Si bridging nanowire MOSFET. (g) Top-view SEM image of a Si bridging nanowire grown between source (S) and drain (D). (h) Perpendicular and (i) longitudinal, cross-sectional TEM images of a nanowire bridging source and drain. Reprinted with permission from (Quitoriano & Kamins, 2008), © 2008 American Chemical Society.

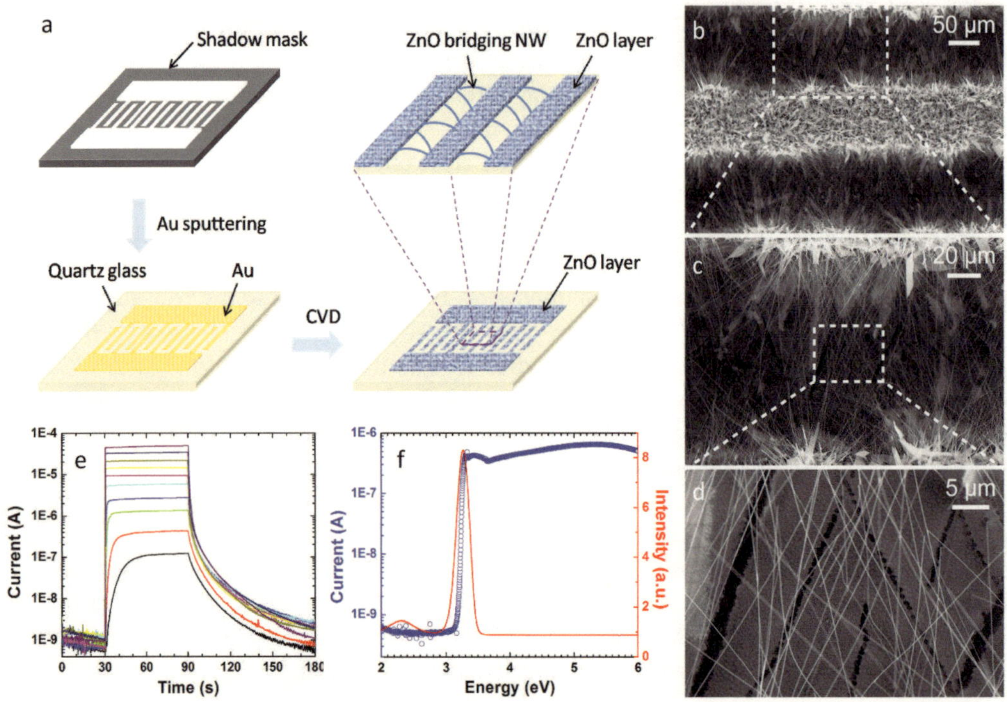

Fig. 5. (a) Schematic process flow of the single-step fabrication of ZnO bridging nanowires. (b-d) SEM images of the ZnO thick layer electrodes and the ZnO bridging nanowires. (e) Time-dependent UV photoresponse of the ZnO bridging nanowire UV photodetector. (f) Cathodoluminescence (CL) and spectral photoresponse of the ZnO bridging nanowires.

The conventional bridging method, as shown in the above examples, requires the fabrication of micro-trenches and electrode posts before the growth of nanowires. The process can be further simplified for some device applications. Li et al. have proposed a single-step bridging method to fabricate ZnO bridging nanowires without resorting to any microfabrication process (Li et al., 2008; Li et al., 2009a, b). In this method, the electrodes and the bridging nanowires are synthesized simultaneously in a CVD process. The schematic fabrication process is shown in Fig. 5a. First, a 2-nm-thick gold catalyst layer is sputtered on a quartz glass through a comb-shaped metal shadow mask. Then, ZnO is grown on the Au patterned substrate in a CVD process. After the CVD process, the Au pattern areas, i.e., the comb pads and fingers, are covered by a thick ZnO layer. SEM observation reveals that the thick layer consists of dense ZnO nanowires and nanosheets (Fig. 5b). The gap (~100 μm in width) between the comb fingers is bridged by many ultra-long nanowires, as shown in Fig. 5c-d. The key to this single-step bridging method is to achieve lateral growth of ultra-long nanowires at the edge of the thick layer, so that physical masking can be applied to pattern the Au catalyst instead of lift-off processes. The as-made structure can be directly used for photoresponse test by contacting the comb pads with In or Al which form ohmic contacts with ZnO. Time-dependent photoresponse of the device in Fig. 5e shows that the current increases drastically and rapidly when exposed to UV illumination. The current increases by

2 to 5 orders under an irradiance of 180 nW/cm^2 to 48 mW/cm^2. The photocurrent decay is also very fast. The decay time is in the order of a few seconds. With the current being insensitive to photon energy lower than ~3.2 eV (Fig. 5f), the device exhibits visible-blind spectral photoresponse. By eliminating lift-off from the fabrication process of the device, the nanowires are free of contamination, which is one of the major advantages of bridging technique as discussed above. Besides, the nanowires are lying over the substrate instead of being in contact with the substrate, rendering their surfaces fully exposed to the ambient atmosphere. This can be very beneficial to nanowire devices such as gas/chemical sensors, in which the surfaces of the nanowires play a central role in the sensing mechanism. Therefore, this single-step bridging method is suitable for the mass-production of low-cost and high-performance nanowire photodetectors and gas/chemical sensors.

4. Conclusion

The remarkable progress made on the synthesis of nanowires over recent years offers a wide selection of building blocks for future nanodevices which are deemed to change our life fundamentally. However, the assembly of these building blocks in reliable and economical ways should be addressed before the flourish of nanotechnology. In this chapter, technologies developed for the assembly of nanowire devices were reviewed. We focused on two main strategies for tackling this problem, that is, transfer with alignment of pre-grown nanowires onto a surface and direct growth of nanowires onto a substrate at desired locations. Nanowire devices demonstrated by these assembly techniques were introduced. By combining the current stage top-down techniques with the nanowire assembly techniques, the fabricated nanodevices show properties that motivate us on carrying on the research. It also points out future research directions for nanotechnology. Nevertheless, many practical problems have to be solved for integration of nanowires into devices. A lot more effort has to be put into the development of new nanowire assembly techniques, as we have been doing for the synthesis of nanowires. It is impossible to find a universal way that solves all the problems. Understanding the weaknesses of each technique is as important as understanding their advantages, for it can help us select the right one for a specific device application. A combination of two or more assembly techniques will be useful as illustrated by some of the cases given above. Through this review, we hope that readers not only learn the state-of-the-art nanowire assembly technologies, but also gain more confidence on the future of nanotechnology despite of all the difficulties.

5. References

Acharya, S.; Panda, A. B.; Belman, N.; Efrima, S. & Golan, Y. (2006). A semiconductor-nanowire assembly of ultrahigh junction density by the Langmuir-Blodgett technique. *Advanced Materials,* Vol. 18, No. 2, (Jan 2006) 210-213

Acharya, S.; Hill, J. P. & Ariga, K. (2009). Soft Langmuir-Blodgett Technique for Hard Nanomaterials. *Advanced Materials,* Vol. 21, No. 29, 2009) 2959-2981

Au, F. C. K.; Wong, K. W.; Tang, Y. H.; Zhang, Y. F.; Bello, I. & Lee, S. T. (1999). Electron field emission from silicon nanowires. *Applied Physics Letters,* Vol. 75, No. 12, (Sep 1999) 1700-1702

Baliga, J. (2004). Chips go vertical. *Ieee Spectrum,* Vol. 41, No. 3, (Mar 2004) 43-47

Bruel, M. (1995). Silicon-on-insulator material technology. *Electronics Letters,* Vol. 31, No. 14, (Jul 1995) 1201-1202

Chen, M.; Guo, L.; Ravi, R. & Searson, P. C. (2006). Kinetics of receptor directed assembly of multisegment nanowires. *Journal of Physical Chemistry B,* Vol. 110, No. 1, (Jan 2006) 211-217

Chen, R. S.; Wang, S. W.; Lan, Z. H.; Tsai, J. T. H.; Wu, C. T.; Chen, L. C.; Chen, K. H.; Huang, Y. S. & Chen, C. C. (2008). On-chip fabrication of well-aligned and contact-barrier-free GaN nanobridge devices with ultrahigh photocurrent responsivity. *Small,* Vol. 4, No. 7, (Jul 2008) 925-929

Chou, S. Y.; Krauss, P. R. & Renstrom, P. J. (1996). Nanoimprint lithography. *Journal of vacuum science & technology. B, Microelectronics and nanometer structures processing, measurement and phenomena,* Vol. 14, No. 6, (Nov-Dec 1996) 4129-4133

Conley, J. F.; Stecker, L. & Ono, Y. (2005). Directed integration of ZnO nanobridge devices on a Si substrate. *Applied Physics Letters,* Vol. 87, No. 22, (Nov 2005) 223114

Cui, Y. & Lieber, C. M. (2001). Functional nanoscale electronic devices assembled using silicon nanowire building blocks. *Science,* Vol. 291, No. 5505, (Feb 2001) 851-853

Cui, Y.; Wei, Q. Q.; Park, H. K. & Lieber, C. M. (2001). Nanowire nanosensors for highly sensitive and selective detection of biological and chemical species. *Science,* Vol. 293, No. 5533, (Aug 2001) 1289-1292

Dayeh, S. A.; Chen, P.; Jing, Y.; Yu, E. T.; Lau, S. S. & Wang, D. L. (2008). Integration of vertical InAs nanowire arrays on insulator-on-silicon for electrical isolation. *Applied Physics Letters,* Vol. 93, No. 20, (Nov 2008) 203109

Deegan, R. D. (2000). Pattern formation in drying drops. *Physical Review E,* Vol. 61, No. 1, (Jan 2000) 475-485

Duan, X. F.; Huang, Y.; Cui, Y.; Wang, J. F. & Lieber, C. M. (2001). Indium phosphide nanowires as building blocks for nanoscale electronic and optoelectronic devices. *Nature,* Vol. 409, No. 6816, (Jan 2001) 66-69

Duan, X. F.; Niu, C. M.; Sahi, V.; Chen, J.; Parce, J. W.; Empedocles, S. & Goldman, J. L. (2003). High-performance thin-film transistors using semiconductor nanowires and nanoribbons. *Nature,* Vol. 425, No. 6955, (Sep 2003) 274-278

Endoh, T.; Kinoshita, K.; Tanigami, T.; Wada, Y.; Sato, K.; Yamada, K.; Yokoyama, T.; Takeuchi, N.; Tanaka, K.; Awaya, N.; Sakiyama, K. & Masuoka, F. (2003). Novel ultrahigh-density flash memory with a stacked-surrounding gate transistor (S-SGT) structured cell. *IEEE Transactions on Electron Devices,* Vol. 50, No. 4, (Apr 2003) 945-951

Fan, H. J.; Lee, W.; Scholz, R.; Dadgar, A.; Krost, A.; Nielsch, K. & Zacharias, M. (2005). Arrays of vertically aligned and hexagonally arranged ZnO nanowires: a new template-directed approach. *Nanotechnology,* Vol. 16, No. 6, (Jun 2005) 913-917

Fan, H. J.; Werner, P. & Zacharias, M. (2006). Semiconductor nanowires: From self-organization to patterned growth. *Small,* Vol. 2, No. 6, (Jun 2006) 700-717

Fortuna, S. A.; Wen, J. G.; Chun, I. S. & Li, X. L. (2008). Planar GaAs Nanowires on GaAs (100) Substrates: Self-Aligned, Nearly Twin-Defect Free, and Transfer-Printable. *Nano Letters,* Vol. 8, No. 12, (Dec 2008) 4421-4427

Fuhrmann, B.; Leipner, H. S.; Hoche, H. R.; Schubert, L.; Werner, P. & Gosele, U. (2005). Ordered arrays of silicon nanowires produced by nanosphere lithography and molecular beam epitaxy. *Nano Letters,* Vol. 5, No. 12, (Dec 2005) 2524-2527

Gangloff, L.; Minoux, E.; Teo, K. B. K.; Vincent, P.; Semet, V. T.; Binh, V. T.; Yang, M. H.; Bu, I. Y. Y.; Lacerda, R. G.; Pirio, G.; Schnell, J. P.; Pribat, D.; Hasko, D. G.; Amaratunga, G. A. J.; Milne, W. I. & Legagneux, P. (2004). Self-aligned, gated arrays of individual nanotube and nanowire emitters. *Nano Letters,* Vol. 4, No. 9, (Sep 2004) 1575-1579

Govender, K.; Boyle, D. S.; O'Brien, P.; Binks, D.; West, D. & Coleman, D. (2002). Room-temperature lasing observed from ZnO nanocolumns grown by aqueous solution deposition. *Advanced Materials,* Vol. 14, No. 17, (Sep 2002) 1221-1224

Greyson, E. C.; Babayan, Y. & Odom, T. W. (2004). Directed growth of ordered arrays of small-diameter ZnO nanowires. *Advanced Materials,* Vol. 16, No. 15, (Aug 2004) 1348-1352

Gwyn, C. W.; Stulen, R.; Sweeney, D. & Attwood, D. (1998). Extreme ultraviolet lithography. *Journal of vacuum science & technology B, Microelectronics and nanometer structures processing, measurement and phenomena,* Vol. 16, No. 6, (Nov-Dec 1998) 3142-3149

Hangarter, C. M.; Rheem, Y.; Yoo, B.; Yang, E. H. & Myung, N. V. (2007). Hierarchical magnetic assembly of nanowires. *Nanotechnology,* Vol. 18, No. 20, (May 2007) 205305

Hangarter, C. M. & Myung, N. V. (2005). Magnetic alignment of nanowires. *Chemistry of Materials,* Vol. 17, No. 6, (Mar 2005) 1320-1324

Haraguchi, K.; Hiruma, K.; Katsuyama, T.; Tominaga, K.; Shirai, M. & Shimada, T. (1996). Self-organized fabrication of planar GaAs nanowhisker arrays. *Applied Physics Letters,* Vol. 69, No. 3, (Jul 1996) 386-387

Haraguchi, K.; Hiruma, K.; Hosomi, K.; Shirai, M. & Katsuyama, T. (1997). Growth mechanism of planar-type GaAs nanowhiskers. *Journal of Vacuum Science & Technology B,* Vol. 15, No. 5, (Sep 1997) 1685-1687

He, R. R.; Gao, D.; Fan, R.; Hochbaum, A. I.; Carraro, C.; Maboudian, R. & Yang, P. D. (2005). Si nanowire bridges in microtrenches: Integration of growth into device fabrication. *Advanced Materials,* Vol. 17, No. 17, (Sep 2005) 2098-2102

He, R. R. & Yang, P. D. (2006). Giant piezoresistance effect in silicon nanowires. *Nature Nanotechnology,* Vol. 1, No. 1, (Oct 2006) 42-46

Henry, T.; Kim, K.; Ren, Z.; Yerino, C.; Han, J. & Tang, H. X. (2007). Directed growth of horizontally aligned gallium nitride nanowires for nanoelectromechanical resonator Arrays. *Nano Letters,* Vol. 7, No. 11, (Nov 2007) 3315-3319

Huang, M. H.; Mao, S.; Feick, H.; Yan, H.; Wu, Y.; Kind, H.; Weber, E.; Russo, R. & Yang, P. (2001). Room-temperature ultraviolet nanowire nanolasers. *Science,* Vol. 292, No. 5523, (Jun 2001) 1897-1899

Huang, Y.; Duan, X. F.; Wei, Q. Q. & Lieber, C. M. (2001a). Directed assembly of one-dimensional nanostructures into functional networks. *Science,* Vol. 291, No. 5504, (Jan 26 2001) 630-633

Huang, Y.; Duan, X. F.; Cui, Y.; Lauhon, L. J.; Kim, K. H. & Lieber, C. M. (2001b). Logic gates and computation from assembled nanowire building blocks. *Science,* Vol. 294, No. 5545, (Nov 9 2001) 1313-1317

Islam, M. S.; Sharma, S.; Kamins, T. I. & Williams, R. S. (2004). Ultrahigh-density silicon nanobridges formed between two vertical silicon surfaces. *Nanotechnology,* Vol. 15, No. 5, (May 2004) L5-L8

Jensen, L. E.; Bjork, M. T.; Jeppesen, S.; Persson, A. I.; Ohlsson, B. J. & Samuelson, L. (2004). Role of surface diffusion in chemical beam epitaxy of InAs nanowires. *Nano Letters,* Vol. 4, No. 10, (Oct 2004) 1961-1964

Jin, S.; Whang, D. M.; McAlpine, M. C.; Friedman, R. S.; Wu, Y. & Lieber, C. M. (2004). Scalable interconnection and integration of nanowire devices without registration. *Nano Letters,* Vol. 4, No. 5, (May 2004) 915-919

Ju, S. Y.; Facchetti, A.; Xuan, Y.; Liu, J.; Ishikawa, F.; Ye, P. D.; Zhou, C. W.; Marks, T. J. & Janes, D. B. (2007). Fabrication of fully transparent nanowire transistors for transparent and flexible electronics. *Nature Nanotechnology,* Vol. 2, No. 6, (Jun 2007) 378-384

Kamins, T. I.; Sharma, S.; Yasseri, A. A.; Li, Z. & Straznicky, J. (2006). Metal-catalysed, bridging nanowires as vapour sensors and concept for their use in a sensor system. *Nanotechnology,* Vol. 17, No. 11, (Jun 2006) S291-S297

Keem, K.; Jeong, D. Y.; Kim, S.; Lee, M. S.; Yeo, I. S.; Chung, U. I. & Moon, J. T. (2006). Fabrication and device characterization of omega-shaped-gate ZnO nanowire field-effect transistors. *Nano Letters,* Vol. 6, No. 7, (Jul 2006) 1454-1458

Kind, H.; Yan, H. Q.; Messer, B.; Law, M. & Yang, P. D. (2002). Nanowire ultraviolet photodetectors and optical switches. *Advanced Materials,* Vol. 14, No. 2, (Jan 2002) 158-160

Könenkamp, R.; Word, R. C. & Schlegel, C. (2004). Vertical nanowire light-emitting diode. *Applied Physics Letters,* Vol. 85, No. 24, (Dec 2004) 6004-6006

Lai, E.; Kim, W. & Yang, P. (2008). Vertical nanowire array-based light emitting diodes. *Nano Research,* Vol. 1, No. 2, (Jun 2008) 123-128

Latu-Romain, E.; Gilet, P.; Noel, P.; Garcia, J.; Ferret, P.; Rosina, M.; Feuillet, G.; Levy, F. & Chelnokov, A. (2008). A generic approach for vertical integration of nanowires. *Nanotechnology,* Vol. 19, No. 34, (Aug 2008)

Law, M.; Greene, L.; Johnson, J. C.; Saykally, R. & Yang, P. (2005). Nanowire dye-sensitized solar cells. *Nature Materials,* Vol. 4, No. 6, (Jun 2005) 455-459

Lee, J. S.; Islam, M. S. & Kim, S. (2006). Direct formation of catalyst-free ZnO nanobridge devices on an etched Si substrate using a thermal evaporation method. *Nano Letters,* Vol. 6, No. 7, (Jul 2006) 1487-1490

Lee, J. S.; Islam, M. S. & Kim, S. (2007a). Photoresponses of ZnO nanobridge devices fabricated using a single-step thermal evaporation method. *Sensors and Actuators B: Chemical,* Vol. 126, No. 1, (Sep 2007) 73-77

Lee, J.; Wang, A. A.; Rheem, Y.; Yoo, B.; Mulchandani, A.; Chen, W. & Myung, N. V. (2007b). DNA assisted assembly of multisegmented nanowires. *Electroanalysis,* Vol. 19, No. 22, (Nov 2007) 2287-2293

Li, Y. B.; Della Valle, F.; Simonnet, M.; Yamada, I. & Delaunay, J. J. (2009a). High-performance UV detector made of ultra-long ZnO bridging nanowires. *Nanotechnology,* Vol. 20, No. 4, (Jan 2009) 045501

Li, Y. B.; Della Valle, F.; Simonnet, M.; Yamada, I. & Delaunay, J. J. (2009b). Competitive surface effects of oxygen and water on UV photoresponse of ZnO nanowires. *Applied Physics Letters,* Vol. 94, No. 2, (Jan 2009) 023110

Li, Y. B.; Nagatomo, I.; Uchino, R.; Yamada, I. & Delaunay, J. J. (2008). Fabrication of ZnO bridging nanowire device by a single-step chemical vapor deposition method. *Materials Research Society Symposium Proceedings,* 1144-LL17-03, Boston, Dec 2008

Liu, M.; Lagdani, J.; Imrane, H.; Pettiford, C.; Lou, J.; Yoon, S.; Harris, V. G.; Vittoria, C. & Sun, N. X. (2007). Self-assembled magnetic nanowire arrays. *Applied Physics Letters,* Vol. 90, No. 10, (Mar 2007) 103105

Liu, Y. L.; Chung, J. H.; Liu, W. K. & Ruoff, R. S. (2006). Dielectrophoretic assembly of nanowires. *Journal of Physical Chemistry B,* Vol. 110, No. 29, (Jul 2006) 14098-14106

Luo, L.; Zhang, Y.; Mao, S. S. & Lin, L. (2006). Fabrication and characterization of ZnO nanowires based UV photodiodes. *Sensors and Actuators A,* Vol. 127, No. 2, (Mar 2006) 201-206

Mai, L. Q.; Gu, Y. H.; Han, C. H.; Hu, B.; Chen, W.; Zhang, P. C.; Xu, L.; Guo, W. L. & Dai, Y. (2009). Orientated Langmuir-Blodgett Assembly of VO2 Nanowires. *Nano Letters,* Vol. 9, No. 2, (Feb 2009) 826-830

Martensson, T.; Carlberg, P.; Borgstrom, M.; Montelius, L.; Seifert, W. & Samuelson, L. (2004). Nanowire arrays defined by nanoimprint lithography. *Nano Letters,* Vol. 4, No. 4, (Apr 2004) 699-702

Mbindyo, J. K. N.; Reiss, B. D.; Martin, B. R.; Keating, C. D.; Natan, M. J. & Mallouk, T. E. (2001). DNA-directed assembly of gold nanowires on complementary surfaces. *Advanced Materials,* Vol. 13, No. 4, (Feb 2001) 249-254

Myung, S.; Lee, M.; Kim, G. T.; Ha, J. S. & Hong, S. (2005). Large-scale "surface-programmed assembly" of pristine vanadium oxide nanowire-based devices. *Advanced Materials,* Vol. 17, No. 19, (Oct 2005) 2361-2364

Ng, H. T.; Han, J.; Yamada, T.; Nguyen, P.; Chen, Y. P. & Meyyappan, M. (2004). Single crystal nanowire vertical surround-gate field-effect transistor. *Nano Letters,* Vol. 4, No. 7, (Jul 2004) 1247-1252

Nie, Z. H.; Fava, D.; Kumacheva, E.; Zou, S.; Walker, G. C. & Rubinstein, M. (2007). Self-assembly of metal-polymer analogues of amphiphilic triblock copolymers. *Nature Materials,* Vol. 6, No. 8, (Aug 2007) 609-614

Nikoobakht, B. (2007). Toward industrial-scale fabrication of nanowire-based devices. *Chemistry of Materials,* Vol. 19, No. 22, (Oct 2007) 5279-5284

Nikoobakht, B. (2009). A Scalable Platform for Integrating Horizontal Nanochannels with Known Registries to Microchannels. *Chemistry of Materials,* Vol. 21, No. 1, (Jan 2009) 27-32

Nilsson, L.; Groening, O.; Emmenegger, C.; Kuettel, O.; Schaller, E.; Schlapbach, L.; Kind, H.; Bonard, J. M. & Kern, K. (2000). Scanning field emission from patterned carbon nanotube films. *Applied Physics Letters,* Vol. 76, No. 15, (Apr 2000) 2071-2073

Ooi, C. & Yellen, B. B. (2008). Field gradients can control the alignment of nanorods. *Langmuir,* Vol. 24, No. 16, (Aug 2008) 8514-8521

Ou, F. S.; Shaijumon, M. M. & Ajayan, P. M. (2008). Controlled manipulation of giant hybrid inorganic nanowire assemblies. *Nano Letters,* Vol. 8, No. 7, (Jul 2008) 1853-1857

Park, J.; Shin, G. & Ha, J. S. (2008). Controlling orientation of V2O5 nanowires within micropatterns via icrocontact printing combined with the gluing Langmuir-Blodgett technique. *Nanotechnology,* Vol. 19, No. 39, (Oct 2008) 395303

Patolsky, F.; Zheng, G. F.; Hayden, O.; Lakadamyali, M.; Zhuang, X. W. & Lieber, C. M. (2004). Electrical detection of single viruses. *Proceedings of the National Academy of Sciences of the United States of America,* Vol. 101, No. 39, (Sep 2004) 14017-14022

Qin, Y.; Yang, R. S. & Wang, Z. L. (2008). Growth of Horizonatal ZnO Nanowire Arrays on Any Substrate. *Journal of Physical Chemistry C,* Vol. 112, No. 48, (Dec 2008) 18734-18736

Quitoriano, N. J. & Kamins, T. I. (2008). Integratable Nanowire Transistors. *Nano Letters,* Vol. 8, No. 12, (Dec 2008) 4410-4414

Quitoriano, N. J.; Wu, W. & Kamins, T. I. (2009). Guiding vapor-liquid-solid nanowire growth using SiO_2. *Nanotechnology,* Vol. 20, No. 14, (Apr 2009) 145303

Sander, M. S.; Prieto, A. L.; Gronsky, R.; Sands, T. & Stacy, A. M. (2002). Fabrication of high-density, high aspect ratio, large-area bismuth telluride nanowire arrays by electrodeposition into porous anodic alumina templates. *Advanced Materials,* Vol. 14, No. 9, (May 2002) 665-667

Schmidt, V.; Riel, H.; Senz, S.; Karg, S.; Riess, W. & Gosele, U. (2006). Realization of a silicon nanowire vertical surround-gate field-effect transistor. *Small,* Vol. 2, No. 1, (Jan 2006) 85-88

Smith, P. A.; Nordquist, C. D.; Jackson, T. N.; Mayer, T. S.; Martin, B. R.; Mbindyo, J. & Mallouk, T. E. (2000). Electric-field assisted assembly and alignment of metallic nanowires. *Applied Physics Letters,* Vol. 77, No. 9, (Aug 28 2000) 1399-1401

Soci, C.; Zhang, A.; Xiang, B.; Dayeh, S. A.; Aplin, D. P. R.; Park, J.; Bao, X. Y.; Lo, Y. H. & Wang, D. (2007). ZnO nanowire UV photodetectors with high internal gain. *Nano Letters,* Vol. 7, No. 4, (Apr 2007) 1003-1009

Tanase, M.; Felton, E. J.; Gray, D. S.; Hultgren, A.; Chen, C. S. & Reich, D. H. (2005). Assembly of multicellular constructs and microarrays of cells using magnetic nanowires. *Lab on a Chip,* Vol. 5, No. 6, 2005) 598-605

Tao, A.; Kim, F.; Hess, C.; Goldberger, J.; He, R. R.; Sun, Y. G.; Xia, Y. N. & Yang, P. D. (2003). Langmuir-Blodgett silver nanowire monolayers for molecular sensing using surface-enhanced Raman spectroscopy. *Nano Letters,* Vol. 3, No. 9, (Sep 2003) 1229-1233

Vieu, C.; Carcenac, F.; Pepin, A.; Chen, Y.; Mejias, M.; Lebib, A.; Manin-Ferlazzo, L.; Couraud, L. & Launois, H. (2000). Electron beam lithography: resolution limits and applications. *Applied Surface Science,* Vol. 164, (Sep 2000) 111-117

Wagner, R. S. & Ellis, W. C. (1964). Vapor-liquid-solid mechanism of single crystal growth. *Applied Physics Letters,* Vol. 4, No. 5, (Mar 1964) 89-90

Wang, D. Q.; Zhu, R.; Zhou, Z. Y. & Ye, X. Y. (2007). Controlled assembly of zinc oxide nanowires using dielectrophoresis. *Applied Physics Letters,* Vol. 90, No. 10, (Mar 2007)

Wang, D. W.; Tu, R.; Zhang, L. & Dai, H. J. (2005a). Deterministic one-to-one synthesis of germanium nanowires and individual gold nanoseed patterning for aligned arrays. *Angewandte Chemie-International Edition,* Vol. 44, No. 19, 2005) 2925-2929

Wang, D. W.; Chang, Y. L.; Liu, Z. & Dai, H. J. (2005b). Oxidation resistant germanium nanowires: Bulk synthesis, long chain alkanethiol functionalization, and Langmuir-Blodgett assembly. *Journal of the American Chemical Society,* Vol. 127, No. 33, (Aug 2005) 11871-11875

Wang, J.; Polizzi, E. & Lundstrom, M. (2004). A three-dimensional quantum simulation of silicon nanowire transistors with the effective-mass approximation. *Journal of Applied Physics,* Vol. 96, No. 4, (Aug 2004) 2192-2203

Wang, M. C. P. & Gates, B. D. (2009). Directed assembly of nanowires. *Materials Today,* Vol. 12, No. 5, (May 2009) 34-43

Wei, W.; Bao, X. Y.; Soci, C.; Ding, Y.; Wang, Z. L. & Wang, D. (2009). Direct Heteroepitaxy of Vertical InAs Nanowires on Si Substrates for Broad Band Photovoltaics and Photodetection. *Nano Letters,* Vol. 9, No. 8, (Aug 2009) 2926-2934

Whang, D.; Jin, S.; Wu, Y. & Lieber, C. M. (2003). Large-scale hierarchical organization of nanowire arrays for integrated nanosystems. *Nano Letters,* Vol. 3, No. 9, (Sep 2003) 1255-1259

Wong, K. W.; Zhou, X. T.; Au, F. C. K.; Lai, H. L.; Lee, C. S. & Lee, S. T. (1999). Field-emission characteristics of SiC nanowires prepared by chemical-vapor deposition. *Applied Physics Letters,* Vol. 75, No. 19, (Nov 1999) 2918-2920

Xu, S.; Ding, Y.; Wei, Y. G.; Fang, H.; Shen, Y.; Sood, A. K.; Polla, D. L. & Wang, Z. L. (2009). Patterned Growth of Horizontal ZnO Nanowire Arrays. *Journal of the American Chemical Society,* Vol. 131, No. 19, (May 2009) 6670-6671

Yoo, B. Y.; Rheem, Y. W.; Beyermann, W. P. & Myung, N. V. (2006). Magnetically assembled 30 nm diameter nickel nanowire with ferromagnetic electrodes. *Nanotechnology,* Vol. 17, No. 10, (May 2006) 2512-2517

Yu, G. H.; Cao, A. Y. & Lieber, C. M. (2007). Large-area blown bubble films of aligned nanowires and carbon nanotubes. *Nature Nanotechnology,* Vol. 2, No. 6, (Jun 2007) 372-377

Zhang, D. H.; Liu, Z. Q.; Li, C.; Tang, T.; Liu, X. L.; Han, S.; Lei, B. & Zhou, C. W. (2004). Detection of NO2 down to ppb levels using individual and multiple In2O3 nanowire devices. *Nano Letters,* Vol. 4, No. 10, (Oct 2004) 1919-1924

Zhang, H. Z.; Kong, Y. C.; Wang, Y. Z.; Du, X.; Bai, Z. G.; Wang, J. J.; Yu, D. P.; Ding, Y.; Hang, Q. L. & Feng, S. Q. (1999). Ga2O3 nanowires prepared by physical evaporation. *Solid State Communications,* Vol. 109, No. 11, (Mar 1999) 677-682

Zhou, H. J.; Fallert, J.; Sartor, J.; Dietz, R. J. B.; Klingshirn, C.; Kalt, H.; Weissenberger, D.; Gerthsen, D.; Zeng, H. B. & Cai, W. P. (2008). Ordered n-type ZnO nanorod arrays. *Applied Physics Letters*, Vol. 92, No. 13, (Mar 2008)

Zhou, R. H.; Chang, H. C.; Protasenko, V.; Kuno, M.; Singh, A. K.; Jena, D. & Xing, H. (2007). CdSe nanowires with illumination-enhanced conductivity: Induced dipoles, dielectrophoretic assembly, and field-sensitive emission. *Journal of Applied Physics*, Vol. 101, No. 7, (Apr 2007) 073704

20

Technology CAD of Nanowire FinFETs

T K Maiti and C K Maiti
Indian Institute of Technology, Kharagpur
India

1. Introduction

The potential applications of semiconductor nanowire (NW) field-effect transistors as potential building blocks for highly downscaled electronic devices with superior performance are attracting considerable attention. In the area of Technology CAD (TCAD) of nanowire FinFETs, it is fair to say that there have been very little or no reports are available in the literature on TCAD modeling of nanowire FinFETs. This chapter presents a detailed framework for virtual wafer fabrication (VWF) including both the device design and the manufacturing technology.

Technology computer aided design is now an indispensable tool for the optimization of new generations of electronic devices in industrial environments. In recent years, non-classical MOS devices such as nanowire FinFETs have received considerable attention owing to their capability of suppression of short channel effects, reduced drain-induced barrier lowering and excellent scalability. Different methods have been reported for the fabrication of silicon nanowires. The novel device designs need three-dimensional (3-D) process and device simulations. Nanowire FinFETs being nonplanar are inherently three-dimensional (3-D) in nature. Sentaurus process tool (3-D process simulation) has been used to study suitability of technology CAD for FinFET process development and the simulation results will be presented in this Chapter. TCAD predictability FinFETs having 25 nm gate lengths with nitride cap layer and fabricated using a conventional CMOS like process flow for novel strain-engineered (process-induced strain) nanowire have been described in detail. Effects of process-induced strain on the performance enhancement of nanowire FinFETs have been discussed.

Sentaurus Device simulator has been used to study electrical characteristics by solving self-consistently five partial differential equations (Poisson equation, electron and hole continuity equations, electron energy balance equation, and the quantum potential equation).The detail of device simulation for FinFETs will be presented. A compact model serves as a link between process technology and circuit design. As FinFETs are predicted to be used in RFIC applications, flicker noise in FinFET becomes very important and modeling of noise in FinFETs will also be described. Hot-carrier induced degradation behavior of nanowire FinFETs and extraction of SPICE parameters essential for circuit analysis and design will be described.

2. Importance of TCAD

For the past four decades, the performance of very large scale integrated (VLSI) circuits and cost-per-function has been steadily increased by geometric downscaling of MOSFETs.

Advanced transistor structures, such as multiple-gate (MuG) or ultra-thin-body (UTB) silicon-on-insulator (SOI) MOSFETs, are very promising for extending MOSFET scaling because short channel effect (SCE) can be suppressed without high channel doping concentrations, resulting in enhanced carrier mobilities. In recent years, non-classical MOS devices such as FinFETs have received considerable attention owing to their capability of suppression of short channel effects, reduced drain-induced barrier lowering and excellent scalability.

The Ω-FinFETs have unique features such as high heat dissipation to the Si substrate, no floating body effect, and low defect density, while having the key advantages of the silicon-on-insulator (SOI)-based FinFETs characteristics. The Ω-FinFET has a top gate like the conventional UTB-SOI, sidewall gates like FinFETs, and special gate extensions under the silicon body. The Ω-FinFET is basically a field effect transistor with a gate that almost covers the body. However, the manufacturability of this type of device structures is still an issue. Many different methods have been proposed to fabricate these devices but most of them suffer from technical challenges mainly due to the process complexity.

Strained-Si technology is beneficial for enhancing carrier mobilities to boost Ion. Both electron and hole mobilities can be improved by applying stress to induce appropriate strain in the channel, e.g., tensile strain for n-channel MOSFETs and compressive strain for p-channel MOSFETs (Maiti et al., 2007). Effect of strain on mobility can be understood by considering the stress induced changes in the complicated electronic band structures of Si. The novel device designs increase the need for three-dimensional (3-D) process and device simulations. FinFET is a nonplanar device and are inherently three-dimensional (3-D) in nature. Therefore, for FinFETs, any meaningful process or device simulation must be performed in three dimensions. Synopsys tools SProcess/SDevice address this need (aSynopsys & bSynopsys, 2006).

2.1 Process simulation

The FinFET process flow used in process simulation is similar to the flow presented by F.-L. Yang *et al.* (Yang et al., 2002). Process flow used is summarized as follows:

1. Fin patterning and threshold voltage implant
2. Gate oxidation, In-situ N+poly gate deposition and patterning
3. nMOS S/D extension implantation
4. Spacer formation
5. S/D implantation

i. Contact formation

As the Sentaurus Process ([a]Synopsys, 2006) is designed as a simulator that works independently of the dimensionality of the structures to be simulated, the process flow was executed by a sequence of three tool instances as discussed above. During the process flow execution, intermediate structures are saved for subsequent use by Sentaurus Process. Sentaurus Process loads the intermediate geometries generated by Sentaurus Structure Editor and performs the implantation and diffusion steps associated with each intermediate structure. The starting material is a (100) FDSOI wafer with a top silicon layer thickness of 25 nm. The top silicon layer has initially a uniform boron concentration of ~5.5x1018 cm^{-3}. The Fin height used in simulation was up to 50 nm, and with 12 nm cap oxide. Channel doping is performed to adjust nMOS Vt using ion implantation. To relieve the etch damage; a sacrificial oxide is removed before gate oxidation. 25 Å thermal oxide is grown and in-situ heavily doped N+ poly-silicon is deposited. After gate plasma etch, the source/drain extension implantation is simulated followed by annealing simulation.

The activation/annealing are simulated by Sentaurus Process using the pair diffusion model. For the 3-D Ω-FinFET simulation, analytic implantation model is used. For the point defects, the appropriate model is activated. For the activation, the solid solubility model is used. To capture the influence of the nonequilibrium point-defect concentrations on the diffusivity of the dopants (phosphorus, arsenic, and boron), the advanced pair-diffusion model is used. It has been observed that during the annealing process, boron is redistributed in a complex fashion. The arsenic extension implant introduces a large amount of interstitials, which during the annealing diffuse quickly to the surface, where they recombine. As boron diffuses only as boron interstitial pairs, boron is transferred to the surface in this process. Due to small volume in the channel Fin, a limited amount of boron is available and, therefore, the boron pileup at the surface in the extension area is accompanied by a corresponding depletion of boron in the center of the channel fin in the extension area. Composite spacer of silicon oxide and nitride is deposited and etched anisotropically with final thickness of ~ 32 nm. Heavily doped N^+ and P^+ junction are made with Phosphorous

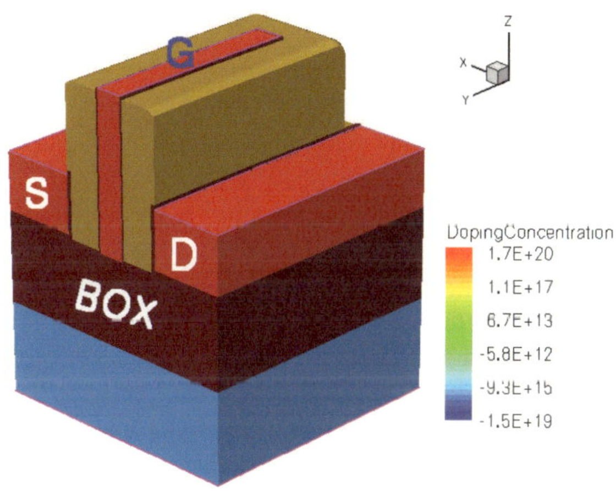

Fig. 1a. Device structure after process simulation.

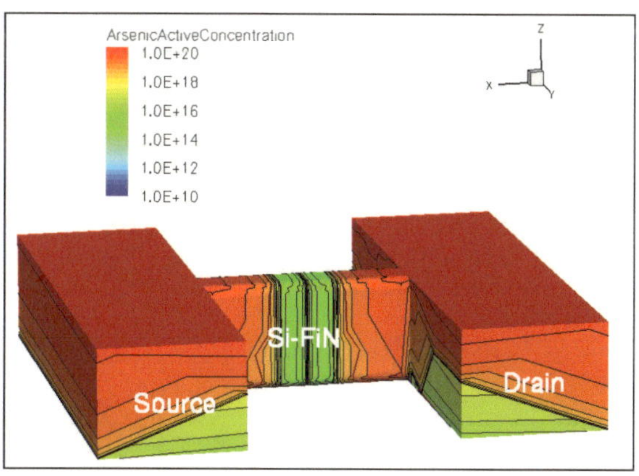

Fig. 1b. Arsenic distribution inside the Strained-Si Fin.

and Boron implantation. Thermal anneals above 1000°C are used for dopants activation. The process was completed by standard metal contact formation. Fig. 1-2 show some process simulated results for 25 nm gate length FinFETs. Silicon-Fin is fabricated on buried-oxide (BOX) and consecutively capped with nitride (Si_3N_4). A tensile process induced strain has been evolved in the Fin, during thermal process.

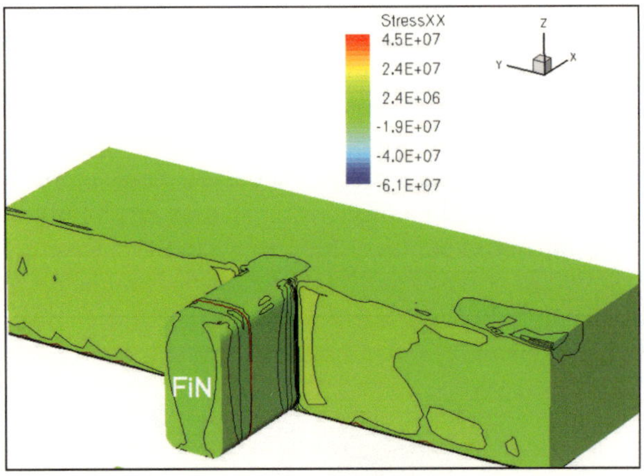

Fig. 2. Stress (ε_{xx}) distribution in channel for Ω-FinFET.

2.2 Device simulation

In advanced deep submicron MOS device structures, transport models must be used. Also the channel-gate oxide interface must be resolved to a very high level of accuracy. The finite-element mesh must also resolve the very steep gradients of the inversion layer. Accordingly, Noffset3D is used to remesh the Ω- FinFET for the device simulations. Also for the Ω-FinFET structure considered here, the body may not be fully depleted. Therefore, the continuity equations for both electrons and holes must be solved simultaneously. The very short gate length of 25 nm mandates the use of the hydrodynamic transport model. Further, the thin oxide thickness (2.4 nm) and relatively high body doping level (~5.5x1018 cm^{-3}) require the consideration of quantization effects. Unlike other approaches to model quantization effects such as 1-D Poisson–Schrödinger and the modified local-density approximation (MLDA), the density gradient model is also applicable to nonplanar 3-D structures. Thus, in simulation, advanced quantization model (density gradient model) is used. Within the density gradient model, an additional partial differential equation is solved to determine the effective quantum potential. For the 3-D FinFET, Sentaurus Device solves self-consistently five partial differential equations (Poisson equation, electron and hole continuity equations, electron energy balance equation, and the quantum potential equation). In device simulation, the I_{ds}–V_{gs} characteristics for a low-drain bias and high-drain bias are simulated and relevant electrical parameters, such as threshold voltages and drain current are extracted.

2.3 Results and discussion

Fig. 3 shows the I_{ds}-V_{gs} characteristics for the device with the strained cap layer and relaxed cap layer. The ON-state Ids of~ 0.42 and ~0.38 mA/μm are obtained for strained and unstrained-Si Ω-FinFET, with OFF-state Ids < nA at an operating voltage of 1.0 V ($|V_{gs}$ –

V_{th}| = 0.96 V and |V_{ds}| = 1.0 V). A slight V_t shift of approximately 15 mV is due to the strain-induced effect on bandgap. One can expect much higher drive currents by further reducing the channel length and gate oxide thickness. Despite the use of a 2.4 nm thick gate oxide, strained-Si Ω-FinFET exhibits near ideal S.S. (~64 mV/dec) and low DIBL (~20 mV/V).

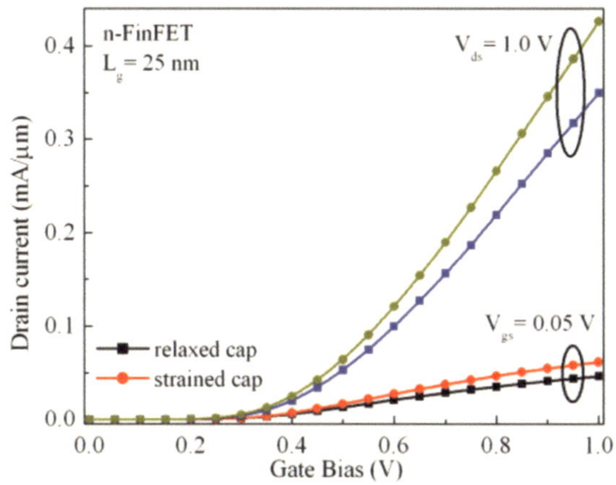

Fig. 3. I_d-V_{gs} characteristic of FinFETs at V_{ds} = 0.05 V and 1.0 V with stressed nitride cap layer and relaxed cap layer.

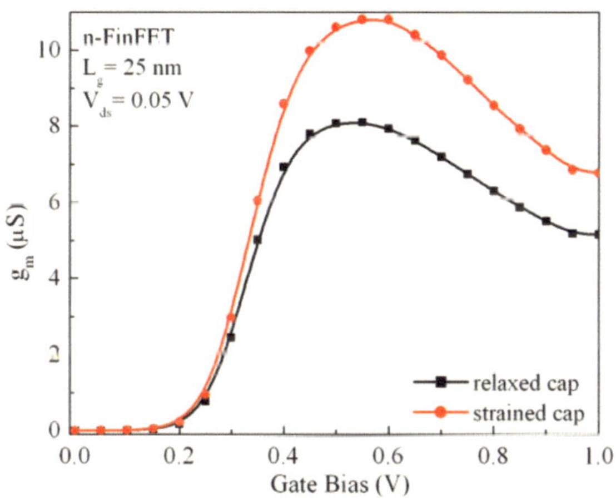

Fig. 4. Comparison of simulated transconductance for both FinFET.

Fig. 4 shows the $g_m(dI_{ds}/dV_{gs})$ variation as a function of gate bias for both devices. Ω-FinFETs shows the peaking in gm value due to mobility dependence on gate-field which is well known for the bulk FET devices. The main reason is the electron confinement due to strain near surface where surface roughness scattering of the carriers at high gate fields (with reduced carrier mobility) leads to the peaking in gm. For strained-Si Ω-FinFET, a 30% higher g_m than reference transistor is observed.

Fig. 5 compares the I_{ds}-V_{ds} characteristics for the devices with a highly tensile cap layer and a relaxed cap layer. The Ω-FinFETs with a highly tensile stress cap layer shows improvement of the saturation current I_{dsat} of approximately 13% compared to the device with the relaxed cap layer. The stress that originates from oxidation and other processing steps has a minor effect on the carriers transport.

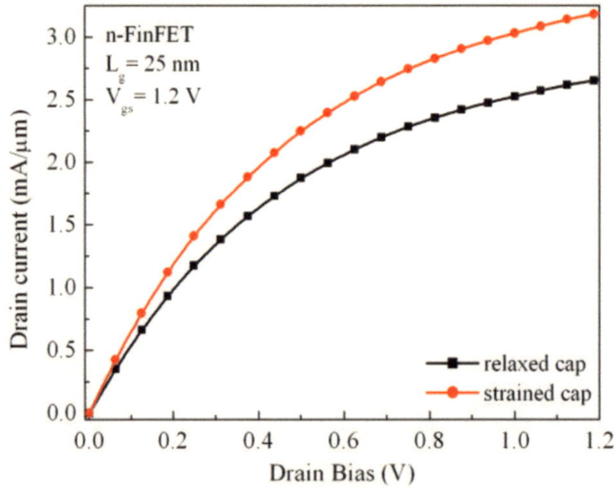

Fig. 5. I_{ds}-V_{ds} characteristics of FinFETs at $V_g = 1.25$ V with strained cap layer and relaxed cap layer

Summary of transistor parameters in comparison with conventional (Si only) FinFET is shown in Table 1. It is observed that simulated transistor performances (subthreshold swing and gate delay) of the Ω-FinFETs are significance.

Type	Bulk-Si FinFET	Strained-Si FinFET
Threshold voltage (mV)	434	381
Ion (μA/μm)	700	853
Ioff (nA/μm)	1.13	2.02
Subthreshold swing (mV/decade)	62.77	64.55
DIBL (mV/V)	26	20

Table 1. Comparison of transistor parameters with conventional FinFET.

3. Hot carrier degradation in nanowire (NW) FinFETs

Hot-carrier induced phenomena are of great interest due to their important role in device reliability ([b]Maiti et al., 2007). High energy carriers (also known as hot carriers) are generated in MOSFETs by high electric field near the drain region. Hot carriers transfer energy to the lattice through phonon emission and break bonds at the Si/SiO_2 interface. The trapping or bond breaking creates oxide charge and interface traps that affect the channel carrier mobility and the effective channel potential. Interface traps and oxide charge also affect the transistor parameters, such as, the threshold voltage and drive currents. Several workers have reported the results of their investigation on hot-carrier effects on the performance of p-MOS transistors (Pan., 1994; Heramans et al., 1998). It has been shown that the degradation of p-MOS transistors is caused by the interface state generation and hole trapping in the gate oxide from the hot-carrier injection. Reliability assurance of analog circuits requires a largely different approach than for the digital case. It is generally accepted that injected and trapped electrons dominate the degradation behavior. In this work, we describe a physics based coulomb mobility model developed to describe Coulomb scattering at the Si-SiO_2 interface and implement in device simulator. Hot-carrier induced current and subsequent degradation in nanowire (NW) Ω-FinFETs are investigated using simulation and validation with reported experimental data. The influence of the hot carriers on the threshold voltage and drive currents is examined in detail for nanowire Ω-FinFETs.

3.1 Quasi-2D coulomb mobility model

The silicon (Si)–silicon dioxide (SiO_2) interface in nanowire (NW) Ω-FinFETs shows a very large number of trap states. These traps become filled during inversion causing a change of conduction charge in the inversion layer and increase the Coulomb scattering of mobile charges. Owing to the large number of occupied interface traps, Coulomb interaction is likely to be an important scattering mechanism in nanowire (NW) Ω-FinFET device operation, resulting in very low surface mobilities and may be described by a quasi-2D scattering model. The coulomb potential due to the occupied traps and fixed charges decreases with distance away from the interface. So, mobile charges in the inversion layer that are close to the interface are scattered more than those further away from the interface; therefore, the Coulomb scattering mobility model is required to be depth dependent. We assume that the electron gas can move in the x-y plane and is confined in the z direction. Electrons are considered confined or quantized if their deBroglie wavelength is larger than or comparable to the width of the confining potential. The deBroglie wavelength of electrons, given by $\lambda = \hbar / \sqrt{2m^* k_B T}$, is approximately 150Å at room temperature, where as the thickness of the inversion layer is typically around 50Å to 100Å. Thus, one may justify treating the inversion layer as a two dimensional electron gas. The scattering from charged centers in the electric quantum limit has been formulated by Stern and Howard (1967). We consider only the p-channel inversion layer on Si (100) surface where the Fermi line is isotropic and calculate the potential of a charged center located at (r_i, z_i). Using the image method, we get

$$V_i(r,z) = \frac{e^2}{4\pi\varepsilon_0 \tilde{k} \sqrt{(r - r_i)^2 + (z - z_i)^2}} \tag{1}$$

where $r^2 = x^2 + y^2$, z = 0 corresponds the Si/SiO_2 interface. z > 0 is in silicon whereas z < 0 is in the oxide. Where $\tilde{k} = (k_{Si} + k_{ox})/2$ for z < 0, and ε_o is the permittivity of free space. We assume parabolic sub bands with the same effective heavy-hole mass, m^*. Since inversion layer electrons are restricted to move in the x-y plane, they would only scatter off potential perturbations that they see in the x-y plane. Therefore, we are only interested in determining the potential variations along that plane. To do so, one needs to calculate the two dimensional Fourier transforms of the potential appearing in Eqn. (1). The hole wave functions are then given by

$$\psi_{i,k}(r,z) = \frac{1}{\sqrt{A}} \xi(z) e^{ik.r} \tag{2}$$

where i represent the subband index and $k=(k_x,k_y)$ is the two-dimensional wavevector parallel to the interface. $\xi(z)$ is the quantized wave function in the direction perpendicular to the interface, E_i its corresponding energy and $r=(x,y)$. We denote the area of the interface by A. The effective unscreened quantum potential for holes in the inversion layer in the electric quantum limit in terms of the 2D Fourier transform is given by

$$v(q,z_i)=\frac{e^2}{2\tilde{k}\varepsilon_0 q}\iint \xi_i(z)\xi_j(z)e^{-q.|z-z_i|}dz \tag{3}$$

We now consider the effect of screening due to inversion layer electrons on Coulombic scattering. Screening is actually a many-body phenomenon since it involves the collective motion of the electron gas. Using the Coulomb screening we get,

$$v(q,z_i)=\frac{e^2}{2\tilde{k}\varepsilon_0 (q+q_s)}\iint \xi_i(z)\xi_j(z)e^{-q.|z-z_i|}dz \tag{4}$$

Where $q_s=\frac{e^2}{2\tilde{k}\varepsilon_0}\iint \xi_i(z)\xi_j(z)e^{-q.|z-z_i|}dz$ one obtains the scattering rate using Fermi's golden rules,

$$S(q,z_i)=\frac{2\pi}{\hbar^2}\left(\frac{e^2}{2\tilde{k}\varepsilon_0 (q+q_s)}\iint \xi_i(z)\xi_j(z)e^{-q.|z-z_i|}dz\right)^2 \delta(E_k-E_{k'}) \tag{5}$$

where $\hbar$ is Planck's constant. E_k and $E_{k'}$ denote the initial and final energies of the mobile charge being scattered. Scattering of inversion layer mobile charges takes place due to Coulombic interactions with occupied traps at the interface and also with fixed charges distributed in the oxide. Defining the 2D charge density $N_{2D}\delta(z_i)$ at depth z_i inside the oxide as the combination of the fixed charge N_f and trapped charge N_{it} as

$$N_{2D}(z_i)=\begin{cases} N_{it}+N_f(0), & z_i=0 \\ N_f(z_i), & z_i<0 \end{cases} \tag{6}$$

Using the above approximation, one obtains the total transition rate. Since, Coulombic scattering is an elastic scattering mechanism, the scattering rate or equivalently the inverse of the momentum relaxation time is then calculated as

$$\frac{1}{\tau_m}=\frac{1}{(2\pi)^2}\cdot\frac{2\pi}{\hbar^2}\left(\frac{e^2}{2\tilde{k}\varepsilon_0}\right)^2 \int\left(\frac{1}{(q+q_s)}\iint \xi_i(z)\xi_j(z)e^{-q.|z-z_i|}dz\right)^2 \delta(E_k-E_{k'})(1-\cos\theta)\delta k \tag{7}$$

Using the above relaxation time, one obtains the mobility of the i-th subband as,

$$\mu_i=\frac{e}{m^*}\frac{\int\sum_i \tau_m \varepsilon \frac{\partial f_0(\varepsilon)}{\partial\varepsilon}d\varepsilon}{\int \varepsilon\frac{\partial f_0(\varepsilon)}{\partial\varepsilon}d\varepsilon} \tag{8}$$

The average mobility, $\bar{\mu}$, is then given by

$$\bar{\mu} = \frac{\sum_i p_i \mu_i^2}{\sum_i p_i \mu_i} \tag{9}$$

where p_i is the hole concentration in the i th subband. Taking into the different scattering mechanism and using the Matthiessen's rule one obtains the total mobility μ.

3.2 Mobility model implementation

The Coulomb scattering mobility model has been implemented in Synopsys Sentaurus Device simulator. To activate the mobility model appropriate mobility values were defined in the fields of the parameter file. Simulation data for the drain current (I_{ds}) versus gate voltage (Vgs) curves match the experimentally measured results very well (Singh et al., 2005). Fig. 6 shows the I_{ds}-V_{gs} characteristics of the simulated p-type nanowire Ω-FinFET with a 10 nm-thick, and 100 nm-long Si-fin as the channel body. At room temperature, the devices show high ON-current (I_{ds} at V_{ds} = Vgs = 1.1 V) of ~0.68mA/μm, V_{th} ~ 0.2 V, and subthreshold swing (SS) of ~68 mV/dec. Low drain-induced barrier lowering (DIBL) of ~10 mV/V is obtained, with $I_{ON}/I_{OFF} > 10^7$ at room temperature. These results are similar to those reported for nanowire Ω-FinFETs by Singh et al. (Singh et al., 2005).

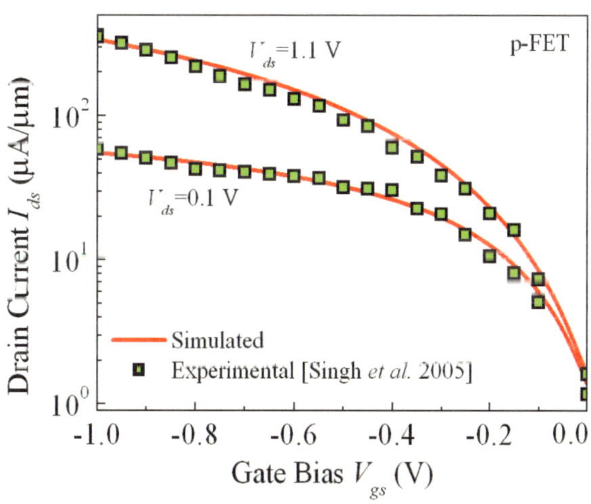

Fig. 6. Gate bias dependence of drain current for nanowire Ω-FinFETs (both simulated and experimental) (after [b]Maiti et al., 2008).

3.3 Results and discussion

Fig. 7 shows a lower drain current for Ω-FinFETs which underwent hot carrier stressing (compared to unstressed devices). Degradation in drain current indicates that hot-carrier induced positive charges are localized near the drain end.

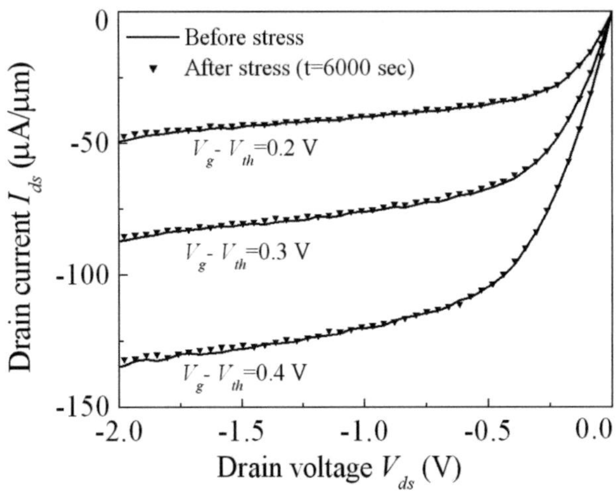

Fig. 7. Degradation of drain current under DC stress (after [b]Maiti et el., 2008).

Fig. 8 shows the threshold voltage V_{th} shift with increasing stress time. The threshold voltage V_{th} shift indicates that net positive charges exist at the gate dielectric interface as a result of hole trapping. As the lateral electric field near the drain increases in short channel devices, electron-hole pairs are generated by impact ionization. These generated holes have energies far greater than the thermal-equilibrium value and are the hot holes. In surface-channel of Ω-FinFETs, hot holes are injected into the gate oxide via hot-carrier injection (HCI), resulting in the formation of dangling silicon bonds due to the breaking of silicon-hydrogen bonds and lead to the interface traps generation (Hu et al., 1985). The charge trapping in interface states causes a shift in threshold voltage and the decrease of transconductance, which degrades the device properties over a period of time.

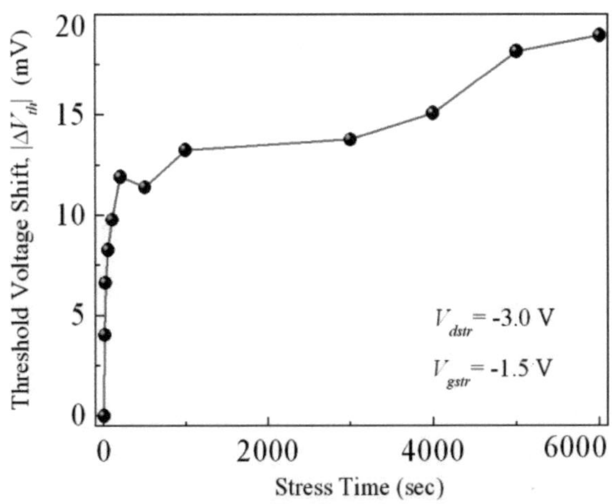

Fig. 8. Threshold voltage V_{th} shift with increasing stress time indicating an accumulation of negative charges due to electron trapping at the Si/SiO_2 interface (after [b]Maiti et el., 2008).

The hot-carrier lifetime measurements were performed and the typical I_{dsat} degradation as a function of stress time is plotted in Fig. 9. The I_{dsat} degradation is consistent with V_{th} shift.

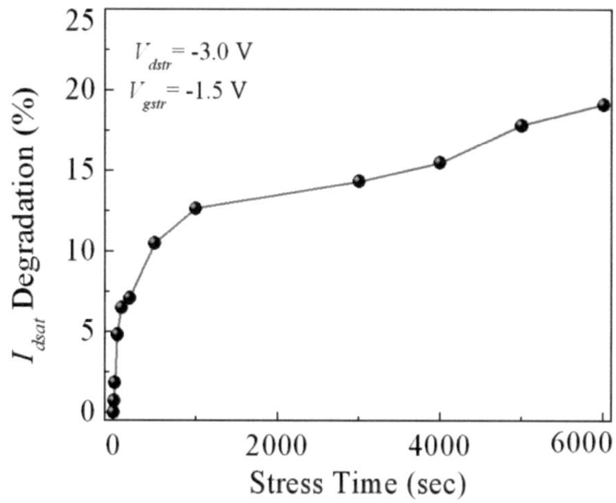

Fig. 9. I_{dsat} degradation as a function of stress time. Hot carrier lifetime in nanowire Ω-FinFETs after stressing for a given I_{sub}/I_d (after [b]Maiti et el., 2008).

4. Spice modeling of silicon nanowire FETs

In this section we will discuss the spice model of silicon nanowire FETs. This section presents the fully depleted BSIMSOI modeling of low power n- and p-MOS nanowire surrounding gate field-effect transistors (SGFETs), extraction of distributed device parasitics, and measuring the capabilities of these FETs for high-speed analog and RF applications.

4.1 Intrinsic SPICE modeling of nanowire FETs

SPICE models of n- and p-MOS SGFETs are created by fully depleted BSIMSOI parameters and are listed in Table 2. These parameters are optimized to ensure input and output I–V characteristics of 10 nm channel length and 2 nm radius SGFETs (Hamedi-Hagh & Bindal, 2008).

The distributed parasitic RC components across the intrinsic SGFET transistor are modeled for n and p-MOS transistors, as shown in Fig. 10 and (b), respectively.

C_{gsx} is the parasitic capacitance between metal gate and the concentric source and C_{gsy} is the parasitic capacitance between metal gate and the source contact. The resistor rg accounts for the effective gate resistance at high frequencies caused by the distributed gate-oxide channel. The resistance R_g accounts for two parallel gate contacts. C_{dsx} is the parasitic capacitance between intrinsic drain and source contacts and C_{dsy} is the parasitic capacitance between drain and source interconnects. Resistors R_{sx} and R_{sy} represent source contacts and resistors R_{nw} and R_{pw} represent overall concentric n-well and p-well resistances from intrinsic source to extrinsic source contacts of n- and p-MOS SGFETs, respectively. C_{gdx} is the parasitic capacitance between gate contact and the intrinsic drain and C_{gdy} is the parasitic capacitance between gate and drain interconnects. The resistor R_d represents the drain contact of the transistor. The effective resistor r_g is given by

$$r_g = \frac{1}{12}\left(R_s \frac{2\pi R}{L} \right) \tag{10}$$

Parameters	Values
Channel Length (L)	10 nm
Channel Radius (R)	2 nm
Gate Oxide Thickness (t_{ox})	1.5 nm
Channel Doping Concentration (n_{ch})	1.5e+19 cm^{-3}
Substrate Doping Concentration (n_{sub})	1.0e+11 cm^{-3}
Threshold Voltage (V_{th0})	0.26 V (nMOS) -0.28 V (pMOS)
Mobility (U_0)	1000 $cm^2/V.s$ (nMOS) 300 $cm^2/V.s$ (pMOS)
Parasitic Resistance Per Unit Area (R_{dsw})	130Ω.μm (nMOS) 360Ω.μm (pMOS)
Saturation Velocity (V_{sat})	≈2e+06 cm/s
Subthreshold Region Offset Voltage (V_{off})	0.06 V
Channel Lenth Modulation (P_{clm})	25
Primary Output Resistance DIBL Effect (P_{diblc1})	1.02e-06
Secondary Output Resistance DIBL Effect (P_{diblc2})	1
Primary Short Channel Effect on V_{th} (D_{vt0})	3.8
Secondary Short Channel Effect on V_{th} (D_{vt1})	2.75
Short Channel Body Bias Effect on V_{th} (D_{vt2})	0 V^{-1}
Primary Narrow Width Effect on V_{th} (D_{vt0w})	0
Secondary Narrow Width Effect on V_{th} (D_{vt1w})	7.25e+07
Narrow Width Body Bias Effect on V_{th} (D_{vt2w})	0.34 V^{-1}
Subthreshold Region DIBL Coefficient (Eta_0)	0.008
Subthreshold Body Bias DIBL Effect (Eta_b)	0.174 V^{-1}
DIBL Coefficient Exponent (D_{sub})	1
Source/Drain to Channel Coupling Capacitance (C_{dsc})	1.373e-10 F/cm^2

Table 2. List of BSIMSOI model parameters of SGFETs (after Hamedi-Hagh & Bindal, 2008).

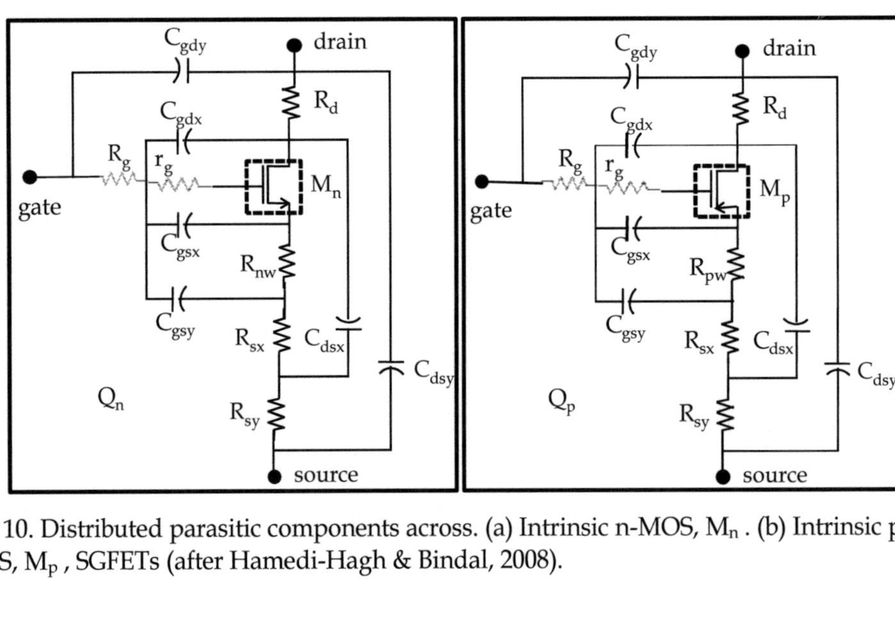

Fig. 10. Distributed parasitic components across. (a) Intrinsic n-MOS, M_n . (b) Intrinsic p-MOS, M_p , SGFETs (after Hamedi-Hagh & Bindal, 2008).

which is equal to the effective gate resistance of the planar transistors with signals applied to both ends of the gate. The distributed SGFET parasitic components are listed in Table 3(a) and (b) for resistors and capacitors, respectively.

Resistors	Values
r_g	10 Ω
R_g	110 Ω
R_{nw} (R_{pw})	2.3 (3.4) k Ω
R_{sx}	100 Ω
R_{sy}	100 Ω
R_d	70 Ω

(a)

Capaqcitors	Values
C_{gsx}	3 aF
C_{gsy}	1 aF
C_{gdx}	0.5 aF
C_{gdy}	0.8 aF
C_{dsx}	0.5 aF
C_{dsy}	0.8 aF

(b)

Table 3. List of SGFET parasitic (a) resistors (b) capacitors (after Hamedi-Hagh & Bindal, 2008).

4.2 Extrinsic SPICE modeling of nanowire FETs

S parameters are obtained by sweeping the frequency from 1 MHz to 10^3 THz and using ports with Z_0 = 1 kΩ internal resistances to ensure stability. The transistors are biased with V_{ds} = 1 V and V_{gs} = 0.5 V to yield the maximum transconductance and to ensure a high power gain. The S_{22} (output return loss) is a measure of the transistor output resistance and S_{21} (forward gain) is a measure of the transistor voltage gain. Due to similar dimensions, n- and p-MOS SGFETs have very similar parasitic components, while the g_m and rout of n- and p-MOS transistors differ from each other. Therefore, it is expected that S_{22} and S_{21} of the n- and p-MOS transistors deviate from each other, while S_{11} (input return loss) and S_{12} (reverse gain) of transistors match more closely. The two important figure of merits for RF transistors are the maximum frequency of oscillation (f_{max}) and the unity current- gain cut-off frequency (f_T). The f_{max} is obtained when the magnitude of the maximum available power gain (G_{max}) of the transistor becomes unity and f_T is obtained when the magnitude of the current gain (H_{21}) of the transistor becomes unity. The G_{max} and H_{21} of the transistor, under simultaneous conjugate impedance-matching conditions at input and output ports, are expressed in terms of S-parameters as (Hamedi-Hagh & Bindal, 2008)

$$G_{max} = \frac{S_{21}^2}{\left(1-S_{11}^2\right)\left(1-S_{22}^2\right)} \tag{11}$$

and

$$H_{21} = \frac{S_{21}}{\left(1-S_{11}\right)\left(1+S_{22}\right)+S_{12}S_{21}} \tag{12}$$

The f_{max} and f_T of n- and p-MOS SGFETs are 120 THz, 36 THz and 100 THz, 25 THz, respectively. All SPICE results indicated the potential use of nanowire FETs in high-speed and low-power next-generation VLSI technologies.

5. Process-compact SPICE modeling of nanowire FETs

In this section, we present a simulation methodology for nanowire FinFETs which allow the flow of pertinent information between process and design engineers without the need for disclosing details of the process. Compact SPICE model parameters are obtained using parameter extraction strategy as a polynomial function of process parameter variations. As a case study, SPICE models are used to identify the impacts of process variability in inverter circuit with nanowire FinFETs.

In advanced technology nodes (< 45 nm), process variations and defects are largely dominating the ultimate yield. The sources of the process variations and defects must be identified and controlled in order to minimize the yield loss. Technology CAD (TCAD) is a powerful tool to identify such root causes for yield loss. TCAD tools are used to study device sensitivities on process variations. Currently, TCAD is heavily used in device research and process integration phases of technology development. However, a major trend in the industry is to apply TCAD tools far beyond the integration phase into manufacturing and yield optimization. In this section, linking of process parameter variations (via DoE) with the electrical parameters of a device through Process Compact Model (PCM) is also demonstrated.

Towards extended TCAD, in process modeling, generally a systematic design of experiments (DoE) run is performed. DoE experiments can be systematically set up to study the control over process parameters and arbitrary choice of device performance characteristics. The models developed from DoE are known as process compact models (PCMs) which are analogous to compact models for semiconductor devices and circuits. PCM may be used to capture the nonlinear behavior and multi-parameter interactions of manufacturing processes (Maiti et al., 2008). SPICE process compact models (SPCMs) can be considered as an extension of PCMs applied to SPICE parameters. By combining calibrated TCAD simulations with global SPICE extraction strategy, it is possible to create self consistent process-dependent compact SPICE models, with process parameter variations as explicit variables. This methodology brings manufacturing to design, so that measurable process variations can be fed into design [borges06]. To design robust circuits using strain-engineered MOSFETs, the effect of process variability on the circuit model parameters examined in detail.

5.1 Process compact models: An overview

Process compact models (PCMs) methodology consists of TCAD simulations, using the process and device models that are calibrated to silicon, and process-dependant compact SPICE model extraction (see Fig. 11). The parameter extraction is performed using the parameter extraction tool Paramos (cSynopsys, 2008), which interfaces with TCAD or experimental data and directly generates process-aware SPICE models. The process-aware SPICE models allow designers to account for process variability and to develop more robust designs.

Process compact models:

Capture the process–device relationships between the process parameters and device performance of a semiconductor manufacturing process.

1. Are robust, fast to evaluate, and can be embedded into other environments such as PCM Studio, spreadsheet applications, and yield management systems.
2. Are analogous to device compact models, which capture electrical behavior and can be derived from measurements or simulations.

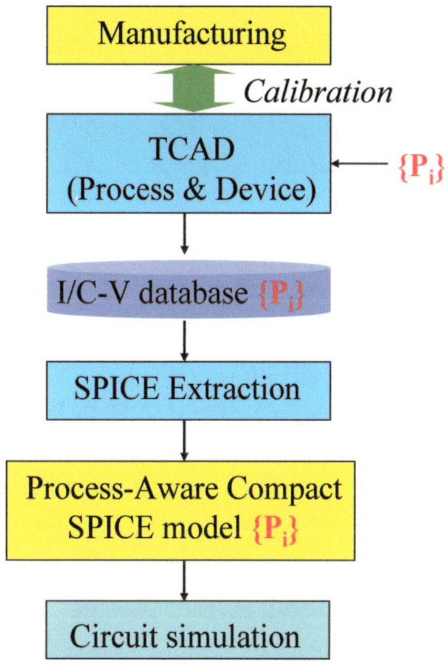

Fig. 11. Compact SPICE model extraction and validation methodology.

SPICE process compact models (SPCMs) can be considered as an extension of PCMs applied to SPICE parameters. Using a global extraction strategy, available from the Synopsys tool Paramos, pertinent compact SPICE model parameters are simultaneously obtained as a polynomial function of process parameter variations. The extraction procedure is performed using Paramos, which will deliver an XML file containing the extracted SPICE model parameters. This methodology brings manufacturing to design, so that measurable process variations can be fed into design. Additionally, design sensitivity to process can be fed back to manufacturing so that product dependent process controls can be performed. Here the chosen SPICE model parameters (Y_i) are extracted as an explicit polynomial function of normalized process parameter variations ($\tilde{P}_j$) as shown in Eqn. 13. Process parameter variations are normalized with respect to the corresponding standard deviation of the parameter as shown in Eqn. 14. Such a normalization process enables the encryption of proprietary information like the absolute values of the process parameters.

$$y_i = y_i^0 + \sum_j \sum_{n=1}^{N} a_{ij}^n \tilde{p}_j^n \tag{13}$$

Where, Y_i - Nominal value of the *i-th* model parameter, j is the *j-th* process parameter, N is the highest order of polynomial, a_{ij}^n is the process coefficient of *j-th* process parameter for the *i-th* SPICE model parameter and for order n of the polynomial, $\tilde{p}_j$ is the normalized process parameter defined as,

$$\tilde{p}_j = \frac{p_j - p_j^0}{\sigma_j} \tag{14}$$

Where, p_j is the *j-th* value of the process parameter, p_j^0 is the nominal value of the *j-th* process parameter, σ_j is the standard deviation of the *j-th* process parameter. Here we represent the BSIM4 SPICE model parameters as quadratic function of process parameters.

This model is easily scalable to higher orders of polynomial (N) for higher accuracy of extraction (Tirumala et al., 2006). Current extraction strategy of the SPICE model parameters involves extraction of nominal SPICE parameters $\left(y_0^i\right)$ followed by extraction of process coefficients $\left(a_{ij}^n\right)$ and re-optimized nominal values of SPICE parameters $\left(y_0^i\right)$.

5.2 Process-aware SPICE parameter extraction

To extract the model parameters, process and device simulations were first performed using typical CMOS process flow. The model parameters extracted are for the nominal process conditions and various drawn gate lengths. One of the SPICE parameters, namely voltage (V_{th}), as a function of process parameters has been extracted. In order to validate the compact SPICE model, for a given set of process conditions and device bias states, I-V curves obtained from TCAD simulations are compared with those obtained from Paramos using process-dependant compact SPICE model card. Fig. 12 shows the current-voltage characteristics. The dots show the TCAD simulation data, and the solid lines show the electrical characteristics generated by global SPICE model.

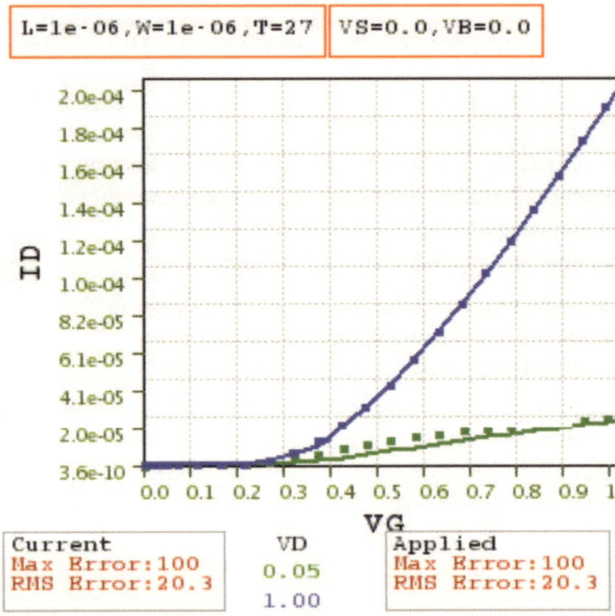

Fig. 12. Current-voltage characteristics.

As an example, SPICE model parameter of threshold voltage (V_{th}) extracted as an explicit polynomial function of normalized process parameter variations (P_i^n) as shown in Eqn. 14.

$$V_{th} = V_{th0} + \sum\sum a_i^n P_i^n \tag{15}$$

Where, V_{th0} is the nominal value of threshold voltage, i is the ith value of the process parameter, a_i^n process coefficient of ith process parameter for the SPICE model parameter and for order n of the polynomial, and P is the normalized process parameter. Such a normalization process (P) enables the encryption of proprietary information like the absolute values of the process parameters. BSIMSOI SPICE model parameters as quadratic function of process parameters have been considered. This model is easily scalable to higher orders of polynomial (n) for higher accuracy of extraction. The SPICE model parameter such as threshold voltage (V_{th}) involves extraction of nominal SPICE parameters (V_{th0}) followed by extraction of process coefficients a_i^n and re-optimized nominal values of SPICE parameters (V_{th0}). Threshold voltage model for strain-engineered nMOSFETs have been obtained using first order polynomial as function of gate length (L_g) and oxide thickness (T_{ox}) as

$$V_{th} = V_{th0} + \left(L_g - \alpha_1\right).\beta_1 + \left(T_{ox} - \alpha_2\right).\beta_2$$

and corresponding threshold voltage Eqn.

$$V_{th} = 0.27+(Lg-90)/45*0.088032+(t_{ox}-0.055)/0.01*0.0231 \quad (16)$$

Here, spice parameters are represented as first order polynomial function of process parameter variations. Threshold voltage parameter generated by the global SPICE model, shows the maximum error is approximately 12% and the root-mean-square (RMS) error is approximately 5%. These results show that the global model can be used to predict the electrical behavior of the devices in the absence of process variability.

6. Noise in silicon nanowire Fin-FET

This section deals with the noise in silicon nanowire FinFETs (SNWFinFETs). The noise of a device is the result of the spontaneous fluctuations in current and voltage inside the device that are basically related to the discrete nature of electrical charge. Noise imposes limits on the performance of amplifiers and other electronic circuits. Si nanowire transistors (SNWTs) have also been widely studied as chemical and biochemical sensors ([a]Zhang, 2007; [b]Zhang, 2008 & Stern, 2007). Biosensing by SNWTs is based on the pronounced conductance changes induced by the depletion of charge carriers in the silicon body when the charged biomolecules are bound to its surface. The high noise level in the depletion (subthreshold) region may lead to reduced signal-to-noise ratios in these sensors (Wei, 2009). In this section Low-frequency noise (LFN) in SNWTs has been demonstrated in the subthreshold region.

6.1 Low frequency noise measurements

The standard noise measurement set-up included an E5263A 2-channel high speed source monitor unit, a SR 570 low noise amplifier (LNA) and a 3570A dynamic signal analyzer. Here, E5263A 2-channel high speed source monitor unit provided the necessary gate-source and drain-source biases as shown in Fig. 13. The minute fluctuations in the drain-source voltage were amplified to the measurable range using low amplifier. The output of the amplifier is fed to 35670A dynamic signal analyzer that performs the fast Fourier transform on the time domain signal to yield the voltage noise power spectral density (S_V) in the 1-100 kHz range after correcting for amplifier gain. In order to obtain a stable spectrum, the number of averages was set at 40 and a 90% sampling window overlap was used for optimal real time processing. A computer interface was provided to control the dynamic signal analyzer and automate the noise data collection.

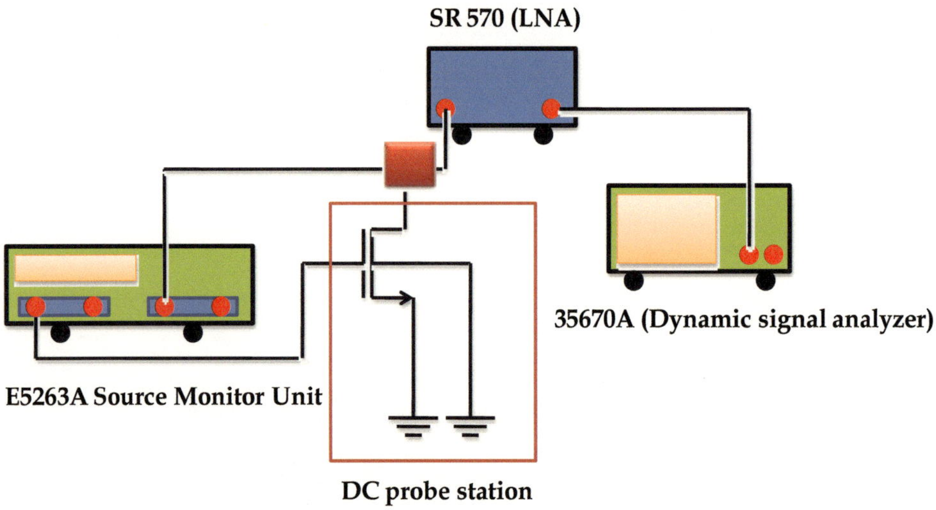

Fig. 13. Low frequency noise measurement system.

6.2 Low-frequency noise

Fig. 14 shows the frequency dependence of the measured drain-current noise spectral density S_v of 100 nm p-type SNWFinFETs, biased at V_{ds} = −50 mV at different gate bias. The S_v extracted at f = 600 Hz of each curve is shown in the inset. The dispersion of the noise spectral density is due to randomly distributed oxide traps, the lattice quality and mobility variations of the ultrascaled dimension of SNWFinFETs.

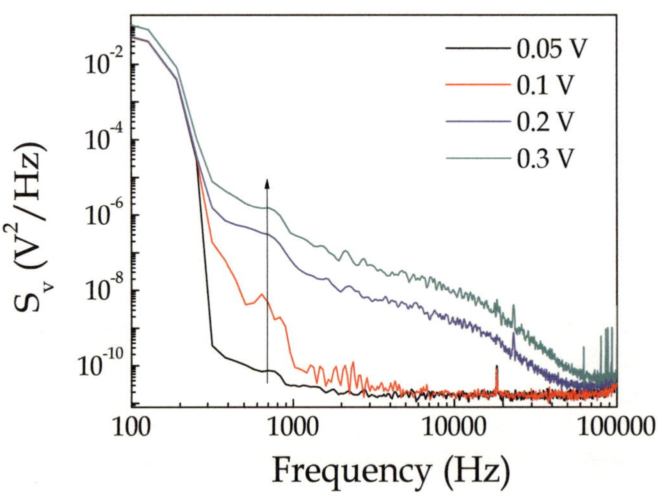

Fig. 14. Drain-current noise spectral density S_v of p-type SNWFinFETs with L = 100 nm biased at V_{ds} = −50 mV at different gate bias (V_{gs}).

7. References

[a]Maiti, C. K.; Chattopadhyay, S. & Bera, L. K. (2007). *Strained-Si Heterostructure Field Effect Devices*, ISBN: 0750309938, CRC Press, Boca Raton.

[a]Synopsys, Inc, Mountain View, California, (2006). Sentaurus Process User Manual, Version Y-2006.06, June 2006.

[b]Synopsys, Inc, Mountain View, California, (2006). Sentaurus Device User Manual, Version Y-2006.06, June 2006.

Yang, F. -L.; Chen, H.-Y.; Chen, F.-C.; Huang, C.-C.; Chang, C.-Y.; Chui, H.K.; Lee, C. C.; Chen, C. C., Huang, H. T.; & Chen, C. J. (2002). 25 nm CMOS Ω FETs, *IEDM Technical Digest*, San Francisco, CA, USA, December, 2002. pp. 255–258.

[b]Maiti, T. K.; Mahato, S. S.; Chakraborty, P.; Sarkar, S. K. & Maiti, C. K. (2009). Impact of Negative Bias Temperature Instability on Strain-engineered p-MOSFETs: A Simulation Study, *Journal of Computational Electronics*, 2009. pp. 1-7, ISSN: 1569-8025 (Print) 1572-8137 (Online)

Pan, Y. (1994). A physical-based analytical model for the hot carrier induced saturation current degradation of p-MOSFETs, *IEEE Transactions on Electron Devices*, vol. 41, no.1, January, 1994, pp.84-89, ISSN: 0018-9383

Heramans, P.; Bellens, R.; Groeseneken, G. & Meas, H. E.(1998). Consistent model for the hot carrier degradation in n-channel and p-channel MOSFETs, *IEEE Transactions on Electron Devices*, December, 1998, vol. 35, no. 12, pp.2194-2209, ISSN: 0018-9383.

Singh, N.; Agarwal, A.; Bera, L. K.; Kumar, R.; Lo, G. Q.; Balasubramanian, N. & Kwong, D. L.(005). Gate-all-around MOSFETs: lateral ultra-narrow (≤10 nm) fin as channel body, IEEE Electronics Letter, vol. 41, no. 4, 2005, pp.1353-1354, ISSN: 0013-5194.

[b]Maiti, T.K.; Bera, M.K.; Mahato, S.S.; Chakraborty, P.; Mahata, C.; Sengupta, M.; Chakraborty, A. & Maiti, C.K. (2008). Hot carrier degradation in nanowire (NW) FinFETs, *15th International Symposium on the Physical and Failure Analysis of Integrated Circuits (IPFA)*, 2008, pp.1–4.

Hu, C.; Tam, S. C.; Hsu, F-C.; Ko, P. K.; Chan, T. Y. & Terril, K. W. (1985). Hot-electron-induced MOSFET degradation- Model, monitor, and improvement, *IEEE Transactions on Electron Devices*, vol. 32, no. 2, February, 1985, pp.375-385, ISSN: 0018-9383.

Hamedi-Hagh, S. & Bindal, A. (2008). Spice Modeling of Silicon Nanowire Field-Effect Transistors for High-Speed Analog Integrated Circuit, *IEEE transactions on nanotechnology*, vol. 7, no. 6, November, 2008. pp. 766–775, ISSN : 1536-125X.

Maiti, C. K. ; Maiti, T. K. & Mahato, S. S. (2008). Strain-Engineered MOSFETs, Semiconductor India, Magazine, June 2008.

[c]Synopsys, Inc., (2008). California, Paramos User Guide, Version B-2008.06, Mountain View, June, 2008.

Tirumala, S. ; Mahotin, Y.; Lin, X.; Moroz, V.; Smith, L.; Krishnamurthy, S.; Bomholt, L. & Pramanik, D. (2006). Bringing Manufacturing into Design via Process-Dependent SPICE Models, *Proc. of the 7th International Symposium on Quality Electronic Design (ISQED'06)*, 2006, pp.806-810, ISBN:0-7695-2523-7.

[a]Zhang, G. J.; Agarwal, Buddharaju, D.; Singh, N. & Gao, Z. (2007). Highly sensitive sensors for alkali metal ions based on complementary-metal-oxide-semiconductor-compatible silicon nanowire, *Applied Physics Letter*, vol. 90, no. 23, June, 2007, pp.233-903,

[b]Zhang, G. J.; Zhang, G.; Chua, J. H.; Chee, R. E.; Wong, E. H.; Agarwal, A. & Buddharaju, K. D., (2008). DNA sensing by silicon nanowire: Charge layer distance dependence, *Nano Letter*, vol. 8, no. 4, April, 2008, pp. 1066–1070.

Stern, E.; Klemic, J. F.; Routenberg, D. A.; Wyrembak, P. N. & Turner-Evans, D. B. (2007). Label-free immunodetection with CMOS compatible semiconducting nanowires, Nature, vol. 445, no. 7127, February. 2007, pp. 519–522.

Wei, C.; Xiong, Y-Z.; Zhou, X.; Singh, N.; Rustagi, S. C.; Lo, G. Q. & Kwong, D. L. (2009). Investigation of Low-Frequency Noise in Silicon Nanowire MOSFETs in the Subthreshold Region, *IEEE Electron Device Letter*, vol. 30, no. 6, June, 2009, pp. 668–671.